PRINCIPLES AND METHODS IN SUPRAMOLECULAR CHEMISTRY

PRINCIPLES AND METHODS IN SUPRAMOLECULAR CHEMISTRY

Hans-Jörg Schneider

Universität des Saarlandes, Saarbrücken, Germany

and

Anatoly K. Yatsimirsky

Universidad Nacional Autónoma de México, México

JOHN WILEY & SONS, LTD
Chichester · New York · Weinheim · Brisbane · Singapore · Toronto

Other Wiley Editorial Offices

John Wiley & Sons, Inc., 605 Third Avenue,
New York, NY 10158-0012, USA

WILEY-VCH Verlag GmbH, Pappelallee 3,
D-69469 Weinheim, Germany

Jacaranda Wiley Ltd, 33 Park Road, Milton,
Queensland 4064, Australia

John Wiley & Sons (Asia) Pte Ltd, 2 Clementi Loop #02-01,
Jin Xing Distripark, Singapore 129809

John Wiley & Sons (Canada) Ltd, 22 Worcester Road,
Rexdale, Ontario M9W 1L1, Canada

Library of Congress Cataloging-in-Publication Data

Schneider, Hans–Jörg, 1935 –
    Principles and methods in supramolecular chemistry/Hans–Jörg
Schneider and Anatoly K. Yatsimirsky.
        p.   cm.
    Includes bibliographical references and index.
    ISBN 0-471-97370-X (hb : alk. paper). — ISBN 0-471-97253-3 (pb:
alk. paper)
    1. Macromolecules.   I. Yatsimirsky, Anatoly K.   II. Title.
QD381. S35   1999   99-24125
547'. 7—dc21   CIP

British Library Cataloguing in Publication Data

A catalogue record for this book is available from the British Library

ISBN 0 471 97370 X (HB)   0 471 97253 3 (PB)

Typeset in 10/12pt Palatino by Thomson Press (India) Ltd, New Delhi
Printed and bound in Great Britain By Bookcraft (Bath) Ltd
This book is printed on acid-free paper responsibly manufactured from sustainable forestry,
in which at least two trees are planted for each one used for paper production.

CONTENTS

FOREWORD

Although supramolecular chemistry has by now become a major field of chemistry, one may wonder why it emerged and developed so late, in the last quarter of this century. There are three main reasons for this: first, supramolecular chemistry requires a solid basis of synthetic methodologies of molecular chemistry for producing the building blocks of the supramolecular entities; second, the supramolecular entities are in principle of greater complexity and lability than molecular species, so that their study presents novel challenges; third, the development of supramolecular chemistry requires the availability of powerful methods for the investigation of the structural, dynamic and physicochemical features of supramolecular entities.

The latter factor certainly played a major role in setting the stage. It is thus of prime importance that all those entering the field have at their disposal a clear and through presentation of the principles and methods available for the study of supramolecular species and processes. This is the purpose of the present book, which fills the need for such a text in a highly competent and attractive fashion. It provides an expert entry into the numerous and varied methodologies at hand and puts them into perspective. The inclusion of exercises (and answers) and of documents on an internet website further extends the value of this book.

The authors have to be warmly congratulated for the quality of their product, for the service they render to the community of supramolecular chemists and to those who intend to join it, and for their contribution to the vitality and further expansion of supramolecular chemistry.

Jean-Marie Lehn
Strasbourg, December 1998

PREFACE

In recent years supramolecular chemistry has established itself as one of the most actively pursued fields of science. Its implications now reach from the basis of molecular recognition in natural and artificial complexes to exciting new applications in chemical technologies, in new materials and in biology or medicine. In consequence, there are already a number of monographs in the field, which leads to the question why one should add another volume to these impressive series.

We hope that the present book distinguishes itself by a systematic and condensed overview of a field which for a novice must often look bewildering in view of its diversity. Particular emphasis is given to the underlying physical principles and to methods which play an increasingly important role in the design, characterization and applications of supramolecular complexes. An effort is made to present essential strategies in a down-to-earth way, which should provide practical useful advice and answers to "frequently asked questions". The contents should help the reader to design host compounds for specific targets, to check their performance, and to understand and to optimize supramolecular systems for a manifold of applications. Exercises, with answers, mostly taken from recent literature, should provide some training ground. As a new feature, access to typical supramolecular structures is available on the internet (http://www.uni-sb.de/matfak/fb 11/schneider/) in connection with specific points and problems in the book. These new and usually freely available techniques allow a much more detailed and, of course, colorful insight into the non-covalent forces that bring synthetic and biological supramolecular systems into action.

One of the difficult, but at the same time exciting, aspects of supramolecular chemistry is that even a single project most often requires modern techniques of organic, inorganic, and physical chemistry, and frequently also applications of biochemical principles and methods. Obviously one cannot cover all aspects and methods of supramolecular chemistry in a single volume. In particular, the intriguing synthetic strategies for artificial receptors are only mentioned where non-covalent interactions play a decisive role. For these and other fields like solid state associations, supramolecular photochemistry, sensor technology, surfactants etc. there are already a number of reviews and monographs that can be consulted. Also we had to concentrate on methods that usually are not only in the hands of specialists but are available in most chemical laboratories. In view of the limited space particularly painful decisions had to be made in the selection of illustrative supramolecular systems and results, which necessarily was often governed by subjective choice. An attempt was made to consider the literature until the end of 1998; this should provide further guidelines to a field in which there is a virtual explosion in the number of relevant papers.

It would have been impossible for us to write a comprehensive monograph on so many aspects of supramolecular chemistry without the support of many coworkers and colleagues, who also helped by supplying relevant material. In particular Alexey Eliseev has read large parts of the manuscript and made very valuable suggestions, as did Helga Schneider-Bernlöhr.

Frank Eblinger has provided the entries on our website. We sincerely thank all of them, as well as Jean-Marie Lehn for a thoughtful foreword, but assume of course the full responsibility for all mistakes.

Hans-Jörg Schneider, Saarbrücken
Anatoly Yatsimirsky, Mexico City
August 1999

A   BASIC CONCEPTS OF HOST–GUEST COMPLEXATION WITH EXAMPLES FROM IONOPHORE CHEMISTRY

A1. GENERAL PRINCIPLES OF MOLECULAR RECOGNITION, COMPLEX FORMATION AND HOST DESIGN

Associations between host and guest molecules, H and G, are usually based on *simultaneous* non-covalent interactions between single binding sites, A (acceptor) and D (donor), which can be combinations like cation–anion, hydrogen-bond-acceptor–donor, etc. Exceptions are solvent-driven equilibria (Section B6) and enforced guest encapsulations within closed host cavities (Section H1). The need for several binding sites is quite evident: non-covalent interactions are usually weak, and concerted interplay between many sites is the only way to achieve strong and specific complexation (*recognition*) of a guest molecule. The principle of multi-site complexation is very general in living systems, where it ensures the efficiency of replication, of enzyme–substrate and of antigen–antibody interactions, as well as of other important biological functions. On the other hand, one can view multi-site complexation as a generalized *chelate effect*, which is of course well known from coordination chemistry. An important requirement for multi-site binding is *complementarity* between binding sites of host and guest molecules. In other words, complexation will be most efficient when the shapes and arrangements of binding sites in host and guest molecules fit each other. This is the general lock and key

principle of Emil Fischer, who had already explained the remarkable specificity of enzyme catalysis a century ago.

An example of extremely efficient biological multi-site binding ($K = 2.5 \times 10^{13}\,\text{M}^{-1}$)[1a] of a low molecular weight guest (biotin) by a protein host (streptavidin) is shown in Fig. A1(a). A very large binding free energy of $-76\,\text{kJ}\,\text{mol}^{-1}$ results here from simultaneous action of more than 10 weak pairwise van der Waals, electrostatic, H-bonding and some lipophilic interactions[2] (see also Sections B2–B4, and website). A synthetic receptor for biotin, shown in Fig. A1(b) for comparison, uses only four hydrogen bonds for the guest recognition and therefore exhibits only $K = 9.3 \times 10^3\,\text{M}^{-1}$ even in the less polar solvent $CDCl_3$, which favors hydrogen bonds.[3] In some cases, however, synthetic host compounds approach the affinity of biological receptors. For example, a protonated azacrown ether[4a] can bind ATP (adenosine triphosphate) in water with an association constant of $K = 10^{11}\,\text{M}^{-1}$; ethyl adenine can be bound by an artificial receptor in chloroform essentially by multiple hydrogen bonds with $K = 4 \times 10^5\,\text{M}^{-1}$.[4b] Some cyclophanes[5] and calixarenes[6] bind choline and acetylcholine with association constants 10^4–$10^5\,\text{M}^{-1}$; crown ethers and cryptands can bind alkali cations with stability constants and selectivities similar or even higher than natural ionophores.[7] Some synthetic siderophores have binding constants for Fe(III) ions

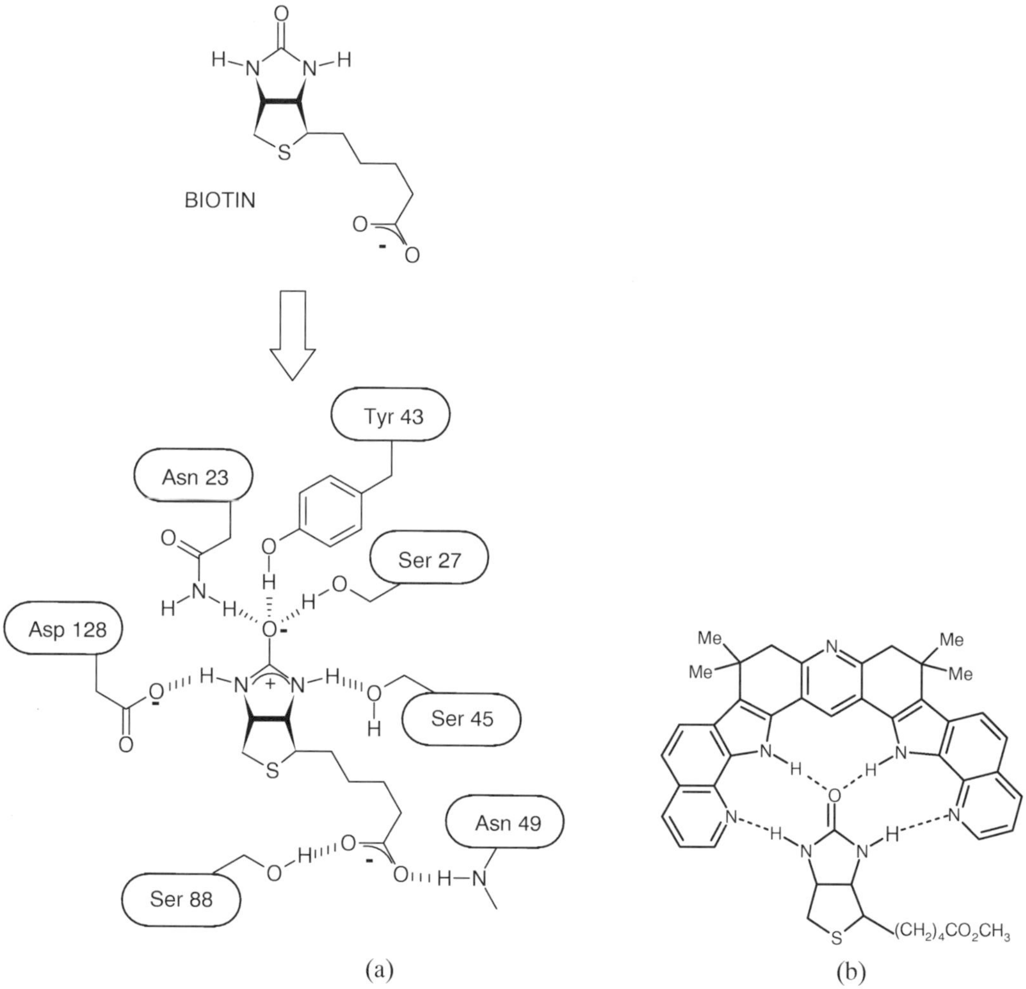

Figure A 1.   (a) Schematic illustration of biotin binding to streptavidin;[1a] Reprinted with permission from *J. Am. Chem. Soc.*, Hegde *et al.* 1993, **115**, 872. Copyright 1992 American Chemical Society (b) Binding of biotin methyl ester to a synthetic host.[3]

around 10^{60} M^{-1} (Section A 5), generating serious problems to determine such values.

A 1.1.   THERMODYNAMICS OF MULTI-SITE HOST–GUEST COMPLEXATION

The simultaneous action of all A $\cdots$ D combinations within one H–G complex, Fig. A 2, can be described to a first approximation by the sum, ΔG_t, of the single free energies, ΔG_{ij}. In consequence the total binding constant, K_t, will be the product of the participating single equilibrium constants, K_{ij}.

Such constants can be defined as the association constants between pairs of species, each bearing only one binding group

$$R\text{–}A_1 + R'\text{–}D_1 \rightleftharpoons R\text{–}A_1 \cdots D_1\text{–}R' \quad (K_{11})$$
$$R\text{–}A_1 + R'\text{–}D_2 \rightleftharpoons R\text{–}A_1 \cdots D_2\text{–}R' \quad (K_{12})$$
$$R\text{–}A_2 + R'\text{–}D_2 \rightleftharpoons R\text{–}A_2 \cdots D_2\text{–}R' \quad (K_{22})$$
$$\text{etc.} \qquad\qquad\qquad\qquad\qquad\qquad (A\,1\text{-}1)$$

One cannot, however, simply write that $K_t = K_{11}K_{12}K_{22}\ldots$ because the total constant is expressed in M^{-1} and the product of single

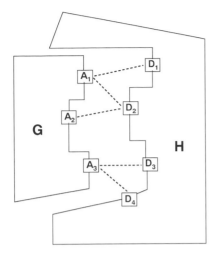

Figure A 2.   Schematic illustration of pairwise multisite host–guest interactions.

constants has the dimension of M^{-n}, where n is the total number of pairwise interactions. In principle, it is also incorrect to sum the single free energies to obtain ΔG_t because each ΔG_{ij} value can involve a sizable entropic term which refers to conversion of two species into one (complex) in the standard 1 M solution, while ΔG_t involves only one such term. A solution of this problem, proposed originally for interpretation of the classical chelate effect,[8] consists in conversion of all equilibrium constants to a dimensionless form, whereby concentrations of components are expressed in molar fractions instead of molarities. Thus a 1 M aqueous solution of a given solute is approximately 1/55.6 in molar fractions. The respective expression for the equilibrium constant will then take the form

$$
\begin{aligned}
K_{ij}(M^{-1}) &= \frac{[R{-}A_i \cdots D_j{-}R']}{[R{-}A_i][R'{-}D_j]} \\
&= \frac{([R{-}A_i \cdots D_j{-}R']/55.6)(1/55.6)}{([R{-}A_i]/55.6)([R'{-}D_j]/55.6)} \\
&= \frac{x_{R{-}A_i \cdots D_j{-}R'}(1/55.6)}{x_{R{-}A_i} x_{R'{-}D_j}} = K_{ij}^x/55.6
\end{aligned}
$$

$$(A 1\text{-}2)$$

where x is molar fraction

and K_{ij}^x is a dimensionless equilibrium constant.

Now one can write

$$
K_t^x = K_{11}^x K_{12}^x K_{22}^x \ldots = (55.6)^n K_{11} K_{12} K_{22} \ldots
$$

On the other hand

$$
K_t^x = 55.6\, K_t
$$

Therefore

$$
K_t = (55.6)^{n-1} K_{11} K_{12} K_{22} \ldots \qquad (A 1\text{-}3)
$$

In terms of free energy this change of the concentration scale implies the change of the standard state of reaction components: instead of a 1 M solution one refers now to a hypothetical pure substance with the same structure and properties, including solvation, as it would have in solution. All free energies of complexation reactions now must be corrected by $-RT \ln 55.6$. This is, of course, the entropic correction (the partial molar entropy of an ideal solution of i-th component is $\bar{S}_i = \bar{S}_i^0 - R \ln x$)[9], which was proposed also for the analysis of thermodynamics of protein association.[10]

An important conclusion which follows from equation (A 1-3) is that even very weak pairwise interactions lead to a strong host–guest binding provided the total number of binding sites is sufficiently large. For example, if all single binding constants are close to $1\,M^{-1}$ (single-site association equilibria (A 1-1) will even be difficult to measure in this case) three pairwise interactions will already give $K_t = 3 \times 10^3\,M^{-1}$. Note that equation (A 1-3) was obtained assuming complete *additivity* of pairwise interactions between host and guest molecules.

Equation (A 1-3) quantitatively explains the classical chelate effect: it allows one to calculate the stability constant of a polydentate ligand (e.g. $H_2NCH_2CH_2NHCH_2CH_2NH_2$) from known values of the binding constants of the respective monodentate ligands (NH_3 in this case), see Section A 2. However, this approach has limitations because generally pairwise interactions are weak and single equilibria like (A 1-1) often cannot be studied experimentally. In another approach, widely used in interpretation of protein–ligand (enzyme–substrate, enzyme–inhibitor, etc.) interactions[11–13] and more recently applied to host–guest equilibria,[14] the free energy of each single interaction is represented by an increment calculated as a difference of the binding free energies in the host–guest systems with and without given pair of sites.[15] Thus, the equilibrium constant of a host–guest complexation reaction (A 1-4)

$$G-(A)_{i-1}-A_i + H-(D)_{j-1}-D_j \overset{K_{ij}^{HG}}{\rightleftharpoons} G-(A)_{i-1}-A_i$$

$$\vdots \quad \vdots$$

$$H-(D)_{j-1}-D_j$$

$$(A 1-4)$$

is compared to that of reactions (A 1-5) or (A1-6)

$$G-(A)_{i-1} + H-(D)_{j-1} \overset{K_{(i-1)(j-1)}^{HG}}{\rightleftharpoons} G-(A)_{i-1}$$

$$\vdots \quad (A 1-5)$$

$$H-(D)_{j-1}$$

$$G-(A)_{i-1} + H-(D)_{j-1}-D_j \overset{K_{(i-1)(j-1)}^{HG}}{\rightleftharpoons} G-(A)_{i-1}$$

$$\vdots \quad (A 1-6)$$

$$H-(D)_{j-1}-D_j$$

Alternatively D and A single sites can be modified chemically to prevent their possible interaction. For example, if ion pairing between $R-COO^-$ and $R'-NH_3^+$ is expected, the carboxylate can be substituted by an alkyl group, or can be esterified, a amino group can, for example, be acylated. Such an approach is frequently used in exploration of interactions with and in proteins, either by covalent modification of the biopolymer or of a ligand with subsequent measurement of, for example, inhibition or of Michaelis–Menten constant. Similarly, the analysis of biological activity variation as a function of structural modification provides a classical technique for the

identification of a pharmacophor with drug receptors (also Section E1).

The ratio

$$K_{ij}^{HG}/K_{(i-1)(j-1)}^{HG} = K_{ij} \qquad (A 1-7)$$

is the incremental dimensionless binding constant between sites A_i and D_j and

$$\Delta\Delta G_{ij} = -RT \ln K_{ij} \qquad (A 1-8)$$

is the free energy increment.

The thus defined experimental increments sometimes can be positive. This happens when host and guest molecules instead of additional complementary D and A sites contain new groups which hinder binding due to steric repulsion, or for other reasons. Such repulsive interactions can be also important in determining the binding specificity[11a] (for an example see Section I1 on chiral recognition). Again one cannot simply multiply K_{ij} to get K_t or sum $\Delta\Delta G_{ij}$ to get ΔG_t. The problem is that the incremental parameters by their meaning refer to an *intra*molecular reaction (A 1-9)

$$\begin{array}{ccc} G-(A)_{i-1}-A_i & & G-(A)_{i-1}-A_i \\ \vdots & \overset{K_{ij}}{\rightleftharpoons} & \vdots \quad \vdots \\ H-(D)_{j-1}-D_j & & H-(D)_{j-1}-D_j \end{array}$$

$$(A 1-9)$$

while the total host–guest complexation process is *inter*molecular. Consequently the total multi-site complexation free energy ΔG_t should be presented as

$$\Delta G_t = \Sigma\Delta\Delta G_{ij} + \Delta G_{ass} \qquad (A 1-10)$$

and the total complexation constant as

$$K_t = K_{ass}K_{11}K_{12}K_{22}\ldots \qquad (A 1-11)$$

Equations (A 1-10) and (A 1-11) contain additional terms ΔG_{ass} and K_{ass} (M^{-1}) which are introduced to account for the bimolecular nature of the association process, namely, to correct for entropy disadvantages which invariably must accompany associations between H and G. Thus, $\Delta G_{ass} = -T\Delta S_{ass}$ and $K_{ass} = \exp \Delta S_{ass}/R$. There is some uncertainty, however, both in the meaning and size of ΔS_{ass} and K_{ass}.

The principal entropic contributions to an association process are due to changes in translational, rotational and (low frequency) vibrational entropy.[16] For gas-phase reactions all these contributions can be calculated from corresponding equations of statistical mechanics,[17] if structures and vibrational frequencies of H, G and their complex are known. Solvation of reaction components makes such calculations practically impossible, but gas-phase parameters can be used at least for illustrative purposes and can be considered as an upper limit (by absolute values) of real solution ΔS_{ass}. For an alternative approach by so-called free energy perturbation calculations see Section E7.

The gas-phase *translational* entropy (S_{trans}) equals approximately $140 \, J \, mol^{-1} K^{-1}$ for a species of molecular weight 100 with three degrees of freedom (standard state 1 M at room temperature). This entropic term does not depend on molecular structure, but depends logarithmically, that is rather weakly, on molecular weight. In the case of complexation of a large host with a small guest, S_{trans} of the host and its complex will as a result be approximately the same, and ΔS_{trans} will be due mainly to the loss of the guest translational entropy. When host and guest molecules have similar weights, ΔS_{trans} is close to S_{trans} of both interacting species. For example, formation of a complex from two species of molecular weight 100 for each gives $\Delta S_{trans} = -132 \, J \, mol^{-1} K^{-1}$. Thus, an estima-tion of $\Delta S_{trans} \approx -135 \, J \, mol^{-1} K^{-1}$ ($T\Delta S_{trans} \approx 40 \, kJ \, mol^{-1}$ at room temperature) can be used for association of species with molecular weights around 100.

The gas-phase *rotational* entropy (S_{rot}) of a polyatomic molecule depends (logarithmically) on the product of molecular moments of inertia around three mutually perpendicular axes, and therefore depends on both weight and shape of molecules. For molecules like substituted benzenes or aliphatic hydrocarbons of similar weight, S_{rot} equals at room temperature approximately $100 \, J \, mol^{-1} K^{-1}$ (three degrees of freedom). Thus a complete loss of rotation of a guest molecule upon complexation will give $\Delta S_{rot} \approx -100 \, J \, mol^{-1} K^{-1}$. Evidently, when the host–guest interaction is weak, some of the rotational degrees of freedom of the interacting species can persist in the complex. Entropies of internal rotations around common single bonds (C–C, C–N, C–O, etc.) are typically around 13–$21 \, J \, mol^{-1} K^{-1}$.[16, 18] *Vibrational* entropy is generally small with exclusion of low-frequency vibrations ($\omega < 200 \, cm^{-1}$), which are typical for weak bonds (as visible in vibration spectra, Section E6, and in the increased heat capacity, Section A 4) and thus can give a considerable positive contribution to total entropy.

It is evident from the discussion above that ΔS_{ass}, which equals the sum $\Delta S_{trans} + \Delta S_{rot} + \Delta S_{intern.rot} + \Delta S_{vib}$, essentially depends also on the strength of host–guest interaction. Table A 1 shows some typical values for different types of interactions. Evidently, only covalent

Table A 1. Selected association entropies ($J \, mol^{-1} K^{-1}$) for different types of interactions compiled from published data.[16, 19–21]

Reaction	Interaction	Phase	ΔS_{ass}	Ref.
Dimerization of cyclopentadiene	covalent	gas	-167	16
		liquid	-154	16
Complexation of MeOH with Et_3N	H-bond	gas	-81	19
		CCl_4 solution	-39	20
Complexation of *p*-xylene with tetracyanoethylene	charge transfer	gas	-65	21
		CH_2Cl_2 solution	-31	21

association leads to experimental association entropy approaching the quantity expected from complete loss of translational and rotational entropies (ca. $-240\,\mathrm{J\,mol^{-1}K^{-1}}$). The entropic cost of weaker interactions is much smaller due to conservation of some rotational degrees of freedom and low-frequency vibrations of weak intermolecular bonds. This interrelation between bond strength and entropy loss is manifested also in numerous observations of compensation behavior of ΔH and ΔS values (Section A 4) for various non-covalent interactions, e.g. for hydrogen bonding in series of related donor–acceptor pairs.[22]

One also can see from Table A 1 that the experimental association entropy becomes considerably less negative on passing from gas phase to solution, especially for weak interactions. This is expected since dissolution of gases in liquids is always accompanied by a negative entropy change. Although there is no simple general relation between gas-phase and solution entropies[23] one can get an idea about the order of magnitude of gas–liquid transition entropy from Trouton's rule, which states that the entropy of evaporation of non-hydrogen-bonded liquids at the pressure 1 atm and the temperature of boiling is nearly constant[9] and equals $88\,\mathrm{J\,mol^{-1}K^{-1}}$. Correction of this number to 1 M standard state in both liquid and vapor phase for a hypothetical substance with molecular weight 100 and density in the liquid state $1\,\mathrm{g\,cm^{-3}}$ gives for ΔS of evaporation approximately $45\,\mathrm{J\,mol^{-1}\,K^{-1}}$, which is in qualitative agreement with observed decrease in the association entropy.

The role of the solvent cannot be over-emphasized (Section C 2). In almost all solvents solvation and desolvation for the participants H, G and the H–G complex have to be considered in the thermodynamic cycle which involves free energy of complexation in the gas phase (ΔG_{HG}^{gas}), solvation free energy of the complex (ΔG_{HG}^{solv}) and desolvation free energies of host and guest molecules

(ΔG_H^{desolv} and ΔG_G^{desolv})

$$\Delta G_t = \Delta G_{HG}^{gas} + \Delta G_H^{desolv} + \Delta G_G^{desolv} + \Delta G_{HG}^{solv}$$

$$(A\,1\text{-}12)$$

The same consideration applies also to the incremental free energies; their additivity will be observed only if solvation–desolvation effects for a given A_i–D_j pair will be at least nearly independent of other pairwise interactions. In addition, these effects should be the same for a given pair in different H and G molecules in order to have universal estimation of the incremental free energy, a condition which can be fulfilled only approximately.

The main problem with the experimental determination of incremental constants and respective free energies from equations (A 1-7) and (A 1-8) is that reactions (A 1-4) or (A 1-9) involve not only the *intrinsic* contributions from A_i–D_j interactions, but also some additional contributions, such as restrictions of internal rotations, repulsions from non-bonded atoms, strain due to structural changes by induced fit, solvation changes, etc. Such contributions modify to a greater or smaller extent the intrinsic values of K_{ij} and $\Delta\Delta G_{ij}$, which can be viewed as maximum possible values for A_i–D_j pairs perfectly fitting each other.[11,12,14] The knowledge of intrinsic incremental constants and free energies is, of course, highly desirable because they can serve for universal evaluation for each type of non-covalent interactions. The additional contributions can be evaluated, in principle, for any particular H–G system by deviation from simple additivity of single ΔG increments.

Depending on the number of free rotations around single bonds in H and G, and depending on the degree these are frozen upon complexation, there can be additional entropy losses. In particular, a two-site interaction like that which occurs in reactions (A 1-4) and (A 1-9) leads to formation of a cycle and, consequently, to loss of internal rotations. Estimation of $\Delta S_{\mathrm{intern.rot}}$ from the entropies of cyclization of linear hydrocarbons[16a] gives an increment of -5 to -6 kJ mol^{-1} for the $T\Delta S$ contribution from the loss of each free rotation (the respective value of $\Delta S_{\text{intern.rot}} = -(17\text{–}20)$ J mol^{-1}K^{-1} lies within the range of entropies of internal rotations around common single bonds given above). This corresponds to a attenuation factor of approximately 0.1 to the binding constant due to the complexation-induced loss of each free rotation in H and G molecules. Estimation of $T\Delta S_{\text{intern.rot}}$ from experimental entropies of freezing of organic substances containing a variable number of single bonds[24] gives an increment ranging from -1.6 to -3.6 kJ mol^{-1} per rotor. It was suggested to correct incremental free energies for the number of free rotations removed on complexation using $T\Delta S_{\text{intern.rot}}$ between -3.5 and -5 kJ mol^{-1} per rotor as a mean 'typical' interval.[24]

On this basis one can conclude that if free rotations will be already frozen in the H and G molecules before complexation, each frozen rotation will increase K by factor of 4 or even 10. This leads to the general assumption that the efficiency of host–guest complexation can be greatly enhanced by rigidification of interacting molecules. However, weak interactions create low-frequency vibrations which give

considerable positive contribution to binding entropy (see above) and can compensate the loss of entropy of internal rotations. Scheme A 1 compares effects of freezing of internal rotations in covalent anhydride formation and non-covalent (actually, partially covalent) metal complex formation cyclization reactions. In these systems internal rotations, which exist in succinic acid and ethylenediamine, are restricted on going to their cyclic derivatives. For the anhydride formation reaction the effect is even higher than one would expect from only restrictions of internal rotations, but it practically disappears for coordination of diamines.

Experimental data for organic H–G complexes even speak for almost undetectable rotatory entropy disadvantages. The ΔG values in Scheme A 2 are selections from series of compounds, in which the number of single bonds was systematically increased; it was secured that at the same time geometric match between the complementary binding sites was possible without the built-up of strain due to formation of non-ideal torsions.[27a] Complexation ΔG values of six hydrogen bonded complexes with the number n of single bonds separating binding sites varying between $n = 5$ and $n = 18$ shows a

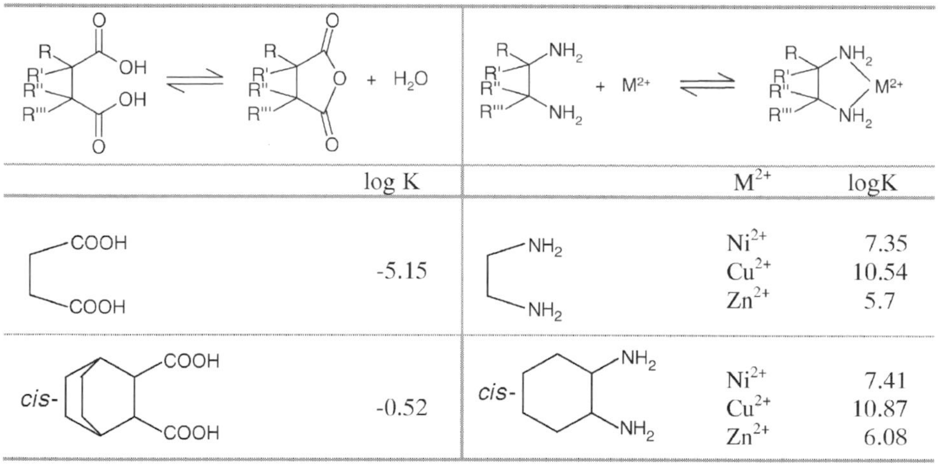

	log K		M^{2+}	logK
(succinic acid)	-5.15	(ethylenediamine)	Ni^{2+} Cu^{2+} Zn^{2+}	7.35 10.54 5.7
cis- (cyclic diacid)	-0.52	cis- (cyclic diamine)	Ni^{2+} Cu^{2+} Zn^{2+}	7.41 10.87 6.08

Scheme A1. Equilibrium constants of cyclic anhydride formation and diamine coordination reactions, from published data.[25,26]

H-Bonded complexes (in CDCL$_3$)[27a]

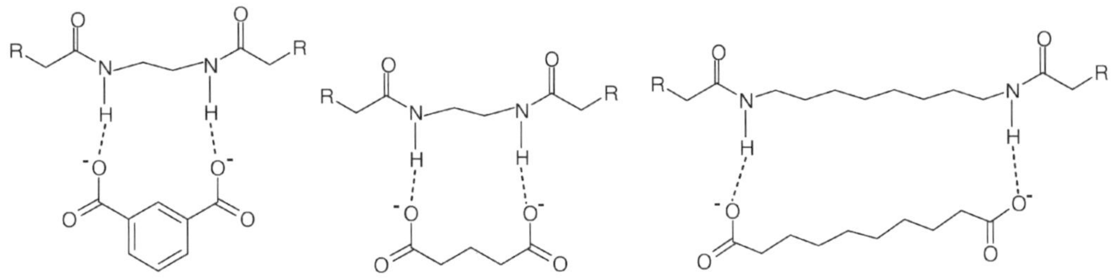

-12.0   -10.9   +(3-4)

Ion pairs (in water at zero ionic strength) [27b]

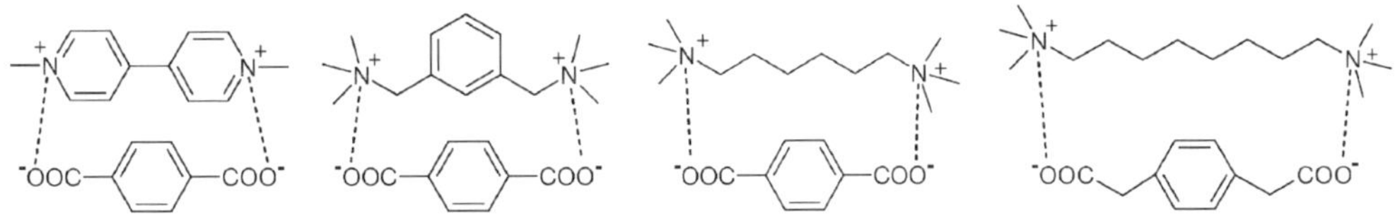

-17.6   -16.3   -14.7   -12.6

Scheme A 2.   Association free energies ΔG (kJ mol^{-1}) of selected host–guest complexes with a varying number of single bonds between binding sites, from published data.[27a,b]

linear correlation with n, indicating a ΔG disadvantage introduced by the presence of one single bond of only around 1.3 kJ mol^{-1} in the case of complexation by hydrogen bonding to anions. For ion pair complexes between dications and dianions with rigid and flexible species one observes an even smaller difference in the binding free energies.[27a] Obviously, entropy contributions either are small, or/and they are compensated (Section A 4) by enthalpy advantages with the more flexible systems; these may allow better contacts between the interacting sites. The so-called macrocyclic effect (Section A 2) is less due to $\Delta S_{intern.rot}$ disadvantages of the conformationally not preorganized host, but rather to the predisposition of cyclic hosts to provide all binding sites in a convergent manner towards the guest compound. The enthalpic reasons for the macrocyclic effect will be discussed in Section A 2, see also literature demonstration of enthalpically driven chelate effect in organic supramolecular complexes.[28]

In view of the above-mentioned problems with the consideration of entropy contributions, one may prefer to use uncorrected empirical increments as 'apparent binding energies'.[14] In such an approach one will need different series of incremental binding energies for different types of functional groups (not only for different types of interactions), in analogy to other empirical linear free energy relationships which always must be based on a sufficiently large data set. Let us discuss now the contribution of $\Delta G_{ass} = -T\Delta S_{ass}$ in equation (A 1-10). From the discussion above one can conclude that for a weak non-covalent association in solution ΔS_{ass} is considerably more positive than is expected from the total loss of translational and rotational entropies of gas phase molecules, and that its estimation is rather uncertain. In an attempt to give a realistic evaluation of $T\Delta S_{ass}$ for a bimolecular liquid-phase non-covalent association, it was concluded that $T\Delta S_{ass}$ may lie 'anywhere in the range -9 to -45 kJ mol^{-1}' (ΔS_{ass} between -30 and -150 J mol^{-1}K^{-1}).[24a] It may be, therefore, more logical to leave ΔG_{ass} and K_{ass} in equations (A 1-10) and (A 1-11) as empirical parameters.

A possible approach to evaluation of ΔS_{ass} is to substitute the molar fraction concentration scale for the molarity scale, as it was done above with the analysis of chelate effect. Such a substitution will change K_t by a factor $1/55.6$ and ΔG_t by a term $RT \ln 55.6$ (in aqueous solution), but it will not change the dimensionless incremental constants K_{ij} and the respective free energy increments $\Delta\Delta G_{ij}$. However, expressions (A 1-10) and (A 1-11) will now not need any correction terms for the association entropy since they will describe the association between hypothetical pure substances. Therefore, $-RT \ln 55.6 = -10$ kJ mol^{-1} can be considered as an equivalent of $T\Delta S_{ass}$, which accounts for non-specific dilution entropy contribution and often is included in the total association free energy as a 'generic entropic contribution'.[13c] This evaluation coincides with the lowest limit of the interval of $T\Delta S_{ass}$ values considered elsewhere[24] (from -9 to -45 kJ mol^{-1}, see above). It should be emphasized that more negative $T\Delta S_{ass}$ values correspond to more tightly bound complexes, but in terms of an empirical incremental approach all entropic contributions related to tightening of the complex by additional interactions are included in incremental experimental ΔG_{ij} values, and should not be included also in $T\Delta S_{ass}$. In summary, we see that 'the entropic cost' of non-covalent association is considerably smaller than one might expect on assuming complete or essential partial loss of translational and rotational entropies of gas-phase molecules. In particular, for evaluation of the binding constants with the use of incremental approach, one needs only minor correction for the adverse entropy of association with a $T\Delta S$ value around -10 kJ mol^{-1}.

On the other hand, experimental ΔG_t values with several hundreds of complexes of different nature show that single $\Delta\Delta G_{ij}$ contributions remain strikingly constant, as long as donor and acceptor site are in the same environment and geometrically match each other. This allows one to extract common $\Delta\Delta G_{ij}$ increments for a given $A_i \cdots D_j$ interaction.[14] The experimental procedure is to plot the observed ΔG_t values against the number n of interactions between A and D occurring in each given complex (provided all interactions are of a similar type). The validity of the ΔG additivity principle is checked by the – in fact frequently observed – linear correlation (for a first example see Fig. A3). The slope of the linear correlation (in this case -5 ± 1 kJ mol^{-1} per ion pair, if extrapolated to zero ionic strength: -8 ± 1 kJ mol^{-1}),[27b] if observed, can then be used to extract an average $\Delta\Delta G_{ij}$ increment for the interaction in question.

Deduction and applicability of general $\Delta\Delta G_{ij}$ increments and their additivity is of

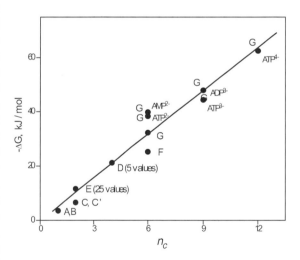

Figure A 3. Experimental complexation free energies ($-\Delta G$, kJ mol^{-1}) *vs.* number n of salt bridges in ion pairs, in water.[14a] A, B and C,C', complexes of a tetraphenolate cyclophane $(4-)$ with Me_4N^+ and an azoniacyclophane $(4+)$ with mono- and dianionic naphthalene derivatives; D, anionic (sulfonate or carboxylate) with cationic (ammonio) triphenylmethane derivatives; E, organic dianions with organic dications; F, cationic azamacrocyle $(6+$ charges) with aliphatic dicarboxylates; G, cationic azacrowns with adenosine mono-, di- and triphosphates.

central significance for the analysis of molecular recognition mechanisms in biological and synthetic complexes as well as for the design of new supramolecular systems, and it is important to define now their limitations. In practice more reliable increments are obtained with a larger number of pairwise interactions because even small association-induced entropic loss may produce a contribution to the experimental binding free energy comparable with that of one or two binding single site D–A interactions. In consequence, increments deduced from complexes with only one or two pairwise interactions may exhibit smaller $\Delta\Delta G_{ij}$ values as they must bear already the full burden of the $T\Delta S_{ass}$ disadvantage, which must be paid only once, in principle after formation of only one non-covalent bond.

How long can one expect the ΔG additivity principle, and therefore an undiminished chelate effect to hold?

(a) Disadvantages due to rotational entropy losses can be negligible as long as rotations are not severely hindered upon complex formation; this is so particularly for weaker, less tightly bound complexes, and with not too many single bonds between the binding sites A_1 and A_2 (see Scheme A 2 for examples).

(b) Solvation and desolvation contributions, equation (A 1-12), should be the same for each binding site, independent of their numbers. This will be so to a lesser degree, if the binding sites are separated by only few bonds and are exposed to solvent, such as for instance in a ethylenediamine unit.

(c) Sufficient contact between binding sites A and D must be possible in the H–G complex without the build-up of excessive strain within the partners H and G. Computer aided molecular modeling and molecular mechanics calculations offer a good way to quantify this (Section E 7). Alternatively, or for preliminary checks,

one may use mechanical models to evaluate these central geometrical requirements for effective molecular recognition; Corey–Pauling–Koltun (CPK) models are appropriate for controlling the steric fit between host and guest, whereas wire (framework) models at least allow one to study whether or how many unfavorable *gauche*-like fragments are necessary to build up a matching geometry between H and G. One may then count approximately $2.5\,kJ\,mol^{-1}$ for each of such *gauche* C–C–C–C interactions, which has to be balanced against the free energy of the complexation process. Thus, a single *gauche* fragment will not lower considerably, for example, a transition metal–ethylenediamine, or an iron–siderophore association energy, but the same lowering effect will be significant (in terms of relative effect) for the stability, for example, of a weaker hydrogen bond-based complex. The requirements of conformational preorganization differ not only as function of the A-D strength, but also as function of the directional nature of the individual non-covalent interactions.

(d) Removal, addition or modification of single site groups can induce conformational, solvation or other changes in host and guest molecules which perturb interactions between other single-site pairs, thus leading to a non-additivity in the binding. In order to correct for such effects in protein–protein interactions the double mutant cycle method was developed,[29] which can be applied also for host–guest complexation. According to this method the incremental free energy of interactions between single sites D and A is calculated as

$$\Delta\Delta G_{AD} = \Delta G_{AD} - \Delta G_{A\to X,D} \atop - \Delta G_{A,D\to Y} + \Delta G_{XY} \qquad (A\,1\text{-}13)$$

where ΔG_{AD} is the complexation free energy for the host–guest pair containing sites A and D, $\Delta G_{A \rightarrow X, D}$ and $\Delta G_{A, D \rightarrow Y}$ are the complexation free energies for the host–guest pairs, where either the A or the D site is removed or modified, ΔG_{XY} is the complexation free energy for the host–guest pair where both A and D sites are removed or modified. If removal or modification of A and D sites produces independent effects on the total complexation free energy without any interaction between the sites, $\Delta\Delta G_{AD}$ should be zero. If there is a complete additivity and the sole result of mutation is the removal of pairwise A–D interaction all three values: $\Delta G_{A \rightarrow X, D}$, $\Delta G_{A, D \rightarrow Y}$ and ΔG_{XY} should then coincide.

A1.2. MACROCYCLES, CLEFTS, AND OPEN CHAIN HOST STRUCTURES

If the ΔG additivity principle holds and entropic disadvantages are small, what can be the possible reasons not to use simple open chain hosts with appropriately placed binding functions A and D, but either cleft-like or macrocyclic hosts (Scheme A 3)? In view of the necessary synthetic effort to prepare suitable macrocycles, which then will be only useful for a special guest molecule, this is also an issue of great practical importance. If we want to complex small, spherical guest molecules like metal ions, only macrocyclic hosts can provide many binding sites directed towards the guest in a convergent manner. One other way to minimize the build-up of strain energies upon complexation with the use of flexible hosts is, to supply along a chain additional substituents which render *gauche–trans* differences small.

Generally, spherical receptor shapes offer a maximum of contact between host and spherical guest, which in comparison to contacts between flat surfaces is almost an order of magnitude larger.[30] This bears in particular on dispersive interactions to be discussed later (Section B 5). A further advantage of macrocyclic hosts is that solvation inside the cavity is absent or weaker than with hosts where the binding sites are more exposed to the solvent. Not only can the desolvation term ΔG_{H}^{desolv} in equation (A 1-12) thus be minimized, the restricted cavity size can also produce high-enthalpy solvent inside, which by uptake of a guest will then be liberated, and thus enhance the association constant (Section B 6). Finally, self-association of hosts, which can block their supramolecular function, is more easily prevented by arranging the binding elements within a cavity. Also for these reasons nature has developed binding pockets inside globular proteins. Cleft-like, concave-shaped hosts,[31–33] offer a convenient way to construct receptors with convergent binding sites without the need to build up *gauche* fragments during complexation. In principle one can provide a U-shaped cavity with variable distances, accommodating host molecules of different shapes as shown with some examples in Scheme A 3.[31,32] To a smaller degree clefts may also be more favorable than open chain receptors in view of the solvation effects discussed above for macrocyclic hosts.

A2. IONOPHORES FOR CATIONS: CHELATE, MACROCYCLIC AND CRYPTATE EFFECTS

Natural and synthetic ionophores are ligands (receptors) capable of forming stable, most often lipophilic complexes with charged hydrophilic species.[34] They can transfer ions into lipophilic phases and for this reason are also used as transmembrane carriers (Section I 6) and as (although expensive) phase-transfer catalysts. As a rule the complexation is fairly specific, and ionophores can discriminate between metal cations of different charge and different size. Ionophores behave as host molecules for the simplest guests like alkali and alkaline earth cations, which are of spherical shape and utilize for binding only electrostatic forces. This makes it relatively easy to follow the energetic and structural

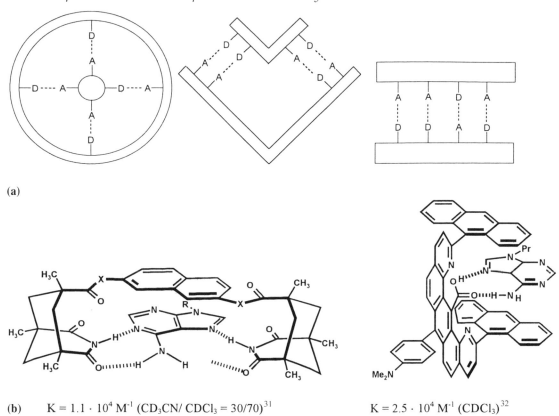

(a)

(b) $K = 1.1 \cdot 10^4 \, M^{-1}$ (CD$_3$CN/ CDCl$_3$ = 30/70)[31] $K = 2.5 \cdot 10^4 \, M^{-1}$ (CDCl$_3$)[32]

Scheme A3.   (a) Macrocyclic, cleft-like, and open chain host structures; (b) Complexes with cleft-like host structures.

requirements for recognition. The study of complexation properties of ionophores, which has been comprehensively reviewed,[7,35,36] created a basis for understanding the main principles of host–guest complexation. We will discuss here mostly complexation of alkali and alkaline earth cations, as the binding of other metal ions is dominated by more covalent-type interactions, and is the subject of classical coordination chemistry. Structures of typical ionophores, (**A-1** to **A-9**) which exemplify the most important classes of these compounds, are shown in Fig. A4. Selected values of the stability constants of ionophore complexes with alkali cations are given in Table A2.

The data in Table A2 demonstrate the ability of cyclic ionophores to form fairly

stable complexes with poorly coordinating species as alkali cations even in highly polar aqueous medium. The remarkable selectivities will be discussed later (Section A3); first we shall concentrate on the explanation of the complex stability.

All ligands mentioned above, both linear and macrocyclic, are polydentate. Therefore, one must attribute at least part of their complexing ability to the *chelate effect*,[37a,b] which refers to the increased stability of complexes with polydentate ligands as compared to those with chemically equivalent monodentate ligands. Quantitatively the chelate effect can be evaluated as discussed in Section A1 assuming the equivalence of the stability constant K of a n-dentate ligand (K_1) and β_n for the coordination of n monodentate

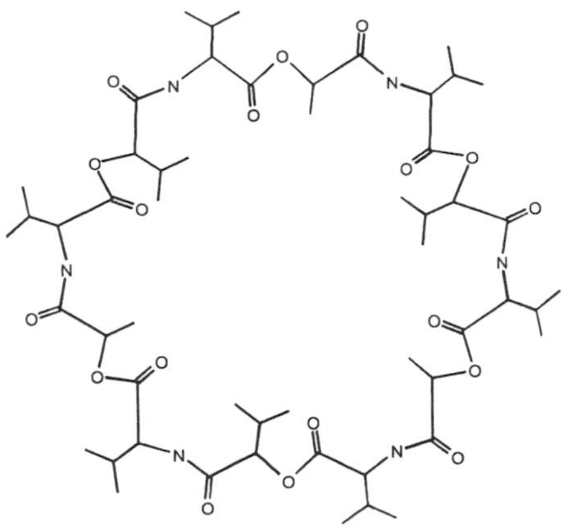

Figure A 4. Typical ionophores (structures **A-1** to **A-9**).

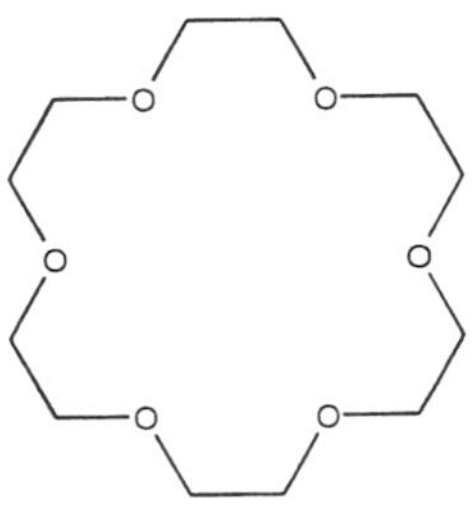

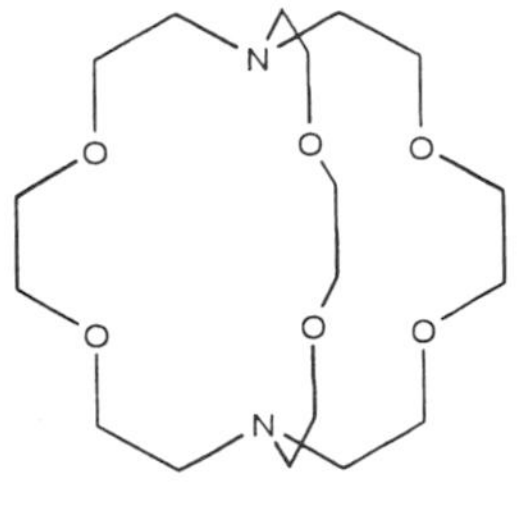

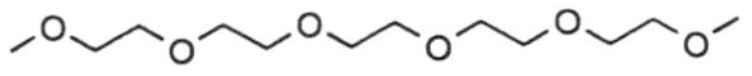

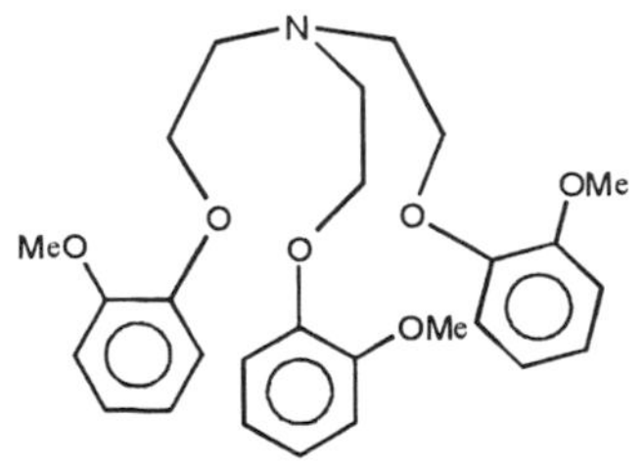

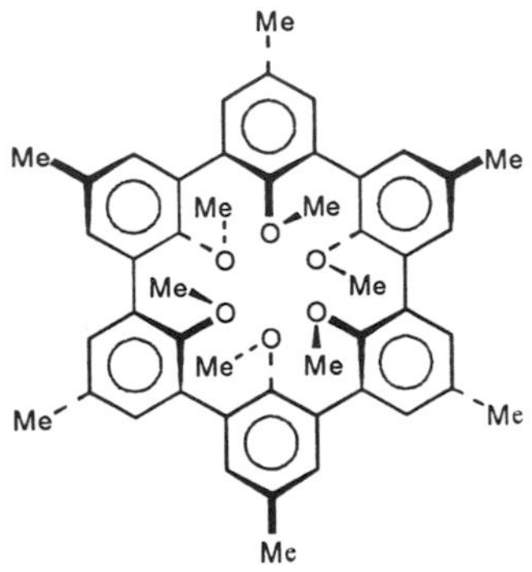

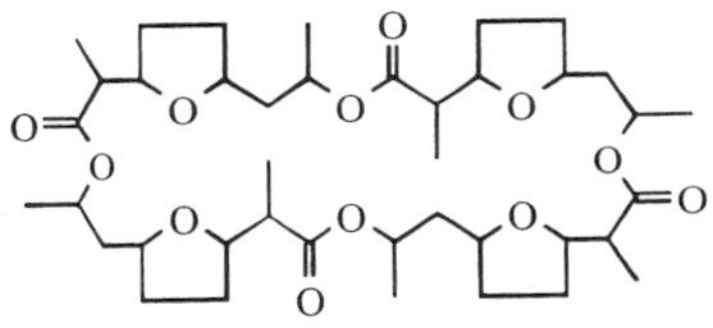

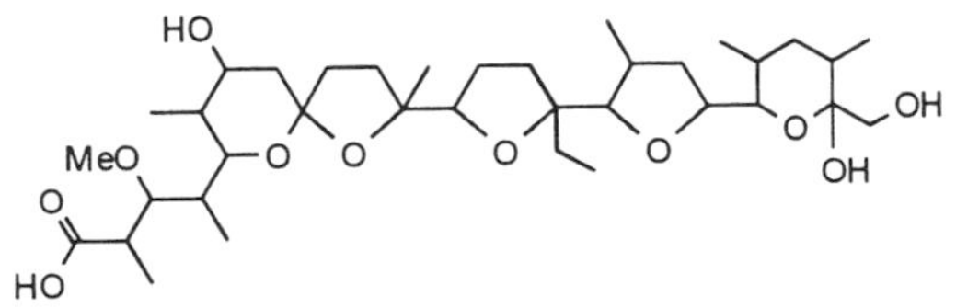

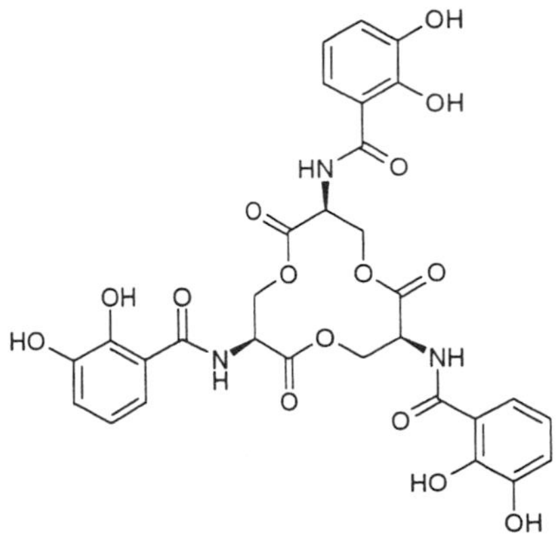

Table A 2.   Logarithms of stability constants (M^{-1}) of complexes of alkali metal cations with some ionophores, compiled from published data.[7,35]

Ionophore	Solvent	Li^+	Na^+	K^+	Rb^+	Cs^+
Valinomycin (**A-1**)	methanol	<0.7	0.9	4.7	5.2	4.4
18-Crown-6 (**A-2**)	methanol	ca.0	4.4	6.1	5.4	4.7
	water		0.8	2.0	1.6	1.0
[2.2.2]-Cryptand (**A-3**)	methanol	2.6	8.0	10.8	9.0	4.4
	water	1.25	3.9	5.3	4.3	
Pentaglyme (**A-4**)	methanol		1.5	2.2		
Tripod (**A-5**)	methanol/water (88/12)	<2	2.2	2.3	<2	

ligands to the same metal cation, if the standard state is chosen as unity mole fraction instead of 1 M.[37c,d] This equation has the form

$$\log K_1 = \log \beta_n + (n-1)\log[\text{solvent}] \quad (A\,2\text{-}1)$$

where [solvent] is the molar concentration of the solvent employed, and corrects the change of the standard state (Section A 1); for example the molar concentration of methanol in methanol is [solvent] $= 24.7\,M$. In the case of polyether ligands β_n refers to the overall stability constant of the complex of a given metal cation with a monodentate ether (e.g. diethyl ether) of the stochiometry $M(ROR)_n$. Of course, the value of β_n is unknown because of the unmeasurably low stability of such a complex in methanol. One can see from equation (A 2-1), however, that with $n = 6$ the value of $\log K = 2.3$ for pentaglyme, or of 2.2 for a tripod ligand would be observed even with a very low $\log \beta_6 = -4.6$ $(\beta_6 = 2 \times 10^{-5}\,M^{-6})$.

More rigorously the chelate effect was quantified for polyamines with ammonia as a reference monodentate ligand in aqueous solutions. The following equation was proposed[37c]

$$\log K_1 = 1.152 \log \beta_n + (n-1) \log 55.5$$
$$(A\,2\text{-}2)$$

where 1.152 is a correction factor which takes into account the higher basicity of a polyamine amino group as compared to NH_3, and 55.5 M is the molar concentration of water in

water. Table A3 (rows 1–4, 6 and 7) illustrates the successful application of equation (A 2-2) to a series of Ni^{2+} complexes of some polyamines. Similar results were obtained also for the complexes of divalent Cu, Fe and Pb.[37c,d] However, as has been pointed out[37c,d] the success of this simple approach in the *quantitative* prediction of stability constants of polydentate ligands may be to some extent fortuitous.

Complexes with the cyclic polyether 18-crown-6 are much more stable than with acyclic ligands, Table A 2, and the same effect is observed for azamacrocycles such as **A–11** as compared with linear polyamines such as **A–10**.[37b,d] Table A 4 illustrates a quantitative comparison of complexation thermodynamics by acyclic and cyclic ligands. The generally observed increased affinity with macrocyclic ligands is referred to as *macrocyclic effect*.

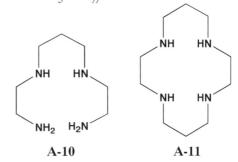

A-10   **A-11**

In terms of equations (A 2-1) and (A 2-2), the chelate effect results principally from a more favorable entropy of complexation with a polydentate ligand. For polyamine ligands

Table A 3. Thermodynamics of complexation of Ni^{2+} with ammonia and polyamines in aqueous solution (all equilibrium constants in M^{-1} and thermodynamic parameters in $kJ\,mol^{-1}$).[37d]

	Ligand	$log\,\beta_n$[a]	$log\,K_1$[a]	$log\,K_{1,calc}$[b]	ΔG	ΔH	$T\Delta S$
1	$(NH_3)_2$	5.06			−28.8	−32.6	−3.77
2	$H_2NCH_2CH_2NH_2$		7.35	7.59	−41.9	−37.7	4.18
3	$(NH_3)_4$	8.09			−46.5	−65.3	−18.8
4	$CH_2NH(CH_2)_2NH_2$ | $CH_2NH(CH_2)_2NH_2$		14.4	14.6			
5	$(H_2NCH_2CH_2NH_2)_2$	13.48			−77.4	−76.6	0.84
6	$(NH_3)_6$	9.05			−51.7	−100	−48.3
7	$CH_2N(CH_2CH_2NH_2)_2$ | $CH_2N(CH_2CH_2NH_2)_2$		19.1	19.2			
8	$(H_2NCH_2CH_2NH_2)_3$	17.64			−101.1	−117	−15.9

[a] experimental values; [b] calculated with equation (A 2-2); Reprinted with permission from the *Journal of Chemical Education*, **69**, 1992, pp. 615–621; Copyright 1992, Division of Chemical Education, Inc.

Table A 4. Differences in thermodynamic parameters ($kJ\,mol^{-1}$) of complexation with cyclic and acyclic ligands, compiled from published data.[7, 35,37d]

Acyclic ligand	Cyclic ligand	Solvent	Metal ion	$\Delta\Delta G$	$\Delta\Delta H$	$T\Delta\Delta S$
Pentaglyme	18-crown-6	MeOH	Na^+	−16.1	−15.5	0.63
			K^+	−23.0	−2.50	20.1
			Rb^+	−19.5	−2.89	16.4
			Cs^+	−15.3	−11.0	3.89
			Ba^{2+}	−25.4	−23.3	2.01
			Pb^{2+}	−27.2	−18.6	8.37
A-10	A-11	H_2O	Cu^{2+}	−18.8	−19.7	−0.84
			Ni^{2+}	−15.9	−23.0	−7.11
			Zn^{2+}	−16.3	−12.1	−4.18

equation (A 2-2) allows one to relate ΔH and ΔS values for ammonia and polydentate ligands: $\Delta H(n\text{-dentate}) = 1.152\,\Delta H(n\text{ monodentate})$ and $T\Delta S\,(n\text{-dentate}) = 1.152T\,\Delta S(n\text{ monodentate}) + 2.3\,RT(n-1)\,log\,55.5$ (neglecting a small decrease in water concentration at increased temperatures). The data in Table A 3 show (compare rows 1 and 2; 3 and 5; 6 and 8) that indeed ΔH values for polydentate ligands are about 15% larger by absolute values than for ammonia and that $T\Delta S$ values are more positive (e.g. for the mono-ethylenediamine complex the expected values are $\Delta H = 37.6\,kJ\,mol^{-1}$ and $T\Delta S = 5.7\,kJ\,mol^{-1}$, in reasonably good agreement with experimental values).

Intuitively the macrocyclic effect also is expected to be associated with the more favorable entropy of complexation because upon coordination a linear ligand loses to some degree free rotations around its single bonds while in a macrocycle these rotations are already partially lost (although macrocycles possess a considerable degree of conformational flexibility, see Section A 5). Thus the presence of 15 single bonds in pentaglyme around which free rotation may be lost upon complexation could result in a ΔG disadvantage of about $17\,kJ\,mol^{-1}$, if one uses the increments discussed in Section A1. The experimental values of ΔH and ΔS of complexation (Table A 4), however, show that the macrocyclic effect is both of entropic and in fact more often of enthalpic origin. For the polyamines the effect is completely enthalpic; $T\Delta S$ values are even more negative for the macrocycle. For the polyethers the difference between binding affinity is predominantly enthalpic for Na^+, Cs^+, Ba^{2+} and Pb^{2+}, but predominantly entropic for K^+ and Rb^+. The enthalpic difference is the result of the build-up of *gauche* conformations along open chains, which is necessary to bind the ion by simultaneous contact with all oxygen atoms.

It should be noted that there are many additional factors which can contribute to the observed values of ΔH and ΔS in solution, and their interpretation is rather complicated, see Section A 4. The enthalpy of complexation by a polydentate ligand may change as result of inductive effects of bridging groups which connect donor atoms, and of the change of transoid to *gauche* O–C–C–O conformations in the open and cyclic structures, respectively. Large and frequently compensating effects in ΔH and ΔS can arise from solvation or desolvation of metal cation, ligand and the complex. With macrocycles one can expect relatively poor solvation of donor groups inside the cavity and generally the solvation or desolvation contribution is much less for the ligand (as long as it is neutral) compared to the ion, see Section C 2. Important information about the role solvents play in the complexation reactions comes from the study of host–guest complexation equilibria in the gas phase (see below).

Further increase in the complex stability is observed on passing to bicyclic cryptand ligands; the phenomenon is referred to as *cryptate effect*. A comparison of complexation properties of mono- and polycyclic ligands must take into account first of all the fact that bicyclic ligands allow a larger number of donor atoms to be in a contact with the metal cation in a three-dimensional cavity. Thus, higher stability constants for the [2.2.2]-cryptand as compared to 18-crown-6 (Table A 2) may be due simply to the fact that the former

involves two additional nitrogen donor atoms. This aspect was analyzed in terms of the additivity principle.[38] The complexation free energy was expressed as a linear function of group electron donor parameters (ED) which stem from independent measurements of stability of hydrogen bonded associates (Section B 3 and Table I in the Appendix).

$$-\Delta G = \text{const} + \Sigma ED \qquad \text{(A 2-3)}$$

When *all* available stability constants of crown ethers and cryptands are used, a very poor correlation in the coordinates of equation (A 2-3) is observed. However, when macrocycles which were either too small or too large for a given cation, and therefore cannot materialize all possible metal cation–donor atom interactions simultaneously, are excluded, the correlation is very good, Fig. A 5. In terms of this approach there is no special 'cryptate effect' in complex stability: donor atoms of both monocyclic and bicyclic ligands possess the same binding ability manifested in general linear correlation of ΔG values for both crown ethers and cryptands with the same ED parameters. Stronger binding of metal cations to cryptands results from larger number of pairwise metal cation–donor atom interactions.

The gas-phase thermodynamic parameters for complexation are of obvious importance for the understanding of the 'effects' discussed in this chapter as well as of the hole-size concept (Section A 3), since they are free of any solvation contributions. The absolute values of these parameters are rather inaccurate owing to serious experimental difficulties in measuring the partial pressures of involatile ligands (Section D 3.5), but relative values and even indirect kinetic results provide already valuable insights. The gas-phase macrocyclic effect was evaluated from so-called reaction efficiencies (rate constants divided by the collision rate constant calculated theoretically) of formation of 1 : 1 and 1 : 2 alkali cation–glyme or crown ether complexes in the gas phase, which appeared to be

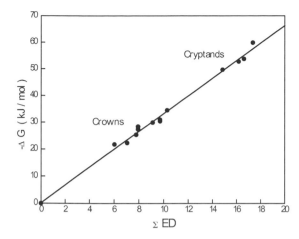

Figure A 5. Correlation of experimental complexation free energies (in methanol) for crown ether and cryptand complexes with the sum of group electron donor parameters (ED).[38] Reprinted with permission from *J. Am. Chem. Soc.*, Schneider *et al.*, 1993, **58**, 3648. Copyright 1993, American Chemical Society.

several times higher for macrocyles.[39] The available data on the gas phase cryptate effect refer to free energies of cation transfer reactions between several macrocyclic and macrobicyclic ligands such as

$$M(18\text{-crown-6})^+ + [2.1.1]$$
$$\rightleftharpoons M[2.1.1]^+ + 18\text{-crown-6} \qquad (A\,2\text{-}4)$$
$$M(21\text{-crown-7})^+ + [2.2.1]$$
$$\rightleftharpoons M[2.2.1]^+ + 21\text{-crown-7} \qquad (A\,2\text{-}5)$$

at 350 K.[40] Representative values for pairs of macrocyclic or macrobicyclic ligands possessing the same number of donor atoms are given and compared with the same values calculated for the methanol solution in Table A 5.

The trend is quite clear and is essentially the same in the gas phase and in solution: small cations capable of inclusion into the three-dimensional cavity of a cryptand form more stable complexes with a bicyclic ligand. With large cations binding to cryptate is thought to occur outside the macrocycle; the more flexible crown ethers then provide a larger number of pairwise contacts with the cation. Apparently the cryptate effect is principally due to more efficient interactions of the guest cation with donor atoms rather than to a desolvation of the receptor interior cavity.

A 3. COMPLEXATION SELECTIVITY, THE HOLE-SIZE CONCEPT AND ITS LIMITATIONS

The remarkable selectivity of metal cation binding to macrocycles illustrates the principle of molecular recognition and is the basis for many applications. Data shown in Table A 2 and Fig. A 6 are classical examples of selective complexation with 18-crown-6 and cryptands, respectively. Apparently the

Table A 5. Selected standard free energies for cation transfer reactions between macrocyclic and macrobicyclic ligands in the gas phase and in methanol at 350 K.[40] Reprinted with permission from *J. Am. Chem. Soc.*, Chen *et al.*, 1996, **118**, 6335. Copyright 1996, American Chemical Society.

	ΔG (kJ mol^{-1})			
	Reaction A2-4		*Reaction A2-5*	
M^+	*Gas phase*	*Methanol*[a]	*Gas phase*	*Methanol*[a]
Li$^+$	− 15.34	− 47.2	− 10.18	
Na$^+$	− 12.26	− 16.5	− 9.58	− 46
K$^+$	6.10	19.4	− 8.58	− 24
Rb$^+$	10.45	11.8	2.01	− 14.3
Cs$^+$	7.77	5.5	14.54	4.0

[a] Calculated for 350 K from values of thermodynamic parameters given in refs. 7 and 35 assuming ΔH and ΔS to be independent of temperature and $\Delta G = 0$ for complexation of Li$^+$ by 18-crown-6 in methanol.

macrocycles select cations primarily by their sizes, and the most stable complexes are formed with the closest match between host cavity and cation size.

Cavity sizes can be obtained from CPK models, or better calculated by computer aided molecular modeling (Section E 7),

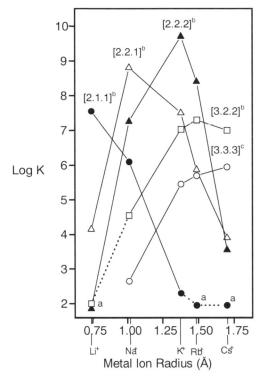

Figure A 6. Size selectivity of cryptands; affinity (lg K) *vs.* ion diameters; (a), values with lg K < 2.0, (b) in 95% MeOH, (c) in MeOH.[41] Reprinted with permission from *J. Am. Chem. Soc.*, Lehn *et al.*, 1975, **97**, 6700. Copyright 1975 American Chemical Society.

which usually agrees with X-ray data. For many macrocycles crystal structures both for free ligands and their metal complexes are available,[42] as are the crystallographic radii of cations[43] (see Table in the Appendix). For example, the diameter of the cavity in 18-crown-6 from crystallographic data equals 2.67–2.86 Å. Thus, the best match among alkali cations is with K^+ (diameter 2.76 Å) and among alkaline earth cations with Ba^{2+} (diameter 2.70 Å), which indeed form the most stable complexes[35b] with 18-crown-6 (Section A 2, Table A 2). Figure A 7 visualize the relation between macrocycle and cation size, showing structures of a series of complexes of the same cation with ligands of different sizes.

The concept of optimal complementarity between guest size and host cavity (hole) is known as *'ion-cavity'* or, more frequently, as *'hole-size'* concept. Although it seems very attractive due to its simplicity, and because it can be considered as a manifestation of famous 'lock and key' principle (Section A 1), this concept has some important limitations, especially for less rigid systems.[44] Let us consider first the selectivity of crown ethers. Fig. A 8 shows that precise size matching exists in only two cases: 18-crown-6 and K^+, 21-crown-7 and Cs^+. In both cases the expected selectivities are observed. However, the smaller 15-crown-5 prefers K^+ in spite of expected better match with Na^+. Even the smaller 12-crown-4 has no selectivity and very little affinity to the Li^+ cation, which is closest in size.

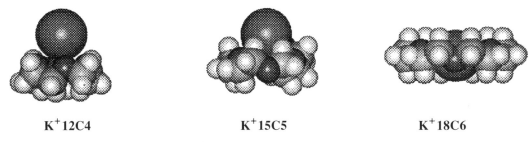

K⁺12C4   **K⁺15C5**   **K⁺18C6**

Figure A 7.   Structures of K^+ complexes of 12-crown-4, 15-crown-5 and 18-crown-6.

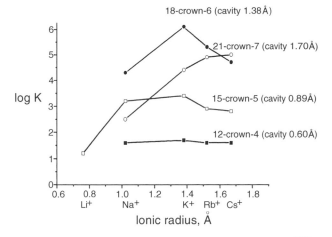

Figure A 8. Logarithms of the binding constants (average of published results),[7,35] of alkali cations by crown ethers in methanol *vs.* ionic radii. In case of Li$^+$ with majority of crown ethers log $K < 1$.

It is interesting to compare binding selectivities of crown ethers, cryptands with natural ionophores such as valinomycin. From the data given in Table A 2, Fig. A 6 and Fig. A 8 one can conclude that the valinomycin displays the highest K$^+$/Na$^+$ selectivity: $\Delta \log K_{K/Na} = 3.8$ for valinomycin, 1.7 for 18-crown-6 and 2.8 for cryptand [2.2.2] (all in methanol). The search for enhanced binding selectivity leads to development of some macrocycles which are even significantly superior to valinomycin in K$^+$/Na$^+$ discrimination, or possess high selectivity to other cations. An important family of such highly selective macrocycles involves calix-crown ethers **A-12**, **A-13** (other highly selective ligands will be discussed in Section A 5 in connection with the pre-organization principle). The selectivity of these ligands is illustrated in Fig. A 9. As one can see, ligand **A-12** in its alternate conformation exhibits $\Delta \log K_{K/Na} > 6$.[45]

Deviations from the hole-size concept were found also for the complexes of transition metal cations with azamacrocycles.[46] The destabilization by torsional strain in *gauche*-like fragments is particularly evident in the sharp drop of the complexation $-\Delta G$ value if one tries to complex, for example transition

metal ions with α,ω-diamines with more than two or three carbon atoms separating the binding sites: longer chains require many unfavorable torsional arrangements, and the in principle strong N–M^{n+} interaction requires the N–M^{n+} distance to be kept within small margins. It has been shown by molecular mechanics calculations that with ethylene and propylene diamines one views the ensuing metal complexes best as metalla-cyclanes, which accommodate large metal ions better with the seemingly smaller ethylenediamine ligand, whereas smaller ions can form with propylenediamine cyclohexane-like structures with almost ideal torsional angles around the ring: here the N–M^{n+} distance is almost the same as the C–C distance in cyclohexane itself (Fig. A 10).[37d,46] Therefore, in contradiction to the simple hole-size concept one observes, for example, that the larger 14-aneN$_4$ binds better small cations such as Ni^{2+}, whereas the smaller 12-aneN$_4$ favors large cations such as Pb^{2+}, Fig. A 11.

The hole-size concept apparently works better for more rigid bicyclic molecules which display a size-match selectivity for both small and large cations, Fig. A 6. The peak selectivity of cryptands manifested in K (or ΔG) is also seen in ΔH values,[47] although maxima in ΔH

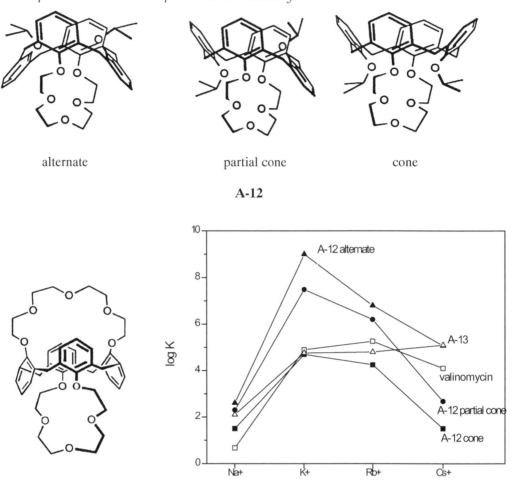

Figure A 9.   Selectivities of calix-crown ethers **A-12** and **A-13** as compared to valinomycin.[45]

for [2.2.1] and [2.2.2] cryptands correspond to larger cations than the maxima in ΔG. Nevertheless, the dominating role of the enthalpy of complexation in cation selection by cryptands is quite evident from these data.

Other factors to be considered are relative Lewis acid strength and solvation of cations, both decreasing on going from Li$^+$ to Cs$^+$. *Ab initio* quantum mechanical calculations suggest that in the absence of any solvent, 18-crown-6 should form most stable complexes with Li$^+$ among alkali cations, and with Mg^{2+} among alkali earth cations.[48] However, inclusion of only few water molecules in the calculations changes the selectivity to K$^+$ and Ba^{2+}, owing to very strong solvation of smaller cations.[48] Thus the actual selectivity is by a balance between solvation and complexation energies rather than only by complementarity. One can expect therefore the selectivity to vary in different solvents as exemplified by data in Table A6 for the

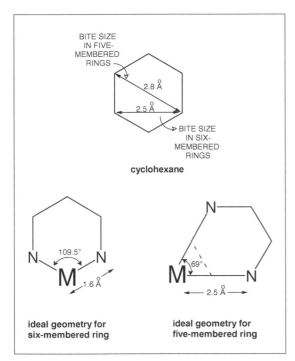

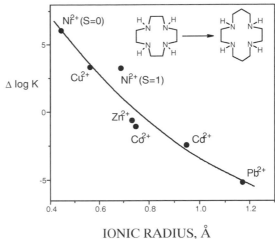

Figure A 11. The effect of increasing chelate ring size.[46] Reprinted with permission from *Progr. Inorg. Chem.*, Hancock, R.D. 1989, **37**, 187. Copyright 1989 American Chemical Society.

Figure A 10. Optimum geometry of six-membered and five-membered rings.[37a,d] Reprinted with permission from *Chem. Rev.*, Hancock, *et al.*, 1989, **89**, 1875. Copyright 1992 American Chemical Society. Reprinted with permission from *J. Chem. Ed.*, Hancock, *et al.*, 1992, **69**, 615. Copyright 1992 American Chemical Society.

binding of alkali cations in acetonitrile. Evidently 15-crown-5 changes its selectivity in this solvent and now prefers Na^+ over K^+ while 18-crown-6 is still selective for K^+, although not as distinctly as in methanol. Both crowns form in acetonitrile considerably more stable complexes with Li^+.

These observations can be rationalized if the free energies of cation transfer from methanol to acetonitrile, given in Table A 6, are taken into account: acetonitrile destabilizes small, and stabilizes large cations, thus favoring the binding of the former. The solvation energies of larger cations Rb^+ and Cs^+ are smaller than those of $Li^+–K^+$. This explains the applicability of the hole-size concept for large macrocycle 21-crown-7. Gas-phase cation complexation results refer to free energies of transfer reactions (A 3-1), Table A 7, and also shed light on the hole-size concept.[39] Evidently, transfer to larger macrocycle, which provides more pairwise interac-

Table A 6. Logarithms of the binding constants of alkali cations by crown ethers in acetonitrile[7,35]; compare with data in methanol from Figure A8.

Cation	15-crown-5	18-crown-6	ΔG of cation transfer from MeOH to MeCN, (kJ mol⁻¹) (see Table C2-1),
Li^+	3.6	3.7	25.9
Na^+	5.0	4.5	5.4
K^+	2.9	5.6	− 2.1
Rb^+	4.0	5.2	− 3.35
Cs^+	3.1	4.9	− 4.6

Table A 7. Standard free energies for cation transfer reactions between different macrocycles, reaction (A 3-1). Reprinted with permission from *J. Am. Chem. Soc.*, Chu *et al.*, 1993, **115**, 5736. Copyright 1993 American Chemical Society.

	$\Delta G \ (kJ \, mol^{-1})$	
Cation	Gas phase	Methanol[a]
Li^+	-7.5	
Na^+	-9.3	15.1
K^+	-3.9	10.4
Rb^+	-11.4	3.0
Cs^+	<-16	-1.3

[a] Calculated from data of refs. 7 and 35

tions with metal cations, is always favorable, although in methanol the transfer ΔG is negative only for the largest cation Cs^+ in agreement with the hole-size concept. However, by its absolute value the transfer ΔG is smallest for K^+; this can be interpreted as by the better fit to the 18-crown-6 macrocycle.[39]

$$M(\text{18-crown-6})^+ + \text{21-crown-7}$$
$$\rightleftharpoons M(\text{21-crown-7})^+ + \text{18-crown-6}$$
$$(A 3-1)$$

A 4. ENTHALPY AND ENTROPY CONTRIBUTIONS AND COMPENSATIONS; HEAT CAPACITY CHANGES

The components of the complexation free energy: ΔH and $T\Delta S$ of complexation are mechanistically important parameters which were already touched upon in connection with chelate and other effects in the preceding chapters. These parameters can be determined either by calorimetry (Section D 5), or from the temperature dependence of the formation constant K according to the van't Hoff isochore

$$d \ln K / dT = \Delta H / RT^2 \qquad (A 4-1)$$

The usual assumption is that in a small temperature interval ΔH can be considered as temperature-independent and, accordingly,

the integral form of the equation (A 4-1) can be presented as

$$R \ln K = -\Delta H / T + \Delta S \qquad (A 4-2)$$

Then ΔH is calculated as the slope of a plot of $R \ln K$ *vs.* $1/T$ and ΔS from ΔG (usually at 298 K) and ΔH.

The temperature dependence of ΔH is given by

$$(d\Delta H / dT)_P = \Delta C_P \qquad (A4-3)$$

and the aforementioned assumption of a constant ΔH value implies that $\Delta C_P = 0$. The available data on ΔC_P values for various host–guest equilibria show, however, that this parameter is often as high as 300–400 J $mol^{-1}K^{-1}$.[35c,49] With such values of ΔC_P the variation of ΔH in a temperature interval of, for example, 50°C is 15–20 kJ mol^{-1}. Unfortunately the temperature dependence of ΔH is often simply ignored.

Since ΔC_P itself also depends on temperature and there are many different expressions which describe this, there are also many equations describing the temperature dependence of K.[50] Often the fitting of experimental results in commonly used temperature intervals of less than or equal to 50°C is equally good within the experimental errors with the simplest equations, which assume a constant ΔC_P, as well as with more complicated equations which take into account the temperature dependence of this parameter. If $\Delta C_P = $ constant, integration of equation (A 4-3) gives

$$\Delta H = \Delta H_i + \Delta C_P T \qquad (A 4-4)$$

where ΔH_i is an integration constant. Then integration of equation (A 4-1) leads to the following expression, known as the Valentiner equation[50]

$$R \ln K = -\Delta H_i / T + \Delta C_P \ln T + I \qquad (A 4-5)$$

where I is another constant of integration.

For the entropy

$$T(d\Delta S / dT)_P = (d\Delta H / dT)_P \qquad (A 4-6)$$

and with the same assumption of $\Delta C_P =$ constant

$$\Delta S = \Delta S_i + \Delta C_P \ln T \qquad (A\,4\text{-}7)$$

Now, taking into account that $\Delta G = -RT \ln K$, one can easily obtain an expression for I

$$I = \Delta S_i - \Delta C_P \qquad (A\,4\text{-}8)$$

Equation (A 4-5) contains three parameters to be calculated from the fitting of the experimental results and, of course, one needs sufficient curvature and more experimental points than for a linear regression with equation (A 4-2). One gains, however, the possibility of calculating ΔH at any temperature as well as the additional important parameter ΔC_P. Equation (A 4-5) was applied, for example to the analysis of complexation thermodynamics of some arene molecules with cyclophanes (Section B 4).[49]

The values of ΔH and ΔC_P found by calorimetry and from the temperature dependence of K sometimes differ considerably, and preference is always given to those determined by calorimetry. First of all, this is because precise determination of ΔH and, especially, ΔC_P requires very high precision in equilibrium constants. Error analysis shows,[51] that for a 1:1 association reaction a relative error of $\pm 5\%$ in K (± 0.02 in log K), which by ordinary standards is considered to be very good, leads to errors in ΔH and ΔC_P at 25°C equal to $\pm 0.8\,\text{kJ}\,\text{mol}^{-1}$ and $\pm 120\,\text{J}$ $\text{mol}^{-1}\,\text{K}^{-1}$, respectively, if the measurements are made in the temperature interval 5–50°C; or $\pm 4\,\text{kJ}\,\text{mol}^{-1}$ and $\pm 2800\,\text{J}\,\text{mol}^{-1}\text{K}^{-1}$, respectively, if the measurements are made in a temperature interval of only 20–30°C. Perhaps even more importantly than the error propagation, the determination of the thermodynamic parameters from the temperature dependence of K often suffers greatly from inconstant reaction conditions over wide temperature intervals. These effects may be larger than statistical errors, and involve temperature-dependent contributions of side reactions, shifts in pH values of buffers, of

medium properties and/or calibration parameters, etc. In general, the temperature dependence of an equilibrium constant rarely allows a reliable determination of ΔC_P and this parameter preferably should be determined by calorimetry. Determination of ΔH in the absence of significant systematic errors is more precise but, of course, gives a mean value of ΔH over a temperature interval used rather than ΔH at a single temperature (as it is often interpreted).

The interpretation of experimental ΔH, ΔS and ΔC_P values of complexation at a molecular level is rarely straightforward. All changes in energetics and geometry of interacting species, which accompany host–guest complexation, such as formation of new non-covalent bonds between host and guest molecules, and breaking of previously existing bonds with solvent molecules in the first solvation shell, changes in the second sphere solvation, change of inter-binding site interactions, complexation-induced strain, etc., may have both enthalpic and entropic contributions. Nevertheless, the knowledge of thermodynamic parameters can be helpful in diagnostics of the dominant type of interactions in a given system. In Chapter B the reader will find typical values of ΔH, ΔS and ΔC_P for different types of interactions responsible for the host–guest complexation. It should be noted that the less popular parameter ΔC_P is fairly sensitive to structural changes upon complexation and by itself provides insight into the binding mechanism. For example, water release upon complexation usually leads to a characteristic decrease of the heat capacity.[52] On the other hand, formation of new bonds can lead to increased heat capacity. The apparently trivial dependence of standard thermodynamic parameters on chosen standard conditions is often overlooked. While ΔH depends only on the standard temperature, ΔG and ΔS depend also on the concentration. In Sections A 1 and A 2 we saw already how the change of standard concentrations from 1 M to unity mole fraction modified the association constants (reaction ΔG). Negative standard ΔS, usually considered as 'unfavorable', can change its sign and can give a 'favorable' contribution to ΔG on going to another standard concentration. In particular, for a bimolecular association reaction in water ΔS increases by $R \ln 55.6 = 33.6 \, \mathrm{J \, mol^{-1} \, K^{-1}}$ on going from 1 M to unity mole fraction standard state ($T\Delta S$ at 298 K increases by 10 kJ $\mathrm{mol^{-1}}$).

The values of ΔH and ΔS for a series of reactions, such as the complexation reactions with structurally related ligands (crown-ethers, glymes, cryptands, etc.) or with the same ligand, but in different solvents, often show a 'compensation behavior', that is a mutual linear relationship

$$\Delta H = T_{\mathrm{iso}}\Delta S + h_0 \qquad \text{(A 4-9)}$$

or, at a constant temperature

$$T\Delta S = a\Delta H + T\Delta S_0 \qquad \text{(A 4-10)}$$

where

$$a = T/T_{\mathrm{iso}}.$$

Combining equations (A 4-9) and (A 4-10) with the expression for ΔG one obtains

$$\Delta G = T_{\mathrm{iso}}\Delta S + h_0 - T\Delta S = (T_{\mathrm{iso}} - T)\Delta S + h_0 \qquad \text{(A 4-11)}$$

and

$$\Delta G = \Delta H - a\Delta H - T\Delta S_0 = (1 - a)\Delta H - T\Delta S_0 \qquad \text{(A4-12)}$$

It is evident from these equations that at temperature $T = T_{\mathrm{iso}}$ (or $a = 1$) the value of ΔG becomes constant for the whole series: $\Delta G = h_0 = T_{\mathrm{iso}}\Delta S_0$. This leads to the commonly used term 'isoequilibrium relationship'.

The related phenomenon in kinetics, known as 'isokinetic relationship', has attracted considerable attention and its theory, applicable also to thermodynamics, is quite developed.[53] In general terms it seems reasonable that a more exothermic interaction will produce

more tight binding and, consequently, will be accompanied by a more negative entropy, as expressed in equation (A 4-10). However, a difficult and often ignored problem is the confirmation of statistical reliability of the isoequilibrium or isokinetic relationship. It has been stressed[53b] that the use of ΔH and ΔS values for the correlation is statistically incorrect, because these parameters are not independent variables, and the analysis of numerous literature data showed, that the majority of published isokinetic relationships based on equations (A 4-9) and (A 4-10) were statistical artifacts. A correct statistical treatment of the temperature-variable data is based on the least-squares calculation of the coordinates of the mutual intercrossing point of the lines in Arrhenius or van't Hoff coordinates.[53b] Indeed, if as required the isoequilibrium temperature really exists, all lines in the coordinates $\ln K$ *vs.* $1/T$ must intercross at the point $1/T = 1/T_{\mathrm{iso}}$ where the equilibrium constants of all reactions of a given series coincide: $\ln K_{\mathrm{iso}} = -h_0/RT_{\mathrm{iso}}$. The principal advantage of this treatment is, that it is applied directly to the independently measured equilibrium or rate constants, and not to the secondary, already mutually dependent parameters. It was suggested to use the term 'isoequilibrium relationship' only for the cases for which a correct statistical analysis has been indeed applied; all statistically non-proven simple correlations between ΔH and ΔS are referred then as a 'compensation effect' which may be either real or only apparent.[53c]

The correct treatment of the binding thermodynamics is exemplified with some α-cyclodextrin complexes in Fig. A 12. It is worth noting that these guests belong to a large series of about 50 molecules for which the application of equation (A 4-9) gives a very good compensation correlation between ΔH and ΔS for all guests. The analysis of van't Hoff plots, however, shows the iso-equilibrium relationship to hold only for several groups of structurally related compounds.[53c]

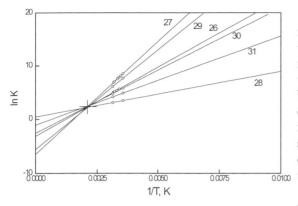

Figure A 12. Van't Hoff plots for the association of α-cyclodextrin with a series of guests derivatives of adamantane and aromatic carboxylic acids ($T_{iso} = 471$ K).[53c] 26 = adamantane-1-ylamine, 27 = L-tryptophan, 28, 29 = hydrocinaminic acids, 30 = L-mandelic acid, 31 = adamantane-1-carboxylic acid. Reprinted with permission from *Chem. Soc. Rev.* Linert, W. 1994, **23**, 429. Copyright 1994, Royal Society of Chemistry.

Numerous compensation effects have been reported for different types of the host–guest complexation, including cation binding to ionophores,[54] and organic host–guest complexation.[55] Many data have been assembled[54a,b,55a] on entropy–enthalpy compensations based solely on equation (A 4-10). The correlation with, for example, less preorganized ligands like glymes was taken as evidence that these suffer from much entropy disadvantage which needs to be 100% compensated by enthalpy contributions. Cryptand and crown ether complexes by this analysis suffer only by 44 or 77% entropic disadvantage. In view of above-mentioned problems, however, it is not clear to which extent these correlations can be considered as true isoequilibrium relationships. The majority of the correlated data are from calorimetric measurements at a single temperature. Evidently, the aforementioned treatment then cannot be applied; in such a case the use of equations (A 4-11) or (A 4-12) is statistically more reasonable than that of (A 4-9) or (A 4-10).[53b]

An interesting question is whether the incremental free energies of pairwise interactions (Section A 1) can be separated to the respective enthalpic and entropic contributions. This has been addressed only recently by an analysis of interaction thermodynamics of mutants of barnase (an extracellular ribonuclease) with its inhibitor barstar. Various substitutions of glutamic acid 73 in the enzyme led to $\Delta\Delta G$ values reasonably close to each other which varied from -7.9 to -12.1 kJ mol^{-1}, while the respective $\Delta\Delta H$ values were very much different and varied form -34.3 to -69.0 kJ mol^{-1}.[29c] Obviously a compensation between $\Delta\Delta H$ and $T\Delta\Delta S$ values leads to nearly constant $\Delta\Delta G$ values in this case; since the nature of this compensation is not clear the authors conclude that only the free energy is a preferable quantity for the analysis of thermodynamics of interactions by incremental approach. The situation is similar to other linear free energy relations with substituent or solvent effects, which to a large degree owe their linearity also to enthalpy–entropy compensations.

A 5. PREORGANIZATION

The preorganization principle states that 'the more highly hosts and guests are organized for binding and low solvation prior to their complexation, the more stable will be their complexes'.[56a] We mentioned already (Section A 1) that multi-site interactions between host and guest molecules are almost invariably accompanied by some energetically unfavorable contributions such as strain, restrictions of internal rotations, desolvation, etc. Evidently, both molecules will be optimally preorganized if (a) all complementary binding sites geometrically match; (b) in the complexed state they are in the same lowest free energy conformation as in the free state; and (c) polar binding sites need not to change solvation. In this case all distortions will be negligible and the complexation will be energetically most favorable, including only the sum of intrinsic binding free energies. One also may expect the interactions of such preorganized molecules to be highly selective since guests of different structures, such as cations of different sizes, will need different optimal host conformations.

Although the general concept seems to be quite simple it is not easy to materialize it for every particular case and there is still only a limited number of highly efficient preorganized hosts. An often applied strategy to preorganize a host molecule is rigidification with the purpose to obtain ideally a single conformation optimal for complexation. As a rule, creation of properly rigidified macrocycles requires complex synthetic routes. On the other hand, as shown in Section A1, effective complexation requires primarily a sufficient number of complementary interaction sites, which in view of the usually minor entropy effects can also be linked by several flexible bonds. It is, however, essential, that the interaction sites are sterically matching without the build-up of considerable strain, like too many *gauche*-conformations. Most non-covalent interactions tolerate a considerable degree of flexibility. The tolerance depends very much on the type of interaction forces; so Coulomb in contrast to dispersive interactions require a much less tight fit (Sections A1 and B2, B5). In addition, complexation with a rigid host can be kinetically slow (Chapter F), which is undesirable for many applications. Perfect biological protein receptors possess considerable conformational mobility. Fischer's lock-and-key hypothesis originally viewed protein binding sites as rigid structures. Only later was the conformational mobility of proteins discovered, and the induced-fit hypothesis was proposed:[57] structures of free and bound proteins often endure considerable conformational changes upon binding.

Inspection of structural data for complexes of proteins with low molecular weight guests (such as sugar, biotin and anion binding proteins, enzymes) shows the primary impor-tance of a very large *number* of pairwise interactions, more than 10 for the small biotin molecule, see Fig. A1a, but barely evidence for rigid preorganization. Interesting observations related to the role of preorganization in a protein host were made by comparing the crystal structures of a germline antibody Fab fragment and its complex with a nitrophenyl phosphonate hapten to those of the affinity-matured antibody.[58] The matured antibody possesses a 3×10^4 times higher affinity to the hapten than the germline antibody and differs from the latter by nine mutations (aminoacid exchanges). None of the amino acid residues in which mutations have occurred are in direct contact with the hapten, but complexa-tion of the hapten with the germline antibody induces considerably larger structural changes in the protein than complexation with matured antibody. It is worth noting that the strong binding of the hapten to matured antibody ($K = 2.2 \times 10^8$ M^{-1}) is still accom-panied by some noticeable conformational changes and involves direct contacts of bound small hapten molecules with seven amino acid residues of the antibody combining site.

Below we discuss examples illustrating relationships between ligand conformation and complex stability. The dramatic macro-cyclic effect in the reaction shown in Scheme A4 ($\Delta \log K = 9.6$ compared with typical values of $\Delta \log K$ in the range 3–4 for poly-amines, Section A2) is due to the strongly 'unorganized' structure of **A-14**.[37b] The piper-azine ring of **A-14** dominates in the chair conformation in the free ligand and must convert to an energetically unfavorable boat or twist boat conformation in the complex. This complexation-induced conformational change strongly reduces the binding constant and is absent in the preorganized macrocycle **A-15**.

The progressively increased binding in the series of macrocyclic ligands in Scheme A5 can be attributed to preorganization effects. Indeed, X-ray analysis shows that the con-formations of the complexed macrocycles

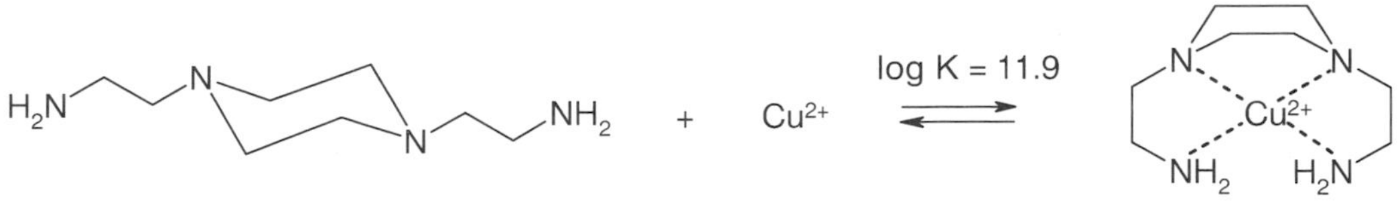

A-14

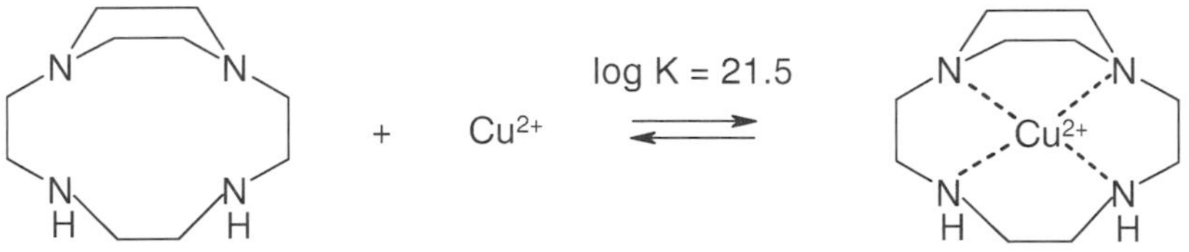

log K = 11.9

A-15

log K = 21.5

Scheme A4. Coordination equilibria of Cu(II) with acyclic and macrocyclic polyamines derived from piperazine.[37b] **A-15**) Reprinted from *Coordination Chemistry Review*, Vol. 133, Martell *et al.* (Factors affecting stabilities of chelate, macrocyclic and macrobicyclic complexes in solution), pp. 39, copyright 1994 with permission from Elsevier Science.

(exemplified by the structure of the complex with unsubstituted macrocycle) are similar for all three ligands, but conformations of the free macrocycles are affected by *gem*-dimethyl substituents in such a way that each pair of the substituents progressively shifts the macrocycle conformation to the one which it adopts upon metal coordination.

Considerable preorganization effects were observed for podands **A-16** and **A-17**, which imitate structural elements of polyether antibiotics like monensin (Section A2, Fig. A4, **A-8**). These and many of other hosts discussed below are insoluble in water, and their metal cation complexation (Table A8) was studied by the picrate extraction method, which limits comparison with other data (Section D4.2). The data in Table A8 show considerable improvement of binding in passing from glymes possessing the same number of oxygen atoms to ligands **A-16** and **A-17**. The latter forms even more stable complexes than 12-crown-4 and approaches the binding properties of 15-crown-5. Molecular mechanics calculations show as the lowest energy conformations of **A-16** and **A-17** structures which resemble a segment of 18-crown-6. The unfavorable *gauche* interactions in the glyme are

in **A-16** and **A-17** balanced out by the additional C–C bonds of the cyclohexyl moiety.

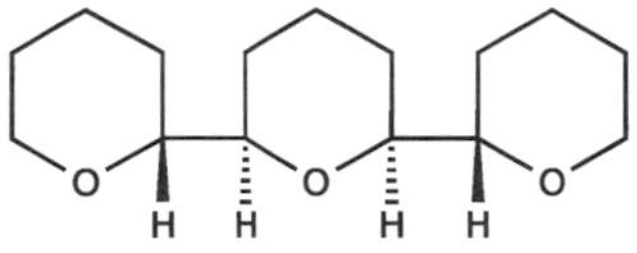

A-16

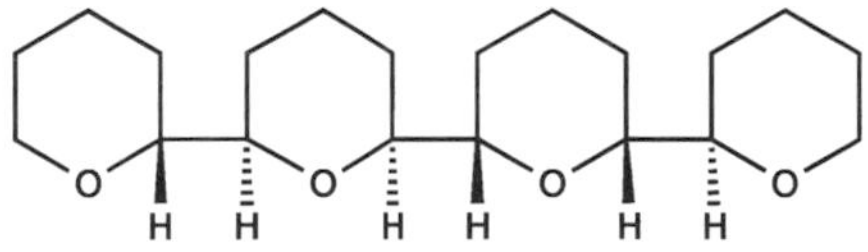

A-17

Considerable efforts were made to reduce the flexibility of 18-crown-6 and to fix it in the optimal binding conformation.[61] This macrocycle is, actually, one of the most 'organized' ionophores which undergoes a minimum conformational change on complexation with its specific guest K^+ (Section A3) and, so, one should not expect a large improvement of its binding properties by further preorganization. Nevertheless, the diequatorial (*ee*) diastereomer of macrocycle **A-18a** shows strongly enhanced binding of Na^+ and reversed Na^+/K^+ selectivity as compared to

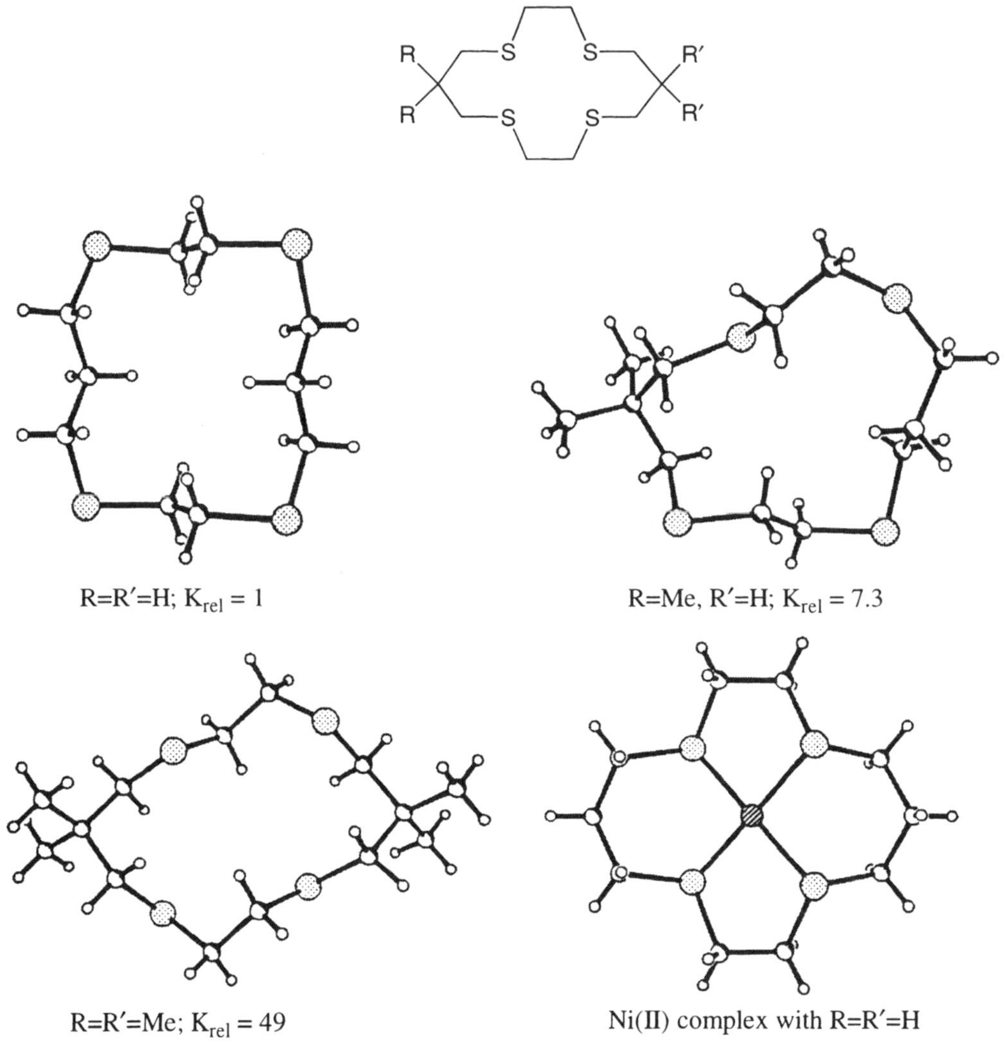

R=R'=H; $K_{rel} = 1$

R=Me, R'=H; $K_{rel} = 7.3$

R=R'=Me; $K_{rel} = 49$

Ni(II) complex with R=R'=H

Scheme A 5. Relative stability constants (in CD_3NO_2) of Ni(II) complexes and conformations of free macrocycles and a Ni(II) complex.[59] Reprinted with permission from *J. Am. Chem. Soc.*, Desper *et al.*, 1991, **113**, 8663. Copyright 1991 American Chemical Society.

18-crown-6, Table A 9. Other diastereomers of **A-18a**, such as axial-equatorial (*ae*) and diaxial (*aa*) combinations are, however, even poorer ionophores than 18-crown-6 itself. The thia macrocycle **A-18b** was obtained only as the diequatorial diastereomer and again shows moderate improvement of binding capacity.

A quite different approach to organized macrocycles was developed with *spherands* as a family of new hosts.[56] These are rigid molecules possessing small spherical cavities with octahedrically arranged oxygen atoms, which are essentially inaccessible for solvent molecules. Typical examples of spherands as well as the 'hybrid' hosts *cryptaspherands* and *hemispherands* are structures **A-6**, **A-23**–**A-26**; **A-19** and **A-20**, respectively.

The general sequence of affinities of various guests towards their most specific guests under conditions of picrate extraction method

Table A 8. Logarithms of association constants (M^{-1}) of alkali metal picrates with **A-16**, **A-17** and other ionophores in $CDCl_3$ saturated with water (picrate extraction method).[60] Reprinted with permission from *J. Am. Chem. Soc.*, Iimori *et al.*, 1989, **111**, 3439. Copyright 1989 American Chemical Society.

Compound	Li^+	Na^+	K^+
A-16	4.83	3.97	< 3.7
$MeO(CH_2CH_2O)_2Me$	< 3.7	< 3.7	< 3.7
A-17	5.48	5.63	4.46
$MeO(CH_2CH_2O)_3Me$	4.08	< 3.7	< 3.7
12-crown-4	4.20	3.86	< 3.7
15-crown-5	5.00	6.61	5.89
18-crown-6(dicyclohexyl)	5.48	6.41	8.20

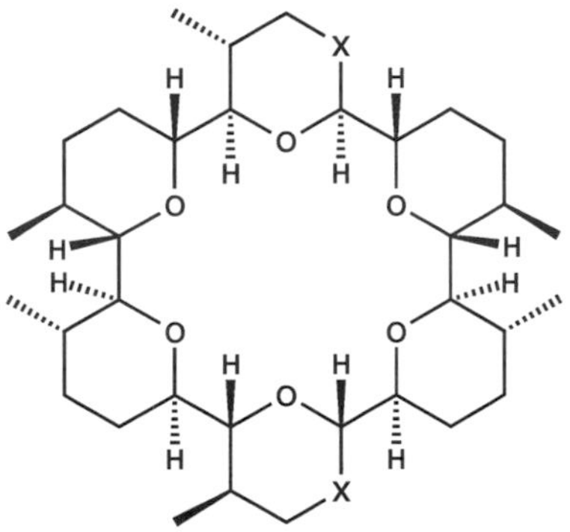

A-18 X = O (a), X = S (b)

is spherands > cryptospherands > cryptands > hemispherands > crowns > podands (see Table A 10). Crystal structures of spherand complexes show that encapsulated Li^+ and

Table A 9. Logarithms of association constants (M^{-1}) of alkali metal picrates with diastereomers of **A-18a,b** $CDCl_3$ saturated with water (picrate extraction method).[61] Reprinted with permission from *J. Am. Chem. Soc.*, Li *et al.*, 1993, **115**, 3804. Copyright 1993 American Chemical Society.

Compound	Li^+	Na^+	K^+
A-18a (ee)	6.30	9.28	8.67
18-crown-6	5.28	6.36	8.30
A-18a (ae)	4.78	4.65	7.50
A-18a (aa)	4.69	4.60	6.95
A-18b (ee)	6.15	7.40	7.97

Na^+ are in contact with six oxygen atoms with distances M^+–O 2.14 Å (Li^+) and 2.28 Å

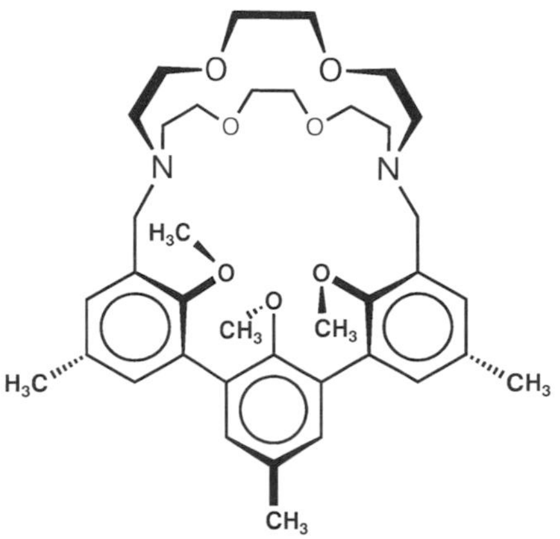

A-19

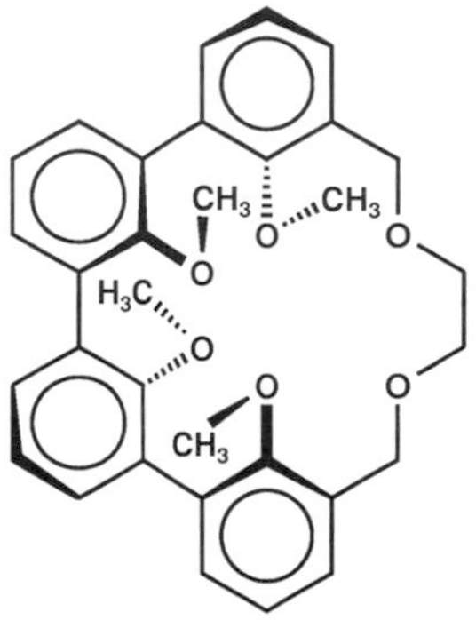

A-20

(Na^+) for **A-7** as a host.[62] The force field calculated K^+–O distance for the potassium complex of the same host equals 2.66 Å.[63] At the same time X-ray analysis of 18-crown-6 complexes of these metals showed Li^+ to be only four-coordinated, with Na^+–O distances between 2.71 and 2.79 Å, and K^+–O distances in the range 2.78–3.05 Å.[42a] Evidently, spherands provide more interaction sites with the small Li^+ cation and bind considerably more tightly with Na^+ and K^+. Together with high affinity these highly preorganized hosts also exhibit very significant selectivity. Thus, spherand **A-6** possesses the highest known Na^+/K^+ selectivity in extraction experiments, and probably also the highest Li^+/Na^+

Table A 10. Negative values of binding free energies ($-\Delta G$, kJ mol^{-1}) of alkali metal picrates to preorganized ionophores in CDCl$_3$ saturated with water (picrate extraction method).[56c] Reprinted with permission from *Angew. Chem., Int. Ed. Engl.*, Cram, D.J., 1983, **25**, 1039.

Compound	Li$^+$	Na$^+$	K$^+$	Rb$^+$	Cs$^+$
A-6	> 96	80	< 25		
A-19	41.4	56.4	79.4	84.8	90.7
A-3		60.2	75.2	70.2	43
A-20	30.1	56.4	44.7	35.1	29.7
A-21	26.3	35.1	47.6	41.4	35.5
A-22	< 25	< 25	< 25	< 25	< 25
A-23	50.6	64.4	65.2	59.4	54.8
A-24	76.5	68.1	51.8	49.3	49.3
A-25	69.4	64.4	45.1	39.3	43.9
A-26	42.9	27.2			

selectivity (it binds Li$^+$ so strongly that only lower a limit of the binding free energy was determined). The cryptospherand **A-19** shows higher K$^+$/Na$^+$ selectivity than cryptand [2.2.2], **A-3**, and **A-20** is more efficient than **A-21**. An obvious tendency of increased selectivity with increased binding exists for these hosts, see also the quantitative treatment below.

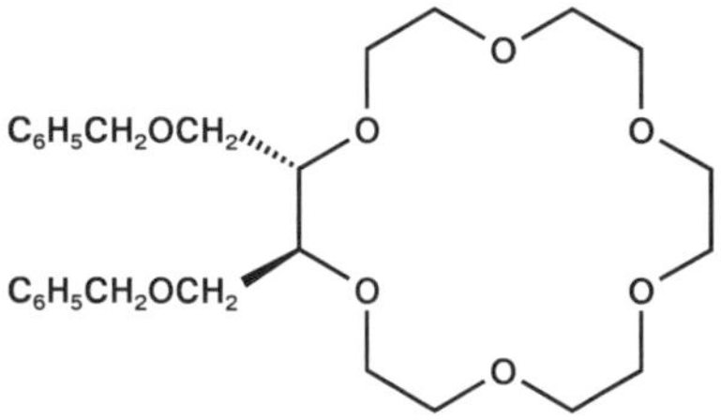

A-21

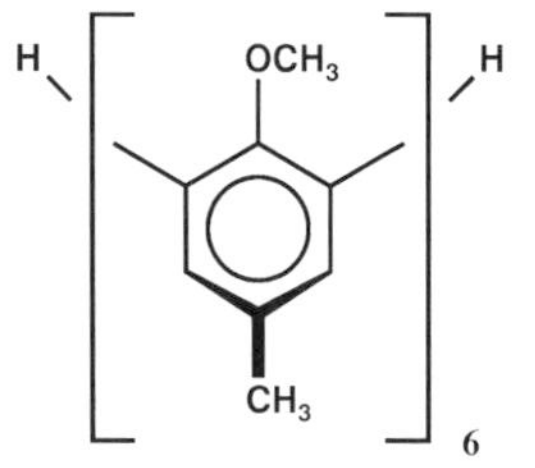

A-22

Comparison of the complexation free energies for **A-6** and **A-26** reveals an unexpected

effect: removal of only one of six oxygen atoms leads to enormous decrease in $-\Delta G$.

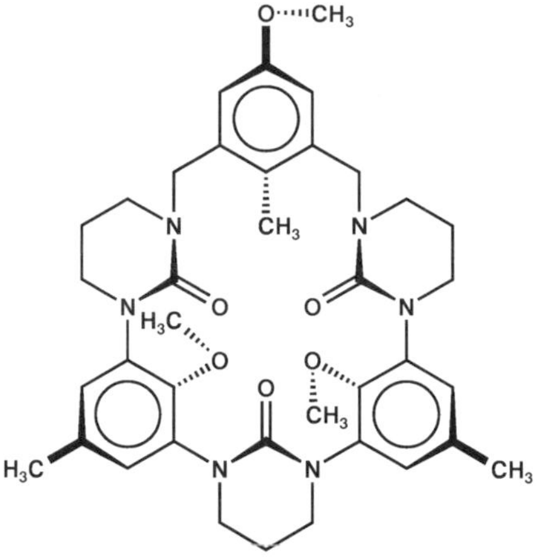

A-23

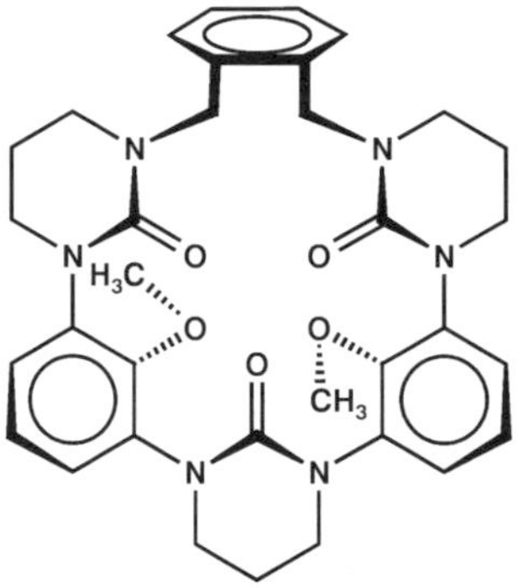

A-24

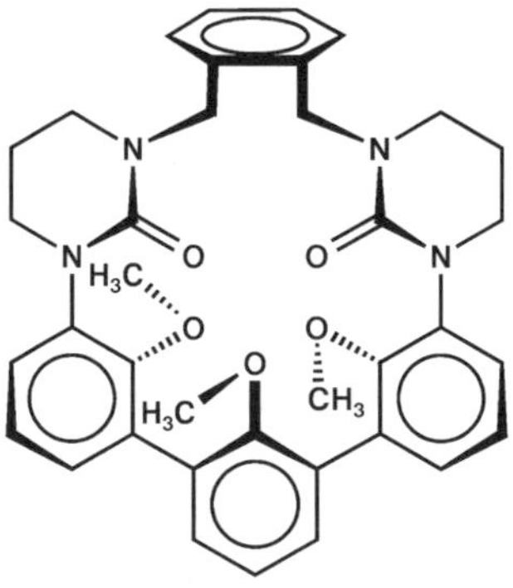

A-25

Similarly, a large difference exists between 18-crown-6 and 18-crown-5,[64] although in 18 C5 only one out of six oxygens is missing. The discrepancy is strongly solvent-dependent:

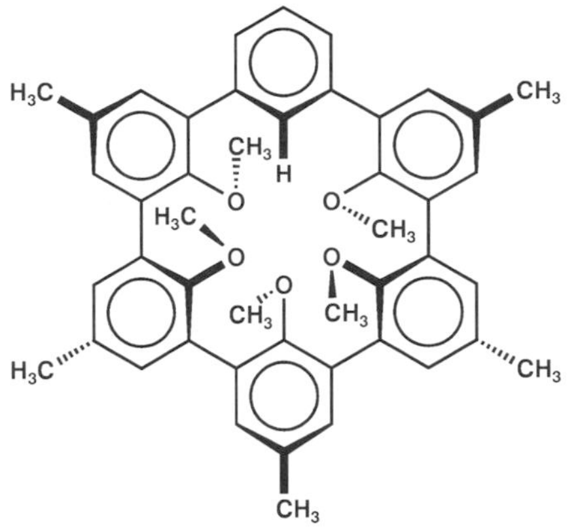

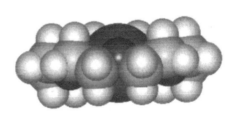

A-26

the differences between $-\Delta G$ values for complexation of K^+ by 18C6 and 18C5 are in MeCN 15.8; in MeOH 18.6 but in water only 4.1 kJ mol^{-1}, respectively. NMR studies and molecular mechanics calculations show that conformations of the two crown ethers in potassium complexes with both ligands are similar, but that with 18C5 the cation is expelled out of the cavity (Fig. A 13) owing to one C–H bond pointing inwards the cavity.[64] The smaller effect in water was explained by a stronger solvation of 18-crown-6 cavity: molecular dynamic water box calculations (Section E 7) show nine H-bonds between water and oxygen instead

of only 4 H-bonds with the smaller 18-crown-5 cavity.

Recently much effort has been directed towards the development of calixarene-based ionophores, which allow sterically fine-tune complexation of many alkali ions.[65,66] A series of conformationally rather rigid 1,3-dialkoxy-*p-tert*-butylcalix[4]arene crown ethers of the general formula **A-27** was prepared, which have cone, partial cone or 1,3-alternate conformation. All ligands are selective for K^+, but the stereoisomers show considerably different affinities, for example cone, partial cone and alternate isomers of **A-27c** bind potassium picrate (by extraction technique) with $\log K = 7.08$, 9.95 and 8.15 respectively. Figure A 14 illustrates a quantitative correlation between selectivity and affinity, which holds for K/Na and K/Cs, but not for K/Rb. As in a case of spherands, rigid calixarenes possess slow complexation kinetics.

The preorganization aspect has been discussed extensively for siderophores as an important class of biological ligands, especially for enterobactin and its models.[67–69] Siderophores are microbial iron transport agents which form strong and selective complexes with Fe(III). In particular, enterobactin (**A-9**, in Fig. A 4) forms a triscatecholate Fe(III) complex with $\log K = 49$.[67b] Of course, one should remember that this huge constant is calculated for the completely deprotonated ligand with a charge of -6; in neutral

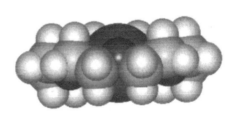

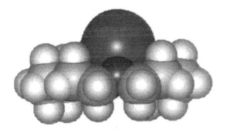

K$^+$18C6 **K$^+$18C5**

Figure A 13. Structures of potassium complexes of 18-crown-6 and 18-crown-5.[64] Reprinted with permission from *J. Org. Chem.*, Raevsky *et al.*, 1996, **61**, 8113. Copyright 1996 American Chemical Society.

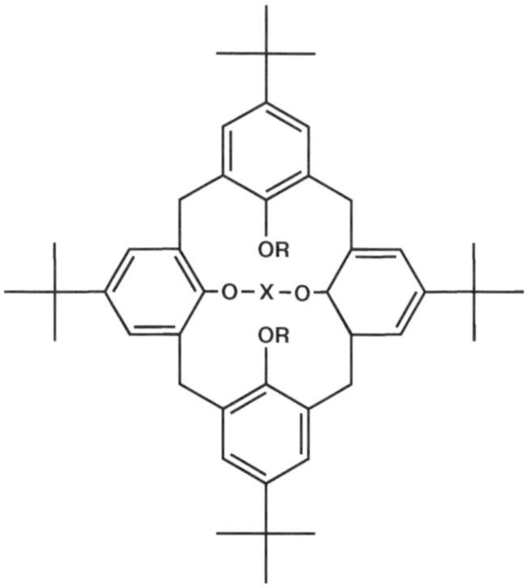

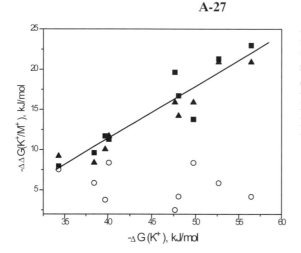

X = $CH_2CH_2(OCH_2CH_2)_n$
a) R = CH_3, n = 3
b) R = $CH_2C_6H_5$, n = 3
c) R = C_2H_5, n = 3
d) R = C_2H_5, n = 4
e) R = n-C_3H_7, n = 3
f) R = i-C_3H_7, n = 3

A-27

Figure A 14. Correlations between differences of complexation free energies of K^+ and Na^+ (squares), K^+ and Cs^+ (triangles), and K^+ and Rb^+ (circles) and free energy of complexation of K^+ by preorganized calixarenes **A-27a–f**.[65a] Reprinted with permission from *J. Am. Chem. Soc.*, Ghidini *et al.*, 1990, **112**, 6979. Copyright 1990 American Chemical Society.

petition experiments (Section D1) using EDTA as the competing ligand with known stability constant.[67d] Many enterobactin models were synthesized amongst which for a long time the best was a triamide ligand known as MECAM (complex A-28) with $\log K = 43$.[67b]

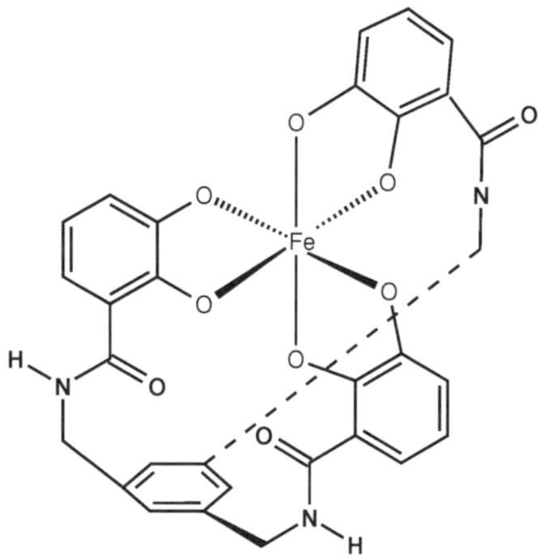

A-28

solutions the observed stability constant is much smaller, but still very large. Thus, at pH 5 when the dominant form of the ligand is neutral, the apparent (conditional) stability constant equals $10^{20.5}$ M^{-1} and cannot be determined by ordinary titration since metal is completely bound even at low µM ligand concentrations; its value was found by com-

Molecular mechanics(MM) calculations of conformations of free and bound enterobactin and MECAM ligands established that enterobactin is far from being perfectly preorganized: its complexation with Fe(III) induces a strain energy of 46 kJ mol^{-1}.[68b] The calculated complexation-induced strain energy for

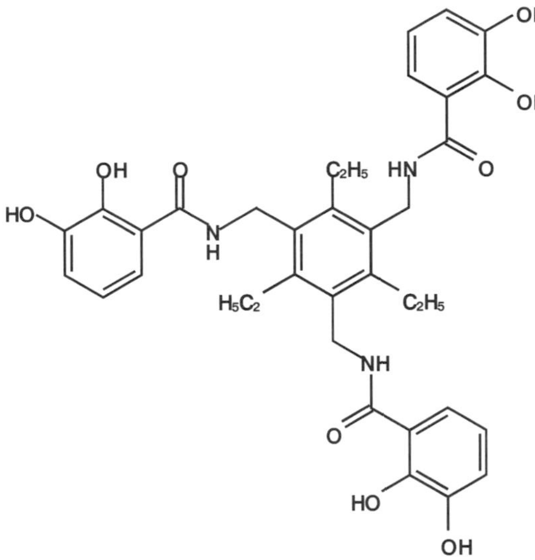

A-29

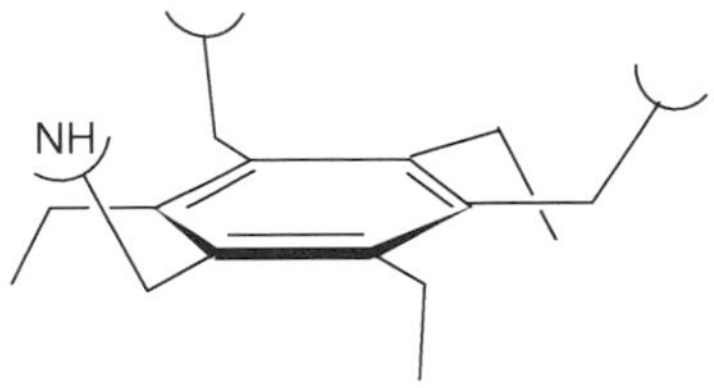

A-30a

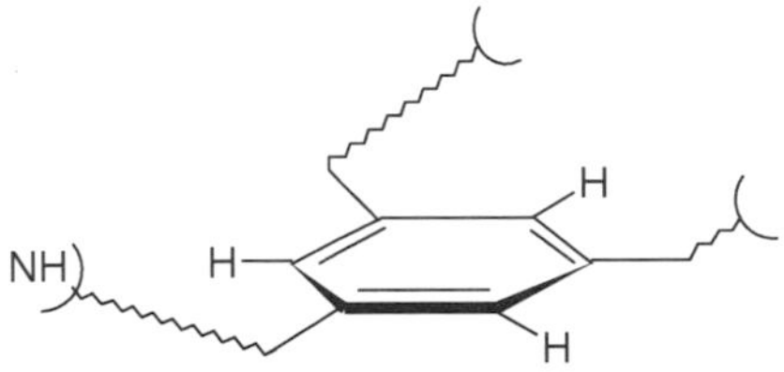

A-30b

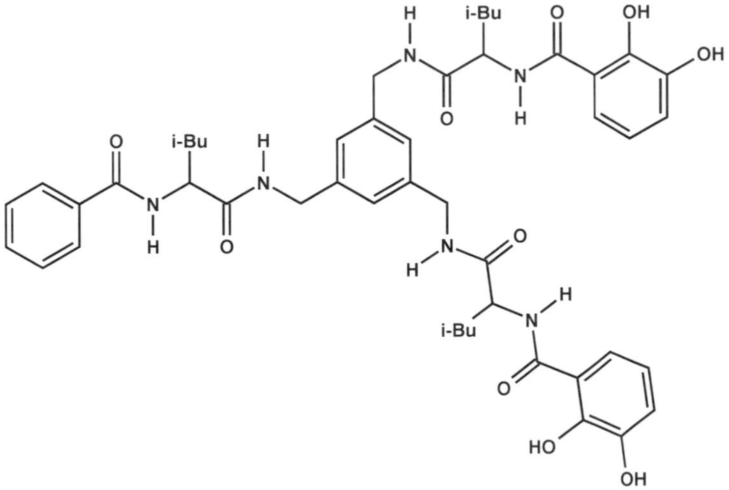

A-31

MECAM was 54.3 kJ mol^{-1}, that is about 8 kJ mol^{-1} higher than for enterobactin. This explains, however, only a small part of observed difference in the binding constants. It was concluded that the principal factor is the higher conformational freedom of MECAM. The higher rigidity of enterobactin was explained by intramolecular hydrogen bonding between amide and catechole groups. Such bonding is possible also in the MECAM molecule, but in this case the methylene groups, which do not participate in H-bonding, give high conformational mobility to the whole molecule. Further improvement of MECAM was achieved by introduction of ethyl groups into the benzene ring, A-29.[67b] Ethyl groups stabilize conformation A-30a in which all catecholeamide groups are preferably directed to one side of the benzene plane, with an also restricted

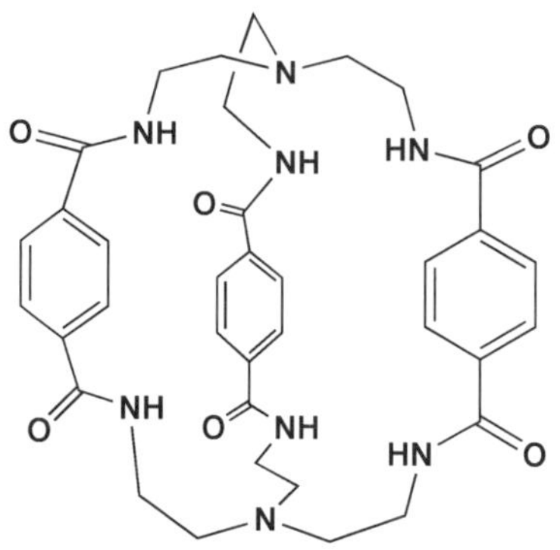

A-32a

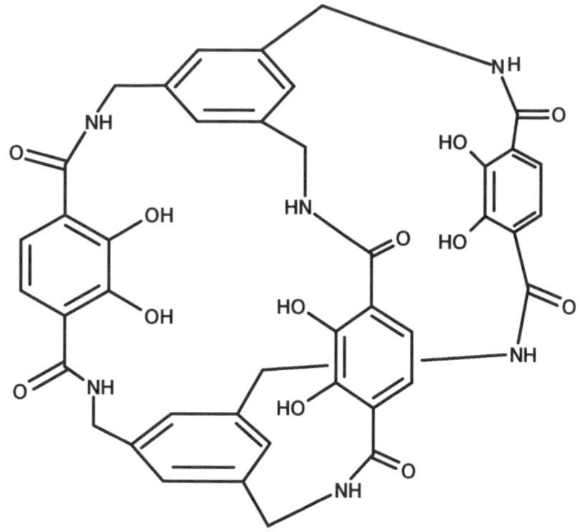

A-32b

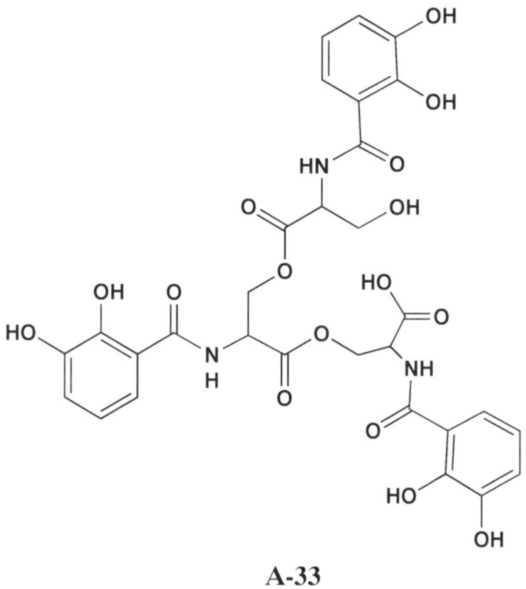

A-33

rotation (compare with MECAM conformation A-30b). As a result the stability constant
for A-29 becomes $\log K = 47$. However, a
significant conformational change of the
ligand is still required for complexation since
catechole groups must rotate 180° to get the
complex conformation like that in A-28.

The possibility of preorganizing model
ligands by intramolecular H-bonding also

was explored. For this purpose amino acid
derivatives like A-31 capable of multiple
intramolecular hydrogen bonding were
prepared.[68a] Stability constants of their
Fe(III) complexes were considerably, sometimes more than 2500 times, higher than for
MECAM. A macrobicyclic cryptand-like ligand
A-32a surprisingly has a binding constant
even smaller than MECAM, $\log K = 43.1$,[67c]

but the structurally related ligand **A-32b** shows a dramatically increased complex stability of log $K = 59$.[68] The results show that efficiencies close to and even higher than the natural receptor can be obtained with synthetic host compounds, which have the additional advantage to be, for example, more stable against hydrolysis.

Although the stability constants of these Fe(III) complexes look extremely large, enterobactin shows most probably nothing more than normal macrocyclic and chelate effects. To evaluate the macrocyclic effect one can compare the stability constant of enterobactin with that of the respective linear ligand – a monohydrolyzed form of native enterobactin **A-33**, log $K = 43$.[67b] The difference of $\Delta \log K = 6$ is explicable in terms of principles discussed in Section A 2 by restriction of rotations around 11 single bonds and the necessity to convert some bonds to the less stable *gauche* conformation upon complexation with the open chain ligand **A-33**. Contribution of the chelate effect can be evaluated from the stability constant of the triscatecholate complex of Fe(III) with pyrocatechole: log $\beta_3 = 43.8$.[70] Using equation A1-3 for the chelate effect (which takes for this case the form log $K = \log \beta_3 + 2 \times \log 55.6$) one obtains log $K = 47.3$, which differs from the enterobactin stability constant by only $\Delta \log K = 1.7$ and exceeds that of **A-33**.

A 6. IONOPHORES FOR ANIONS

In comparison to cations, the area of supramolecular anion complexation has developed more slowly, to some degree restricted to solid state studies, but has attracted increasing attention in recent years.[71–75] Generally, anions possess larger ionic radii and higher solvation energy in protic solvents than cations. They are also of more diverse shapes: spherical, linear, angular etc. Anions receptors can be neutral Lewis acids, neutral proton donors, metal cations or positively charged organic groups, such as ammonium or gua-

nidine centers, which can serve also as proton donors. The closest analogue to macrocyclic hosts for cations, which are based on electron-donor (Lewis base) heteroatoms such as O, N or S, are Lewis acid-type receptors. Hosts of this type (Fig. A 15) are neutral species which involve electron-deficient atoms such as tin,[76] boron,[77] silicon,[78] mercury,[79] mercurocarborands[80] and uranium.[81]

The usually small stability constants of anion complexes have been determined only for a few hosts of this type; in many cases the anion inclusion was inferred only from structural solid state data. The macrocyclic effect is sometimes absent, for example the equilibrium constant for complexation of Cl^- by host **A-34a** in $CDCl_3$, which is the best of a whole series, is equal to that observed with Bu_3SnCl. Anion binding ($Br^- > Cl^- \gg F^-, I^-$) by macrocycle **A-35** was tested by its capacity to accelerate anion transport through a water–organic solvent interface.[78] Acyclic analog Me_4Si did not promote anion transport under the same conditions. ^{199}Hg NMR titration of **A-36** by halide ions in acetone shows formation of 1 : 1 (for Cl^-) and 2 : 1 (for Br^-, I^-) host–guest adducts with stability constants too large to be measured by the method employed.[80] Anion complexes in organic solvents (MeCN, DMSO) by uranyl salophene derivatives like **A-37** reveal a general selectivity to $H_2PO_4^-$.[81] The cleft type host **A-37**, which provides additional H-bonding stabilization of bound anions forms quite stable adducts with binding constants log $K (M^{-1}) => 5$ ($H_2PO_4^-$), 3.2 (Cl^-) and 2.6 (NO_2^-) in MeCN.[81a]

A remarkable feature of the family of receptors which use transition metal cations as anion binding sites is their ability to give readily detectable spectral or electrochemical signals upon binding of anions; they therefore can act as anion sensors[71,82–84] (Section I 4). Here we shall restrict ourselves to the discussion of some general principles of anion recognition by such receptors; several typical examples of which are given in Fig. A 16.

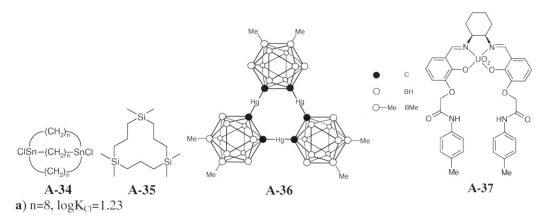

a) n=8, logK$_{Cl}$=1.23

Figure A 15.   Lewis acid-containing macrocycles designed for recognition of anions: **A-34**-tin-based; **A-35**-silicon-based; **A-36**-mercurocarborand; **A-37**-uranyl salophene complex.

The binuclear complex **A-38** binds different phosphate anions better than a respective mononuclear complex.[85] The complex stability increases in the order $HPO_4^{2-} < HP_2O_7^{3-} < HP_3O_{10}^{4-}$, which is in line with the increase of the guest negative charge, but the difference between the first and second anion is 2 logarithmic units while between the second and third it is only 0.8. This and other observations were interpreted as evidence of geometric complementary binding of the pyrophosphate anion to receptor **A-38**.[85] The cationic receptors **A-39** and **A-40** were designed as electrochemical sensors for anions[82] Host **A-40** closely resembles neutral

A-37 and is also specific for $H_2PO_4^-$ ($\log K = 3.9$ in DMSO).[82] A X-ray structure of the chloride complex of **A-40** shows that the anion is stabilized by six hydrogen bonds (two amide and four C–H groups).[82] Thus, although hosts **A-39** and **A-40** are positively charged, hydrogen bonding of anions is also considered to be an important factor. Uptake of anions is characterized by cathodic shifts of reduction potentials of **A-39** and **A-40** by 30–200 mV providing a possibility of measuring anion concentrations electrochemically.

Positively charged polyammonium macrocycles (Fig. A 17) represent the most widely explored and most efficient group of anion

Figure A 16.   Transition-metal receptors for anions: **A-38** – binuclear Cu(II) complex; **A-39** – cobaltocenium receptor; **A-40** – Ru(II)-bipyridyl receptor.

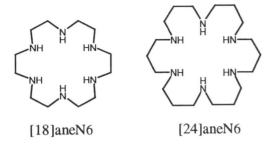

[18]aneN6   [24]aneN6

Figure A17. Typical polyammonium macrocycles (macrocyclic polyazaalkanes).

X-ray data show that Cl^- and NO_3^- are located outside the macrocycle cavity in the salt [18]aneN$_6$H$_4$(Cl$^-$)$_2$(NO$_3^-$)$_2$.[86b] Large anions such as $Co(CN)_6^{4-}$, $Pt(CN)_4^{2-}$ and $PtCl_6^{4-}$,[88,a,b] also do not form inclusion complexes with protonated [30]aneN6; only $PdCl_4^{2-}$, which is planar, is located inside the macrocycle.[88c] It is therefore not unexpected, that mono-macrocyclic polyamines generally do not show size-selective binding. The smaller macrocycle [18]aneN6 binds both small Cl^- and larger SO_4^{2-} better than [24]aneN6 (Table A11, no. 1 and no. 2), perhaps simply due to higher charge density in the former. Indeed, a general decrease in the binding constants on passing to larger polyazamacrocycles of the same total charge was observed for some anions ($Co(CN)_6^{4-}$, PO_4^{3-}, etc.), and interpreted in this way.[88d] Interestingly, in some cases, such as for $Co(CN)_6^{4-}$, a plot of $\log K$ *vs.* number, k, of monomeric units from which the macrocycles [3k]aneN$_k$ are constructed shows a minimum after which the binding constants increase with increasing macrocycle size. For example, such a minimum occurs at $k = 10$ for $Co(CN)_6^{4-}$.[88d] This observation was explained by the ability of large macrocycles to encapsulate even such large anions. Moderate size specificity was observed for the small fluoride anion binding by tetraprotonated macrocycles [16]aneN$_4$, [18]aneN$_4$ and [20]aneN$_4$: logarithms of stability constants were 1.9, 2.0 and 2.8 respectively.[89]

receptors, which function also in aqueous solutions. Anion binding by these macrocycles has some important analogies with cation complexation by macrocycles which are discussed before. Let us examine first anion binding by acyclic and mono-macrocyclic polyazaalkanes in order to analyze chelate and macrocyclic effects in this case. Table A11 shows some representative data for anions of different sizes and charges. Binding of mono-anions in water is generally rather weak and was detected only in a few cases, for example with tetraprotonated hexacyclen (see Fig. A17), Table A11, no. 1. The chelate effect in these systems is evident from the increased binding with higher charged hosts. It can be treated most adequately by the incremental approach in terms of the correlation discussed in Section A1 (see Fig. A3). A moderate macrocyclic effect can be seen on comparison of no. 2 and no. 3 for sulfate binding.

Table A11. Logarithms of binding constants of various anions by protonated polyazaalkanes in water.[86,87]

| | Polyamine | Protonation form | $\log K$ (M^{-1}) | | Ref. |
			Cl^-	SO_4^{2-}	
1	[18]aneN6	H$_4$L^{4+}	1.84	4.12	86
2	[24]aneN6	H$_4$L^{4+}	weak	2.50	87
		H$_5$L^{5+}		3.05	
		H$_6$L^{6+}		4.05	
3	H$_2$N(CH$_2$CH$_2$CH$_2$NH)$_5$H	H$_4$L^{4+}		1.15	
		H$_5$L^{5+}		2.00	
		H$_6$L^{6+}		2.50	

Macrocycles like **A-41** bind monoanions (Cl^-, NO_3^-) when $x = 4$ and larger anions like HgI_3^- and HgI_4^{2-} when $x = 6$[90]. Diprotonated dicationic sapphyrine **A-42**, an expanded porphyrin shows remarkable binding selectivity and high affinity to F^-, however only in non-aqueous media.[91] A X-ray analysis shows that the rigid planar cation **A-42** possesses about a 5.5 Å diameter core and incorporates F^- (ionic diameter about 2.4 Å), affording five nearly equivalent hydrogen bonds with the N to F distance of 2.7 Å; Cl^- with an ionic diameter of about 3.3 Å is therefore not incorporated. The binding constant for F^- in methanol equals 1×10^5 M^{-1} while those for Cl^- and Br^- are less than 100 M^{-1}.

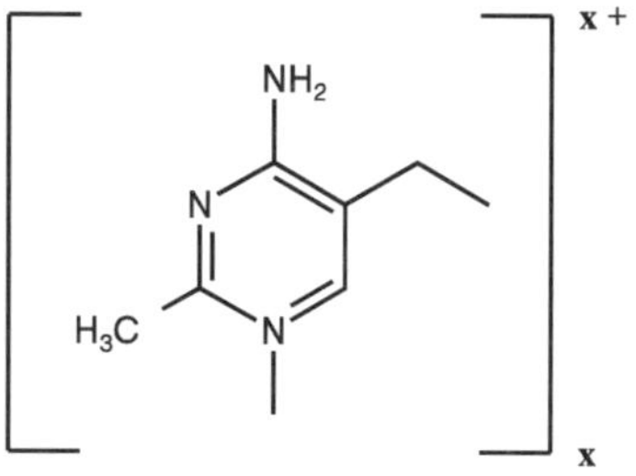

A-41

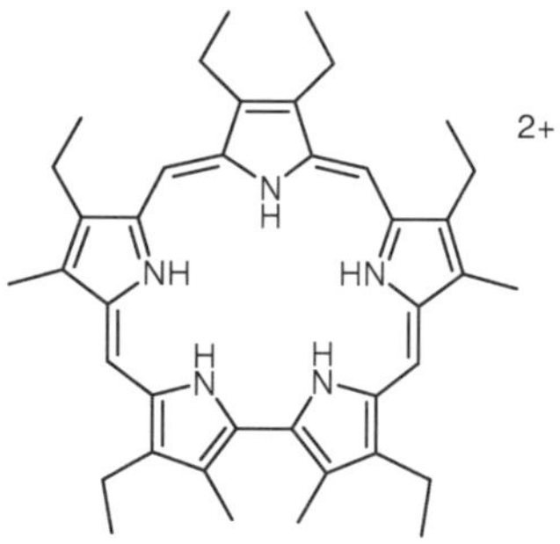

A-42

Macrobi- and macrotricyclic polyamines form inclusion complexes with anions more efficiently than monocyclic rings. The stability of inclusion complexes with macrotricyclic diamines (katapinates), is quite low, for example stability constant of Cl^- katapinate for protonated **A-43** with $n = 9$ only equals $4\,M^{-1}$, measured with however large excess

(50% in water) of CF_3COOD.[92c] Significant improvement of binding and selectivity is achieved with bicyclic and tricyclic hosts possessing four or more nitrogen atoms like in **A-44**–**A-47**. The cavity of $[A\text{-}44 \cdot 6H]^{6+}$ is of

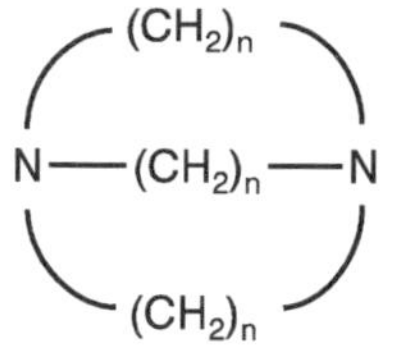

A-43

$[A\text{-}44.6H]^{6+} + X^-$	log K
F^-	4.1
Cl^-	3.0
Br^-	2.6
I^-	2.15
N_3^-	4.3

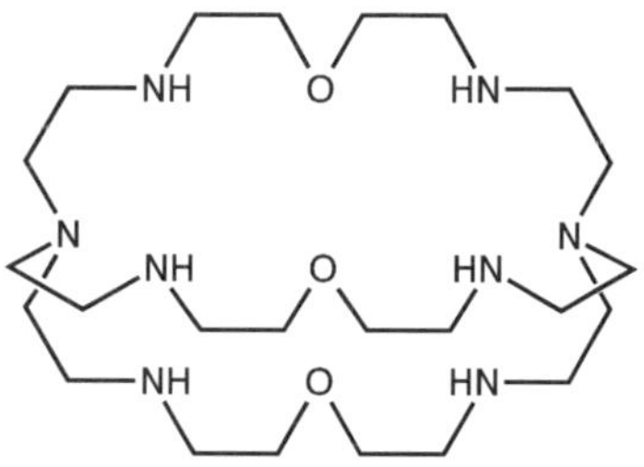

A-44

ellipsoidal shape with dimensions of about 4.5 Å length and 2.5–3.0 Å width; in contrast to halides the azide anion fits well into the cavity and uses a maximum number of H-bonds, see **A-48**, explaining the highest stability of the azide complex.[93] The small F^- anion also forms a stable complex, believed to be due to higher charge density and known tendency of this anion to form stable H-bonds.

The macrotricyclic anion cryptands **A-45a** and **A-46** show remarkable selectivity between Cl^- and Br^- anions with a factor $> 10^3$.[94] The related macrobicycle **A-47**, also in a tetraprotonated form, is much less selective and forms less stable complexes. Cryptands **A-45a** and **A-46** do not form any detectable complexes with larger anions like I^-, NO_3^- and ClO_4^-. Interestingly, full methylation of **A-45a** yielding **A-45b** leads to a thousand-fold decrease of the stability of the chloride complex and inversion of Cl^-/Br^- selectivity.[95] This observation again clearly

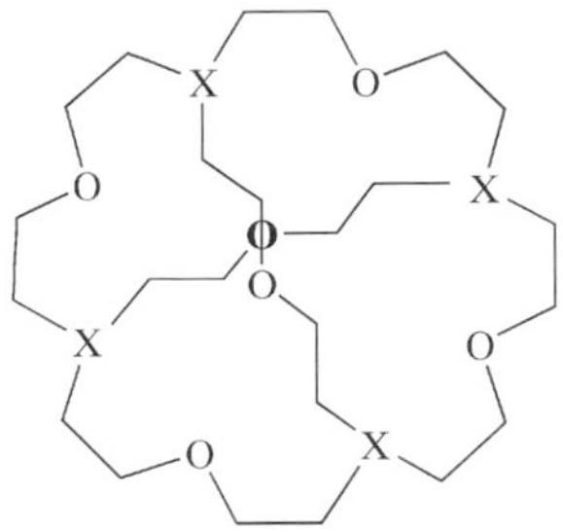

A-45 (a) X=N; (b) X=NCH$_3^+$

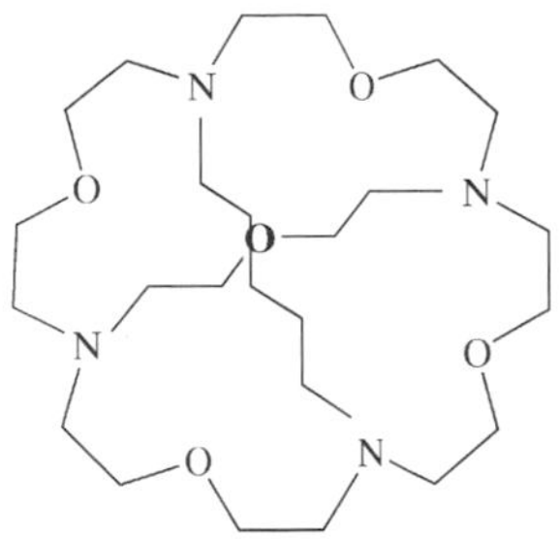

A-46

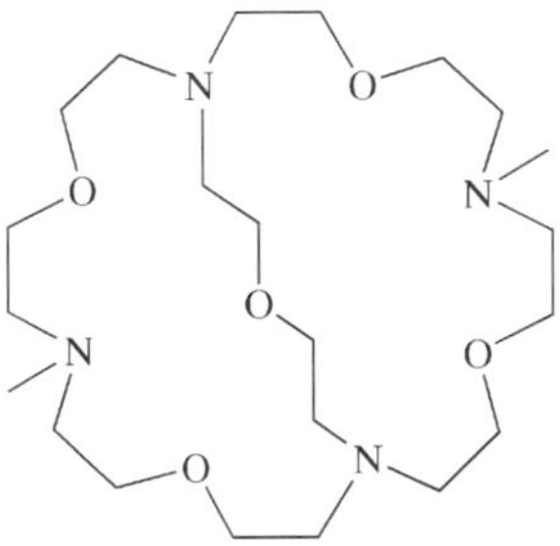

A-47

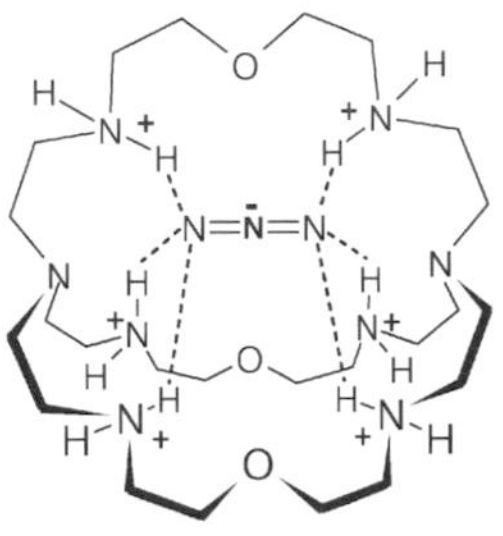

A-48

demonstrates the importance of hydrogen bonding.

In order to achieve also high F$^-$/Cl$^-$ selectivity, the smaller macrotricycles A-49a,b were prepared.[96] The protonated forms of these hosts indeed give very stable complexes with F$^-$ in water: $\log K = 10.55$, 8.40 and 5.50 for [A-49a·6H]$^{6+}$, [A-49a·5H]$^{5+}$, and [A-49a·4H]$^{4+}$ respectively. At the same time, [A-49a·6H]$^{6+}$ binds Cl$^-$ with only $\log K < 2$. Thus a F$^-$/Cl$^-$ record selectivity of $> 10^8$ is achieved.

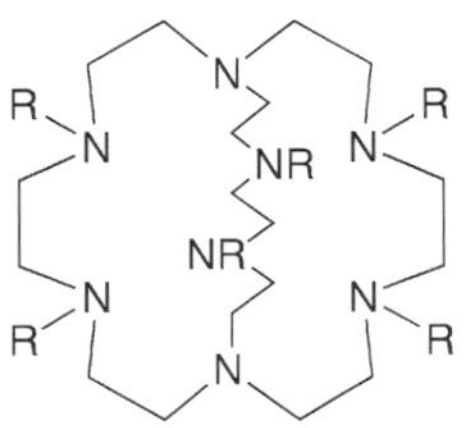

A-49 (a) R=H, (b) R=Me

The quaternary ammonium macrotricycles A-50a bind preferably heavier halide anions and show low affinity to phosphates, for

example the values of $\log K$ for $n = 6$ equal 1.7 (Cl$^-$), 3.01 (Br$^-$), 2.7 (I$^-$), 2.1 (H$_2$PO$_4^-$), 2.54 (HPO$_4^{2-}$) in water.[97] Zwitterionic hosts such as the borate-amine adducts A-50b[98a] and betains A-50c[98b] have the advantage of needing no counterion for the host charges,

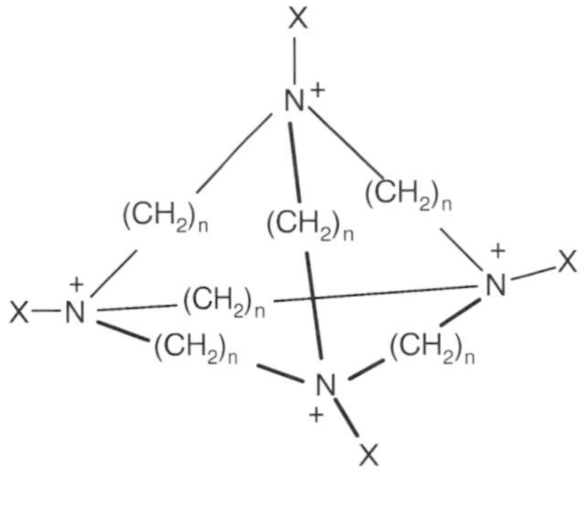

A-50
(a) X=Me; (b) X=BH$_3^-$;
(c) X=CH$_2$C$_6$H$_4$COO$^-$-p

which can compete with desired anion complexation. Host A-50c is capable of anion complexation in water with stability constants of the same order of magnitude as cationic host **A-50a**: for $n = 6$ the values of $\log K$ equal 2.4 (Cl^-), 3.3 (Br^-), 3.8 (I^-)[98b]. The last group of anion receptors considered here are neutral molecules capable of anion binding, in particular by inclusion through hydrogen bonding in absence of competing water as solvent. The macrocycle **A-51** binds halide anions in $CDCl_3$[99] (Table A 12).

A new family of neutral H-bonding macrocyclic receptors are calixpyrroles (or porphyrinogens) like **A-52**,[100] which bind strongly and selectively to F^- (Table A 12) owing to its smaller size and higher basicity. Hydrogen bonding is also favored here by the high acidity of the pyrrole NH group ($pK_a = 17.5$), as compared to usual amines (pK_a in the range 30–35) and amides (pK_a around 25).

The hydrogen bond donor capacities of amide and urea groups have been used as the anion binding elements in several acyclic receptors. These hosts suffer to some extent from inter- and intramolecular self-association, which can be very strong in solvents like chloroform or dichloromethane, but often disappears in more polar solvents like DMSO or acetonitrile. Bis-urea and bis-thiourea hosts such as **A-53** bind different anions even in DMSO with a pronounced specificity for $H_2PO_4^-$.[101] The thiourea host **A-53b** shows no self-association and forms more stable complexes with the anions studied, Table A 13. The stability constants indicate generally better binding between more basic anions and a more acidic urea group (thiourea > urea), as expected for the hydrogen bonding. The anion geometry also seems to be important: dihydrogen phosphate anion is less basic than acetate, but its structure allows better fits between guest H-acceptor and host H-donor sites as illustrated in supposed host–guest complex structures **A-54a,b**. The results with **A-55a,b** in MeCN[102] (Table A 14) again demonstrate the importance of acid–base properties of interacting groups: the most basic dihydrogen phosphate anion always forms more stable complexes than other anions, more acidic sulfonamides form more stable complexes than carboxylic acid amides, and the electron acceptor substituent CH_2Cl enhances the binding in comparison to alkyl and phenyl substituents. In chloroform even monodentate N-methyl acetamide forms complexes with halide anions: $K = 9.9$ (Cl^-), 6.5 (Br^-), and 5.3 M^{-1} (I^-).[103]

Urea derivatives generally provide stronger binding than amides because the E,E conformer (around C(O)=N pseudo-double bond), which contains two N–H bonds converging to the anion, is formed without build-up of considerable strain. Thus, polydentate urea-containing host **A-55c** binds Cl-anion with $\log K = 3.7$ even in such polar solvent as DMSO.[103] The structure of the complex established by computer-aided molecular modeling is shown in **A-56**.

In conclusion let us look briefly at the binding mechanisms of natural anion hosts. Detailed study of phosphate- and sulfate-

Table A 12.   Binding constants of anions with neutral macrocyclic H-bonding hosts.[99,100]

	$\log K$ (M^{-1})		
Anion	**A-51** (in $CDCl_3$)	**A-52a** (in CD_2Cl_2)	**A-52b** (in CD_2Cl_2)
F^-	3.09	4.23	3.56
Cl^-	3.0	2.54	2.07
Br^-	2.4	1	
I^-		< 1	
$H_2PO_4^-$		1.99	< 1
HSO_4^-		< 1	

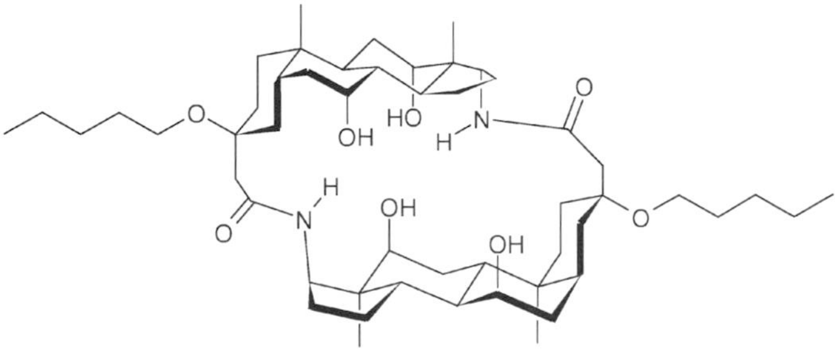

A-51

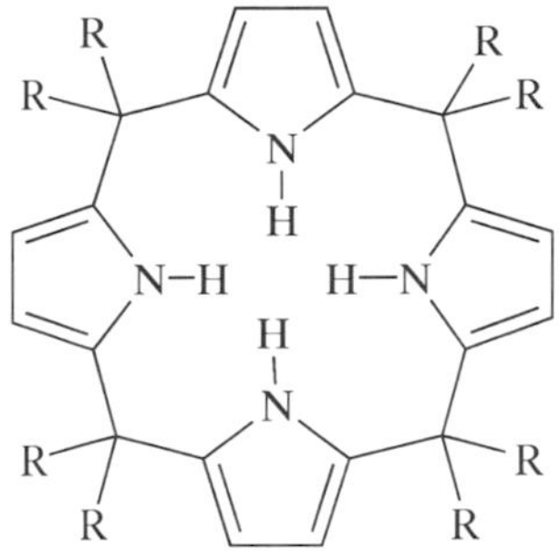

A-52 (**a**) R=Me; (**b**) R,R=*cyclo*-C$_6$H$_{11}$

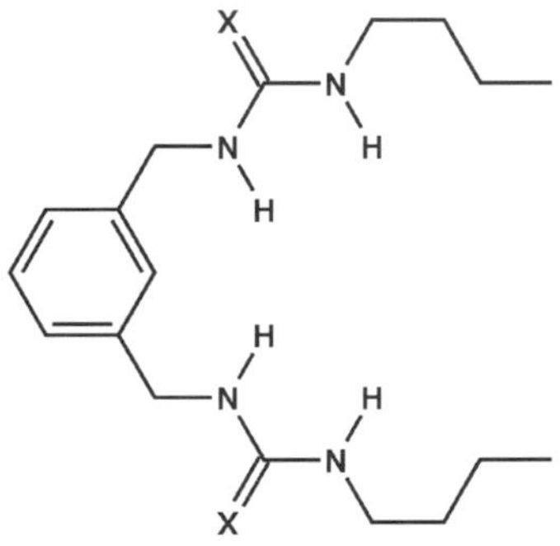

A-53 (**a**) X = O; (**b**) X = S

binding proteins[104] revealed strong and specific binding of the respective anions. Thus, a phosphate-binding protein binds HPO$_4^{2-}$ with $K = 10^6 \, \text{M}^{-1}$ and SO$_4^{2-}$ with $K < 10 \, \text{M}^{-1}$, while a sulfate-binding protein possesses $K = 6 \times 10^6 \, \text{M}^{-1}$ for SO$_4^{2-}$ and does not bind the phosphate anion. The efficiency and selectivity is due to multiple hydrogen bonding of anions by neutral proton-donor groups without any contribution of ion-pairing from cationic sites. One can identify here a large number of pairwise interactions: HPO$_4^{2-}$ is bound through 12, and SO$_4^{2-}$ through seven hydrogen bonds in active sites of the respective proteins. The active site of a phosphate-binding protein involves even one anionic carboxylate residue of aspartic acid which forms a hydrogen bond with a P–OH fragment of the phosphate. There are also two hydrogen bonds between the phosphate anion and a cationic arginine residue, but the latter is salt-linked to an aspartate carboxylate. We see that proteins do not use cationic binding elements, in contrast to the most efficient synthetic supramolecular hosts. The hydrogen bonds used in neutral natural systems require for most of the known

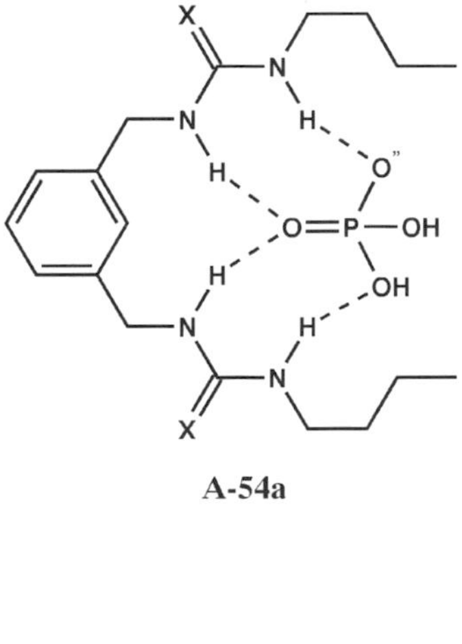

A-54a

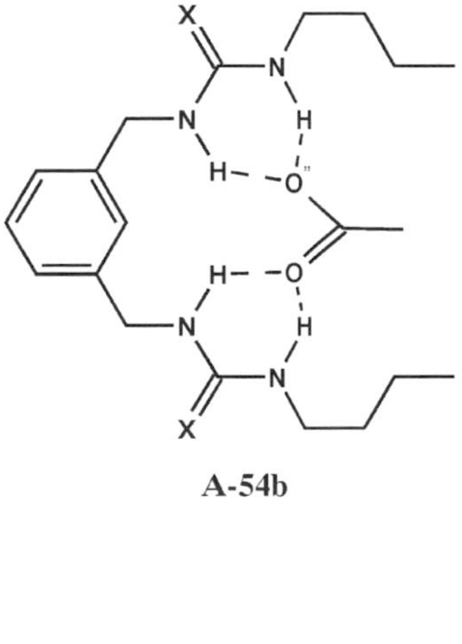

A-54b

(a) X =

(b) X =

(c) X =

A-55

Table A 13. Stability constants of anionic complexes of the hosts **A-53a,b** in DMSO.[101] Reprinted from *Tetrahedron Letters*, Vol. 36, Nishizawa *et al.*, (Anion recognition by urea and thiourea groups: remarkably simple neutralreceptors for dihydrogen phosphate), pp 6483, copyright 1995 with permission from Elsevier Science.

| | \multicolumn{6}{c}{$log K (M^{-1})$} | | | | | |
	$H_2PO_4^-$	CH_3COO^-	Cl^-	HSO_4^-	NO_3^-	ClO_4^-
A-53a	2.08	1.63	0.6	0.0	<0	no binding
A-53b	2.91	2.67	0.95	0.3	<0	no binding

Table A 14. Binding constants (M^{-1}) of anions with hosts **A-55a,b** in MeCN.[102] Reprinted from *Angew. Chem., Int. Ed. Engl.*, Valiyaveettil *et al.*, 1993, **32**, 900. Copyright 1993 John Wiley & Sons.

Anion	Host	R	log K	Host	R	log K
$H_2PO_4^-$	**A-55a**	CH_2Cl	3.78	**A-55b**	4-MeC_6H_4	3.54
		C_6H_5	2.94			
HSO_4^-		CH_2Cl	2.23		4-MeC_6H_4	1.90
		C_6H_5	1.75			
Cl^-		CH_2Cl	3.24		4-MeC_6H_4	2.73
		C_6H_5	2.00			

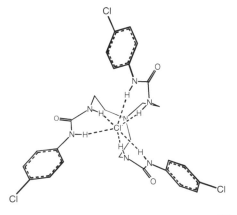

A-56-Structure of Cl⁻ complex of **A-55c**.[103]

synthetic systems the absence of proton donating solvents.

A7.   MACROCYCLES WITH SECONDARY BINDING SITES

The analysis of chelate, macrocyclic and cryptate effects already led us to conclusion that it is the number of interacting single sites which is of primary importance for tight and selective binding. In this chapter we shall consider several approaches to enhance the number of such sites by attachment of secondary sites to an already existing macro-

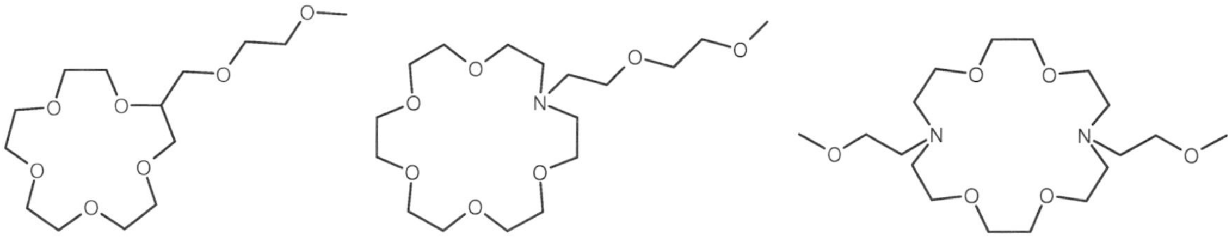

C-pivot N-pivot bibrachial

Figure A 18. Typical macrocyclic ligands with pendent donor groups.[105] Reprinted with permission from *Chem. Soc. Rev.*, Gokel, G.W., 1992, **20**, 39. Copyright 1992 Royal Society of Chemistry.

cycle, or by combining two binding sites of similar or different chemical nature.

A7.1. LARIAT ETHERS

The so-called lariat structures contain one or more pendent donor groups in side arms of macrocycles (see Fig. A 18). These macrocycles occupy an intermediate position between crowns and cryptands.[105] The pendent groups are designed to provide additional binding sites above and/or below the macrocycle 'plane'.

The binding of the pendent donor groups to a cation incorporated into the macrocycle was proven for several systems by X-ray analysis. One usually observes increased stability of lariat ether complexes with alkali cations although they do not reach the stability of cryptates. Thus, stability constants (in methanol) of potassium complexes of N-pivot and bibrachial lariat ethers shown in Fig. A 18, which possess the same number of donor atoms as the cryptand [2.2.2], are of the same order as of 18-crown-6 and are about 10^4 times smaller than those of the cryptand. Figure A 19 illustrates the effects of polyether sidearms on the stability of ammonium complexes of aza-crowns with the general structure **A-57**. With ammonium ion complexes a rather irregular and small affinity enhancement is observed, owing to the need of optimal conformations for hydrogen bonds in these complexes. In contrast carboxylate sidearms lead to a regular increase of binding

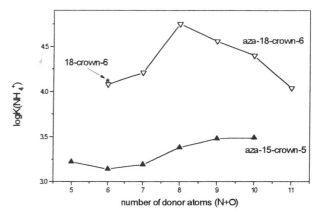

Figure A 19. Binding of ammonium cations by lariat ethers **A-57**. The first point in each plot corresponds to R = Me.[106] Reprinted with permission from *J. Am. Chem. Soc.*, Schultz et al., 1985, **107**, 6659. Copyright 1985 American Chemical Society.

constants (Fig. A 20), due to the dominance of less conformation-dependent Coulomb interactions (Section B 2).

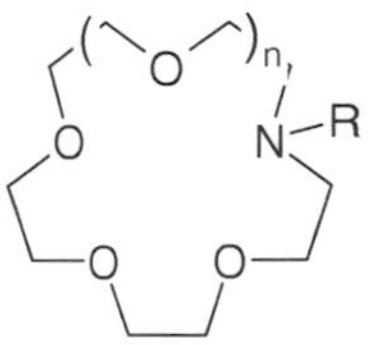

A-57
R = CH₃, CH₃(OCH₂OCH₂)ₓ,
x = 1-8, n = 1; 2

One would expect that attachment of pendent donor groups must always be favorable

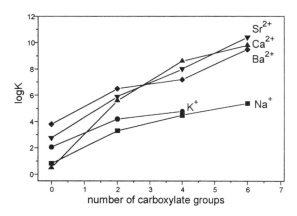

	A-58	A-59
logK (Zn^{2+})	10.51	5.90
logK (Cd^{2+})	10.90	8.84
logK (Pb^{2+})	9.01	10.72

Scheme A 6. Stability constants of different metal cations with a tetraazamacrocyclic ligand and its derivative with four pendent donor groups.[107] Reprinted from *Coord. Chem. Rev.* Vol. **148**, Hancock *et al.* (Macrocyclic ligands with pendent amide and alcoholic oxygen donor groups) 315, copyright 1996 with permission from Elsevier Science.

Figure A 20. Plots of the logarithms of stability constants of complexes of alkali and alkaline earth cations with 18-crown-6 and **A-61**–**A-63** in water *vs.* the number of carboxyl groups.[111] Reprinted with permission from *Can. J. Chem.*, Dutton *et al.*, 1988, **66**, 1097. Copyright 1998 NRC Research Press.

or at least harmless for the complexation. However, analysis of pendent oxygen donor effects on the binding of a wide series of metal cations actually showed a stability *decrease* with metal complexes of smaller, and an increase with larger cations.[107] This is illustrated in Scheme A 6 where the binding constants of metal cations of different sizes to a tetraazamacrocycle **A-59** and its derivative **A-58** are compared: zinc(II) cation (ionic radius 0.74 Å) forms a much less stable

complex with **A-59** than with **A-58**, the difference is smaller for cadmium(II) (ionic radius 0.95 Å) and only lead(II) (ionic radius 1.19 Å) gains an additional stability from four pendent donor groups. In this and other similar cases the main factor is thought to be the cation size because an unstrained five-membered chelate ring with oxygen donor atoms can be formed only with very large cations[107] (Section A 2). When oxygen atoms cannot interact with the cation for steric reasons the pendent groups produce only unfavorable steric crowding. Thus the favorable effects of side arms on the binding of alkali metal and ammonium cations are observed because these cations are sufficiently large to allow binding interactions with pendent donor groups.

Generally N-pivot lariat ethers (Fig. A 18) form more stable complexes than C-pivot ligands.[108] Lariat macrocycles with numerous pendent groups other than polyethers were synthesized (often such groups serve as redox, membrane active, hydrogen bonding, etc. sites rather than additional cation binding sites). Among the latter the highest affinities for cations are observed for anionic pendent groups which provide electrostatic stabilization of bound cations as illustrated in **A-60c**.[109]

logK(Na^+) in MeOH	
A-60 R = COOH (a)	2.22
R = COOMe (b)	2.80
R = COO^- (c)	5.0

Some general trends can be seen by comparison of the binding constants of alkali and alkaline earth cations with a series of anionic crown ethers **A-61**–**A-63** derived from tartaric acid.[110,111] The $\log K$ values for 18-crown-6 and macrocycles **A-61**–**A-63** in water are plotted against the number of carboxyl groups in Fig. A 20. As one would expect for additive electrostatic interactions (Section B 2) the plots are roughly linear: addition of each pair of carboxylates gives a mean $\Delta \log K$ increment of 1.4 for monovalent and 2.6 for divalent cations, not far from the Coulomb increment ($\Delta \log K$ around 1) discussed in Section B 2. The total effect on passing from 18-crown-6 to **A-63** is fairly large: about 10^4 for Na^+ and 10^6–10^9 for Ba^{2+}–Ca^{2+}. Note that the size selectivity of the parent crown ether disappears on passing to ligands **A-62** and **A-63**, which discriminate cations only by charge.

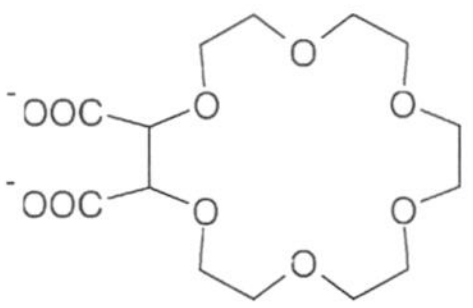

A-61

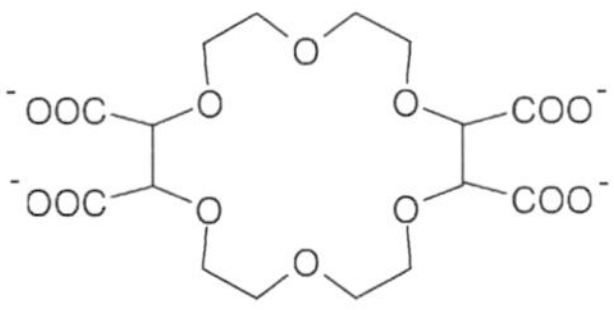

A-62

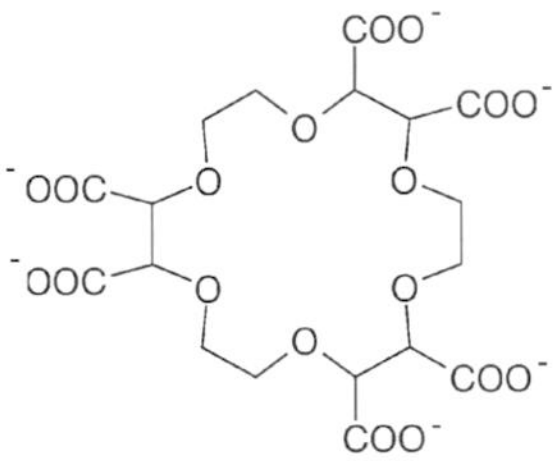

A-63

Data involving carboxylate derivatives of aza-crowns like **A-64** and transition metal cations revealed more complex trends.[110] It has been concluded that the electrostatic effect is generally of minor importance for com-plexation of transition metal cations and the dominant interaction type is chelation of the ion by fragments $N–CH_2–COO^-$ or $O–CH_2–COO^-$, amongst which the former provides the higher stability,[110] whereas in alkali and alkaline earth cations electrostatic interactions always dominate.

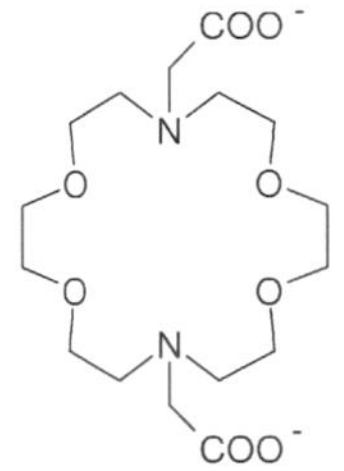

A-64

A 7.2. DITOPIC RECEPTORS

Instead of adding secondary binding sites to a given host molecule, one can achieve a larger number of pairwise interactions with the guest by constructing a host with two or more structurally or energetically similar binding sites. Such hosts are referred as di- or poly-topic receptors, which can be homotopic (con-taining chemically similar binding sites) or heterotopic (containing chemically different sites). Many hosts of this type were designed for recognition of organic guests, which usually offer more binding sites than inor-ganic ions: they will be discussed in later chapters.

In previous sections we already met several receptors of this type. The binuclear macro-cyclic complex of Cu(II) (**A-38**) is representa-tive of a group of metal-based ditopic receptors for anions. It shows moderately increased binding of phosphate anions com-pared to the respective mononuclear complex (Section A 6). Usually metal-based ditopic macrocyclic receptors bind effectively small inorganic anions like OH^-, O_2^{2-}, CN^-, etc.[112] affording bridged binuclear complexes. Besides binuclear assembly inside one macro-cycle, two covalently linked macrocyclic metal complexes can be arranged as a ditopic receptor. For example, the binuclear Zn(II) complex **A-65** binds OH^- and organic dianions: barbiturates and phosphates.[113a] The tritopic receptor **A-66** constructed on the same chemical basis shows about 100 times higher affinity towards phosphate dianions.[113b]

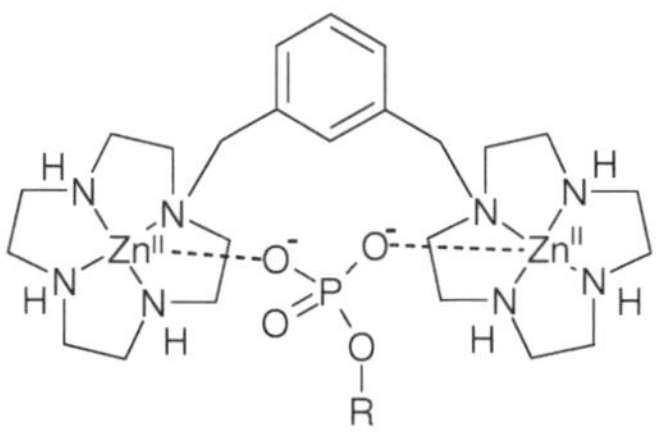

A-65

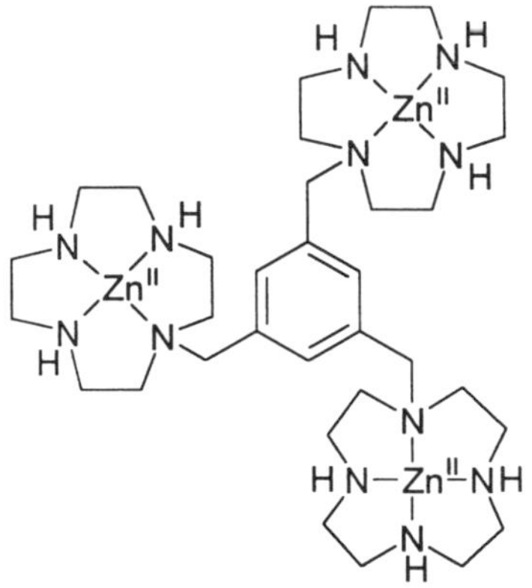

A-66

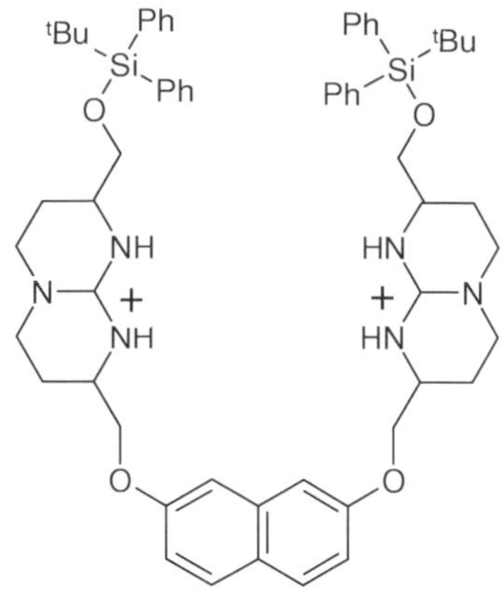

The previously discussed (Section A6) polyammonium cationic macrocycles possessing spatially separated charged fragments (e.g. **A-44**) also can be considered as ditopic receptors for anions (**A-48**). Acyclic ditopic guanidinium receptors like **A-67** which bind halide, sulfate and organic phosphate anions,[114] represent another type of cationic ditopic hosts. Bis crown ethers like **A-68** serve as ditopic receptors for cations which are too large for inclusion inside a given macrocycle cavity and can give sandwich-type complexes with two macrocycle molecules.[115]

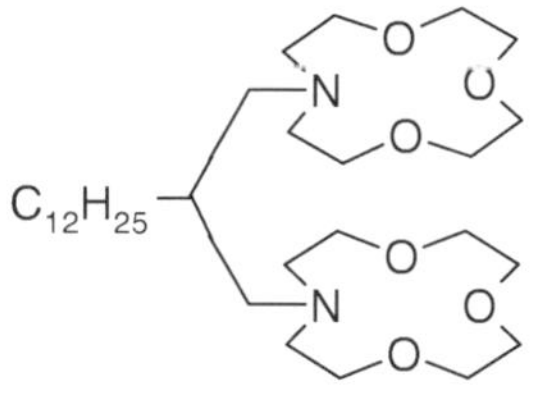

A7.3. CO-COMPLEXATION

Heterotopic receptors can involve both cation and anion binding sites and, therefore, are

A-67

A-68

able to complex ion pairs. The crown ether-boryl derivative **A-69** interacts with dry KF in dichloromethane affording complex **A-70a**.[116a] Here boron acts as a Lewis acid site for a fluoride anion and the polyoxyethylene chain binds the potassium cation. Host **A-69** does not interact with KCl and KBr, but affords usual monotopic cation inclusion complexes with KI and KSCN, where anions remain free. Receptor **A-69** also binds simultaneously amine and alcohol molecules with a 1:1 ratio yielding the ionic complex **A-70b**.[116b] In this case the same combination of interaction types as in **A-70a** is involved, although the guest species are initially neutral molecules.

Chemically different, but similar by the nature of interactions involved, is zinc porphyrin capped with a calixarene **A-71** which operates as a ditopic receptor for alkali metal halides.[117] The highest stability was observed for the complex with KI ($\log K = 5.23$ in $CHCl_3/MeCN$ mixture) where I^- interacts both with Zn(II) cation as an axial ligand and

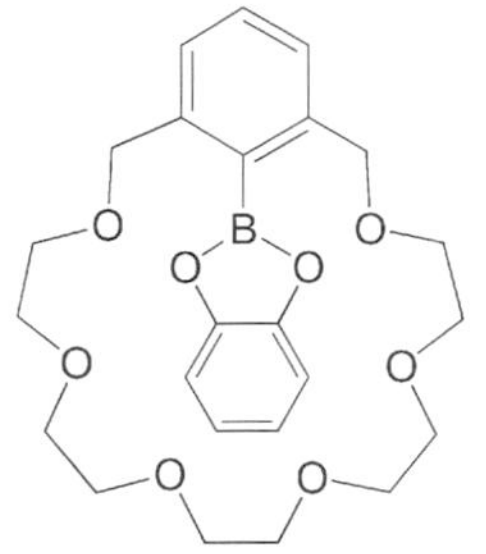

A-69

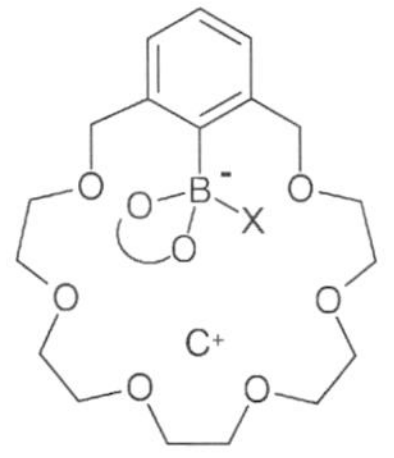

A-70 (a) $C^+ = K^+$, $X = F$
      (a) $C^+ = RNH_3^+$, $X = OR'$

with calixarene-bound K^+ electrostatically. Very weak binding is observed for $KClO_4$ ($\log K < 2$), while NaI and $NaClO_4$ both form complexes of comparable stability ($\log K = 3.91$ and 3.69 respectively). These observations agree with monotopic binding of sodium salts due to inclusion of cation into calixarene component of the receptor, but ditopic binding of potassium salts. Another type of ditopic receptors based on simultaneous first- and second-sphere coordination will be discussed in the next section.

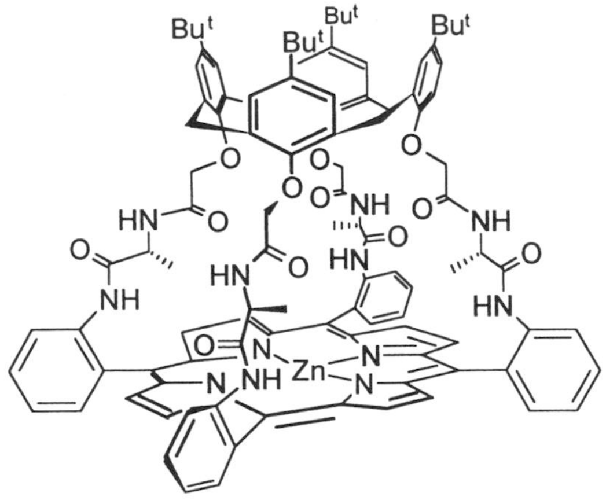

A-71

A 8. SECOND-SPHERE COORDINATION

Non-covalent binding of already coordinatively saturated metal complexes with external ligands represents the area of the second-sphere coordination.[118a,b] Generally, non-covalent interactions of metal complexes with surrounding (outer-sphere) solvent molecules have been studied for a long time as an aspect of solvation phenomena and intermolecular solid-state interactions, and date back to Werner's original concepts of coordination chemistry.[118c] Solvation already affects spectral, electrochemical and other properties of metal complexes, but second-sphere host–guest interactions cause more specific and more effective modification of such properties. Another well known type of outer-sphere interactions is ion-pairing of charged metal complexes. A supramolecular analogue of the latter is the binding of anionic complexes, such as $Co(CN)_6^{4-}$, with protonated aza-macrocycles, as discussed in Section A6. Here we shall discuss second-sphere association processes which occur generally between uncharged species due to hydrogen bonding, hydrophobic and other non-covalent interactions.

Crown ethers were amongst the first macrocycles for which second-sphere coordination of metal complexes was demonstrated. In Section B3 we shall discuss the binding mode of ammonium ions to crown ethers based on hydrogen bonding of the type $NH^+ \cdot O$. There is an evident analogy between an alkylammonium cation ($R-NH_3^+$) and a metal ammine complex (L_nM-NH_3) in terms of availability of three protons for hydrogen bonding to oxygen atoms of the crown ether. Indeed, the expected three-point hydrogen bonding was observed in crystals of many adducts of metal ammine complexes with crown ethers, as exemplified by the structure **A-72** for the second-sphere complexation of *trans*-[Pt(PMe$_3$)Cl$_2$NH$_3$] with 18-crown-6.[119] Related complexation through an ammonium group which is a part of organic ligand was

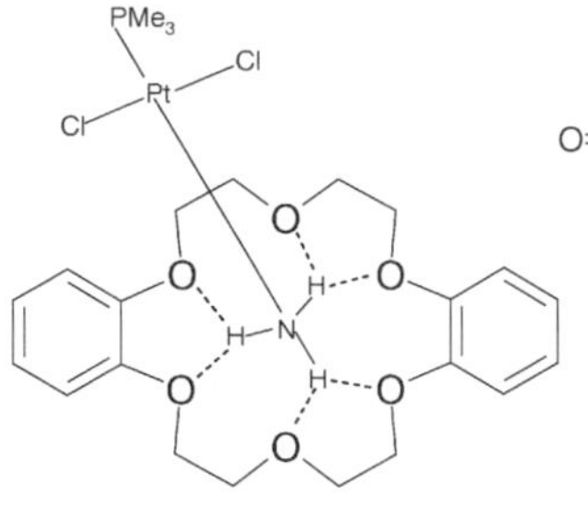

A-72

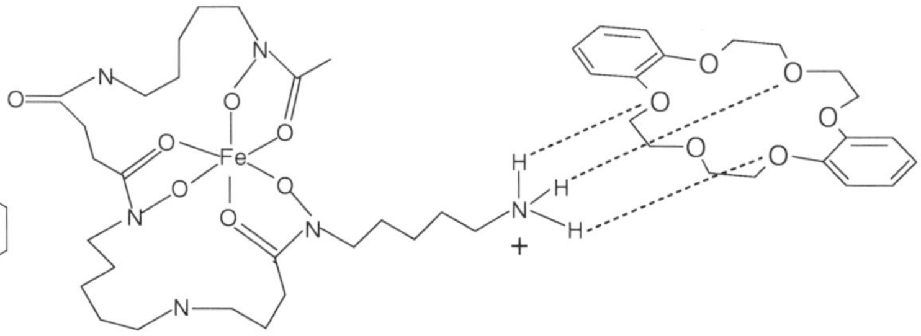

A-73

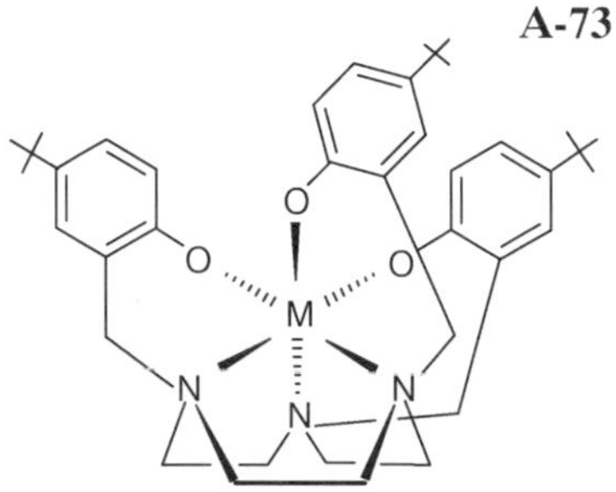

A-74

employed for recognition of ferrioxamine B by crown ethers in solution as shown in structure **A-73**.[120] Besides NH_3, coordinated organic amines, such as ethylenediamine, can act as H-donor groups.

Coordinated water is another protic, first-sphere ligand that can serve as a hydrogen bond donor for the second-sphere complexation. The trisphenolato complex **A-74** (M = Cr^{III} or Fe^{III}) serves as host to bind hexaaquo transition metal ions in the second coordination sphere.[121] Coordinated phenolato oxygen atoms behave in these complexes as sufficiently basic proton acceptor sites which in several cases form hydrogen bonds with aquo ligands of divalent cations (like Zn, Ni, Mn or Co) affording trinuclear sandwich-type complexes [(**A-74**)–$M(H_2O)_6^{2+}$–(**A-74**)]. Even the much less acidic CH_3CN can bring about second-sphere H-bonding with crown ethers as was found, for example, in a bis adduct of {[trans-$Ir(CO)(CH_3CN)(PPh_3)_2$]$_2$·18-crown -6}$^{2+}$ in the solid state.[118b]

Bipyridine and other π-acceptor ligands bound to the metal ion can considerably increase the stability of second-sphere complexes formed by hydrogen bonding of the ammine ligands, due to increased acidity of the protic ligand. Moreover, in complexes with benzocrowns such ligands provide an additional stabilization due to π–π stacking interactions (Section B 5) with catechol rings of the macrocycle as shown in structure **A-75**.[118a,b] The stabilization effect of such interactions in solution is exemplified by data for association of some Ru(II) ammine complexes with ligand **A-76** in nitromethane.[122] No association occurs with $Ru(bipy)_3^{2+}$, but complexes like $Ru(NH_3)_4(bipy)^{2+}$ or $Ru(NH_3)_5(py)^{2+}$ form adducts with K ranging from 200 to 800 M^{-1} indicating the presence of at least one NH_3 ligand which serves as potential H-donor to be indispensable for binding. On the other hand, complexes with the same number of ammine ligands, but different heterocyclic ligands form adducts of different stability: K (M^{-1}) = 260 for $Ru(NH_3)_4(py)_2^{2+}$, 800 for $Ru(NH_3)_4$ (bipy)$^{2+}$ and 2500 for $Ru(NH_3)_4$ (phen)$^{2+}$. This trend can be attributed to stacking interactions between the aromatic host and guest parts in a manner analogueous to that in complex **A-75**.

Variation of the pyridine ligand basicity in a series of pentammine complexes $Ru(NH_3)_5$

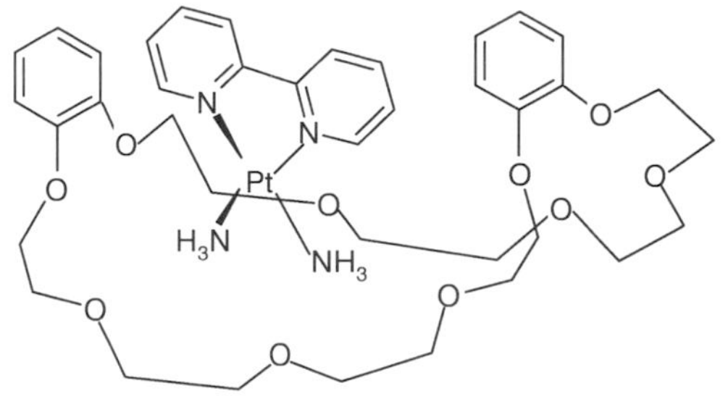

A-75

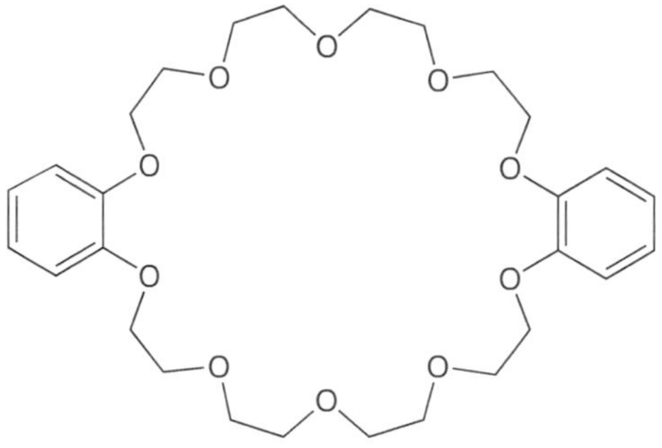

A-76

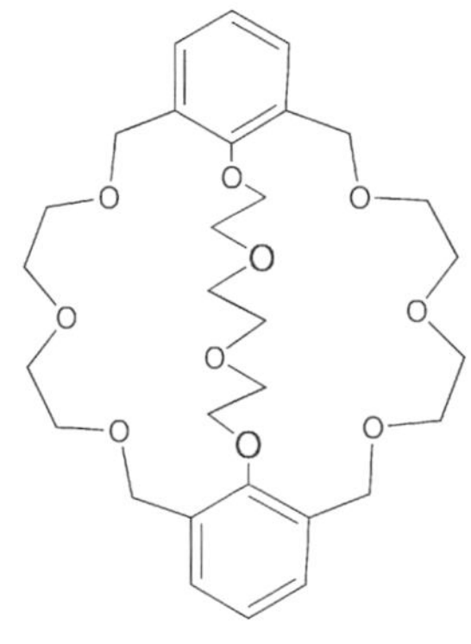

A-77

stants and pK_a of substituted pyridines with a negative slope was demonstrated for this and other related series of second-sphere complexation reactions.[122]

One may expect formation of more stable second-sphere adducts with polycyclic hosts, which in principle can provide more H-bonding contacts with ammine ligands than even large polyethers can. Clear evidence of more than three-point hydrogen bonding was obtained for the adduct of **A-77** with $Rh(cod)(NH_3)^{2+}$;[123] the X-ray structure of the adduct showed eight N·O distances in the range 3.00–3.26 Å, indicative of hydrogen bonding (Section B2). The binding free energy in CD_2Cl_2 solution was estimated as $\Delta G < -13.4$ kJ mol^{-1} ($K > 200$ M^{-1}).

Another important group of second-sphere ligands involves cyclodextrins (CDs) and cyclophanes which utilize predominantly lipophilic interactions for the inclusion of metal complex guests into their large cavities. Complexation of ferrocene with β-CD ($K = 50$ M^{-1} in water)[124] was amongst the first examples of such interactions. Various complexes of general structure $M(L)(NH_3)_n$, where L is a hydrophobic ligand capable of fitting the CD cavity, form supercomplexes with CD based on simultaneous inclusion of L and hydrogen bonding of coordinated ammonia to CD hydroxyl groups. Typical examples are complexations of [Pt(1,1-cyclobutanedi-carboxylate)(NH$_3$)$_2$] with α-CD[125a,b] and [Ru(4-phenylpyridine)(NH$_3$)$_5$]$^{2+}$ with β-CD.[126] In both cases the binding constants are of the order of 200 M^{-1} in water. Inclusion of extended hydrophobic fragments of first-sphere ligands into cyclodextrins have been exploited to construct various self-assembling superstructures, such as rotaxanes and pseudorotaxanes[127] (Section I5).

Second-sphere coordination can occur simultaneously with the first-sphere coordination if the macrocycle contains functional groups capable of both types of interactions. Such receptors can be considered as di-topic (Section A7.2). They bind protic ligands, such

$(X–py)^{2+}$ $(X = 3,5-Cl_2,$ 3-Cl, H, 4-CH$_3$)[122] shows an expected trend: more strongly electron-withdrawing substituents increase stability of the adduct with **A-76** presumably due to acidification of N–H bonds of coordinated NH$_3$ ligands, leading to formation of stronger hydrogen bonds with ether oxygens (Section B3). A reasonably good linear correlation between logarithms of stability con-

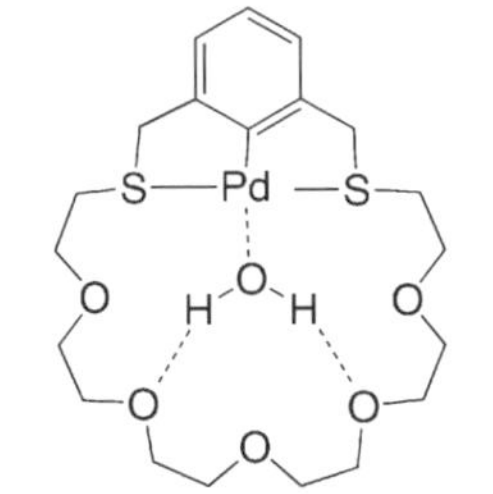

A-78

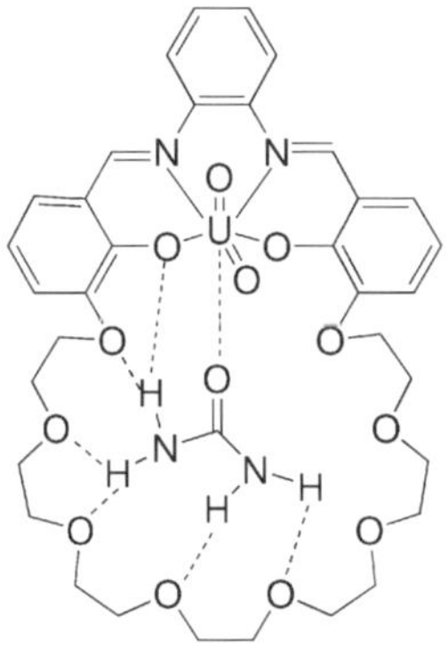

A-79

as H_2O, NH_3, NH_2NH_2, urea, primary and secondary organic amines, through classical first-sphere coordination to a metal site linked to an appropriate chelate fragment of a macrocyclic ligand, and hydrogen bonding to a macrocycle polyether chain. One family of such receptors consists of palladium containing macrocycles exemplified by the complex **A-78**.[128] Other promising receptors of this type are Schiff base uranyl complexes.[129] Urea binding in one of such complexes is shown in **A-79**.

A 9. CONFORMATIONAL COUPLING BETWEEN BINDING SITES: COOPERATIVITY, ALLOSTERIC EFFECTS AND INDUCED FIT

Host compounds containing at least two spatially separated binding sites can show cooperativity (Section D 1.3) due to conformational coupling between sites in the sense, that occupation of one site will affect the affinity of the second site.[130a−c] This should not be confused with often observed situation when the binding at the second site is affected by *direct* interaction between bound guests and not between the sites. Thus, in binuclear metal complexes of, for example the polyamine type ligands occupation by one cation will by electrostatic repulsion weaken the binding of a second cation at a neighboring site (e.g. logarithms of stability constants for first and second Cu^{2+} cations in the complex **A-38** are 13.8 and 9.7 respectively,[85] Section A 6). Inversely, affinity enhancement is expected if at one site a cation, and at the other site an anion is bound, leading to efficient complexation of ion pairs; the same holds for Lewis acid–base pairs. Such a co-complexation has been already used in several synthetic receptors and is discussed in Section A 7.3. In contrast, the interaction between binding sites in allosteric systems is mediated by conformational changes induced in the receptor upon occupation of one site. The need to find an optimal geometric contact between host and guest molecules can of course lead to an *induced fit* with or without affecting any other binding sites. This requires a certain flexibility in host or/and guest molecules and may be accompanied by free energy losses due to such factors as restrictions of internal rotations, induced torsional strain, etc. As discussed in Section A 1, for weak non-covalent interactions these losses fortunately are in most cases not very large; by the induced fit the necessary complementarity is achieved with usually not perfectly preorganized hosts, both in synthetic and biological systems. In proteins the conformational changes induced, for example by the binding of a coenzyme can lead to significantly stronger binding of a substrate. This is illustrated, for example by the quite different conformations of alcohol dehydrogenase in the binary complex with NADH and in the ternary complex with an additional substrate (see website). Conformational coupling between sites required for allosteric effects will be more effective when the sites are connected through a rigid spacer capable to conduct complexation-induced displacements of interacting groups (see below). Thus an optimum balance between flexibility and rigidity is necessary for an efficient allosteric host molecule.

In case of affinity enhancement (positive cooperativity) the binding isotherms will be sigmoidal. In the classical case of hemoglobin[131] the binding constant for the fourth oxygen molecule is about 300-fold increased compared to association of the first oxygen molecule. In the presence of an allosteric effector (2,3-diphosphoglycerate) this ratio can be increased to 1000. If successive ligands are bound with decreasing affinity (negative cooperativity) binding isotherms usually are hyperbolic.

The mathematical treatment of cooperative systems is considered in Section D1. Here we discuss briefly some models which simplify the complicated binding equations. A graphical method is often used to obtain information on the extent of cooperativity. The graph is the Hill plot,[132] which is based on the idea that the ligand binding can be described by a single-step interaction between host and n guests, according to equation

$$H + nG \rightleftharpoons HG_n$$

The equilibrium is defined by a stochiometric binding constant K as

$$K = [HG_n]/([H][G]^n)$$

The apparent value of K is the inverse of the n-th degree of the guest concentration needed for half-saturation. Taking logarithms and introducing the degree of saturation $\bar{Y} = [HG_n]/(n[H]_{tot})$ the Hill equation results

$$\log[\bar{Y}/(1 - Y)] = n\log[G] + \log K \quad (A\,9\text{-}1)$$

The slope of the plot[130b] $\log\{\bar{Y}/(1 - \bar{Y})\}$ versus $\log[G]$ (where $[G]$ = free guest concentration) gives at 50% saturation the Hill coefficient $h \leq n$. The plot is linear with $h = 1$ for one or several non-interacting binding sites, and

with positive cooperativity becomes sigmoidal. For example, oxygen binding to hemoglobin shows at very low and very high degree of saturation a slope of one, corresponding to the non-cooperative binding of the first and fourth oxygen molecule. The intercepts of the straight lines with the abscissa give K_1 and K_n (see Fig. A21). Between 10 and 90% saturation the plot is nearly linear. At 50% saturation one observes as slope $h = 2.8$ for hemoglobin. The Hill coefficient h can never exceed n, but varies between 1 and n for positive cooperativity. Values of $h = 1$ indicate absence of cooperativity, smaller values ($h < 1$) indicate negative cooperativity.

The difference $\Delta\Delta G = -RT \ln(K_1/K_n)$ which was defined as the interaction energy between the binding sites, is the interesting measure of cooperativity *strength*. $\Delta\Delta G$ can also be graphically derived from the distance d in Fig. A21, with

$$\Delta\Delta G = -2.3\,RT\,d\sqrt{2}.$$

Cooperative systems regulate chemical processes in living organisms. Numerous attempts can be found in the literature to understand this phenomenon on a molecular bases as well as to give its mathematical description.[133,134] The mostly used models are: the Koshland–Nemethy–Filmer (KNF)[133a] sequential model, and the Monod–Wyman–Changeux (MWC)[133b] concerted mechanism. Originally they are developed for ligand binding to proteins, which are composed of n subunits with n binding sites, but they can be used too for cases where the n binding sites are on a single molecule. Both models assume an equilibrium between lower and higher affinity conformations of binding sites. The assumptions in the KNF model are: (1) in absence of guests the host exists in one conformation, (2) upon binding, the guest induces a conformational change, which is transmitted to the second binding site. The expression for the binding isotherm for this model is similar to the Adair (Bjerrum) equation described in Section D 1. The mathematically more simple MWC description of multi-site binding systems assumes only symmetrical states (all oligomers are either in the tense (T) or relaxed (R) state), which are in equilibrium with an equilibrium constant L. Here the ratio c between the dissociation constants K'_R and K'_T in different states R and T is used; T has a lower affinity compared to R. R_0 and T_0 refer to the states before ligand binding

$$c = K'_R/K'_T, \text{ with an 'allosteric constant'}$$
$$L = [T_0]/[R_0]$$

The MWC model needs always only three constants: K'_R, K'_1 and L. Using the mass action law and $\alpha = [G]/K_R$ one can derive

$$\bar{Y} = [L\,c\alpha(1 + c\alpha)^{n-1} + \alpha(1 + \alpha)^{n-1}]/$$
$$[L(1 + c\alpha)^n + (1 + \alpha)^n]$$

It is the dissociation constant ratio c which in the MWC model is used to classify allosteric proteins to have weak ($1 > c > 0.1$) or strong ($c \leq 0.1$) cooperativity with large values of L. Sigmoidal binding curves are observed only if $n > 1$, $L > 1$ and $c \ll 1$.

Also by these criteria artificial allosteric systems can be much more effective than proteins. If the apparent binding constants K_1 and K_2 with at least one of two available binding sites are large enough – as can easily be reached with artificial receptors – one can simply take the ratio of apparent binding constants before and after occupation of one binding site as a measure of cooperativity strength.[134] These ratios relate K_1 of a binary complex H–G, and K_2 of a ternary complex H–G_A–G_B. Artificial receptors can be designed with a much stronger and direct conformational control between binding sites than is possible in proteins, in which the very nature of the peptide chains prevents a high conformational rigidity. On the other hand, no artificial receptor has until now been prepared with a cooperativity over such large distances as exist, for example between the hemoglobin subunits. In Scheme A-7 we show some simple principles of conformational coupling between binding sites A and B, which can be either directed towards the *same* kind of guest compounds (*homotropic* systems), or of a *different* kind (*heterotropic*). In example I of a homotropic system the most stable conformation I a (analogous to the so-called tense state in the MWC model) will upon binding of a guest molecule G1 under build-up of some strain convert to conformation Ib, which then is preorganized to take up *without* additional strain energy another G2 to give the H·G_1·G_2 complex Ic. The thus increased affinity in the second binding leads to positive cooperativity, and is also observed in heterotropic systems II and III.

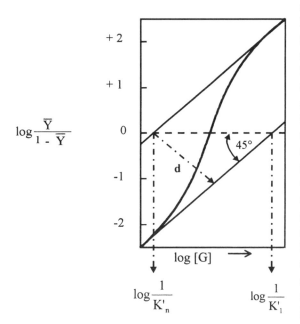

Figure A 21. Hill plot[130b]. The limiting slopes at very low and very high ligand concentrations ([G] → 0, [G] → ∞, respectively) are 1.0. The slope at the midpoint ([G] = [$G_{0.5}$]) is the Hill coefficient h. The straight lines from the limiting slopes 1.0 can be used to obtain the values of the ligand effinity to the first and the last site as dissociation constants K'_1 and K'_n, respectively. Reprinted with permission of Springer Verlag N.Y. (*Quantitative Aspects of Allosteric Mechanism*), A. Levitzki *et al.*, **1978**, Fig. 4, p. 20, copyright 1978, Springer Verlag.

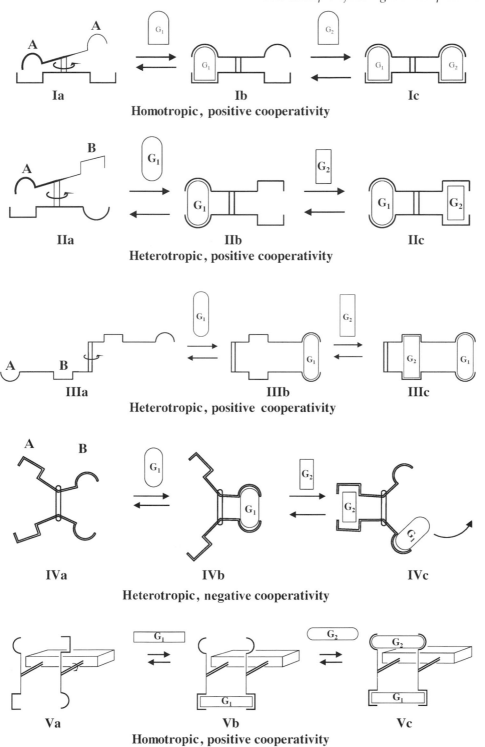

Scheme A 7. The molecular machinery of selected allosteric systems.

Negative cooperativity, like with system IV, has been materialized less frequently until now, but might be even of more interest for applications such as liberations of toxins, catalysts etc. than the positive counterpart. It should be pointed out that the recognition sites A and B can of course also differ in, for example, electric charge instead shape, which is pictured in Scheme A 7. In the case of the allosteric principle V, the system will then switch between large and zero dipole moments. Related molecular two-state systems, switched mostly by light until now, are described in Section I5 and also in some

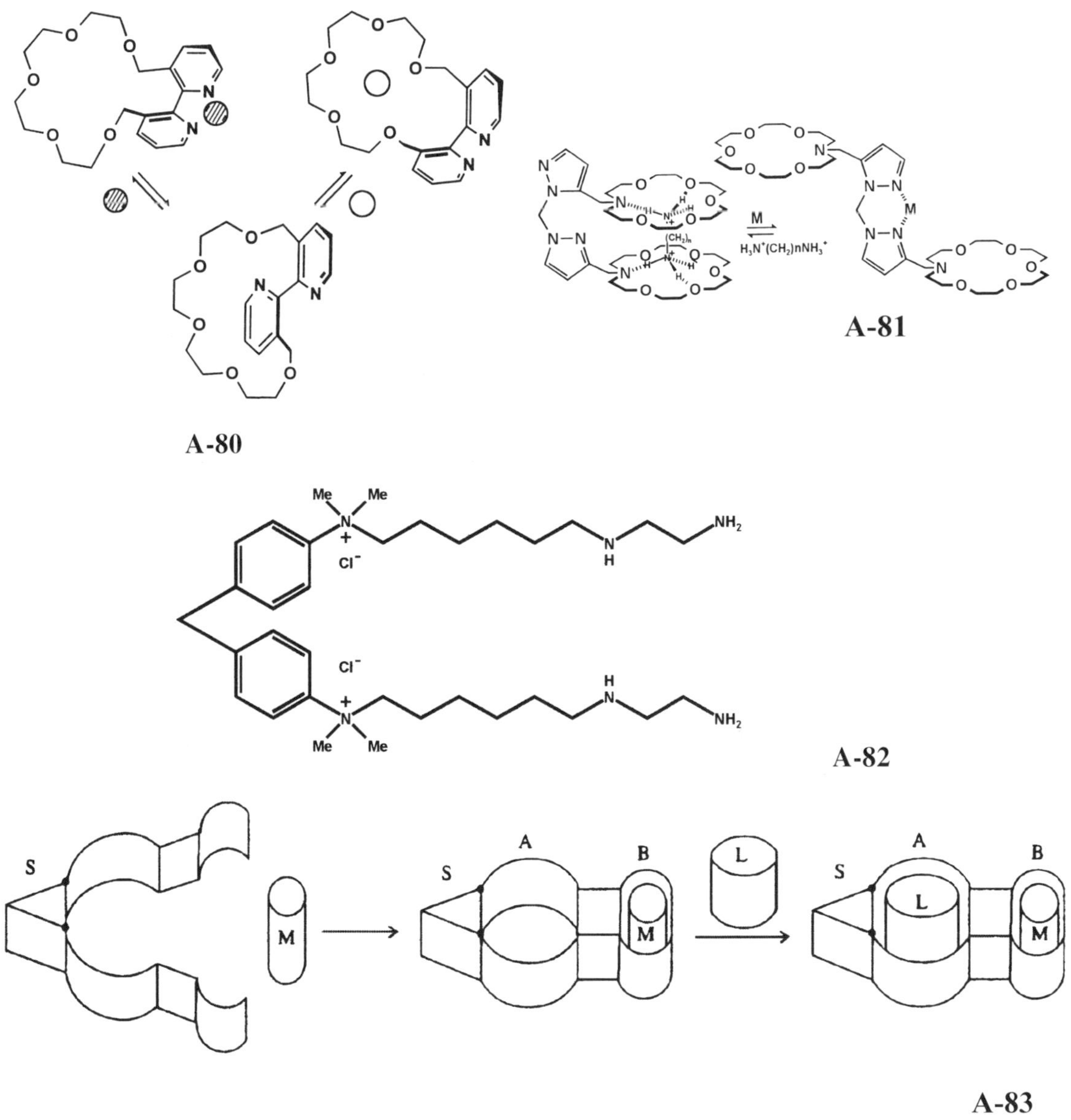

A-80

A-81

A-82

A-83

Scheme A 8.   Selected examples of synthetic allosteric systems with ligands **A-80** to **A-85**.

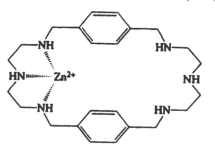

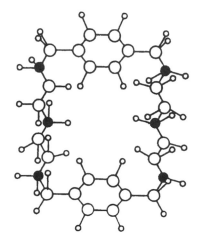

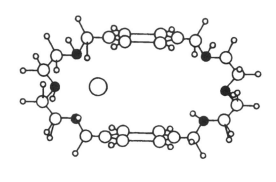

A-84

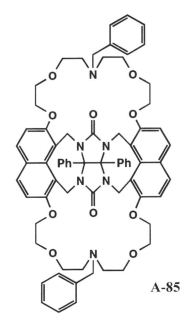

A-85

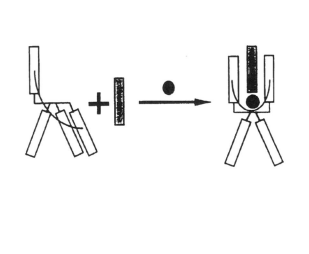

Scheme A8.   (*continued*)

recent reviews.[118c] Obviously, materialization of the molecular machinery sketched in Scheme A 7 requires an intelligently chosen combination of rigid levers or hinges and joints.

The earliest examples of synthetic allosteric complexes involved two heterotropic binding sites for metal ions,[118c,135,136] like in **A-80** (Scheme A 8). These receptors belong to type II in Scheme A 7, only with closed cyclic hinges at one or two sides of the central handle. The one site (A) is a crown ether equipped for complexation of alkali or alkali earth ions, whereas the vicinal nitrogen atoms at the other side (B) interact with transition metal ions. If a copper ion is taken up at the B site, the conformation, which in the ground state has a distorted crown, changes to a favorable geometry for binding potassium ions, and vice versa. Related conformational reorganization upon uptake of metals in such

ene units is the basis of new techniques for self-assembling complexes (Section 12). Receptor **A-81** (Scheme A 8) represents one of the (until now) rare cases of negative cooperativity between binding of a diprotonated diamine with the two crown ether units, and Zn^{2+} with the pyrazole nitrogens.[137]

In some artificial allosteric systems like **A-82/83** (Scheme A 8) the cooperativity is so strong that one can give only upper limits such as K_1/K_2 of <0.01 for the cooperativity parameter: without occupation of the metal binding center B there is not any detectable association with the lipophilic fluorescent dye which is used as second guest molecule.[138a] A similar on/off situation with $K_1/K_2 < 0.01$ exists for the host **A-84**, in which binding of metal ions like Zn^{2+} leads to constriction of the macrocycle with subsequent binding of lipophilic guest dyes.[138b] Such receptors can be used to

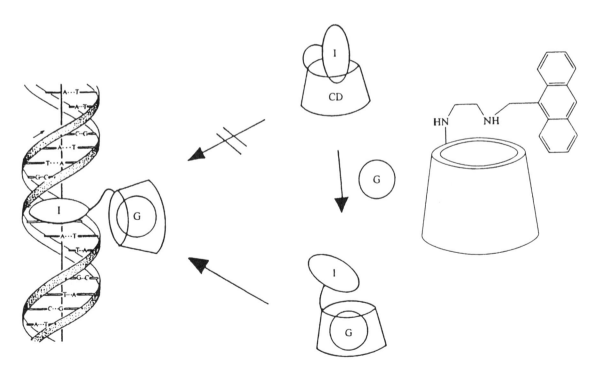

Scheme A 9.   An allosteric system targeting DNA.[140] Reprinted with permission from *J. Am. Chem. Soc.*, Ikeda *et al.*, 1995, **117**, 1453. Copyright 1995 American Chemical Society.

measure metal concentrations in water with highly sensitive fluorescence techniques, for example.[118c] The molecular clip **A-85**[139] shown in Scheme A 8 exists in three conformations *ss*, *sa* and *aa*, interconverting slowly on the NMR timescale. Both the *ss* and the *sa* conformer convert upon uptake of a potassium ion in the crown ether bottom unit to *aa*, which then binds an aromatic guest such as 1,3-dimethoxybenzene about four times better than without K[+].

Low molecular weight allosteric compounds can also target biopolymers, as shown in Scheme A 9. Here the anthracene unit **I** can

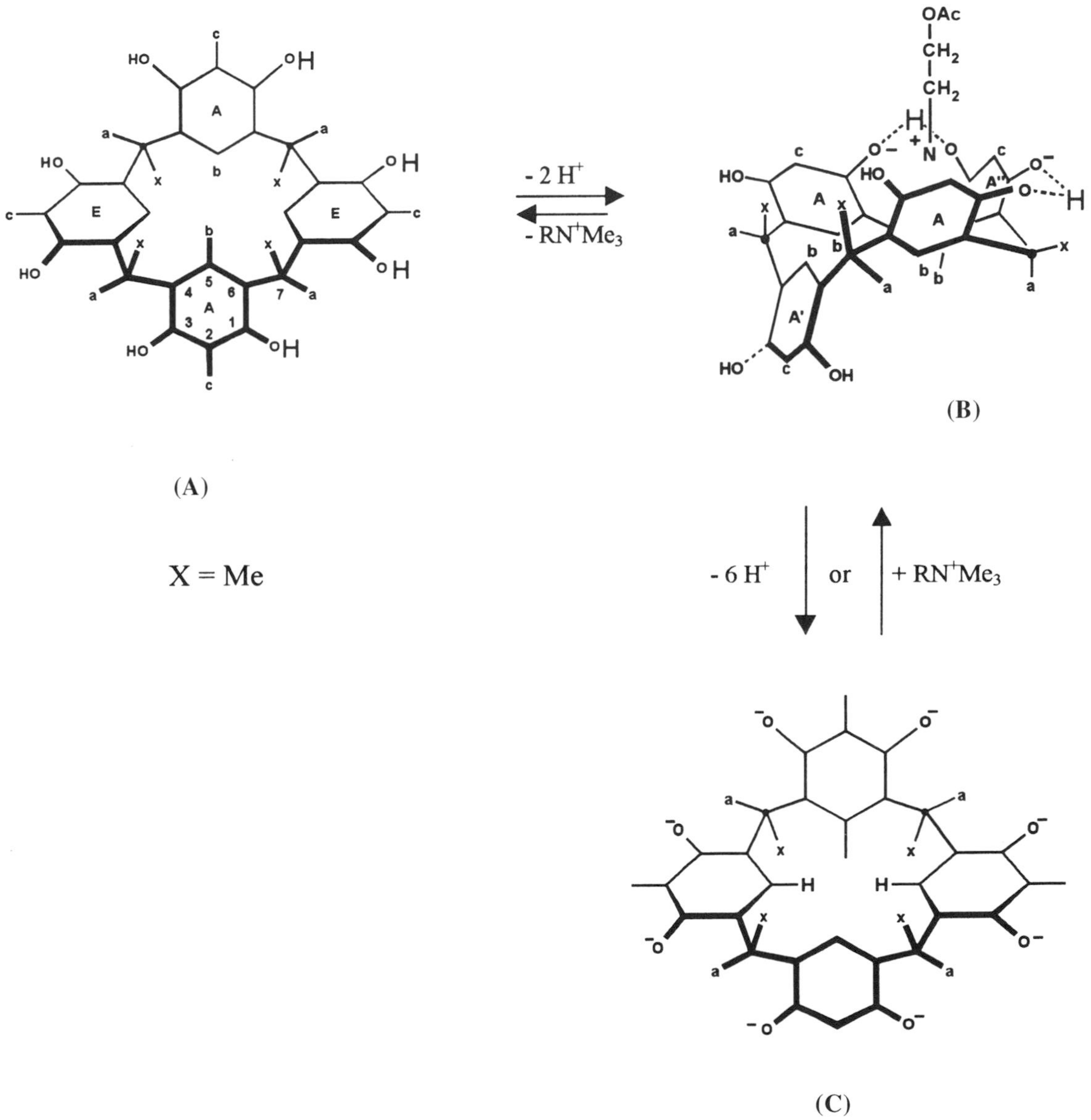

(A)

X = Me

(B)

(C)

Scheme A 10. An allosteric element of a proton pump.[141] Reprinted with permission from *J. Am. Chem. Soc.*, Schneider *et al.*, 1998, **110**, 6449. Copyright 1988 American Chemical Society

only intercalate into double-stranded DNA after it is expelled from the cyclodextrin cavity by a suitable second guest G like adamantanol.[140]

Proton delivery can be conformationally coupled to release or uptake of organic guest molecules if the protons are part of a hydrogen bridge which is opened or closed by the process.[141] In the resorcarene in Scheme A 10 conformation *A* converts to *B* by release of two protons, thus allowing the additional

strain in *B* to be overcome by the formation of two strong hydrogen bonds. The same process happens in basic solution where conformer *C* dominates, by uptake of choline-type tetraalkylammonium compounds, which can only be bound in structure *B*. At the same time, two protons must be taken up from the solution in order to form the hydrogen bonds in *B*. This allosteric system therefore represents a simple element of a proton pump.

A 10.   EXERCISES AND ANSWERS

A 10.1   EXERCISES

1. Scheme A11 shows the binding constants of a series of steroids to a glucocorticoid receptor (published data[13a]). Use data for molecules **A-86** – **A-89** to calculate free energy increments of OH groups in different positions of steroid molecules and then use these increments to calculate the binding constants of steroids **A-90** and **A-91**.

2. Apply equation (A1-3) to the analysis of the linear correlation between complexation free energies and number of ionic pairwise interactions shown in Fig. A3. Which value of the formation constant K of a single 1:1 ionic pair is necessary to get the observed slope of -5 kJ mol^{-1} per ionic pair? Which value does one obtain from the slope of -8 kJ mol^{-1}, which one observes for ionic strength equal to zero? (Section B 2.)

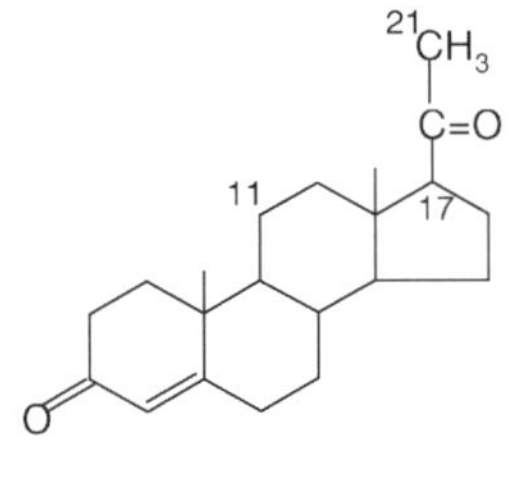

A-86 logK=6.85	**A-87** logK=7.36	**A-88** logK=7.51
A-89 logK=6.48	**A-90** logK=8.18	**A-91** logK=7.81

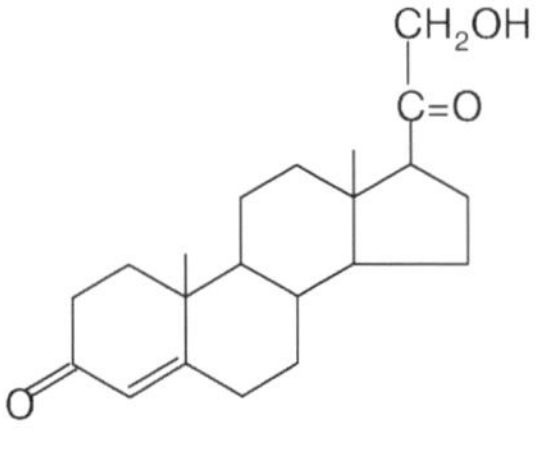

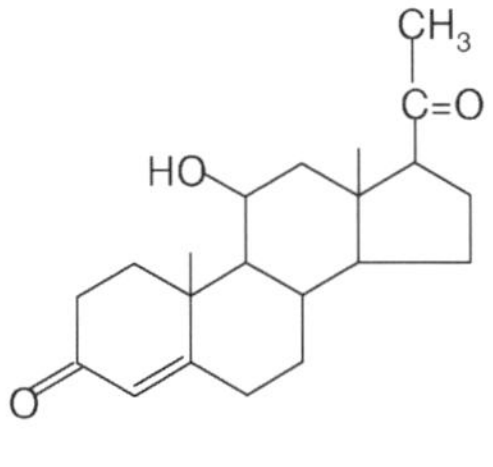

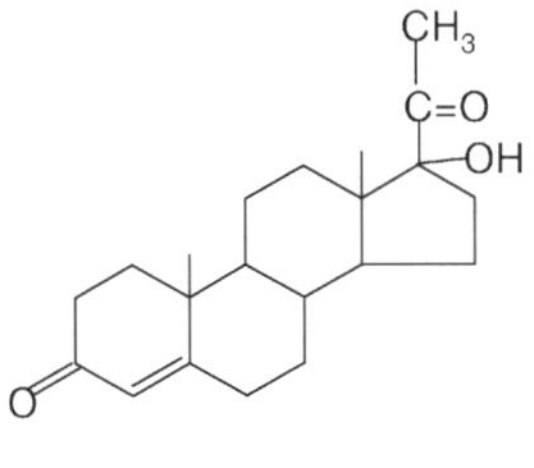

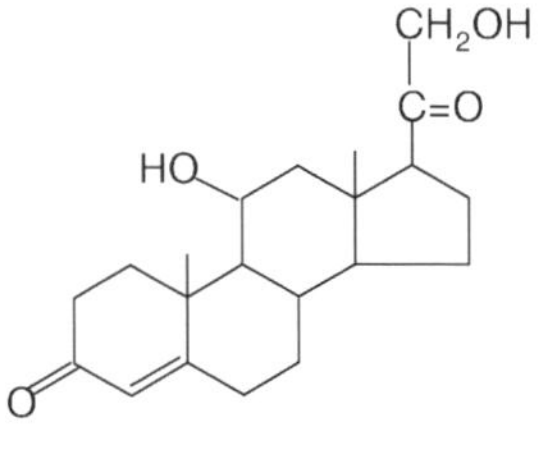

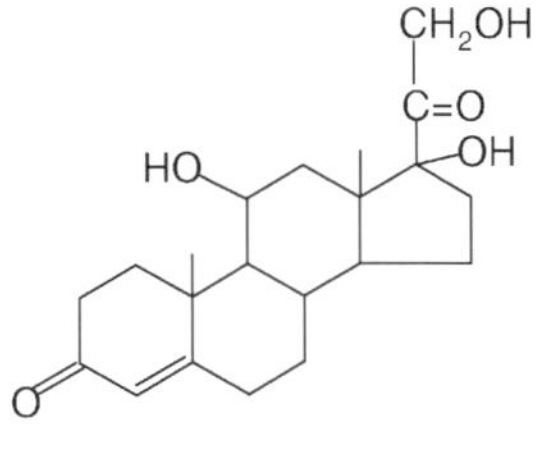

Scheme A 11.   Binding constants of a series of steroids to a glucocorticoid receptor (From published data[13a])

3. Binding of the hydrocinnamate anion to the active site of carboxypeptidase A involves one salt bridge, two hydrogen bonds and inclusion of the inhibitor phenyl group into a hydrophobic cavity of the protein (Scheme A 12). Calculate the binding constant of the hydrocinnamate anion with the free energy increment for H-bonding found in Exercise 1 (use the mean value for 11- and 21-OH), an increment for a salt bridge $-5 \pm 1 \, kJ \, mol^{-1}$ (slope of the line in Fig. A 3), and an increment for hydrophobic binding of phenyl group $-11.2 \, kJ \, mol^{-1}$ (Section B 6) assuming $\Delta G_{ass} = 0$. Compare the result with the experimental $K = 1.6 \times 10^4 \, M^{-1}$. Binding constants of anions of phenylacetic and 3-phenylbutyric acids to carboxypeptidase A equal $2.6 \times 10^3 \, M^{-1}$ and 8.8×10^2 M^{-1} respectively.[142] Propose an explanation of the observed trends in the binding constants.

4. Estimate incremental free energy of ion pairing between guest catecholate and host ammonium ionic groups from the formation free energies of the complexes **A-92 a–f** of different forms of caffeic acid and cinnamic acid with unsubstituted β-cyclodextrins and those bearing ammonium groups, as shown in Scheme A 13.[143] Use data for single 'mutations' of host and guest structures as well as apply the double mutation cycle method, equation (A 1-13).

5. Propose plausible explanations of tendencies in thermodynamic parameters $(kJ \, mol^{-1})$ given in Table A 15 for complexation of ethylenediaminetetraacetate homologs **A-93** with Zn^{2+} yielding complex **A-94**.

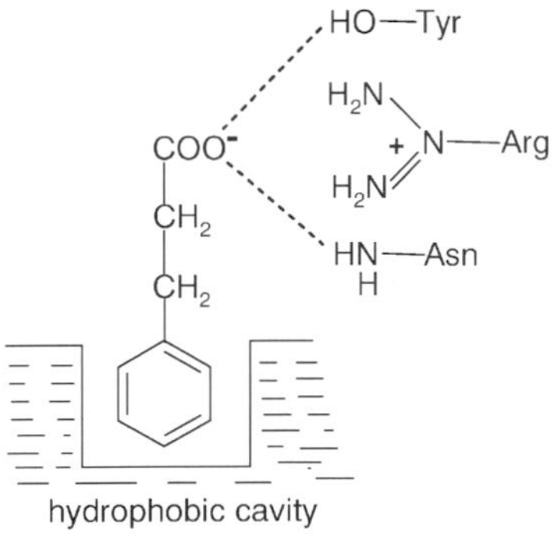

Scheme A 12. Schematic illustration of the hydrocinnamate anion binding to the active side of carboxypeptidase A.

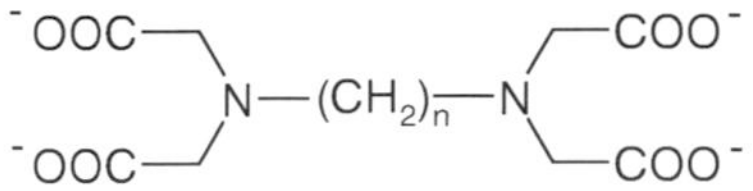

A-93

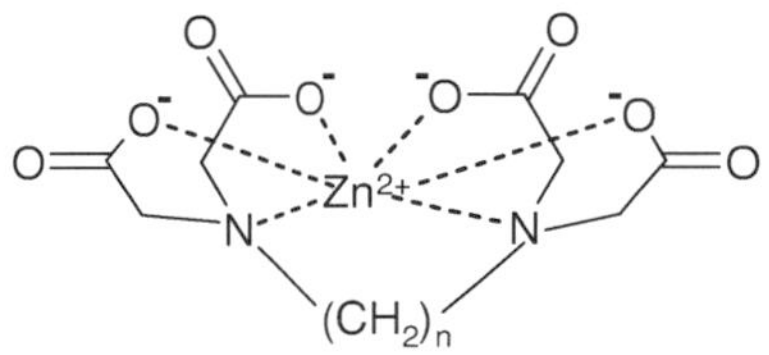

A-94

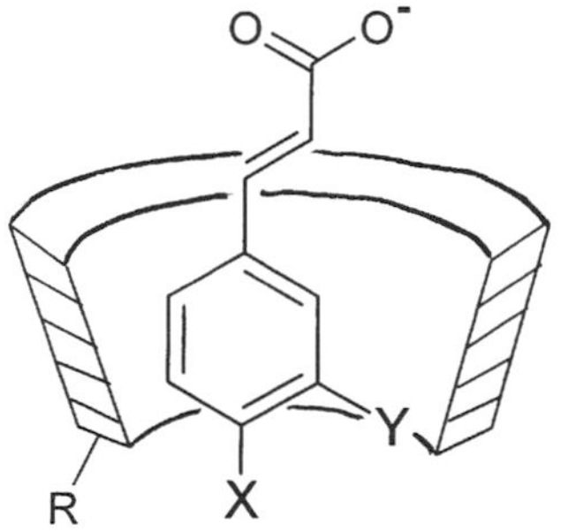

$-\Delta G \, (kJ/mol)$

a) R = H, X = O⁻, Y = OH   11.5
b) R = NH₂Me⁺, X = O⁻, Y = OH   21.8
c) R = NH₂Me⁺, X = Y = H   13.5
d) R = H, X = Y = OH   12.9
e) R = NH₂Me⁺, X = Y = OH   14.3
f) R = X = Y = H   15.6

A-92

Scheme A 13. Schematic illustration of the binding of caffeic and cinnamic acids to cyclodextrins.

Table A 15. Thermodynamic parameters (kJ mol^{-1}) for complexation of ethylenediamine tetra-acetate homologs **A-93** with Zn^{2+} yielding complex **A-94** (from published data[37b]).

n	ΔG (kJ mol^{-1})	ΔH (kJ mol^{-1})	$T\Delta S$ (kJ mol^{-1})
2	-95.04	-19.2	75.8
3	-87.55	-9.6	80.0
4	-86.40	-14.6	71.8
5	-72.58	-11.3	61.5

Table A 16. Formation constants for complexation of 18-crown-6 with cation **A-95** in MeOH/H$_2$O (9/1) at different temperatures.[145]

T (K)	$10^{-3}K$, (M^{-1})
280	2.52
290	1.77
300	1.27
310	0.755
320	0.49

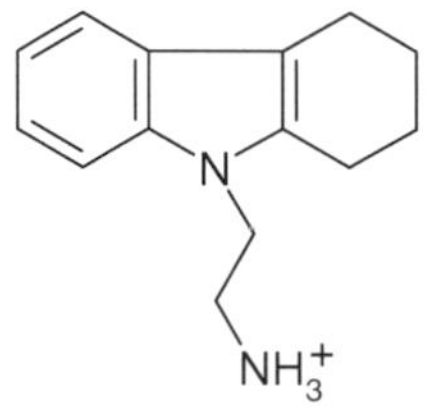

6. Explain the trends in relative stability of oxalate and malonate chelates with the following metal ions:

	log K	
	oxalate	*malonate*
Be^{2+}	4.96	6.18
Mg^{2+}	3.43	2.85
Pb^{2+}	4.91	3.68

7. Figure A 22 contains data on the selectivity patterns of cryptate [2.2.2] and its less symmetric [3.2.1] isomer.[144] Try to explain the difference.

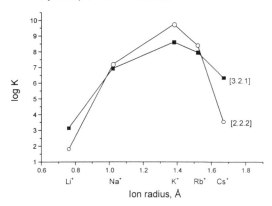

Figure A 22. Binding selectivity of cryptate [2.2.2] and its [3.2.1] isomer[144]

8. Analyze the data given in Table A 16 on the temperature dependence of the formation constant for complexation of 18-crown-6 with cation **A-95** in MeOH/H$_2$O (9/1)[145] in terms of the linear equation (A 4-2) and the Valentiner equation (A 4-5). If possible use a commercially available program for non-linear regression. Interpret the results.

A-95

9. Test the presence or absence of the isoequilibrium relationship in the thermodynamic parameters (kJ mol^{-1}) of complexation of macrocyclic antibiotics with Na$^+$ and K$^+$ cations (Table A 17). Try to correlate the results with equations (A 4-10) and (A 4-12).

A 10.2   ANSWERS

1. Incremental binding free energies equal $\Delta G_{OH} = -2.3\,RT(\log K_{OH-substituted\ steroid} - \log K_1) = -3.8$, -2.9, $+2.1$ kJ mol^{-1} for 11$-$OH, 21$-$OH and 17$-$OH. Positive value for 17$-$OH most probably is due to mismatch. Calculated values of the binding constants for **A-90** and **A-91** are log $K_5 = 8.0$ and log $K_6 = 7.65$ in good agreement with experimental values.

2. The single equilibrium constants can be defined here as the formation constants of 1:1 ionic pairs between small ions bearing the same ionic groups as the host and guest polyions, for example CH$_3$COO$^-$ and NH$_4^+$. Assuming that all ionic pairwise interactions to be equal, only one single equilibrium constant K_{11} is necessary and equation (A 1-3) takes the form

$$K_t = (55.6)^{n-1} K_{11}^n$$

Consequently

$$\Delta G_t = -2.3\,RT(n-1)\log 55.6 - 2.3\,RTn \log K_{11}$$
$$= 2.3\,RT \log 55.6 - 2.3\,RT(\log 55.6 + \log K_{11})n$$

Table A 17. Thermodynamic parameters (kJ mol^{-1}) of complexation of macrocyclic antibiotics with Na$^+$ and K$^+$ cations (from published data[54b]).

Ligand	Solvent	Cation	ΔH	$T\Delta S$	ΔG
Nonactin	MeOH	Na$^+$	− 11.1	4.4	− 15.5
		K$^+$	− 43.5	− 17.9	− 25.6
	EtOH	Na$^+$	− 27.4	− 8.7	− 18.7
		K$^+$	− 52.2	− 22.1	− 30.1
Monactin	MeOH	Na$^+$	− 25.1	− 10.0	− 15.1
Dinactin	MeOH	Na$^+$	− 27.6	− 10.9	− 16.5
Trinactin	MeOH	Na$^+$	− 30.5	− 12.1	− 18.4
Valinomycin	MeOH	K$^+$	− 19.0	9.0	-27.9
	EtOH	K$^+$	− 37.2	− 2.7	− 34.3
Nigericin	MeOH	Na$^+$	7.1	29.3	− 22.2
		K$^+$	− 4.1	27.6	− 32.0
Monensin	MeOH	Na$^+$	− 16.3	18.0	− 34.3
		K$^+$	− 15.6	10.4	− 26.0

Evidently ΔG_t is a linear function of n with the slope $-2.3RT(\log 55.6 + \log K_{11})$ and an intercept $2.3RT \log 55.6$. Introducing the values of parameters at 298 K one obtains the intercept 10 kJ mol^{-1} and the slope $-5.7(1.74 + \log K_{11})$ kJ mol^{-1}. In order to get the slopes of -5 kJ mol^{-1} and -8 kJ mol^{-1}, the values of K_{11} must be equal to 0.14 and 0.46 M^{-1} respectively. Experimental determination of formation constants of 1:1 ionic pairs in water is rather uncertain due to their small values (Sections B 2 and D 2). From the Fuoss equation (Section B 2) the theoretically expected value of K_{11} at zero ionic strength is about 0.8 M^{-1} (taking the Bjerrum distance of closest interionic approach $a = 3.57$ Å for a 1:1 electrolyte). This is reasonably close to $K_{11} = 0.46$ M^{-1}, also at zero ionic strength. The smaller K_{11} value found from the slope of the plot in Fig. A 3 refers to a higher ionic strength of about 0.1 M.

3. The hydrogen bond increment is -3.35 kJ mol^{-1} and $\Delta G_t = -22.9$ kJ mol^{-1} (experimental $\Delta G_t = -24.0$ kJ mol^{-1}), which gives $K = 1.02 \times 10^4$ M^{-1}. Both shortening and extension of the hydrocarbon chain connecting the phenyl group and carboxylate anion leads to a decrease in K. This indicates that the complementarity between binding sites is achieved just when these groups are separated by two methylene groups.

4. Binding of the anion of caffeic acid to amino-cyclodextrin (complex **A-92b**) involves ion pairing between the guest catecholate anion and the host ammonium cation as well as hydrophobic interaction of the hydrocarbon part of the guest with the cavity of the host. In order to evaluate $\Delta\Delta G$ of ion pairing one can consider three 'mutations': removal of the ammonium group (complex **A-92a**), removal of hydroxyl groups of catecholate (complex **A-92c**) and neutralization of catecholate anion by protonation (complex **A-92e**). The three resulting $\Delta\Delta G$ values are -10.3, -8.4 and -7.5 kJ mol^{-1}. Such scatter of increments is quite typical (previously discussed,[13b] Section A 1) and is partially due to different contributions of secondary effects. The mean value of $\Delta\Delta G$ is -8.7 ± 1.4 kJ mol^{-1} and is not too far from the mean electrostatic increment per ionic pair given in Sections A 1 and B 2.

Results for complexes **A-92d** and **A-92f** lacking both anionic and cationic groups allow one to construct two double mutation cycles: one which involves complexes **A-92a**, **A-92b**, **A-92d** and **A-92e**, and another one which involves complexes **A-92b**, **A-92c**, **A-92e** and **A-92f**. Application of equation (A 1-13) gives from the first cycle $\Delta\Delta G = -8.9$ kJ mol^{-1} and from the second $\Delta\Delta G = -9.6$ kJ mol^{-1} in reasonably good agreement with each other. The mean value of $\Delta\Delta G = -9.25$ kJ mol^{-1} is, actually, very close to the mean value from single mutations, but, as expected, the double mutation method gives less scatter.

5. Increased length of a connector between two iminodiacetate fragments leads to a decreased complex stability. The effect is both enthalpic and entropic: ΔH increases by roughly 2 kJ mol^{-1} with every additional methylene group (2.5 kJ mol^{-1} per *gauche* C−⌣−C interaction, Section A1) and $T\Delta S$ decreases by roughly 5 kJ mol^{-1} per additional methylene group, showing an increment typical for covalent cyclization reactions (Section A 1) in accordance with strong metal−ligand interaction in this system.

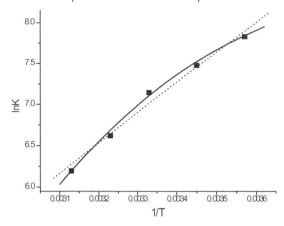

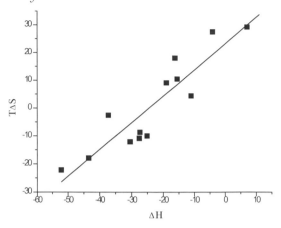

Figure A 23. Linear (dotted line) and non-linear (solid curve) regressions of the data in Table A16 according to equation (A4-2) or equation (A4-5), respectively.

Figure A 24. Data of Table A17 plotted as $T\Delta S$ vs. ΔH.

6. Smaller cations prefer the larger 6-membered malonate cycle, while the larger cations prefer smaller 5-membered oxalate cycle.

7. The less symmetrical structure of the [3.2.1] isomer possessing both smaller and larger areas inside its cavity, as compared to the [2.2.2] isomer, allows both smaller and larger cations to fit better in the respective areas. At the same time K^+ does in this isomer not have the optimum contact with all six donor atoms.

8. Figure A 23 shows the results of linear (dotted line) and non-linear (solid curve) regressions in accordance with equations (A 4-2) and (A 4-5) respectively. The non-linear regression gives a better fit with the parameter values $\Delta H_i = 155 \pm 72\,\text{kJ mol}^{-1}$; $\Delta C_P = -623 \pm 240\,\text{J mol}^{-1}\,\text{K}^{-1}$) and $I/R = 500 \pm 200$. From linear regression: $\Delta H = -30.7 \pm 2.1\,\text{kJ mol}^{-1}$. From equation (A 4-4) one obtains at 298 K $\Delta H_{283} = -30.65\,\text{kJ mol}^{-1}$ in good agreement with the result of the linear correlation, which ignores the temperature dependence of ΔH. As one would expect, ΔH found by linear regression is close to that calculated for the mean temperature of the interval employed, which is 300 K. Enthalpy values at lowest and highest temperatures of the interval are considerably different: $\Delta H_{280} = -19.4\,\text{kJ mol}^{-1}$ and $\Delta H_{320} = -44.4\,\text{kJ mol}^{-1}$. Note a large error in ΔC_P in

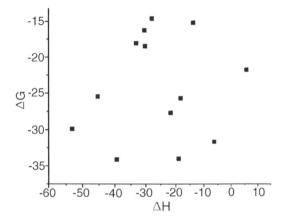

Figure A25. Data of Table A17 plotted as ΔG vs. ΔH.

spite of its large absolute value. Evidently more points (or higher precision) are required to get a more accurate evaluation.

9. The plot of $T\Delta S$ vs. ΔH shows a satisfactory correlation, Fig. A 24, with the slope $a = 0.95$ and $R = 0.90817$, however, the plot of ΔG vs. ΔH, Fig. A 25, shows practically no correlation, $R = 0.1000$. Since the last plot is statistically more appropriate the conclusion is that no isoequilibrium correlation exists in the given system.

REFERENCES

1 (a) Weber, P.C.; Wendoloski, J.J.; Pantoliano, M.W.; Salemme, F.R. (1992) *J. Am. Chem. Soc.*, **114**, 3197; (b) Sano, T.; Cantor, C.R. (1995) *Proc. Natl. Acad. Sci. USA*, **92**, 3180.

2 Through this book 'reaction free energy (enthalpy, entropy)' and symbols ΔG, ΔH, ΔS stand for 'reaction standard free energy (enthalpy, entropy)'.

3 Hegde, V.; Hung, C.-Y.; Madhukar, P.; Cunningham, R.; Höpfner, T.; Thummel, R.P. (1993) *J. Am. Chem. Soc.*, **115**, 872.

4 (a) Hosseini, M.W.; Lehn, J.-M. (1987) *Helv. Chim. Acta*, **70**, 1312.; (b) Lonergan, D.G.; Deslongchamps, G.; Tomás, S. (1998) *Tetrahedron Lett.*, **39**, 7861.

5 Dougherty, D.A. (1996) *Science*, **271**, 163.

6 (a) Lehn, J.-M.; Meric, R.; Vigneron, J.-P.; Cesario, M.; Guilhem, J.; Pascard, C.; Asfari, Z.; Vicens, J. (1995) *Supramol. Chem.*, **5**, 97; (b) Schneider, H.-J.; Güttes, D.; Schneider, U. (1988) *J. Am. Chem. Soc.*, **110**, 6449.

7 Inoue, Y.; Gokel, G.W. (Eds.) (1990) *Cation Binding by Macrocycles*, Marcel Dekker, New York.

8 Adamson, A.W. (1954) *J. Am. Chem. Soc.*, **76**, 1578.

9 Mortimer, R. (1993) *Physical Chemistry*, Benjamin/Cummings Publ., Redwood City.

10 Kauzmann, W. (1959) *Adv. Protein Chem.*, **14**, 1.

11 (a) Fersht, A.R. (1985) *Enzyme Structure and Mechanism*, 2nd edn., Freeman, New York; (b) Fersht, A.R.; Shi, J.-P.; Knill-Jones, J.; Lowe, D.M.; Wilkinson, A.J.; Blow, D.M.; Carter, P.; Waye, M.M.; Winter, G. (1985) *Nature*, **314**, 235; (c) Fersht, A.R. (1987) *Trends Biochem. Sci.*, **12**, 301.

12 Jencks, W.P. (1975) *Adv. Enzymol.*, **43**, 219.

13 (a) Wolff, M.E.; Baxter, J.D.; Kollman, P.A.; Lee, D.L.; Kuntz, I.D.; Bloom, E.; Matulich, D.T.; Morris, (1978) *J. Biochemistry*, **17**, 3201; (b) Morgan, B.P.; Scholtz, J.M.; Ballinger, M.D.; Zipkin, I.D.; Bartlet, P.A. (1991) *J. Am. Chem. Soc.*, **113**, 297; (c) Wallqvist, A.; Covell, D.G. (1996) *Proteins: Str. Func. Genet.*, **25**, 403.

14 (a) Schneider, H.-J.; (1994) *Chem. Soc. Rev.*, **23**, 227; (b) Schneider, H.-J. (1991) *Angew. Chem., Int. Ed. Engl.*, **30**, 1417.

15 Group contribution methods have been used long ago for estimations of thermodynamic quantities of organic compounds, see, e.g. Cohen, N.; Benson, S.W. (1993) *Chem. Rev.*, **93**, 2419; Gianni, P.; Leprot, L. (1996) *J. Solut. Chem.*, **25**, 1; Klotz, I.M.; Rosenberg, R.M. (1994) *Chemical Thermodynamics*, 5th edn., Wiley, New York, Chapter 22.

16 (a) Page, M.I.; Jencks, W.P. (1971) *Proc. Natl. Acad. Sci. USA*, **68**, 1678; (b) Page, M.I. (1973) *Chem. Soc. Rev.*, **2**, 295; (c) Jencks, W.P. (1978) *Proc. Natl. Acad. Sci. USA*, **78**, 4046.

17 Stull, D.R.; Westrum, E.F.Jr.; Sinke, G.C. (1969) *The Chemical Thermodynamics of Organic Compounds*, Wiley, New York.

18 Benson, S.W. (1967) *Thermochemical Kinetics*, 2nd edn., Wiley, New York.

19 Fild, M.; Swinarski, M.F.; Holmes, R.R. (1970) *Inorg. Chem.*, **9**, 839.

20 Gramstad, T. (1961) *Acta. Chem. Scand.*, **16**, 807.

21 Kroll, M. (1968) *J. Am. Chem. Soc.*, **90**, 1097.

22 Joesten, M.D.; Schaad, L.J. (1974) *Hydrogen Bonding*, Marcell Decker, New York.

23 Abraham, M.H. (1981) *J. Am. Chem. Soc.*, **103**, 6742.

24 (a) Searle, M.S.; Williams, D.H; Gerhard, U. (1992) *J. Am. Chem. Soc.*, **114**, 10697; (b) Williams, D.H; Searle, M.S.; Mackay, J.P.; Gerhard, U.; Maplestone, R.A. (1993) *Proc. Natl. Acad. Sci. USA*, **90**, 1172.

25 Kirby, A.J. (1980) *Adv. Phys. Org. Chem.*, **17**, 183.

26 Smith, R.M.; Martell, A.E. (1975) *Critical Stability Constants, Vol. 2*, Plenum Press, New York.

27 (a) Eblinger, F.; Schneider, H.-J. (1998) *Angew. Chem., Int. Ed. Engl.*, **37**, 826; (b) Hossain, M.A.; Schneider, H.-J., (1999) *Chem. Eur. J.* in press.

28 Zhang, B.; Breslow, R. (1993) *J. Am. Chem. Soc.*, **115**, 9353.

29 (a) Horovitz, A. (1987) *J. Mol. Biol.*, **196**, 733; (b) Fersht, A.R.; Matouschek, A.; Serrano, L. (1992) *J. Mol. Biol.*, **224**, 771; (c) Frish, C.; Schreiber, G.; Johnson, C.M.; Fersht, A.R. (1997) *J. Mol. Biol.*, **267**, 696.

30 Boer, J.H.; Custers, J.F.H. (1934) *Z. Phys. Chem.*, **B25**, 225; see also Page, M.I. in Page, M.J., Ed.: (1984) *The Chemistry of Enzyme Action*, Elsevier, Amsterdam, 46.

31 Rebek, J., Jr. (1990) *Angew. Chem., Int. Ed. Engl.*, **29**, 245.

32 Zimmerman, S.C.; Wu, W.; Zeng, Z. (1991) *J. Am. Chem. Soc.*, **113**, 196.

33 For the use of concave hosts for selective reactions see Lüning, U. (1995) *Top. Curr. Chem.*, **175**, 57.

34 Hilgenfeld, R.; Saenger, W. (1982) *Top. Curr. Chem.*, **101**, 1.

35 (a) Christensen, J.J.; Eatough, D.J.; Izatt, R.M. (1974) *Chem. Rev.*, **74**, 351; (b) Izatt, R.M.; Bradshaw, J.S.; Nielson, S.A.; Lamb, J.D.; Christensen, J.J. (1985) *Chem. Rev.*, **85**, 271; (c) Izatt, R.M.; Pawlak, K.; Bradshaw, J.S.; Bruening, R.L. (1991) *Chem. Rev.*, **91**, 1721.

36 Gokel, G.W. (1991) *Crown Ethers and Cryptands*, The Royal Society of Chemistry, Cambridge, UK.

37 (a) Hancock, R.D.; Martell, A.E. (1989) *Chem. Rev.*, **89**, 1875; (b) Martell, A.E. Hancock, R.D.; Motekaitis, R.J. (1994) *Coord. Chem. Rev.*, **133**, 39; (c) Hancock, R.D.; Marsicano, F. (1976) *J. Chem. Soc., Dalton Trans.*, 1096; (d) Hancock, R.D. (1992) *J. Chem. Ed.*, **69**, 615.

38 Schneider, H.-J.; Rüdiger, V.; Raevsky, O.A. (1993) *J. Org. Chem.*, **58**, 3648.

39 Chu, I.-H.; Zhang, H.; Dearden, D.V. (1993) *J. Am. Chem. Soc.*, **115**, 5736.

40 Chen, Q.; Cannell, K.; Nicoll, J.; Dearden, D.V. (1996) *J. Am. Chem. Soc.*, **118**, 6335.

41 Lehn, J.-M.; Sauvage, J.P. (1975) *J. Am. Chem. Soc.*, **97**, 6700.

42 (a) Bajaj, A.V.; Poonia, N.S. (1988) *Coord. Chem. Rev.*, **87**, 55; (b) Goldberg, I. (1984) *Inclusion Compounds*; Atwood, J.L.; Davies, J.E.D.; Mac-Nicol, D.D., Eds. Vol. 2, Academic Press, London, 261 ff.; see also ref.[33]

43 There are several noticably different scales of ionic crystal radii proposed by different authors (see e.g. Marcus, Y. (1988) *Chem. Rev.*, **88**, 1475). Crystal radii depend on the coordination number of cations; currently most used, in particular in cation solution chemistry, are the values for coordination number 6 (for a compilation see, e.g. Müller, U. (1992) *Inorganic Structural Chemistry*, 2nd edn., Wiley; Huheey, J.E.; Keiter, E.A.; Keiter, R.L. (1993) *Inorganic Chemistry: Principles of Structure and Reactivity*, 4th edn., Harper Collins College Publishers, New York).

44 (a) Michaux, G.; Reisse, J. (1982) *J. Am. Chem. Soc.*, **104**, 6895; (b) Gokel, G.W.; Goli, D.M.; Minganti, C.; Echegoyen, L. (1983) *J. Am. Chem. Soc.*, **105**, 6786.

45 Arnaud-Neu, F. personal communication.

46 Hancock, R.D. (1989) *Progr. Inorg. Chem.*, **37**, 187.

47 Kauffmann, E.; Lehn, J.-M.; Sauvage, J.-P. (1976) *Helv. Chim. Acta.*, **59**, 1099.

48 Glendening, E.D.; Feller, D. (1996) *J. Am. Chem. Soc.*, **118**, 6052.

49 Staufer, D.A.; Barrans, R.E.; Dougherty, D.A. (1990) *J. Org. Chem.*, **55**, 2762.

50 Blandamer, M.J.; Burges, J.; Robertson, R.E.; Scott, J.M.W. (1982) *Chem. Rev.*, **82**, 259.

51 King, E.J. (1965) *Acid Base Equilibria*, Pergamon Press, Oxford.

52 Abraham, M.H. (1982) *J. Am. Chem. Soc.*, **104**, 2085; Blokzijl, W.; Engberts, J.B.F.N; Blandamer, M.J. (1990) *ibid.*, **112**, 1197; Privalov, P.L.; Gill, S.J. (1988) *Adv. Protein. Chem.*, **39**, 191.

53 (a) Leffler, J.E.; Grunwald, E. (1963) *Rates and Equilibria of Organic Reactions*, Wiley, New York; (b) Exner, O. (1973) *Progr. Phys. Org. Chem.*, **10**, 411; (c) Linert, W. (1994) *Chem. Soc. Rev.*, **23**, 429; (d) Lumry, R.; Rajender, S. (1970) *Biopolymers*, **9**, 1125.

54 (a) Inoue, Y.; Hakushi, T. Liu, Y.; in ref 7, p.1; (b) Inoue, Y.; Hakushi, T. (1985) *J. Chem. Soc. Perkin Trans. 2*, 935.

55 (a) Inoue, Y.; Hakushi, T.; Liu, Y.; Tong, L.-H.; Shen, B.-J.; Jin, D.-S. (1993) *J. Am. Chem. Soc.*, **115**, 475; (b) Bertrand, G.L.; Faulkner, J.R.; Han, S.M.; Armstrong, D.W.; (1984) *J. Phys. Chem.*, **93**, 6863.

56 (a) Cram, D.J. (1988) *Angew. Chem., Int. Ed. Engl.*, **27**, 1009; (b) Cram, D.J.; Lein, G.M. (1985) *J. Am. Chem. Soc.*, **107**, 3657; (c) Cram, D.J. (1986) *Angew. Chem., Int. Ed. Engl.*, **25**, 1039; (d) Cram, D.J. (1983) *Science*, **219**, 1177.

57 Koshland Jr., D.E. (1995) *Angew. Chem., Int. Ed. Engl.*, **33**, 2475.

58 Wedemayer, G.J.; Patten, P.A.; Wang, L.H.; Schultz, P.G.; Stevens, R.C. (1997) *Science*, **276**, 1665.

59 Desper, J.M.; Gellman, S.H.; Wolf, Jr., R.E.; Cooper, S.R. (1991) *J. Am. Chem. Soc.*, **113**, 8663.

60 Iimori, T.; Still, W.C.; Rheingold, A.L.; Staley, D.L. (1989) *J. Am. Chem. Soc.*, **111**, 3439.

61 Li, G.; Still, W.C. (1993) *J. Am. Chem. Soc.*, **115**, 3804 and references therein.

62 Trueblood, K.N.; Knobler, C.B.; Maverick, E.; Helgeson, R.C.; Brown, S.B.; Cram, D.J. (1981) *J. Am. Chem. Soc.*, **103**, 5594.

63 Kollman, P.A.; Wipff, G.; Singh, U.C. (1985) *J. Am. Chem. Soc.*, **107**, 2212.

64 Raevsky, O.A.; Solov'ev, V.P.; Solotnov, A.F.; Schneider, H.-J., Rüdiger, V. (1996) *J. Org. Chem.*, **61**, 8113.

65 (a) Ghidini, E.; Ugozzoli, F.; Ungaro, R.; Harkema, S.; El-Fald, A.A.; Reinhoudt, D.N. (1990) *J. Am. Chem. Soc.*, **112**, 6979; (b) Casnati, A.; Pochini, A.; Ungaro, R.; Geraci F.; Arnaud, F.; Fanni, S.; Schwing, M.-J.; Egberink, R.J.M.; de Jong, F.; Reinhoudt, D.N. (1995) *J. Am. Chem. Soc.*, **117**, 2767.

66 (a) Geraci, C.; Piattelli, M.; Neri, P. (1995) *Tetrahedron Lett.*, **36**, 5429; (b) Neri, P.; Consoli, G.M.L.; Cunsolo, F.; Geraci, C.; Piattelli, M. (1996) *New J. Chem.*, **20**, 433.

67 (a) Hou, Z.; Sunderland, C.J.; Nishio, T.; Raymond, K.N. (1996) *J. Am. Chem. Soc.*, **118**, 5148; (b) Stack, T.D.P.; Hou, Z.; Raymond, K.N. (1993) *J. Am. Chem. Soc.*, **115**, 6466; (c) Garrett, T.M.; McMurry, T.J.; Hosseini, M.W.; Reyes, Z.E.; Hahn, F.E.; Raymond, K.N. (1991) *J. Am. Chem. Soc.*, **113**, 2965; (d) Harris, W.R.; Carrano, C.J.; Cooper, S.R.; Sofen, S.R.; Avdeef, A.E.; McArdle, J.V.; Raymond, K.N. (1979) *J. Am. Chem. Soc.*, **101**, 6097.

68 (a) Tor, Y.; Libman, J.; Shanzer, A.; Felder, C.E.; Lifson, S. (1992) *J. Am. Chem. Soc.*, **114**, 6661; (b) Shanzer, A.; Libman, J.; Lifson, S.; Felder, C.E.; (1986) *J. Am. Chem. Soc.*, **108**, 7609.

69 Stutte, P.; Kiggen, W.; Vögtle, F. (1987) *Tetrahedron*, **43**, 2065.

70 Martell, A.E.; Smith, R.M. (1977) *Critical Stability Constants, Vol. 4*, Plenum Press, New York.

71 (a) Bianchi, A.; Bowman-James, K.; Garcia-España, E. Eds., (1997) *The Supramolecular Chemistry of Anions*, Wiley-VCH, New York; (b) Fabbrizzi, L.; Licchelli, M.; Pallavicini, P., Parodi, L.; Poggi, A., Taglietti, A. (1996) in: *Physical Supramolecular Chemistry*, Echegoyen, L.; Kaifer, A.E. (Eds.) Kluwer, Dordrecht, 433.

72 Scheerder, J.; Engbersen, J.F.J.; Reinhoudt, D.N. (1996) *Rec. Trav. Chim. Pay-Bas*, **115**, 307; (b) Antonisse, M.M.G.; Reinhoudt, D.N. (1998) *J. Chem. Soc., Chem. Commun.*, 443.

73 Atwood, J.L.; Holman, K.T.; Steed, J.W. (1996) *J. Chem. Soc., Chem. Commun.*, 1401.

74 Steed, B. (1993) *Pure. Appl. Chem.*, **65**, 1457.

75 (a) Schmidtchen, F.P.; Gleich, A.; Schummer, A. (1989) *Pure Appl. Chem.*, **61**, 1535; Schmidtchen, F.P.; Berger, M. (1997) *Chem. Rev.*, **97**, 1609.

76 (a) Newcomb, M.; Horner, J.H.; Blanda, M.T.; Squattrito, P.J. (1989) *J. Am. Chem. Soc.*, **111**, 6294; (b) Blanda, M.T.; Newcomb, M. (1989) *Tetrahedron Lett.*, **30**, 3501; (c) Blanda, M.T.; Newcomb, M.; Horner, J.H. (1989) *J. Org. Chem.*, **54**, 4626.

77 (a) Jacobson, S.; Pizer, R. (1993) *J. Am. Chem. Soc.*, **115**, 11216; (b) Katz, H.E. (1985) *J. Org. Chem.*, **50**, 5027; (c) Katz, H.E. (1986) *J. Am. Chem. Soc.*, **108**, 7640; (d) Katz, H.E. (1987) *Organometallics*, **6**, 1134.

78 Jung, M.E.; Xia, H. (1988) *Tetrahedron Lett.*, **29**, 297.

79 (a) Wuest, J.D.; Zacharie, B. (1987) *J. Am. Chem. Soc.*, **109**, 4714; (b) Shur, V.B.; Tikhonova, I.A.; Dolgushin, F.M.; Yanovsky, A.I.; Struchkov, Y.T.; Volkonsky, A.Yu.; Solodova, E.N.; Panov, S.Yu.; Petrovskii, P.V.; Vol'pin, M.E. (1993) *J. Organomet. Chem.*, **443**, C19.

80 Hawthorne, M.F.; Zheng, Z. (1997) *Acc. Chem. Res.*, **30**, 267.

81 (a) Rudkevich, D.M.; Verboom, W.; Brzozka, Z.; Palys, M.J.; Stauthamer, W.P.R.V.; van Hummel, G.J.; Franken, S.M.; Harkema, S.; Engbersen, J.F.J.; Reinhoudt, D.N. (1994) *J. Am. Chem. Soc.*, **116**, 4341; (b) Rudkevich, D.M.; Verboom, W.; Reinhoudt, D.N. (1994) *J. Org. Chem.*, **59**, 3683; (c) Rudkevich, D.M.; Brzozka, Z.; Palys, M.J.; Visser, H.C.; Verboom, W.; Reinhoudt, D.N. (1994) *Angew. Chem., Int. Ed. Engl.*, **33**, 467.

82 Beer, P.D. (1996) *J. Chem. Soc., Chem. Commun.*, 689.

83 Ahlers, B.; Cammann, K.; Kramer, R. (1996) *Angew. Chem., Int. Ed. Engl.*, **35**, 2141.

84 Beer, P.D.; Fletcher, N.C.; Wear T.J. (1997) *Polyhedron*, **16**, 815.

85 Nation, D.A.; Martell, A.E.; Carrol, R.I.; Clearfield, A. (1996) *Inorg. Chem.*, **35**, 7246.

86 (a) Gelb, R.I.; Lee, B.T.; Zompa, L.J. (1985) *J. Am. Chem. Soc.*, **107**, 909; (b) Cullinane, J.; Gelb, R.I.; Margulis, T.N.; Zompa, L.J. (1982) *J. Am. Chem. Soc.*, **104**, 3048; (c) Gelb, R.I.; Schwartz, L.B.; Zompa, L.J. (1986) *Inorg. Chem.*, **25**, 1527.

87 Hosseini, M.W.; Lehn, J.-M. (1988) *Helv. Chim. Acta*, **71**, 749.

88 (a) Bianchi, A.; Micheloni, M.; Paoletti, P. (1988) *Pure Appl. Chem.*, **60**, 525; (b) Bencini, A.; Bianchi, A.; Dapporto, P.; Garcia-España, E.; Micheloni, M.; Ramirez, J.A.; Paoletti, P.; Paoli, P. (1992) *Inorg. Chem.*, **31**, 1902; (c) Bencini, A.; Bianchi, A.; Micheloni, M.; Paoletti, P.; Dapporto, P.; Paoli, P.; Garcia-España, E. (1992) *J. Incl. Phen. Mol. Recognit. Chem.*, **12**, 291; (d) Bencini, A.; Bianchi, A.;. Paoletti, P.; Paoli, P. (1993) *Pure Appl. Chem.*, **65**, 381.

89 Suck, E.; Handel, H. (1984) *Tetrahedron Lett.*, **25**, 645.

90 (a) Cramer, R.E.; Carrié, M.J.J. (1990) *Inorg. Chem.*, **29**, 3902; (b) Cramer, R.E.; Fermin, V.; Kuwabara, E.; Kirkup, R.; Selman, M.; Aoki, K.; Adeyemo, A.; Yamazaki, H. (1991) *J. Am. Chem. Soc.*, **113**, 7033.

91 Shionoya, M.; Furuta, H.; Lynch, V.; Harriman, A.; Sessler, J.L. (1992) *J. Am. Chem. Soc.*, **114**, 5714.

92 (a) Simmons, H.E.; Park, C.H. (1968) *J. Am. Chem. Soc.*, **90**, 2428; (b) Park, C.H.; Simmons, H.E. (1968) *J. Am. Chem. Soc.*, **90**, 2429; (c) Park,

C.H.; Simmons, H.E. (1968) *J. Am. Chem. Soc.*, **90**, 2431.

93 Dietrich, B.; Guilhem, J.; Lehn, J.-M.; Pascard, C.; Sonveaux, E. (1984) *Helv. Chim. Acta*, **67**, 91.

94 Graf, E.; Lehn, J.-M. (1976) *J. Am. Chem. Soc.*, **98**, 6403.

95 Schmidtchen, F.P. (1977) *Angew. Chem., Int. Ed. Engl.*, **16**, 720.

96 Dietrich, B.; Dilworth, B.; Lehn, J.-M.; Souchez, J.-P.; Cesario, M.; Guilhem, J.; Pascard, C. (1996) *Helv. Chim. Acta*, **79**, 569.

97 Schmidtchen, F.P. (1981) *Chem. Ber.*, **114**, 597.

98 (a) Worm, K.; Schmidtchen, F.P.; Schier, A.; Schafer, A.; Hesse, M. (1994) *Angew. Chem., Int. Ed. Engl.*, **33**, 327; (b) Worm, K.; Schmidtchen, F.P. (1995) *Angew. Chem., Int. Ed. Engl.*, **34**, 65.

99 Davis, A P; Gilmer, J.F.; Perry, J.J. (1996) *Angew. Chem., Int. Ed. Engl.*, **35**, 1312.

100 Gale, P.A.; Sessler, J.L.; Král, V.; Lynch, V. (1996) *J. Am. Chem. Soc.*, **118**, 5140.

101 Nishizawa, S.; Bühlmann, P.; Iwao, M.; Umezawa, Y. (1995) *Tetrahedron Lett.*, **36**, 6483.

102 Valiyaveettil, S.; Engbersen, J.F.J.; Verboom, W.; Reinhoudt, D.N. (1993) *Angew. Chem., Int. Ed. Engl.*, **32**, 900.

103 Werner, F. PhD (1998) Dissertation, Universität des Saarlandes.

104 (a) Luecke, H.; Quiocho, F.A. (1990) *Nature*, **347**, 402; (b) He, J.J.; Quiocho, F.A.; (1991) *Science*, **251**, 1479.

105 Gokel, G.W. (1992) *Chem. Soc. Rev.*, **20**, 39; Gokel, G.W.; Trafton, J.E. in ref. 7, p. 253 ff.

106 Schultz, R.A.; White, B.D.; Dishong, D.M.; Arnold, K.A.; Gokel, G.W. (1985) *J. Am. Chem. Soc.*, **107**, 6659.

107 Hancock, R.D.; Maumela, H.; de Sousa, A.S. (1996) *Coord. Chem. Rev.*, **148**, 315.

108 Schultz, R.A.; Dishong, D.M.; Gokel, G.W. (1981) *Tetrahedron Lett.*, **22**, 2623.

109 Adamic, R.J.; Lloyd, B.A.; Eyring, E.M.; Petrucci, S.; Bartsch, R.A.; Pugia, M.J.; Knudsen, B.E.; Liu, Y.; Desai, D.H. (1986) *J. Phys. Chem.*, **90**, 6571.

110 Anantanarayan, A. Fyles, T.M. (1990) *Can. J. Chem.*, **68**, 1338.

111 Dutton, P.J.; Fyles, T.M.; McDermid, S.A. (1988) *Can. J. Chem.*, **66**, 1097.

112 Motekaitis, R.J.; Martell, A.E.; Dietrich, B.; Lehn, J.-M. (1984) *Inorg. Chem.*, **23**, 1588.

113 (a) Fujioka, H.; Koike, T.; Yamada, N.; Kimura, E. (1996) *Heterocycles*, **42**, 775; (b) Kimura, E.;

Aoki, S.; Koike, T.; Shiro, M. (1997) *J. Am. Chem. Soc.*, **119**, 3068.

114 Stepan, H.; Gloe, K.; Schiessl, P.; Schmidtchen, F.P. (1995) *Supramol. Chem.*, **5**, 273.

115 (a) Bourgoin, M.; Wong, K.H.; Hui, J.Y.; Smid, J. (1975) *J. Am. Chem. Soc.*, **97**, 3462; (b) Sakamoto, H.; Kimura, K.; Koseki, Y.; Motsuo, M.; Shono, T. (1986) *J. Org. Chem.*, **51**, 4974; (c) Kikukawa, K.; He, G.-X.; Abe, A.; Goto, T.; Arata, R.; Ikeda, T.; Wada, F.; Matsuda, T. (1987) *J. Chem. Soc., Perkin Trans. 2*, 135.

116 (a) Reetz, M.T.; Niemeyer, C.M.; Harms, K. (1991) *Angew. Chem., Int. Ed. Engl.*, **30**, 1472; (b) Reetz, M.T.; Niemeyer, C.M.; Harms, K. (1991) *Angew. Chem., Int. Ed. Engl.*, **30**, 1474.

117 Nagasaki, T.; Fujishima, H.; Takeuchi, M.; Shinkai, S. (1995) *J. Chem. Soc., Perkin Trans. 2*, 1883.

118 (a) Raymo, F.M.; Stoddart, J.F. (1996) *Chem. Ber.*, **129**, 981; (b) Colquhoun, H.M.; Stoddart, J.F.; Williams, D.J. (1986) *Angew. Chem., Int. Ed. Engl.*, **25**, 487; (c) Canary, J.W.; Gibb, B.C. (1997) *Progr. Inorg. Chem.*, **45**, 1.

119 Colquhoun, H.M.; Stoddart, J.F.; Williams, D.J. (1983) *J. Chem. Soc., Dalton Trans.*, 607.

120 Crumbliss, A.L.; Batinic-Haberle, I.; Spasojevic, I. (1996) *Pure Appl. Chem.*, **68**, 1225.

121 Auerbach, U.; Stockheim, C.; Weyhermüller, T.; Wieghardt, K.; Nuber, B. (1993) *Angew. Chem., Int. Ed. Engl.*, **32**, 714.

122 Zhang, X.L.; Yoon, D.I.; Hupp, J.T. (1995) *Inorg. Chim. Acta.*, **240**, 285.

123 Alston, D.R.; Slawin, A.M.Z.; Stoddart, J.F.; Williams, D.J.; Zarzycki, R. (1984) *Angew. Chem., Int. Ed. Engl.*, **23**, 821.

124 Siegel, B.; Breslow, R. (1975) *J. Am. Chem. Soc.*, **97**, 6869.

125 (a) Alston, D.R.; Lilley, T.H.; Stoddart, J.F. (1985) *J. Chem. Soc., Chem. Commun.*, 1600; (b) Alston, D.R.; Slawin, A.M.Z.; Stoddart, J.F.; Williams, D.J. (1985) *J. Chem. Soc., Chem. Commun.*, 1602.

126 Johnson, M.D.; Bernard, J.G. (1996) *J. Chem. Soc., Chem. Commun.*, 185.

127 Stoddart, J.F. (1992) *Angew. Chem., Int. Ed. Engl.*, **31**, 846.

128 (a) Kickham, J.E.; Loeb, S.J. (1995) *Inorg. Chem.*, **34**, 5656; (b) Kickham, J.E.; Loeb, S.J.; Murphy, S.L. (1993) *J. Am. Chem. Soc.*, **115**, 7031.

129 van Staveren, C.J.; van Eerden, J.; van Veggel, F.C.J.M.; Harkema, S.; Reinhoudt, D.N. (1988) *J. Am. Chem. Soc.*, **110**, 4994.

130 For reviews directed to biological systems see: (a) Fersht, A. (1977) *Enzyme Structure and Mechanism*, W.H. Freeman, Reading, San Francisco; (b) Levitzki, A., (1978) *Quantitative Aspects of Allosteric Mechanism*, Springer, New York; (c) Wyman, J., (1964) *Adv. Protein Chem.*, **19**, 223.

131 Arnone, A.; Perutz, M.F. (1974) *Nature (London)*, **249**, 34.

132 Hill, R.; (1925) *Proc. Roy. Soc.*, **B100**, 419.

133 (a) Koshland, D.E.; Nemethy, G.; Filmer, D. (1966) *Biochemistry*, **5**, 365; (b) Monod, J.; Wyman, J.; Changeux, J.D.; (1965) *J. Mol. Biol.*, **12**, 88.

134 Czerlinski, G.H.; (1989) *Biophys. Chem.*, **34**, 169.

135 Rebek, J. (1984) *Acc. Chem. Res.*, **17**, 258.

136 Shinkai, S. (1992) *Crown Ethers and Analogous Compounds, Vol. 45.* Hiraoka, M., ed., Elsevier, Amsterdam, p. 335.

137 Rodriguez-Ubiz, J.C.; Juanez, O.; Brunet, E. (1994) *Tetrahedron Lett.*, **35**, 8461.

138 (a) Schneider, H.-J.; Ruf, D. (1990) *Angew. Chem., Int. Ed. Eng.*, **29**, 1159; (b) Baldes, R.; Schneider, H.-J. (1995) *ibid.*, **34**, 321.

139 (a) Sijbesma, R.J.; Nolte, R.M. (1991) *J. Am. Chem. Soc.*, **113**, 6695. (b) Van Nunnen, J.L.M.; Folmer, B.F.B.; Nolte, R.J.M. (1997) *ibid.*, **119**, 283.

140 Ikeda, T.; Yoshida, K.; Schneider, H.-J. (1995) *J. Am. Chem. Soc.*, **117**, 1453.

141 Schneider, H.-J.; Güttes, D.; Schneider, U. (1988) *J. Am. Chem. Soc.*, **110**, 6449.

142 Zollner, H. (1989) *Handbook of Enzyme Inhibitors*, VCH, Weinheim.

143 Yatsimirsky, A.K.; Eliseev, A.V. (1991) *J. Chem. Soc., Perkin Trans. 2*, 1769.

144 Zhang, X.X.; Izatt, R.M.; Krakowiak, K.E.; Bradshaw, J.S. (1997) *Inorg. Chim. Acta.*, **254**, 43.

145 Shizuka, H.; Serizawa, M.; Kobayashi, H.; Kameta, K.; Sugiyama, H.; Matsuura, T.; Saito, I. (1988) *J. Am. Chem. Soc.*, **110**, 1726.

NON-COVALENT INTERACTIONS AND ORGANIC HOST–GUEST COMPLEXES

B1. QUANTIFICATION OF NON-COVALENT FORCES

The quantification of non-covalent forces[1] is of paramount importance for the design of synthetic host–guest complexes, of new drugs and of new materials, of enzyme-analogue catalysts, for active-site directed mutagenesis and for many other applications. Synthetic host–guest complexes also offer a unique opportunity to elucidate the energies and the geometric requirements involved in formation of natural and artificial supramolecular systems. One major advantage of synthetic complexes compared to natural systems is that their conformations are better defined and can be analyzed experimentally in more detail. In particular, they can be designed to bear only those functionalities in appropriate places which are needed to bring about a certain operation, like specific binding of some substrate, or which are needed to answer specific questions about certain non-covalent forces.

Recent advances in biotechnology have made it possible to vary systematically, for example, protein structures with interesting activity changes. The problem with biopolymers is that even with exchange of a single aminoacid in an active site many things (like changes in conformations, pK_a values of neighboring aminoacids etc.) can happen at the same time. Although conformations in the smaller synthetic host–guest complexes are much better defined, it should be realized that in the functional state of solution often they also can be present in a multitude of structures. For this reason spectroscopic and computational methods (Sections D3 and E7) are indispensable tools of supramolecular chemistry. In fact, an optimal method for studying non-covalent forces with well defined conformations experimentally is, to construct *covalent* models in which internal motions are restricted essentially to only two conformations differing by built-in non-covalent interactions. We will illustrate the use of such 'intramolecular balances' for non-covalent forces in Section B5. Analogous to the study of enzyme mechanisms, covalent models of course lack supramolecular functions, but allow us to investigate mechanistic requirements in more detail.

The extreme opposite to this approach are studies of associations between small solutes, such as between simple alcohols, amines, amides, ammonium ions, carboxylates etc. These have the serious disadvantage that they are ill-defined even with respect to their stochiometry, as often they form not only a 1:1 but a whole range of higher aggregates. Nevertheless, associations between small molecules have been studied in very large numbers (thus, a database of hydrogen bonded systems[2] contains currently more than 10,000 different reports). Systematic analysis of such a large number of entries has led to unified 'factor values'[3] for non-covalent interactions, somewhat analogous to substituent constants in Hammett or Taft correlations. These factors (see Table I in the appendix) are useful not only for understanding and prediction of hydrogen bonded systems (Section B3), but even for other polar interactions (Section B4).

The *'structural'* counterpart to databases of intermolecular interaction *energies* are those based largely on X-ray structures (Section E2), such as the Cambridge file for molecules of small and intermediate size, and the Brookhaven Protein Data Bank (bpd files) for biopolymers, which also contain some neutron diffraction- and NMR-derived structures (nuclear magnetic resonance). Careful statistical analyses[4] of these solid state structures lead to a clear picture of preferred geometries, in particular for hydrogen bonded complexes.[5] Although these data stem from the solid state where the lattice may lead to significant deviations from solution state, the large number of structures and their frequently observed congruence with conformations expected for a given interaction makes it likely that at least similar geometric requirements exist in the 'free' state of supramolecular complexes. It should, however, be borne in mind, that non-covalent interactions are generally 'soft' in strength and in directionality, with the exception of partially covalent bonds in transition metal coordination compounds. This, and the presence of many, often similar binding sites and *inter*molecular forces directed towards the guest, leads in solution (or in the gas state) usually to many complex structures of similar energy. The crystal lattice picks up only one (occasionally two) of them in the unit cell. In contrast, in proteins there is a manifold of *intra*molecular interactions stabilizing the conformation. This, and the poor accessibility of the interior of these extremely large moieties for solvation leads to crystal-like densities, and makes differences between solution and solid state structures less likely.

B2.   ION PAIRS

Electrostatic interactions play a nearly ubiquitous role in synthetic and natural supramolecular systems,[6a,b] particularly if one includes hydrogen bonds, which are now often described in this way (Section B3). In this section we discuss interactions between permanent charges, which can be point poles, dipoles, or polypoles, with selected typical examples of host–guest complexes.

In organic ions the charge is heavily delocalized, which is one of the obstacles to a rigorous theoretical treatment. Thus, the positive charge in alkylammonium ions resides completely on the carbon, and on the hydrogen atoms of alkyl groups surrounding the nitrogen atom, which bears no charge at all.[7] Fortunately, it is possible to describe interactions approximately by referring to a point charge of the whole group in these cases, as will be seen below.

For spherical ions A and B with point charges Bjerrum[8] has on the basis of the Debye–Hückel theory described association constants K as a function of the charges z_A, z_B, the dielectric constant ε, and a factor $Q(b)$ which depends on ε, z and on the distance of closest approach a between A and B:

$$K = (4\pi N/1000)(z_A z_B e^2/\varepsilon kT)^3 Q(b) \quad \text{(B2-1)}$$

where $b = z_A z_B e^2/\varepsilon kTa$ (in this and following equations the product $z_A z_B$ is taken by its absolute value). The Q values are tabulated.[8a] Parameter b is related to the so-called critical distance $q = ab/2 = z_A z_B e^2/2\varepsilon kT$, which is defined as the distance at which the mutual electric potential energy of ions A and B equals $2kT$. Bjerrum theory considers that the ion pairing occurs if the centers of ions A and B approach each other at a distance shorter than or equal to q. Therefore this model counts as ion pairs not only the real associates in which the ions are in a physical contact (contact ion pairs) but also all 'pairs' for which the interionic distance lies between a and q. The distance a equals the sum of ionic radii of A and B and the critical distance $q = 3.57z_A z_B$ Å at room temperature in aqueous solution. For 1 : 1 electrolytes a is close to q, but for polyvalent ions q is considerably larger; for 2 : 2 electrolytes $q = 14.3$ Å. Also q becomes very large in solvents of lower dielectric constants. On the other hand, large ions with $a > q$ according to this theory do not associate at all.

For contact ion pairs the Fuoss equation (B 2-2)[8b] gives

$$K = (4\pi N a^3/3000) \exp(z_A z_B e^2/\varepsilon kTa) \quad \text{(B 2-2)}$$

or (at 298 K)

$$K = 0.00252 a^3 \exp(560 z_A z_B/\varepsilon a)$$

In the logarithmic form

$$\log K = -2.6 + 3\log a + 243 z_A z_B/\varepsilon a \quad \text{(B 2-3)}$$

where a is in Å. The equation is applicable also for solvent-separated pairs and for loose solvated complexes provided the solvent molecule size is included in evaluation of a. The experimental results do not allow a clear distinction between Bjerrum and Fuoss equations, however, the latter is often preferred because of its simpler mathematical form and conceptual basis.[8b]

For strongly interacting ions, when association free energies are $\gg kT$, even a simple equation (B 2-4) derived from consideration of ion pairing in terms of the Born cycle gives satisfactory results.[9]

$$K = \exp(z_A z_B e^2/\varepsilon kTa) \quad \text{(B 2-4)}$$

The theoretical expressions for enthalpic and entropic contributions to association free energy can be obtained by differentiation of $\ln K$ by T (see equation (A 4-1), Section A 4). Thus, from the Fuoss equation it follows that[10]

$$\Delta H = -(z_A z_B e^2/\varepsilon kTa)RT^2\left(\frac{d\ln\varepsilon}{dT + 1/T}\right)$$

and

$$\Delta S = R\ln(4\pi N a^3/3000)$$
$$- (z_A z_B e^2/\varepsilon kTa)RT\left(\frac{d\ln\varepsilon}{dT}\right)$$

(expressions which follow from other models can be found in the literature[10]). Using $d\ln\varepsilon/dT = -0.00455$ for water[11], one obtains (at $T = 298$ K) $\Delta H = 6.3(z_A z_B/a)$ kJ mol^{-1} and for a in the range 3–7 Å this equation predicts a small positive value of ΔH ranging from $0.9 z_A z_B$ to $2.1 z_A z_B$ kJ mol^{-1}. Experimental values indeed often are close to zero although larger both positive and negative values also were reported (Table B 1). Thus, ion pairing is expected to be, and often is, an entropy-driven process.

Considerable improvement has been achieved by application of new numerical methods of solution of the Poisson equation (Debye–Hückel theory utilizes its analytical solution for a special simplest case of hard-sphere charges in a uniform medium and small electric potentials), which allow one to take into account also polarization effects and charge delocalizations.[13]

Experimental values for simple ion pairs in water and water–organic mixtures fairly agree with equations (B 2-1)–(B 2-4). In particular, these equations predict an experimentally confirmed rapid increase in K on going to solvents with decreased dielectric constant, $\log K$ being a linear function of $1/\varepsilon$.[8a] Perhaps the most difficult problem lies in uncertainty in the ionic size parameter a. Comparisons of a values calculated from experimental ion-pair association constants by using theoretical equations with sum of crystallographic radii show generally poor agreement. Even qualitative trends in ion-pair stability *vs.* ionic size

Table B 1. Selected thermodynamic ion-pairing parameters (kJ mol^{-1}) at 25°C in water.[9,12]

Ion pair	ΔG	ΔH	$T\Delta S$	Ref.
$Ca^{2+}SO_4^{2-}$	-13.2	6.9	20.1	9
$Zn^{2+}SO_4^{2-}$	-13.6	16.8	30.4	9
$Ca^{2+}(EDTA)^{4-}$	-59.9	-27.0	32.9	9
$La^{3+}Fe(CN)_6^{3-}$	-21.3	8.4	29.7	9
$Co(NH_3)_6^{3+}Cl^-$	-9.7	2.7	12.4	12
$Co(NH_3)_6^{3+}ClO_4^-$	-9.1	-3.6	5.5	12
$Li^+Co(C_2O_4)_3^{3-}$	-5.2	6.4	11.6	12
$Cs^+Co(C_2O_4)_3^{3-}$	-9.3	-1.5	7.8	12

Table B 2. Ion-association constants for ions of different sizes.[7b,14]

Cation	Anion	log K	Ref.
Li^+	$Cr(C_2O_4)_3^{3-}$	0.843	7b
Na^+		1.513	
K^+		1.570	
Rb^+		1.586	
Cs^+		1.617	
trans-			
$Co\,(En)(NO_2)_2^+$	Cl^-	0.20	14
	Br^-	0.30	
	I^-	0.41	
	ClO_4^-	0.43	

often are opposite to theoretical expectations. Equations (B 2-1) and (B 2-4) predict a decrease in K on going to larger a values. According to the Fuoss model there is a minimum (never observed) in K as a function of a at $a_{min} = 2.4\,z_A z_B$ in water. However, experimental values of K often increase on going to larger ions, as illustrated in Table B 2. To explain such a trend one should take into account that, first, smaller ions are more strongly hydrated and participate in ion-pair formation with their hydration shells, which make their sizes even larger than sizes of less hydrated bigger ions, and second, larger ions may give larger non-Coulombic contribution to the binding.

The problems with a rigorous description of ion pairs are, *inter alia*, specific interactions between ions and between ions and solvent molecules, which involve polarizations particularly of oxygen lone pairs. Furthermore, the thermodynamics of ion pairing is characterized by large and variable positive entropy contributions (see above) in particular due to liberation of tightly bound water molecules upon complex formation.[6,15] Another complication is that these associations comprise contact ion pairs, as well as those which are solvent separated. The first ones will be measured largely by spectroscopic methods, like UV/vis techniques, the second together with the first traditionally by conductivity changes (see Section D 2.4).

Given these complications, it is surprising that systematic analyses of several hundreds ion pairs in water, exemplified by a plot (Fig. A 3, Section A 1) of free energies *vs.* the number n of single charge–charge interactions possible in these associations, shows a fairly constant increment of $\Delta\Delta G = -(5 \pm 1)$ kJ mol^{-1} per 1 : 1 interaction.[16] The correlation extends from inorganic salts such as zinc sulfate, up to complicated and very anisotropic systems, like azoniacyclophanes, or even to DNA groove binders (see below). Note that there are two types of charge–charge interactions: one when cation and anion are hydrogen bonded to each other, such as $R–NH_3^+\,^-O_2C–R'$ (such pairs are called salt bridges),[17] and when they are not, such as $Zn^{2+}SO_4^{2-}$. The correlation involves both types of interactions indicating small, if any, difference in their formation free energy in aqueous solution (see also below).

The value of -5 kJ mol^{-1}, corresponding approximately to the incremental constant $K_{ij} \approx 10$ for a single charge–charge interaction, has been found also with many ammonium-carboxylate interactions, and with ions of very different polarizibility and shape, such as with metal, protonated and peralkylated ammonium and sulfonium cations, as well as with sulfate, phosphate, carboxylate and phenolate anions.[16] More recent correlation with K values extrapolated to zero ionic strength gives the increment of -8 ± 1 kJ mol^{-1} (ref.[27b] Chapter A). The increment is not far from the predictions by equations (B 2-1) or (B 2-2), although these should apply only to small spherical ions (Exercise 2 in Chapter A).

In practice it is not always easy to estimate the number of single interactions in multiply bound host–guest complexes, which can adopt numerous low energy conformations in solution. A simplified treatment can be suggested, which correlates the association constant with the product of total charges $z_A z_B$ of guest and host (for other simple correlations between log K and ionic charges

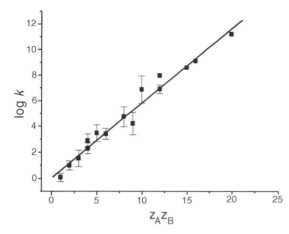

Figure B 1. Ion-pair association constants at zero ionic strength as function of charge product. Data from refs. 18a (alkylammonuim cations with $Fe(CN)_6^{4-}$), 18b (alkylammonuim cations with sulfate, phosphate, pyrophosphate and triphosphate anions), 18c (alkylammonuim cations with pyrophosphate anions), 18d (various organic dications with di- and tetracarboxylate anions) and 19 (inorganic ion pairs). Points show the mean values of $\log K$ calculated from data available for each given charge product; vertical bars are the respective standard deviations. There are two points for $z_A z_B = 4$ (1 : 4 and 2 : 2 ion pairs) and for $z_A z_B = 12$ (2 : 6 and 3 : 4 ion pairs). Data for 203 ion pairs are included.[18a–d,19]

see the literature[18a–d]). Figure B 1 correlates in this way the logarithms of association constants for 203 ion pairs at zero ionic strength (*I*). The dependence can be satisfactorily described by the linear equation

$$\log K(I = 0) = 0.04 + 0.57\,z_A z_B \qquad \text{(B 2-5)}$$

with correlation coefficient $R = 0.98558$ and standard deviation of 0.6. The slope of the dependence corresponds to that of the Fuoss equation (B 2-3) with $a = 5.3\,\text{Å}$. This correlation indicates the primary importance of the charge of interacting ions while their size and charge distribution play a less important role.

Correlations like those in Fig. A 3 (Section A 1) and Fig. B 1 are useful for estimation of a mean 'typical' increment for charge–charge interactions. Of course, any particular system can show substantial deviations from a mean

value leading to a higher or lesser degree of complexation selectivity in addition to the selectivity determined by host and guest total charges (or number of possible single charge–charge interactions). For example, data in Fig. A 3 demonstrate that for a given number of pairwise charge–charge contacts, *n*, the stability constants of host–guest complexation can vary by 10^2–10^4 times. These variations are due to numerous factors such as charge size and delocalization, complementarity of opposite charges, contribution of non-electrostatic interactions, etc. In connection with the recognition problem, the important aspect is possible shape selectivity of interaction due to complementarity of charges. Certain selectivity of this type was reported for binding of anions of dicarboxylic acids possessing different distances between their carboxyl groups to cationic protonated azamacrocycles.[20] Shape selectivity in complexation of isomeric anions or tricarboxylic acids by a tetracationic host **B-1**[21a] and protonated forms of azamacrocycle [21]aneN₇[22] is illustrated in Scheme B 1.

Host **B-1** shows a pronounced specificity to *trans*-aconitic acid; higher protonated forms of [21]aneN₇ discriminate isomers of 1,3,5-trimethyl-1,3,5-cyclohexanetricarboxylic and benzenetricarboxylic acids with factor of 100 and bind rigid trianions up to 10^4 times better than flexible citrate. As expected, both complexation strength and selectivity increase on going to more highly charged host cations.

[21]aneN₇   **B-1**

A noticeable specificity of electrostatic binding of tricarboxylates with **B-2** may find

Log K

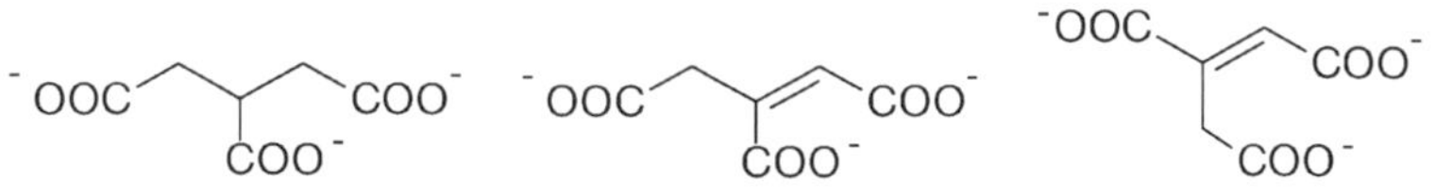

B-1	4.4	5.1	4.3

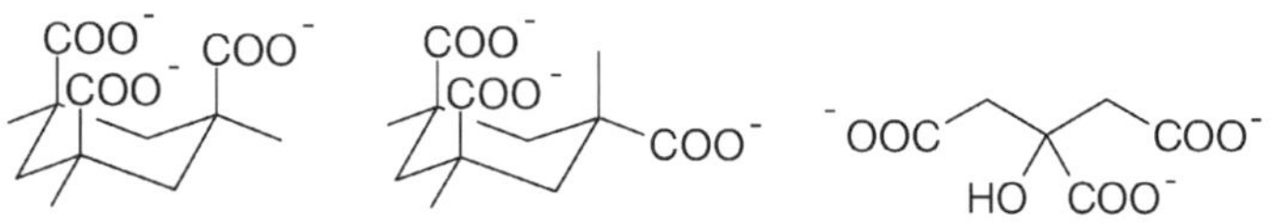

B-1	4.1		
$H_3[21]aneN_7^{3+}$	3.4	2.5	
$H_4[21]aneN_7^{4+}$	5.2	4.0	2.7
$H_5[21]aneN_7^{5+}$	8.6	6.6	4.6

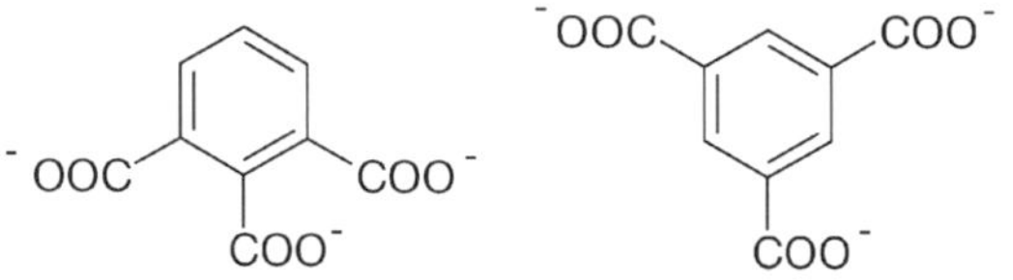

$H_3[21]aneN_7^{3+}$	3.2	3.1
$H_4[21]aneN_7^{4+}$	5.2	4.6
$H_5[21]aneN_7^{5+}$	8.2	6.2

Scheme B 1. Shape selectivity in recognition of tricarboxylates by tetracationic host **B-1**[21a] and differently protonated forms of [21]aneN$_7$ in water.[22]

even application for citrate determination in orange juice.[21b] These examples demonstrate that even directionless interionic interactions can provide noticeable shape selectivity of complexation.

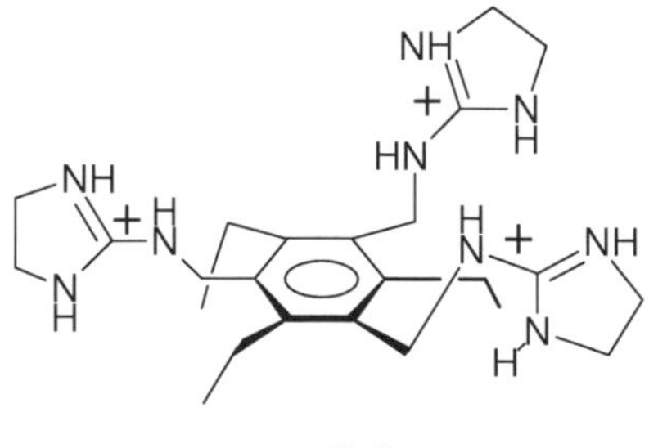

B-2

With protonated ammonium centers the question arises to what degree one ascribes interactions with anions to interionic attraction, or to hydrogen bonds between these exceptionally good donors and acceptors. An experimental answer could be the investigation with permethylated analogs of protonated amines. Complexes with corresponding derivatives of the host **B-3** (Scheme B 2) with, for example, anionic fluorophores have shown the same ΔG value, within experimental error.[23] That interactions with protonated amines can indeed be described largely by Coulomb increments given above is also obvious from the study of polyamine interactions with the phosphate residues in double stranded DNA, which show the same affinity with, for example natural spermine and its permethylated analog.[24] The linear plot of the observed affinities *vs.* the number of charged nitrogen atoms in many polyamines, includ-

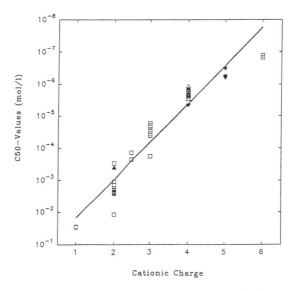

Guest	logK (K in M⁻¹)
Adenine	1.70
GMP²⁻	2.65
Adenosin	1.60
Uridin	1.00
AMP²⁻	3.28
UMP²⁻	2.90
ADP³⁻	4.15
Cytosin	1.23
ATP⁴⁻	4.57
CMP²⁻	2.97
Guanosin	1.00

B-3

Scheme B 2. Association constants for interactions of cyclophane **B-3** with nucleotides, nucleosides and nucleobases in water.[23]

ing macrocycles of different size (Fig. B 2) indicates again additive contributions. Comparison of the number of possible contacts between the ammonium centers and the singly charged DNA phosphate groups with corresponding equilibrium constants suggests again an increment of around 5 kJ mol⁻¹ and per salt bridge,[24] indicating the applicability of the additivity principle also for biopolymer interactions.

However, in other systems peralkylated ammonium cations possess both increased and decreased affinity to anions as compared with protonated amines. For example, permethylated azacryptand **A 45b** (Section A 6) binds Cl⁻ much weaker, but Br⁻ substantially stronger than its protonated analog [**A 45a** · 4H]⁴⁺. There is practically no association between NH₄⁺ and halide anions in water, but cations R₄N⁺ form detectable ion pairs with stability constants increasing on going from R = Me to R = n–Bu.[19] This diversity of peralkylation effects results form a complex combination of different factors: besides removal of possible hydrogen bonding peralkylation changes substantially the ionic size, hydration, charge delocalization and van der Waals contributions.

Crystal structures of solid state complexes involving salt bridges as a rule show the

Figure B 2. Affinities of polyamines to ds-DNA (estimated by a fluorescence competition assay with ethidiumbromide) *vs.* the number of charged nitrogen groups.[24]

presence of zwitterionic hydrogen bonds of the type R₃N–H⁺···X⁻ (for representative examples see the literature[25] and Section A 6). Theoretical simulation of guanidinium acetate and methylammonium acetate ion pairs in water also reveals the hydrogen bonding inside the pairs.[26] These results are quite expectable, however, because the positive charges of ammonium cations are located on hydrogen atoms (see above) and to some extent it seems trivial to say that salt bridges are hydrogen bonded. The point is whether or not this bonding provides an extra stability or leads to some other important features of salt bridges as compared to ion pairs without such bonding.

It is well known that stability of hydrogen bonds correlates with basicity of proton donors and acidity of proton acceptors (Section B 3). Such correlation should be observed also for the salt bridges and its observation was proposed as a test for contribution of hydrogen bonding to the stability of salt bridges in proteins.[27] The problem with interpretation of this and similar correlations is that increased anion basicity, as well as cation acidity, results from higher charge density and, consequently, the Coulomb interaction with a counter-ion should be stronger independently of whether it is protic or aprotic. For example, $\log K$ for ion pairing of lithium fluoroacetates ($CH_{3-n}F_nCOO^-$ Li^+, $n = 0$–3) in DMSO decreases from 2.7 to 1.0 on going form $n = 0$ to $n = 3$ and perfectly correlates with pK_a of the respective fluoroacetic acids.[28] Comparison of ion-pair formation constants of a given anion with simple inorganic cations and with protonated organic polyamines of similar charge shows somewhat higher values for the latter (e.g. for $Fe(CN)_6^{4-}$ at zero ionic strength $\log K = 3.8$ with Mg^{2+}–Ba^{2+} and range from 4.4 to 4.9 with diprotonated ethylendiamine, diethylenetriamine and triethylenetetramine, Fig. B 1), but with large number of ion pairs the differences are in limits of statistical deviations of reasonable good general correlation given by equation (B 2-5). No contribution of hydrogen bonding to stability of phosphate-guanidinuim ion pair may be supposed from practically equal binding constants of 2′,3′-cAMP and 3′,5′-cAMP to a guanidinium-containing receptor, which according to molecular modeling can form up to three addi-

tional hydrogen bonds with the former ligand.[29]

The applicability of additive increments for Coulomb interactions also to supramolecular systems, in which actually other non-covalent forces may vary and dominate, is illustrated with complexes between the azoniacyclophane **B-3** and nucleotides or nucleosides (Scheme B 2). Obviously, the major binding contribution, such as for the adenine derivatives, stems from lipophilic interactions between nucleobase and the host, but the free energy difference between the electroneutral nucleoside and the doubly charged nucleotide remains constant for all compounds, with $\Delta\Delta G \approx 10\,kJ\,mol^{-1}$. Molecular models show, that at a given time the phosphate dianion can be in contact only with one ammonium group of **B-3**[23], which agrees with the presence of two salt bridges.

In general, charged groups often serve as secondary binding sites operating through ion pairing in combination with interactions of other types. We already discussed in Section A 7 crown ethers with pendent ionogenic groups. Numerous cationic cyclophanes like **B-3** have been described in literature.[30] As a rule they bind substrates by a combination of lipophilic and electrostatic interactions similarly to discussed above situation in Scheme B 2. However, the spatial organization of charged and lipophilic subsites should meet some yet not clearly understood requirements. For example, cyclophane **B-4** designed as a mimic of vancomycin carboxylate binding site binds anions only electrostatically without any significant contribution of expected hydrophobic binding (the values of $\log K$ for anions $RCOO^-$ are within the range 1.34–1.82 expected for $z_A z_B = 3$ (equation B 2-5) for R = H, Me, MeCH$_2$, Me$_2$CH, *cyclo*-C$_5$H$_9$, Ph).[31] Computer modeling suggests that the substrates possess necessary complementarity for simultaneous electrostatic and hydrophobic binding.

Another family of such hosts are cyclodextrins modified by charged substituents exem-

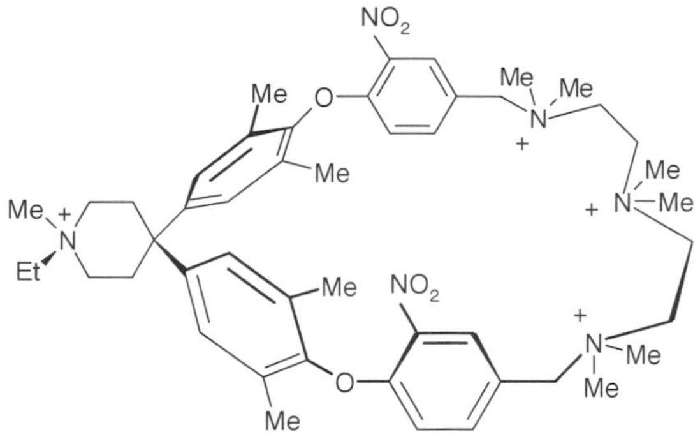

B-4

Table B3. Logarithms of binding constants K (M^{-1}) of nucleotides and related compounds to cyclodextrins at 25°C in water $(I = 0.1M)$.[32]

	$\log K$		
Guest	**B-5**	**B-6**	*β-cyclodextrin*
5'-AMP^{2-}	4.15	5.10	1.95
3'-AMP^{2-}	3.18	4.96	2.40
5'-GMP^{2-}	3.80	4.60	
5'-CMP^{2-}	2.92	4.30	
5'-UMP^{2-}	2.92	4.94	
ribose-5-phosphate^{2-}	4.05	5.93	
ribose		1.41	0.00
PO$_4^{3-}$	2.30	3.57	

plified here by aminocyclodextrins **B-5** and **B-6** employed for recognition of nucleotides[32] (see Scheme B3, Table B3 and structure of such complexes on the *website*).

An interesting feature of hosts **B-5** and **B-6** is their ability to discriminate nucleobases which apparently contribute repulsive, rather than attractive, forces to complex formation, as is evident from comparison of the complexes with ribose-5-phosphate and nucleotides that have the same stabilities. Attractive interactions involve binding of the sugar moiety to the cavity of cyclodextrins and ion pairing between ribose phosphate dianion and positive ammonium charges of the hosts. Separation of contributions shows the host **B-5** to form the expected two salt bridges, but

B-5 : R$_A$ = R$_B$ = NHCH$_3$, other R=OH
B-6 : all R = NHCH$_3$

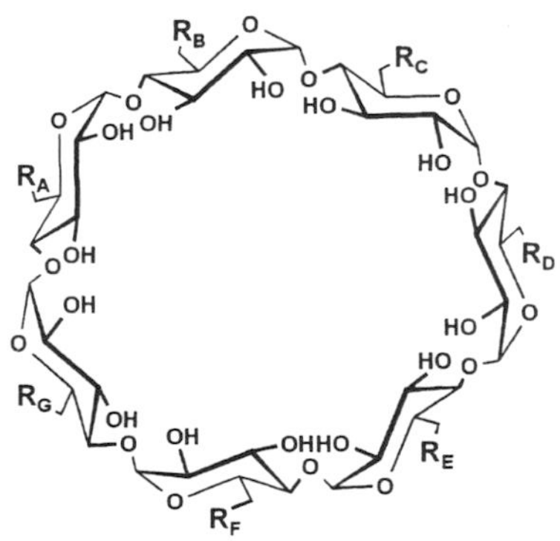

Scheme B3. Structures of amino-cyclodextrins **B-5**, **B-6**; (*β*-cyclodextrin: all R=OH).[32]

B-6 forms only four bridges in spite of the presence of seven ammonium cationic groups, as simultaneous contact between all ionic sites is geometrically not possible.

Cation binding to anionic cyclophanes is exemplified in Scheme B4, which demonstrates that multidentate anionic hosts also bind positively charged guests with an affinity that is predictable by counting the number of possible salt bridges or charge product.[33] As the Coulomb interactions are just the sum of individual charge–charge interactions, it does not make a difference whether the charges are delocalized, like in the tetraphenolate **B-7** (Scheme B4)[33] for which we can count either 4×1, or 8×0.5 or 16×0.25 salt bridges etc. with the same result. This is true, of course, only if the fractional charges are distributed symmetrically at the same distance from the guest charge. Asymmetrical charge delocalization, like that in the case discussed above of lithium fluoroacetates, transfers partial charges at higher distances and leads to decreased binding. The examples in Scheme B4 belong to the strongest known receptors for choline or cholinaceatate; they illustrate how large affinities can simply be obtained by accumulation of charge–charge interactions.

For a correct dissection of the forces within these complexes we also have to take into

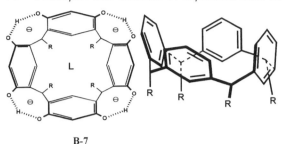

B-7

Scheme B 4. Association free energy of ligand $L =$ (Me$_3$N$^+$–CH$_2$–CH$_2$–OCOMe) (acetylcholine) $\Delta G =$ -27 kJ mol^{-1} in D$_2$O. with resorcarene **B-7** (R = Me); substituents partially omitted for clarity.[33]

account any non-Coulombic interactions. This can be done by comparing the affinities of charged guest molecules to those of electroneutral, but isosteric analogs. Thus, the tetraanion **B-7** shows with tert-butanol an association ΔG which must be subtracted from the ΔG of choline etc. in order to obtain the pure Coulomb contribution, which then (by division by four for four salt bridges) gives the predicted increment (Scheme B 4).

This simplified approach to quantification and prediction of affinities must be modified if there is no simultaneous contact between the cationic and anionic binding sites. Computer aided molecular modeling is the best way to check this, and to eventually extend the prediction to complexes with insufficient fit, based on the simple r^{-1} dependence of Coulomb interactions for example,[33] where r is the distance between the ionic centers. That electrostatic interactions fall off only with r^{-1} makes this potential much softer than that of any other intermolecular force (Section E 7). This is a major reason for the applicability of simple and additive increments, as Coulomb-type forces tolerate considerable mismatch between binding sites. Additive Coulomb interactions can also be used to describe interactions with and between di- and polypoles instead of point charges. Most currently used force fields calculate such forces by attributing partial charges at the atoms which

form such polypoles, and thus offer a convenient method for the evaluation of any electrostatic contributions. The major problem here is the choice of the appropriate partial charges, and of the local dielectrics around them (Section E 7).

In conclusion let us briefly discuss charge–charge interactions in proteins. Salt bridges are formed predominantly between cationic groups of lysine and arginine and anionic groups of aspartic and glutamic acids. Empirical evaluations of interaction free energy by site-specific mutagenesis give values ranging from -4 to -20 kJ mol^{-1}.[17,34] A mean value of about -10 kJ mol^{-1} per protein salt bridge is twice the mean increment typical for host–guest complexation. Stronger electrostatic interactions in proteins most probably are due to low local dielectric constant. Theoretical calculations predict even higher values up to -50 kJ mol^{-1},[35] which of course strongly depend on the chosen dielectric constant. Interestingly, the calculations show that formation of salt bridges in some protein–protein complexes is accompanied by positive ΔG due to uncompensated loss of solvation free energy.[35] This phenomenon is responsible for often observed destabilizing effects of salt bridges on folded protein conformations.[13b,17,34]

It is worth noting that even for large protein macromolecules the binding free energy of charged low molecular weight ligands is proportional to the total protein host charge. This was demonstrated in the study of binding of ligands **B-8**–**B-10** to acetylated derivatives of bovine carbonic anhydrase II.[36] Acetylation of the protein converts its positively charged lysine ε-ammonium groups into neutral amide groups. Thus the protein derivatives with charges ranging from -3.5 (native protein at pH 8.3) to -19.5 were prepared. Binding $-\Delta G$ was independent of the protein charge for neutral ligand **B-8**, but increased for positively charged **B-9** and decreased for anion **B-10**. In both cases the dependencies of $-\Delta G$ on the protein charge were linear with slopes 0.2 and $-0.4\,\text{kJ mol}^{-1}$ respectively. These small slopes are in agreement with the large size of the protein.

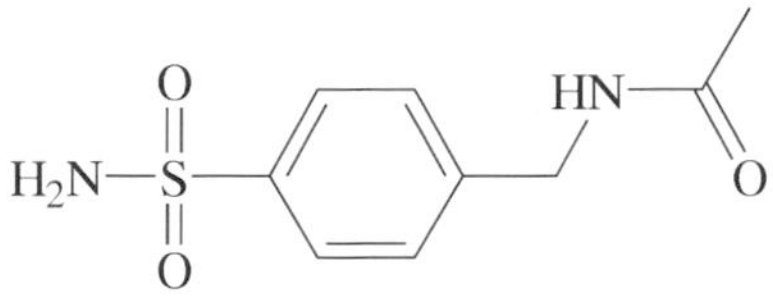

B-8

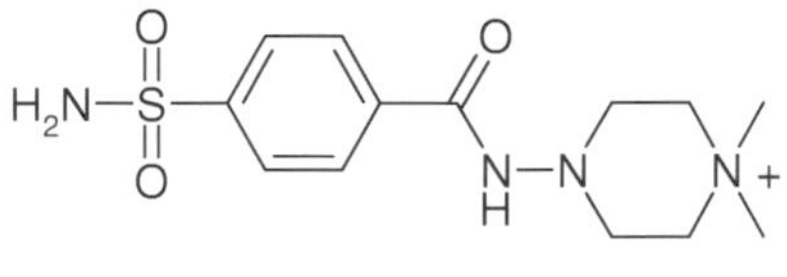

B-9

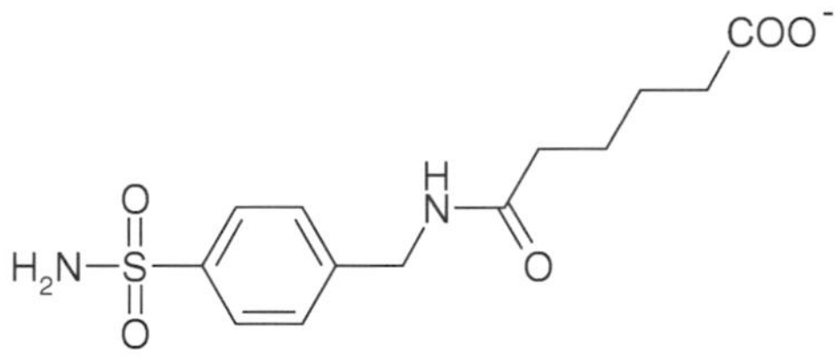

B-10

Affinity of simple inorganic ions to proteins usually follows the Hofmeister or lyotropic series which for anions is $ClO_4^- > SCN^- > I^- > Br^- \sim NO_3^- > Cl^- > BrO_3^- > H_2PO_4^- > F^-$ and is less definitive for cations.[1a] Essentially the same order of affinities follow anion binding to cation-exchange resins[1a] and to micelles of cationic surfactants (Section G2), as well as usually rather weak binding of anions to neutral hosts such as cyclodextrins (e.g. $ClO_4^- > SCN^- > I^- > Br^- > IO_3^- > NO_3^-$ for β-cyclodextrin).[37] This trend in anion affinities seems to be related to different ion properties, in particular to their effects on the structure of water (Section C3), and cannot be explained by simple Coulomb charge–charge interactions.

B3. HYDROGEN BONDS

Probably the most important non-covalent interaction between a proton, and corresponding donor **D** and acceptor **A** atoms[38] can vary between few and hundreds of kJ mol^{-1}. In the case of the bifluoride anion the corresponding interatomic distances are closer to covalent bond length than to the sum of the van der Waals radii (Table B4). Noticeable reductions of the distances between **A** and **D** in comparison to the sum of the van der Waals radii are often taken as evidence for the presence of a hydrogen bond. However, this criterion alone is doubtful in view of the $A\cdots D$ distance being also a function of the A–H–D angle (see below) and often ill-defined van der Waals radii.[38b,c] Stronger bonds are expected to be shorter, but there is only approximate (a sigmoidal) correlation between distance and energy.[38c] Contributions from exchange integrals between overlapping orbitals were earlier believed to be significant, but *ab initio* molecular orbital (MO) calculations with increasingly large basis sets indicate, that the electrostatic forces[39] between the positively charged hydrogen H, and the negatively charged acceptor and donor atoms **A** and **D** are dominating, in particular for so-called weaker bonds like those involving amides, alcohols etc.[40a,b]

As long as the charges of **A** and **D** are centered in these atoms one would predict for an optimal interaction a linear $D\text{-}H\cdots A$ arrangement with an $D\text{-}H\cdots A$ angle close to 180° (Type (a) in Scheme B5). In solid state structures, which offer a rich source of corresponding geometries, this is, however, rarely found: typical $D\cdots A$ distances between hydroxy oxygen atoms are for example 2.72 Å, whereas the sum of the van der Waals radii would be for $O\cdots O$ 3.04 Å, and for a linear $O\cdots H\cdots O$ bond even 5.44 Å. Statistical analysis of directionality in organic crystals gives an average angle of 167° for $O\text{-}H\cdots O$ bonds and 161° for $N\text{-}H\cdots O$ bonds.[41a,b] One hydrogen atom H can also interact with two acceptor **A** atoms in a three-center, often called bifurcated, arrangement (type b), less frequent are other types (types c,

Table B 4.   Some gas phase data for intermolecular hydrogen bonds.[38d]

		Distance ($\mathring{A}$)[a]	E (kJ mol^{-1})
[F-H-F]$^-$ (semi-covalent)	anionic	F-H 1.13 (covalent: 0.92) (van der Waals 2.70) F-F 2.26 (covalent: 1.42) (van der Waals 3.10)	163
HOH$\cdots$OH$^-$	anionic	2.29	135
MeOH$\cdots$H$^+\cdots$MeOH	cationic	2.39	131
DMSO$\cdots$H$^+\cdots$DMSO	cationic	2.42	106
MeOH$\cdots$MeOH	neutral	2.70	32
HOH$\cdots$HOH	neutral	2.98	22

^{a} all distances between oxygen atoms (except for FHF$^-$).

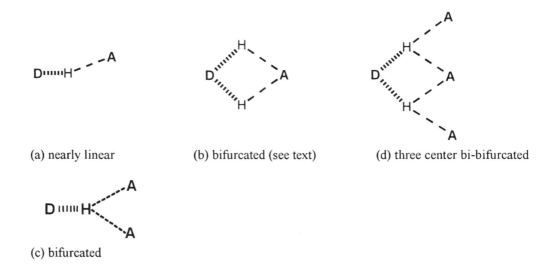

(a) nearly linear   (b) bifurcated (see text)   (d) three center bi-bifurcated

(c) bifurcated

Scheme B 5.   Hydrogen bond types.

and d Scheme B 5). Unfortunately the descriptor 'bifurcated' is in use for both configuration (b) and (c), and for (d) in combination with the term 'three center'.

As far as Coulomb attraction with a very soft r^{-1} dependence on distances r dominates, the conformational requirements are quite liberal as outlined in Section B 2. *Ab initio* calculations on amide-type associations, taking into account also aqueous solvation suggest, that indeed the hydrogen bonding energies are largely orientation independent as long as there is no steric interference between other parts of the structure.[42] Other, more empirical calculations, based largely on structural data from X-ray analyses, try to take into account angular dependencies,[43] but neglect, like almost all methods (Section E 7), orientation effects by lone pairs.

A practically useful empirical approach for the quantification of donor and acceptor abilities in hydrogen bonding rests on the analysis of thousands of binding constants between simple solute molecules, usually

0

taken from spectroscopic measurements in solvents like carbon tetrachloride.[44,45] The thus derived factor values (for a selection see Table I in Appendices) allow one to estimate the binding free energies, or $\log K$ values, for all kinds of bonding between **A** and **D** functions in non-protic solvents, based on a multiplicative combination of the factors C_A and C_D,[44] or acidity and basicity constants α_2^H and β_2^H,[45] in equation (B3-1a) or (B3-1b)

$$\Delta G\,[kJ/mol] = 2.43\,C_A C_D + 5.70 \quad \text{(for CCl}_4\text{)}$$
$$\text{(B 3-1a)}^{44}$$
$$= 1.93\,C_A C_D + 5.7$$
$$\text{(for CHCl}_3\text{)}$$

or

$$\log K\,(K \text{ in M}^{-1}) = 7.354\,\alpha_2^H \beta_2^H - 1.094$$
$$\text{(for CCl}_4\text{)} \quad \text{(B 3-1b)}^{45}$$

with $R = 0.9956$ and SD $= 0.09$ for the latter.

The factor values in Table I (see Appendix) relate to other, less comprehensive, scales of electron accepting and donating power. Gutman's[46] donor and acceptor numbers are based on relative stability of selected Lewis-type complexes measured by formation enthalpy and via ^{31}P-NMR shift changes, respectively. Drago's[47] E and C values are based on enthalpy data for a set of Lewis-type donor–acceptor interactions. Indeed, all these constants can be used not only to estimate the strength of hydrogen bond combinations, but also that of many so-called electron-donor–acceptor (EDA), or even crown ether and cryptand complexes (Section A 2).

For structurally related compound series one observes linear correlations between complexation free energies and pK_a values of hydrogen bond donors or acceptors (Fig. B 3).[45,48] Apart from well established classical steric and electronic substituent effects (see Scheme B 6 for examples), which can be described by Hammett-type equations, there are neighboring group effects, which can exert an enormous influence on the basicity or

acidity of acceptor or donor functions. The dissociation free energies of benzoic acid (in DMSO) drop by 33 kJ mol^{-1} upon introduction of one *ortho*-hydroxy group, a second one has an almost additive effect of 59 kJ mol^{-1} (Scheme B 7).[49]

Related strong hydrogen bonds exist, such as between neighboring carboxyl groups; in these recently-called low-barrier hydrogen bonds (LBHB) the proton H is in the time average centered between the **A** and **D** atoms, but separated by relatively low barriers[38d,e,50] (Scheme B 8). As a consequence the hydrogen atom H is particularly depleted of shielding electrons and is characterized by very large downfield NMR shifts. Only for the symmetric LBHB profile[38d,50b] (Scheme B 8) does one expect a substantial difference between the small hydrogen H, and the larger deuterium atom D, and a corresponding isotope effect on the NMR shieldings. LBHB associations are intermediates between the above-mentioned weaker bonds (with up to

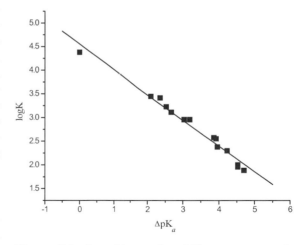

Figure B 3. Logarithms of stability constants of hydrogen-bonded complexes of 3,4-dinitrophenolate anion with substituted phenols in THF *vs.* differences of pK_a values of donor and acceptor molecules.[48] Reprinted with permission from *Science*, Vol. **272**, Shan *et al.* (The energetics of hydrogen bonds in model systems: implications for enzymic catalysis), pp. 97, copyright 1996 American Association for the Advancement of Science.

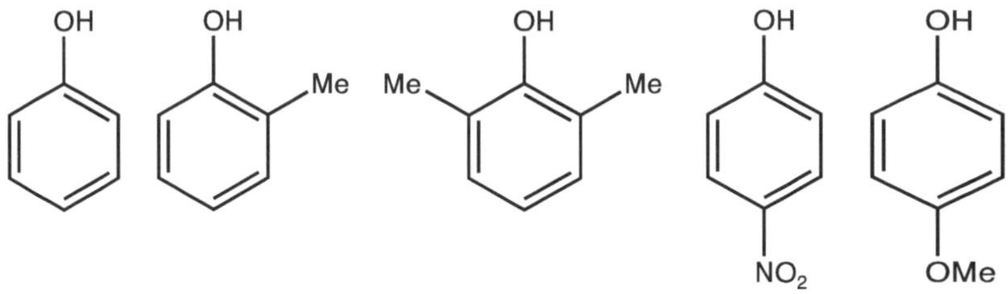

$C_A =$	- 2.50	- 2.27	- 1.58	- 3.45	- 2.41	Ref.44
$\alpha_2^H =$	0.60	0.55	0.39	0.82	0.58	Ref.45

Scheme B 6.   Steric and electronic effects on hydrogen bond strength factor values.,[44,45]

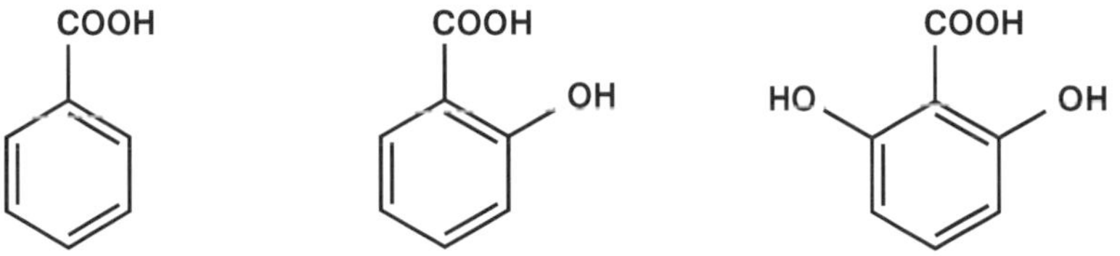

$\Delta\Delta G =$ 0.0   32.9   60.0

Scheme B 7.   Free energy changes $\Delta\Delta G$ [kJ mol^{-1} in DMSO] due to hydrogen bonds by neighboring protic substituents from ΔpK_a measurements.[49] Reprinted with permission from *J. Am. Chem. Soc.*, Shan et al., 1996, **118**, 5515. Copyright 1996 American Chemical Society.

35 kJ mol^{-1} in the gas phase) with a distinct double-well potential, and very strong, close to single well bonds (e.g. the F–H–F anion with 163 kJ mol^{-1}). Such NMR characteristics are also observed for intermolecular hydrogen bonds and in the catalytic triad of serine proteases. However, the existence of LBHB in these systems was seriously questioned on the basis of more detailed studies.[51]

Obviously, the symmetric potential profile depicted in LBHB bonds in Scheme B 8 is expected only if A and D atoms posses similar basicities, and to reach its maximum if the pK_a values of donor and acceptor unit are matching (see Scheme B 9). It has been proposed that such LBHB not only show unusual spectral characteristics, but also are anomaleously strong and their formation explains, in particular, extremely efficient transition state stabilization in active sites of some enzymes.[50b,c,52] Of course, for a given hydrogen bond donor D–H the most stable hydrogen bond will be formed with acceptor A possessing the same basicity as D because interaction free energies correlate with pK_a values of A–H (see above) increasing on increase in pK_a until $\Delta pK_a = 0$; after that the proton transfer occurs from more acidic D–H to more basic A. However, no deviations in linear correlations between log K (or ΔG) and pK_a values at $\Delta pK_a = 0$ were ever reported (see as an example Fig. B 3) indicating that there are no additional energy effects associated with symmetrical bonds of the LBHB type, although they do show specific NMR and IR (infrared) spectroscopic features (Sections E 2, E 4). Also, recent measurements of model compounds indicate that there is only a relatively small, if any, additional gain in free energies in such systems. Thus, the associa-

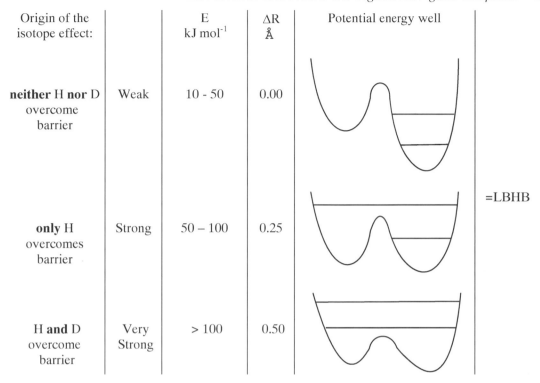

Origin of the isotope effect:		E kJ mol⁻¹	ΔR Å	Potential energy well
neither H **nor** D overcome barrier	Weak	10 - 50	0.00	
only H overcomes barrier	Strong	50 – 100	0.25	=LBHB
H **and** D overcome barrier	Very Strong	> 100	0.50	

Scheme B 8. Typical reaction profiles for weak, intermediate (LBHB) and strong hydrogen bonds. [38d] Free energies, E; $\Delta R = R_{\mathrm{vdw}}\,(A + B) - R_{\mathrm{AB}}$; LBHB = Low barrier hydrogen bond. Reprinted with permission from *Adv. Phys. Org. Chem.*, Hibbert *et al.*, 1990, **26**, 225. Copyright 1990 Academic Press.

tion free energy of the (symmetric) free acid in the Kemp acid derivative in Scheme B 9 is only higher by 1.4–2.4 kcal mol⁻¹ than that of the corresponding asymmetric amide–carboxylate interaction.[50a]

If the pK_a values of the donor and of the conjugate acid of the acceptor are similar, there will be an increase in proton transfer which ultimately (with stronger acids and bases) will lead to a salt bridge with predominantly Coulombic attraction; for borderline cases with cationic donors see the literature.[53] Determination of pK_a values in non-aqueous solvents such as acetonitrile can help to distinguish hydrogen bonding and ion pairing mechanisms.[54]

The role of hydrogen bonding in ion pairs was discussed in Section B 2, as well as the experimental evidence of often minor changes in the binding free energy of amines to strongly acidic groups upon permethylation. This has been observed with azoniacyclophanes as well as with polyamine-DNA complexes, and is in accord with the positive charge being localized mostly on the hydrogen atoms surrounding the nitrogen centers. On the other hand, weak donors such as C–H protons of alkyl ammonium groups[55] interact barely with weak acceptors like neutral oxygen lone pairs. Ammonium cation–crown ether complexes therefore rely entirely on hydrogen bonds, and show a linear decrease of association free energy with the number of possible hydrogen bonds (Fig. B 4). The nature of substituents at the nitrogen atom exerts little influence, if they point away from the binding site (cf. benzylammonium etc. has a similar affinity to 18C6 as the ammonium ion

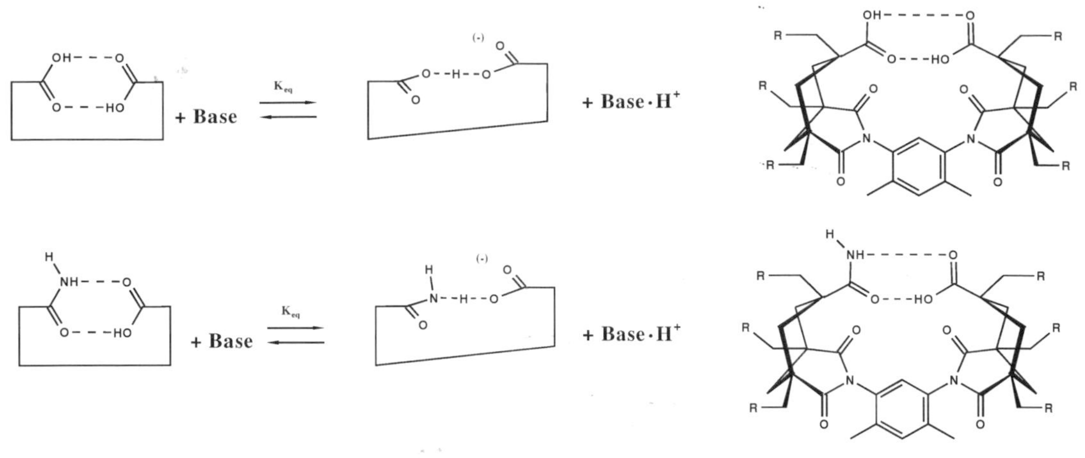

Scheme B 9.   Equilibria in Kemp acid derivatives.[50a] Reprinted with permission from *J. Am. Chem. Soc.*, Kato *et al.*, 1996, **118**, 8575. Copyright 1996 American Chemical Society.

NH_4^+). For the same reason chiral discrimination (Section I 1) requires additional groups interacting above the crown ether plane.

The quantification of non-covalent binding increments from 'simple' solute molecules,[44,45] (Table I in Appendix) suffers, as pointed out already, from problems of ill-defined geometries and even stochiometries in the complexes. Thus, the free energy for *N*-methylacetamide self-association in chloroform, which is cited in textbooks as cornerstone for peptide-like structures, varies considerably between -4 and $-7\,kJ\,mol^{-1}$ as result of the many ill-defined aggregates involved here. Even larger discrepancies exist with values derived from measurements with biopolymers, such as from peptides and proteins where values between -2 and $-6\,kJ\,mol^{-1}$ (uncharged donor and acceptor) and from -15 to $-19\,kJ\,mol^{-1}$ (charged donor or acceptor) from enzymes with aminoacids exchanged in the active site[56a] as well as values from -1 to $-7\,kJ\,mol^{-1}$ from studies of associations of vancomycin group antibiotics[56b] were reported. Such exchanges in the active center can lead to position changes of neighboring groups, which as demonstrated above (Scheme B 6) may exert

a large influence on binding affinities. It should be noted, that free energy perturbation (FEP) calculations of amide–amide interactions in water predict zero free energy values.[57]

Synthetic host–guest chemistry has produced a large array of effective hosts for guest molecules with peptide-like partial structures[58] (Scheme B 10). Other than in the simple solute–solute associations these are usually well-ordered, and allow to assign energetic values to the participating hydrogen bonds on firmer grounds. If we analyze the ΔG values of systems like in Scheme B 10 as a function of the possible number n of hydrogen bonds, we observe a fairly linear correlation.[59] The few outliers can be assigned to unfavorable steric interactions, or to too-flexible frameworks. As with the corresponding analysis with ion pairs we can extract from the correlation a common increment, which for these amide-type of hydrogen bonds amounts to $-(5\pm1)\,kJ$ per mol and hydrogen bond. The extreme dependence of this increments on solvents will be discussed in the framework of medium effects (Section C 2).

Associations of amide-type structures are often accompanied by *secondary* interactions,

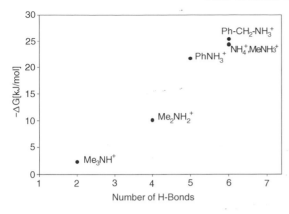

Figure B 4. A plot of complexation free energies *vs.* the number of hydrogen bonds in 18-crown-6 ether complexes with different ammonium cations (from measurements in methanol; assuming two bonds for each hydrogen; the correlation shows only negligible change if one bond is assumed).[55]

Molecular mechanics calculations indicate similar complexation modes either with (a) one linear hydrogen bond between each hydrogen and oxygen (with a distance of 1.88 Å), or (b) bifurcated bonds (then with a H–O distance of 2.06 Å). Tetramethylammonium chloride shows no measurable complexation with 18-crown-6, which indicates the predominance of hydrogen bonding over Coulomb forces in such ionophores.

which have been evaluated first on the basis of charge density calculations.[60] These explain, for example why the Watson–Crick-base pairing free energy (measured with nucleobase derivatives in chloroform) between A and T is only one third of that of the G–C analogue, although G–C has only one more hydrogen bond than A–T (Scheme B 11). The reason is, that in A–T pairs there are only *repulsive secondary* interactions between the atoms vicinal to the primary hydrogen bond,

whereas in the G–C combination there are *attractive* secondary interactions. Systematical analysis of 58 complexes with amide-like partial structures reveals, that one can describe the secondary interactions by a common increment of 2.9 kJ per mol and hydrogen bond, irrespective whether they are repulsive or attractive[61] (see the agreement between calculated and experimental ΔG values in Scheme B 11). The increment then for the primary amide-type hydrogen bond increases to 7.9 kJ; it must be higher than the 5 kJ mentioned above, as this value is derived neglecting the more often repulsive secondary forces. All values are given for the weak hydrogen bond donor chloroform as solvent; they approximately double if one uses carbontetrachloride instead.[59]

The weakening of hydrogen bonds in protic environment is the major reason for the rather small affinity of hydrogen bonding ligands by Hogsteen base pairing in DNA or RNA grooves. and poses severe problems for medicinally interesting so-called triple helix (anti-gene, directed towards DNA) and antisense (directed towards RNA) strategies.[62] Nevertheless, the accumulation of many donor and acceptor functions, such as in pyrroles connected by amide linkers (Scheme B 12)[63] lead to nanomolar affinities to the minor groove of double-stranded DNA even though there is competition with water, other than for the Watson–Crick base pairs inside the helix. The hairpin-like folded polyamides can be made specific for recognition of the four nucleobases also by introduction of additional hydroxy groups and of imidazole units. The design of such hydrogen bonding ligands can be the basis of new gene-based drugs, controlling gene expression.

Weaker hydrogen bond donors such as aliphatic hydroxy groups need much stronger acceptors, and a non-competitive microenvironment or solvent. For this reason carbohydrate receptors in nature[64] are embedded inside globular proteins, and use carboxylic groups of acidic aminoacids as acceptors

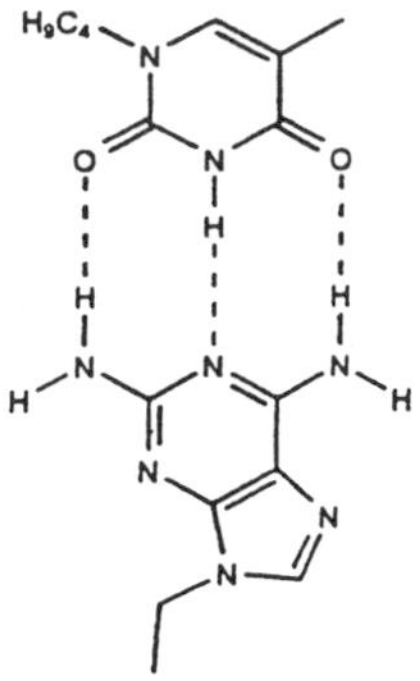

I Rich 1967:   **12**
                     calc.   *12.1*

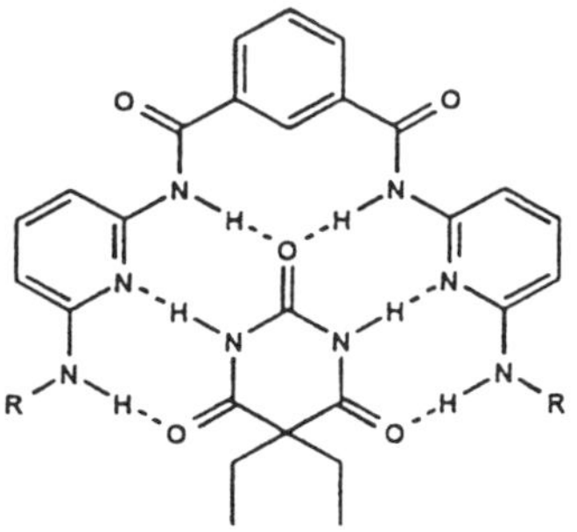

II Hamilton 1988:   **24.5**
                          calc.   *24.2*

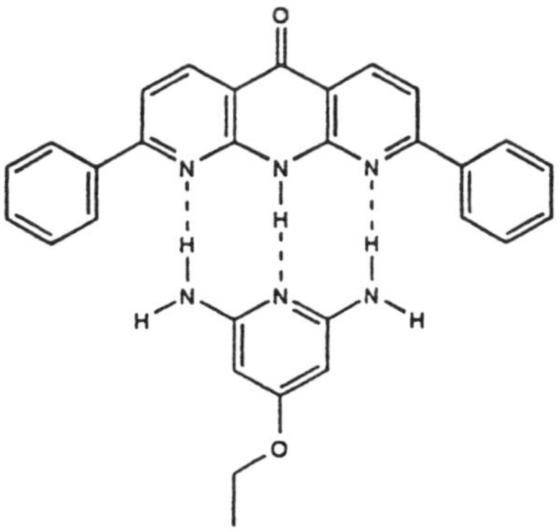

III Zimmermann 1993: **10.5**
                          calc.   *12.1*

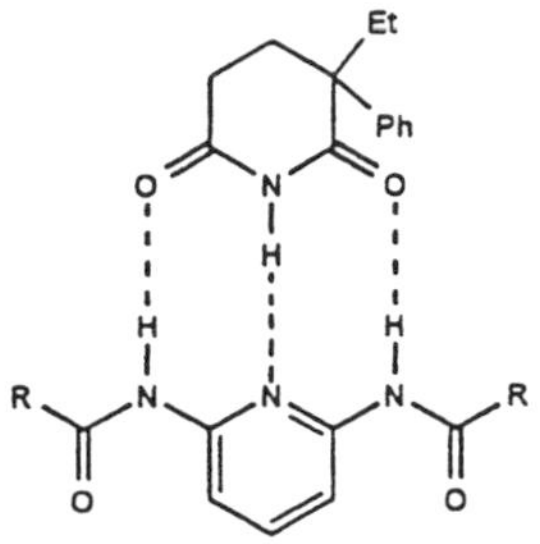

IV Schneider 1989:   **13.0**
                           calc.   *12.1*

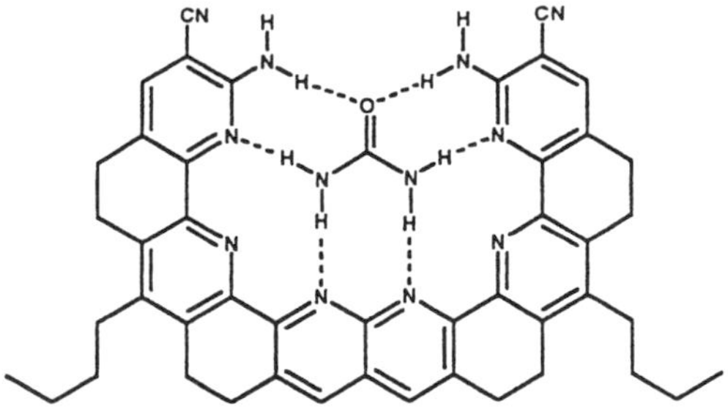

V Bell 1997:   **29.4**

Scheme B 10. Selected synthetic biomimetic hydrogen bonded complexes with experimental association free energy $-\Delta G$ (in CDCl$_3$, [kJ mol^{-1}]) and $-\Delta G$ values (*in italics*) calculated with increments as described in the text.

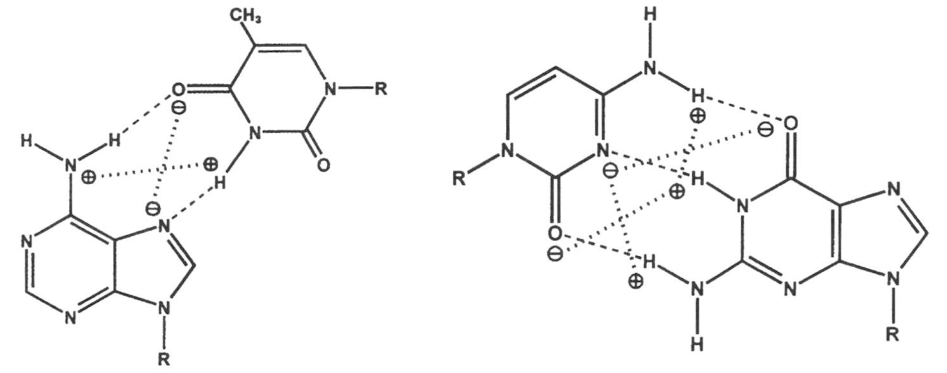

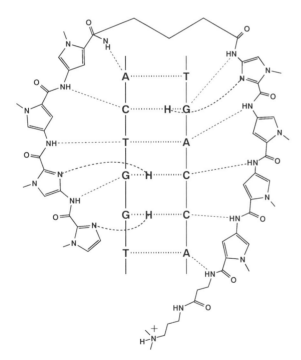

A ≡≡≡ T

-ΔG (CDCl₃) exp.   **8.5**

calc. 10.0

G ≡≡≡ C

24.5   [kJ/mol]

23.8

Scheme B 11.   Watson–Crick base pairs with primary and secondary interactions.[60] Reprinted with permission from *J. Am. Chem. Soc.*, Pranata *et al.*, 1991, **113**, 2810. Copyright 1991 American Chemical Society. Experimental association energies $-\Delta G$ (in CDCl₃, [kJ mol^{-1}], and ΔG values calculated with primary and secondary increments, see text.[61]

Scheme B 12.   A polyamide hairpin wrapping around double-stranded DNA in the minor groove; dotted lines: are hydrogen bonds.[63] Reprinted with permission from *Nature*, Vol. **391**, White *et al.*, pp 468, copyright 1998 Macmillian Magazines Limited.

(Scheme B 13). The few studies on synthetic receptors for carbohydrates[65] show that strong acceptors like carboxylic or phosphonate anions can bind monosaccharides with binding constants up to 10^3 M^{-1} in chloroform, with a ditopic acceptor even with up to 10^4 M^{-1} (Scheme B 14). The presence of several vicinal hydroxy groups in the guest molecule enhance the binding considerably. It has been shown how even less basic or acidic phenolic groups in the host can, in chloroform with moderate binding constants, complex hydroxy compounds if several phenols in a resorcinarene-derived host are properly arranged.[66] As in natural receptors complexation of carbohydrates can significantly increase by simultaneous lipophilic interactions (Sections B 5 and B 6). It is obvious that the development of efficient synthetic receptors for weak hydrogen bond donors will require considerable more effort, in particular if these should work also in aqueous solutions.

Even weaker hydrogen bonds, such as those between C–H and O, have until now been identified only in the gas phase or the solid state.[67] In view of the strong competition

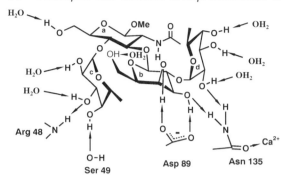

Scheme B 13.   Binding mode of a tetrasaccharide to lectin.[64] Reprinted with permission from., *Acc. Chem. Res.*, Lemieux, R.U., 1996, **29**, 373. Copyright 1996 American Chemical Society.

with all conventional solvents measurements of such weak complexation energies can only be carried out in the gas phase.

That the chelate effect by accumulating a large number of complementary acceptor–donor pairs can overcome the only moderate force of amide-type hydrogen bonds is best illustrated with the streptavidin–biotin complex (Fig. A 1). Here we find (depending on how much we account for bifurcated bonds) up to seven hydrogen bonds, which makes this (with $K = 2.5 \times 10^{13}$ M^{-1} or $\Delta G = -76$ kJ mol^{-1}) one of the strongest associations known until now. The value is not far from the interaction energy expected from seven hydrogen bonds in a very hydrophobic environment (cf. the increment of -10 kJ mol^{-1} per bond in CCl$_4$[59]); recent studies

have shown that interactions between lipophilic aminoacids also contribute to the extraordinary stability of the biotin streptavidin complex.

The strength of hydrogen bonds can, in principle, be increased by neighboring protic groups, manifested, for example in increased stability of hydrogen bonds in long chains of OH groups (so-called cooperative effect in structures like $\cdots$O–H$\cdots$O–H$\cdots$O–H$\cdots$).[41a] An interesting consequence of such an effect is the possibility of strengthening hydrogen bonds by polar solvents, including water, which traditionally are considered exerting only the inhibitory effect due to competition with proton donors and acceptors. *Ab initio* calculations demonstrate that the hydrogen bond between two *N*-acetamide molecules in the dimer **B-11** solvated by three water molecules is approximately 12 kJ mol^{-1} more stable than in the unsolvated dimer.[68] Of course, when a strong acid of Lewis or Bronsted type is added to a proton donor group, this can finally convert the latter to a positively charged H-bonding site with considerably increased affinity. For example, the urea derivative **B-12** binds acetate anion in DMSO with $K = 370$ M^{-1}, but its boronate adducts **B-13** and **B-14** bind acetate under the same conditions with $K = 7 \times 10^3$ and 6×10^4 M^{-1} respectively.[69]

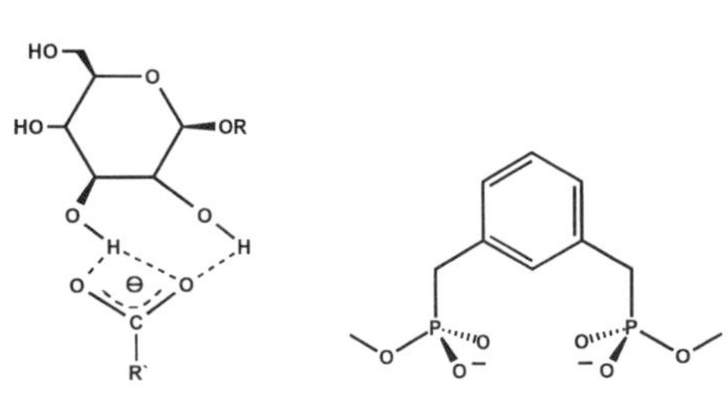

Scheme B 14.   Synthetic receptors for sugar binding (possible hydrogen bonds indicated with dashed lines).[65a,b]

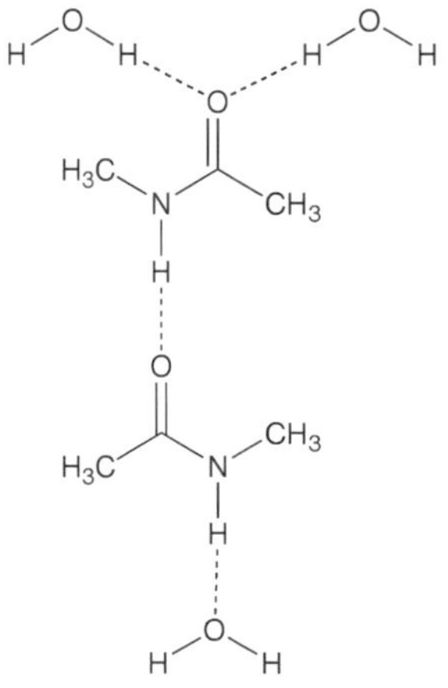

A
with R' = C$_6$H$_5$,   R' = C$_{12}$H$_{25}$
in CDCl$_3$ $K \approx 10^3$ M^{-1}

B
with glucose-OR' :
$K \approx 10^4$ M^{-1}

B-11

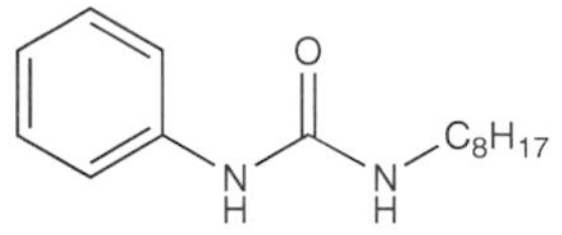

B-12

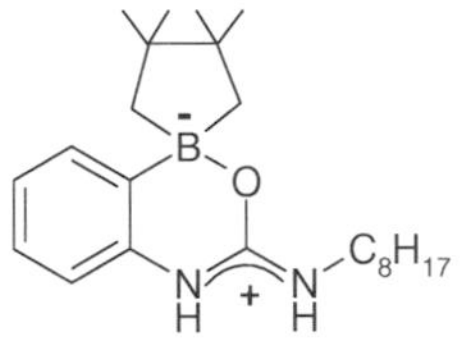

B-13

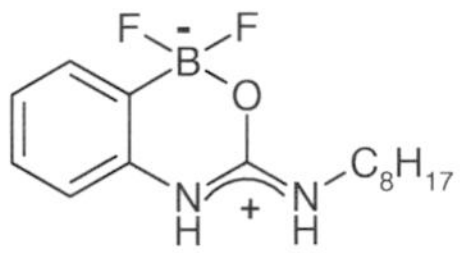

B-14

Table B 5. Gas phase free binding energies, $-\Delta G$, of cations with selected bases.[a]

	$Li^{+\,b}$	$NR_4^{+\,c}$
H_2O	114	10
C_6H_6	124	14
Et_2S	122	
MeOH	130	12
Me_2O	135	
Et_2O	150	
$MeNH_2$	143	15
MeCN	153	
Me_2CO	159	30
THF	148	
MeCOOMe	158	
$MeCONH_2$	176	
$MeCONMe_2$	185	48
DMSO	180	

[a] in kJ mol^{-1} at ambient temperature.
[b] from ref. 71
[c] for R=Me, calculated from ΔH and ΔS values for about 300 K with data from ref. 72

B 4. CATION–π-ELECTRON AND RELATED INTERACTIONS

Forces between cations and electroneutral acceptor parts such as π-electrons and lone pair electrons play an important role in synthetic and biological supramolecular complexes. They are the basis of most ionophores of the crown ether type discussed in earlier chapters, and hydrogen bonds to neutral acceptors also belong to this category.

Affinities (ΔG) of simple cations to such acceptors have been measured early in the gas phase[70] by mass spectrometry and cyclotron resonance (Table B 5).[71–73] They demonstrate, that in the absence of competing solvents interactions of small cations with π-systems can be even stronger than those with for instance water. The $-\Delta G$ values increase with the increasing electron donicity of the neutral (Lewis-type) base (the values do not correlate, however, with the hydrogen bond acceptor increments discussed in Section B 3), and represent valuable numbers for the design of receptors. Even large peralkylammonium ions show significant attraction, in

particular with the relatively large negative point charges of amides. This is of obvious importance for cation recognition and transport in proteins. In an electroneutral host like cucurbituril (**B-15**) both strong dipoles and a delocalized π-system lead to strong complexation of alkali metal[74] and organic ammonium ions,[75] in the latter case with substantial contributions of hydrophobic interactions as evident from the effects of alkyl substituents.

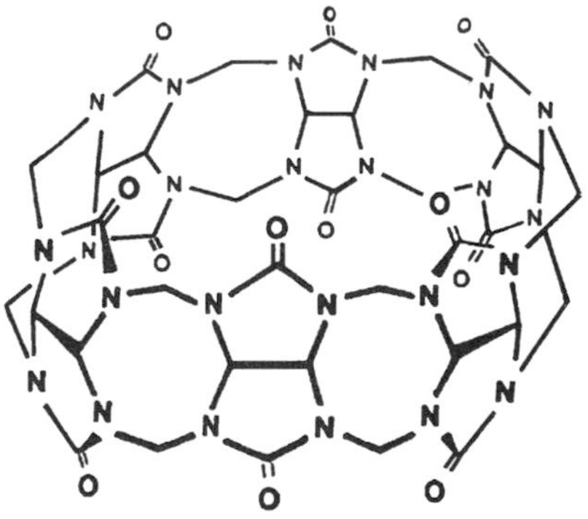

B-15

A distinct decrease of affinity with increasing ion diameter is seen in the sequence

Li > Na > K ≫ NMe$_4$ cations with, for example benzene;[70,76] this points to electrostatic forces as the dominant factor, in agreement with high level *ab initio* calculations of such systems.[72,73,76] The partial charge in ether oxygen lone pairs might be too small for purely electrostatic binding, as obvious from the undetectable complexation of tetramethylammonium ions[55] to crown ethers in solution. In contrast, the surface of phenyl rings provide an excellent fragment for binding cationic guests of the choline type (see below).

All aromatic moieties possess a permanent negative potential above the plane, which can exert a Coulomb-type attraction towards permanent full or partial charges, the latter being the basis of the edge-to-face benzene dimers discussed below. The charge of the π-cloud is set off by positive charges in the aromatic C–H bonds, as seen in Fig. B 5.[73] Benzene is also characterized by a permanent quadrupole moment, composed of the dipoles within the ring system. The calculational difference between the Coulomb-type interactions and those with the quadrupole is the steeper distance dependence of the latter (Section E 7), requiring tighter fit between cation and π-system. Another binding contribution, demanding even shorter distances between cation and host moiety, is the polarization induced by the ion within the acceptor system, producing that way an always favorable ion–dipole or ion–quadrupole attraction. Recent *ab initio* MO calculations at the MP2 level indicate also significant contributions of electron correlation effects in complexes between tetramethylammonium cations and aromatic moieties.[77]

Besides the different geometric requirements there is also a distinct difference in solvent effects (Section C 2): water will always weaken complexes which are dominated by electrostatic interactions, but the extremely low polarizability of water (mostly due to its small molecular volume) will make it less competitive for interactions dominated by polarization effects. Another factor leading to relatively strong benzene–NMe$_4$ cation interactions in water is the smaller dehydration energy compared to protonated ammonium ions.

Associations of cations with electroneutral bases are characterized by substantial entropic disadvantages, which generally take away at least half of the enthalpy gain. The data assembled in Table B 6[73] illustrate the increasing entropic disadvantage with the increasing number of bases surrounding one highly charge density cation like lithium; nevertheless such higher complexes were shown to exist in the gas phase (Fig. B 5). Notice, that the entropic disadvantage in the complexes is less pronounced with water than with benzene, which partially offsets the enthalpic

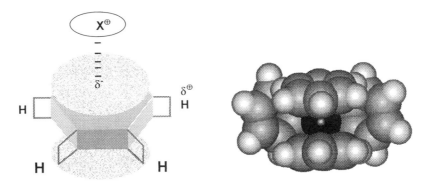

Figure B 5.   Visualization of a cation–π complex (upper part) and of four benzene molecules wrapped around one lithium cation.[73]

Table B6. Thermodynamic data[a] for complexes between potassium cations (K$^+$) and benzene (B) or water (W) in the gas phase.[73] Reprinted with permission from *J. Phys. Chem.*, Sunner *et al.* 1981, **85**, 1814. Copyright 1981 American Chemical Society.

	$-\Delta H$	$-\Delta S$	$-\Delta G$
K$^+$ + B = K$^+$B	80	103	50
K$^+$B + B = K$^+$B$_2$	79	142	37
K$^+$B$_2$ + B = K$^+$B$_3$	61	137	20
K$^+$B$_3$ + B = K$^+$B$_4$	53	173	1
K$^+$ + W = K$^+$W	75	90	48
K$^+$W + W = K$^+$W$_2$	68	101	38
K$^+$W$_2$ + W = K$^+$W$_3$	55	96	26
K$^+$W$_3$ + W = K$^+$W$_4$	49	103	18
K$^+$W$_4$ + W = K$^+$W$_5$	45	105	13
K$^+$W$_5$ + W = K$^+$W$_6$	42	107	10

[a] Enthalpies ΔH and free energies ΔG in kJ mol^{-1}; entropies ΔS in J K^{-1} mol^{-1}; standard state 1 atm.

advantage of aromatic moieties for cation solvation. The association constants with metal cations can be multiplied by the chelate effect if the surrounding aromatic moieties are covalently linked by suitable spacers. Thus,

the paracyclophane **B-16** (Fig. B6) forms in the gas phase particularly stable complexes[78] with cesium and rubidium cations, less stable ones with the too small Li or Na ions, as evident from electrospray ionization mass spectrometry (ESI-MS) measurements (Section D3.5). In solution, ammonium ion complexation by calixarenes[79] **B-17** (Fig. B6) is also substantially increased by the surrounding aryl rings.

Most of the artificial receptors utilizing cation–π interactions work in aqueous medium, which can lead also to stronger polarization contributions. These host compounds allow to dissect the binding contributions and show the way to design effective host systems potentially of practical importance, as will be illustrated with several examples. The cyclophanes[76] **B-18** etc. with $X = Cs^+$ (in water carboxylate anion) (Fig. B7) bear permanent charges only outside the cavity; these anionic groups can make small electrostatic contributions to the complexation, but essentially serve to make the host water-soluble. Hosts **B-18** with $X = Cs^+$ binds compounds **B-19** and **B-20** with high affinities, which are

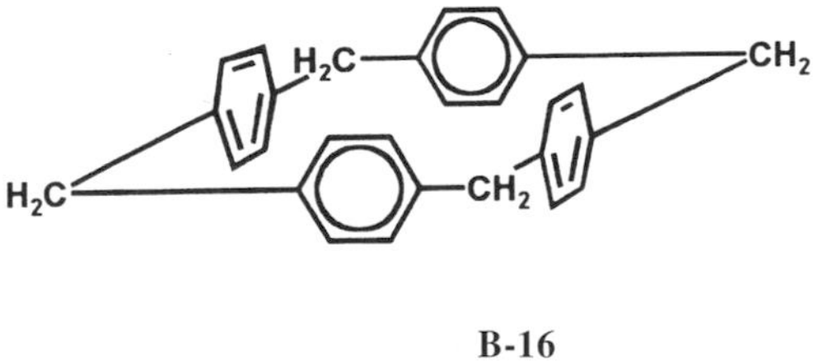

B-16 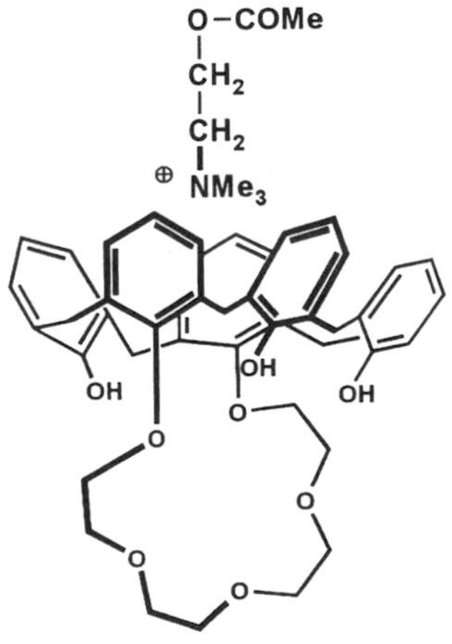   B-17

$\Delta G° = -13.4$ kJ/mol (CDCl$_3$)

Figure B6. Aromatic hosts for cations. **B-16**[78] for Li$^+$; **B-17**[79] for ammonium compounds.

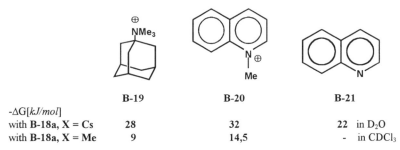

B-18-a: Sp = 1,4-CH$_2$-Ph-CH$_2$, X = Cs
B-18-b: Sp = 1,4-Cyclohexyl, X = Cs

-ΔG[*kJ/mol*]	B-19	B-20	B-21	
with **B-18a**, X = **Cs**	28	32	22	in D$_2$O
with **B-18a**, X = **Me**	9	14,5	-	in CDCl$_3$

Figure B 7.   Complexation free energies ($-\Delta G$) with electron-rich cyclophanes **B-18a,b**, **B-19**, **B-20**, **B-21**.[76]

increased considerably if, as in the case of **B-18a**, the cavity walls are made completely aromatic. The affinities are substantially lower in chloroform instead of water as solvent, which is the opposite of what one would expect if only electrostatic interactions would prevail and indicates polarization effects, but also may be due to hydrophobic contributions. In contrast to the expectation, that the lipophilic cavity of **B-18** would bind less hydrophilic substrates better, much smaller affinities are found with electroneutral guest compounds of similar shape, such as **B-21**, since the cation–π effect is missing.

Host structures with aromatic units in a approximately perpendicular orientation towards ammonium ion centers of guest molecules provide a very efficient ligand to complex, for example the neurotransmitter acetylcholine. Host compounds such as **B-22**[80] are relatively open and complex for instance acetylcholine with only $K = 5 \times 10^2$ M^{-1}. The ligand **B-18a**[76b] binds acetylcholine with $K = 2 \times 10^4$ M^{-1}, that is as strongly as acetylcholine esterase ($K \approx 10^3$ M^{-1})[76b], which contains a tryptophane indole unit in contact with the choline NMe$_3$ group.[81] That electro-

B-22

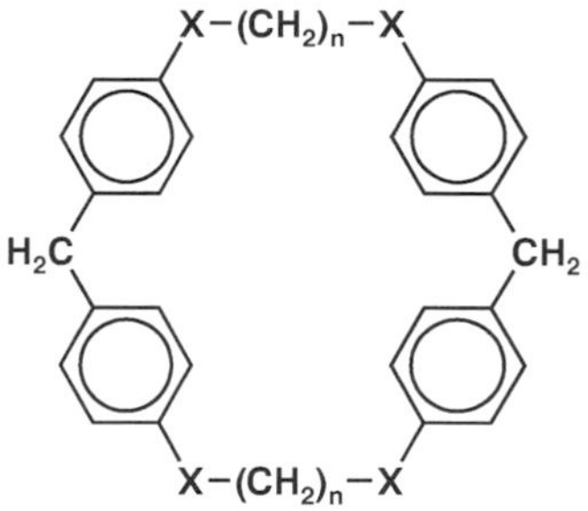

B-23a: X = $^+NMe_2$

B-23b: X =

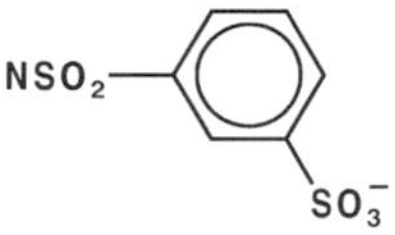

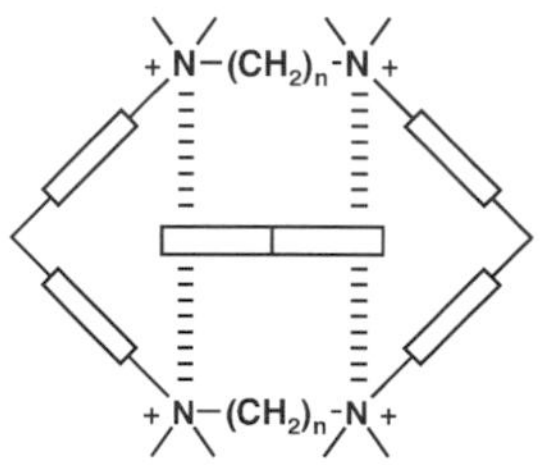

B-23a

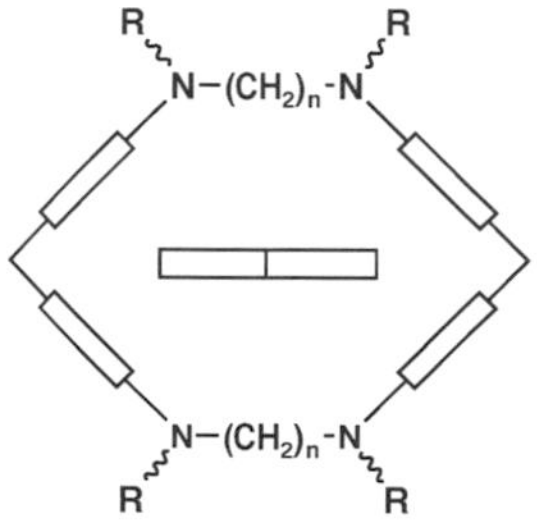

B-23b

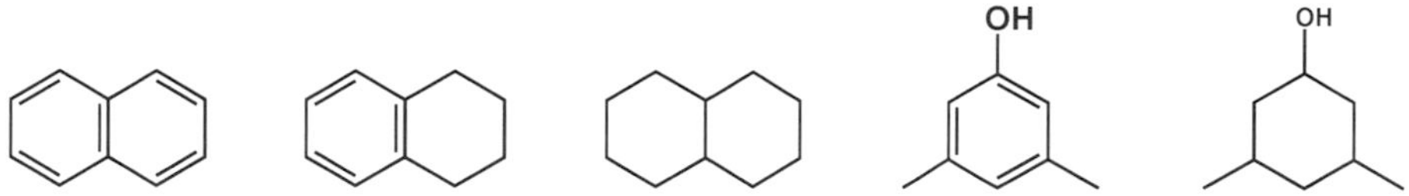

B-23a:	17	11	7	11	~3
B-23b:	13	-	-	6.5	~4

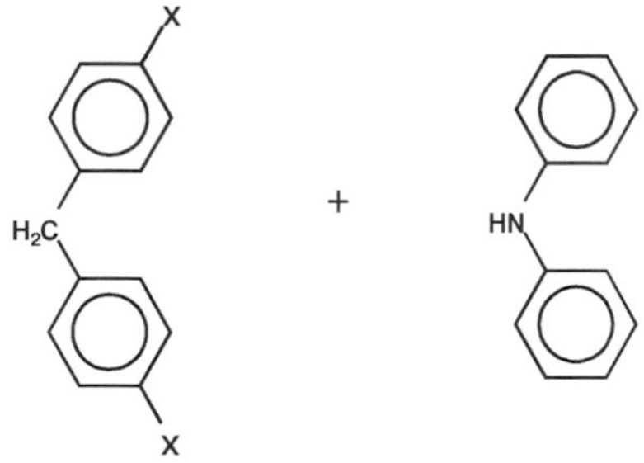

B-24

$X = NMe_3^{\oplus}$ 7.0   (in 5% MeOH)

$X = SO_3^{\theta}$ 5.3   (in 20% MeOH)

Figure B 8. Complexation free energies $-\Delta G$ [kJ mol^{-1}] with the cyclophane **B-23a,b** and cleft hosts **B-24** in D$_2$O/CD$_3$OD (80/20) unless noted otherwise, showing the magnitude of cation–π effects.[83]

static interactions with Glu 202 contribute little to the binding is borne out by studies with mutants, in which this aminoacid was exchanged to glutamine. This represents a nice example of using site-directed mutagenesis to identify non-covalent interactions in proteins. Interactions of aromatic parts of aminoacids with positively charged nitrogen groups from basic aminoacids has been found to be a frequent and important motif in many proteins.[17]

Comparisons to binding of acetylcholine in biological systems needs, however, some caution. Enzymes are designed for strong binding of transition states rather than ground states of their substrates. Thus, binding constant of acetylcholine to acetylcholine esterase is about 10^4 times smaller than that to the high affinity site of the nicotinic acetylcholine receptor.[76b] There is evidence that distant anionic groups in the active site of the nicotinic receptor create an electrostatic potential of about $-80\,mV$ at the binding site,[82] which provides a binding constant increase of about 25 times.

The complexation free energies ΔG observed with host **B-23** in Fig. B 8, which binds naphthalene and nucleobases only well, as long as the cationic charge lining the cavity is retained (**B-23a**), provide numbers for the aryl–cation interaction increments $\Delta\Delta G$.[83] If the positive charge is removed (host **B-23b**), aromatic guest molecules are bound only weakly. Moreover, lipophilic saturated guest molecules are complexed much less than their aromatic counterparts as long as the latter can interact with N^+ charges of the host (**B-23a**), but show better binding with the host **B-23b** lacking the positive charge (Fig. B 8). Ligands like the cleft **B-24** also have the ion above the aromatic guest centers; the ΔG observed with **B-24** ($X = SO_3^-$) demonstrate that the anion–π interaction can have almost the same size as the cation–π interaction, in line with strong contributions from polarization of the aromatic electron cloud. For the cation–π interaction there are more data available (Fig. B 8), which allow one to extract an approximate number of $\Delta\Delta G = -3\,kJ\,mol^{-1}$ in water per phenyl group. A similar binding increment emerges from affinities between aromatic ion pairs, which show additive contributions of Coulomb interactions.[83c]

There are now almost countless examples of synthetic hosts, in which interactions between ionic or polar groups with neutral Lewis bases are used for specific binding of a large variety of substrates. We close this short overview by pointing out that many hydrogen bond-like associations also belong to this category, potentially also including for instance C–H–π interactions. They are, however, difficult to separate from often dominating dispersive and other high order interactions, and will therefore be dealt with in the next chapter. If aromatic substrates can be efficiently bound

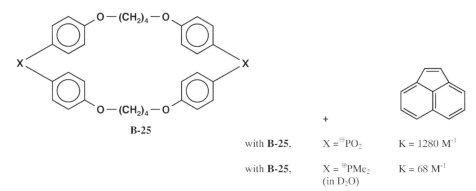

with **B-25**,	$X = {}^\ominus PO_2$	$K = 1280\ M^{-1}$
with **B-25**,	$X = {}^\oplus PMe_2$	$K = 68\ M^{-1}$
	(in D_2O)	

Figure B 9.   Charge-mediated binding selectivity with isosteric hosts of opposite charge.[84]

by interactions with the negatively charged π-clouds, they also should exert attractive forces with the aromatic hydrogen atoms, which bear balancing positive charges. Such interactions have been identified with cyclophanes like **B-25** bearing anionic groups in close vicinity of aromatic guest protons (Fig. B 9).[84]

B 5.   VAN DER WAALS INTERACTIONS, DISPERSIVE FORCES, STACKING, CHARGE TRANSFER COMPLEXES AND RELATED INTERACTIONS

In this Section we summarize a number of weak interactions, which in most cases cannot be distinguished clearly with respect to their mechanistic origin. Dispersive forces in the classical concept of London and Eisenschitz are described as electrostatic attractions between fluctuating weak dipoles like neighboring C–H bonds, which at some time period can assume local charge distributions which differ from the permanent time-averaged moments and can assume a energetically favorable mutual orientation.[85] These forces fall off with r^{-6} between interacting bonds (Section E 7) and therefore require an extremely tight fit between molecular surfaces. In the molecular orbital concept the corresponding lipophilic interaction can be described as electron correlation, which even for the cation–π effect (see preceding Section) has been assumed to contribute significantly.[77b]

The edge-to-face benzene dimer[86] (Scheme B 15) also has been shown by *ab initio* molecular orbital calculations to be stabilized primarily by correlation energy.[87] Characteristic thermodynamic parameters such as ΔH, ΔS and ΔC_P for interactions between non-polar substrates in water will be discussed in the next Section with solvophobic forces, from which attractive lipophilic forces are often difficult to distinguish. Upfield NMR shifts in the naphthyl units of compound **B-26** (Scheme B 15) indicate formation of edge-to-face conformers, which, however, is similar in water and in chloroform or benzene as solvent, indicating a weak non–hydrophobic attraction between the aromatic rings.[88] Such edge-to-face orientations between aromatic moieties are frequently found also in proteins[17] and will be discussed in the following.

B 5.1.   C–H–π AND RELATED INTERACTIONS

Generally C–H bonds have a small dipole moment (around 0.4 Debye for aliphatic systems) with a positively charged hydrogen atom; this can exert an attractive force with any molecular fragments bearing a negative partial charge, or a polarizable electron cloud.[89] Obviously these interactions are mechanistically related to Coulomb effects, to hydrogen bonds and to the cation–π effects discussed earlier. They are dealt with here in view of their frequently concomitant occur-

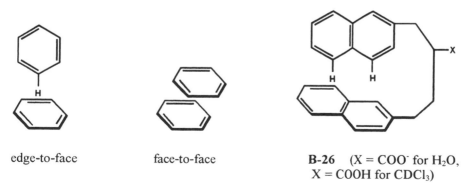

edge-to-face   face-to-face   **B-26**   (X = COO⁻ for H_2O,
                                                           X = COOH for $CDCl_3$)

Scheme B 15.   Types of arene–arene interactions.

Scheme B 16. Intramolecular C–H–π interactions.

rence with dispersive and other interactions. Particularly in water they are often also difficult to separate from solvophobic forces. In most cases C–H–π interactions are much smaller than other non-covalent forces; they are visible particularly in solid state structures or in changes of conformational equilibria. Thus, the population of the rotamer **B-27a** in Scheme B16 increases from $K_{rel}=1.10$ for X= NO_2 to 1.65 for X=NH_2.[90]

Similarly, the preference for the dibenzo-diazocine conformer **B-28a** in Scheme B16 increases from $1\,kJ\,mol^{-1}$ for X=H to $2.9\,kJ\,mol^{-1}$ for X=NO_2.[91] That an increasing acidity of the C–H group also leads to a distinct increase of these interactions is well known from the NMR shielding of, for instance, $CHCl_3$ protons in aromatic solvents (the so-called aromatic solvent induced shift or ASIS), which again increases with the electrondonating power of substituents at the phenyl ring. It is obviously difficult to disentangle the various mechanistic contributions to the C–H–π effect, and to obtain from experiments a reliable quantity for its size in solution. A recent NMR analysis of the conformers **B-28a/b** yielded after correction for solvent induced changes a ΔG value of around $2\,kJ\,mol^{-1}$,[92] which for intermolecular interactions will be even smaller, but never-

theless have been shown to contribute significantly.[93]

B 5.2.   STACKING OR FACE-TO-FACE COMPLEXES

The permanent negative charges of π-electron clouds will lead to repulsion in face-to-face complexes unless the aromatic units are so displaced that the correspondingly more positively charged aromatic hydrogen atoms can exert an attractive force towards the π-system (Scheme B15). The high polarizibility of the π-cloud can enhance dispersive forces also in less displaced orientations. Molecular dynamic (MD) and *ab initio* molecular orbital (MO) calculations for benzene dimers indicate a predominance of edge-to-face complexes over the stacked sandwich arrangement[94] by $3\,kJ\,mol^{-1}$ in the gas phase and by $6\,kJ\,mol^{-1}$ in water;[94c] with toluene the preference seems to be reversed in the gas phase, but not in water. Experimental data on benzene association constants[86] vary considerably, but show in water only values around $1\,M^{-1}$. Magnetic field alignment NMR studies with benzene in water speak for a edge-to-face preference of up to $3\,kJ\,mol^{-1}$.[95] Substituents X at phenyl rings change the association pattern significantly.[96] Thus, with 1,8-diarylnaphthalenes one observes an increase of rotational barriers with the number of fluorine groups in one of the phenyl rings as the result of a decreased electron density, leading to less unfavorable orientation in parallel stacks in the ground state. On the other hand, the quadrupoles of the protonated and fluorinated benzene rings have a similar magnitude, but opposite signs with an inversion of the C–X bond polarity, which favors even parallel stacks.[96]

The association free energies ΔG in supramolecular complexes of **B-29a–d** and *N*-ethyl adenine (Scheme B17) show a systematic increase by $\Delta\Delta G=1.85 \pm 0.15\,kJ\,mol^{-1}$ for each additional benzene moiety interacting with the adenine nucleobase.[97] Noticeably, this is observed in chloroform, where solvophobic

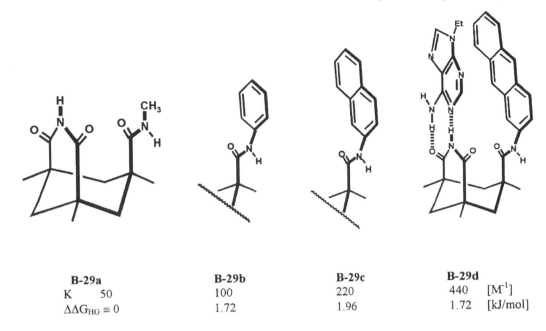

	B-29a	**B-29b**	**B-29c**	**B-29d**	
K	50	100	220	440	[M⁻¹]
$\Delta\Delta G_{HG}$	$\equiv 0$	1.72	1.96	1.72	[kJ/mol]

Scheme B 17. Association constants for complexes between a series of diamide ligands successively substituted with aryl groups of growing size and *N*-ethyl adenine in chloroform. For **B-29b** and **B-29c** the left hand side imide is omitted. The $\Delta\Delta G$ values refer to differences between **B-29b** and **B-29a**; **B-29c** and **B-29b**; **B-29d** and **B-29c**, respectively.[97]

forces will be diminished and where the polarizibility of the competing solvent also is much larger than in water.

Stacking can be exploited for nucleotide recognition even in chloroform if, for instance hydrogen bonds provide for the primary binding force between host and guest. As illustrated with the structure **B-30**[98]; a lipophilic thymine derivative is bound with $K = 290\,M^{-1}$. Much higher affinities with up to $K = 1.5 \times 10^4\,M^{-1}$ with adenosine derivatives are the consequence of providing not only one π-surface in the host for stacking, but two as in the cleft (or tweezer) compound **B-31**[99]; the smaller pyrimidine derivatives of uridine and cytidine are as consequence of the smaller π-system associated only with affinities around $10^2\,M^{-1}$. The host **B-32** formed from proflavine (Scheme B18, 'Flav') linked to thymine T assumes a folded stacked structure, and then is able to bind, for example the adenine compound **B-33 (A)** with $K=300\,M^{-1}$;

B-30

shielding effects on the NMR shifts indicate a stacked entity, likely with the common base pairing between A and T.[100] The dominance of stacking interactions is also visible in complexes of nucleotides to azapyrenium structures like **B-34**, for which one observes surprisingly no significant difference between AMP, ADP and ATP. Favorable orientations for stacking seems to preclude here essentially ion pairing with the positively charged nitrogen which is buried by the π-system in the stacked dimer.

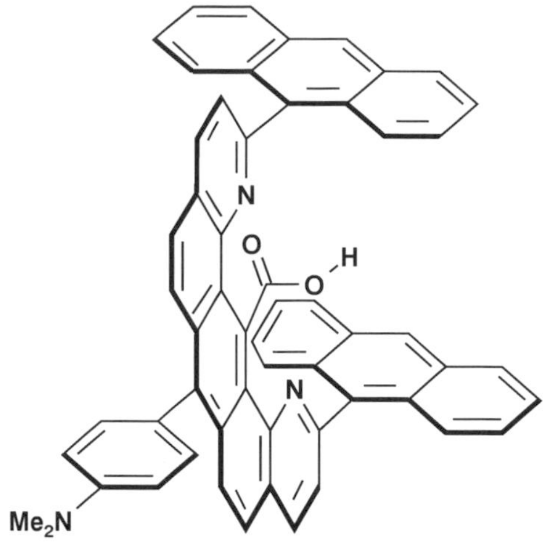

B-31

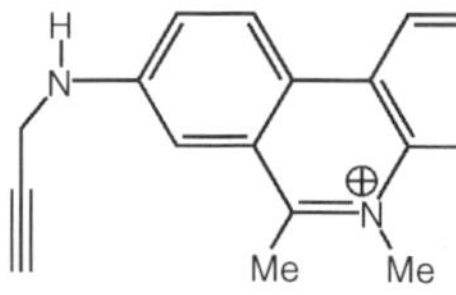

B-34

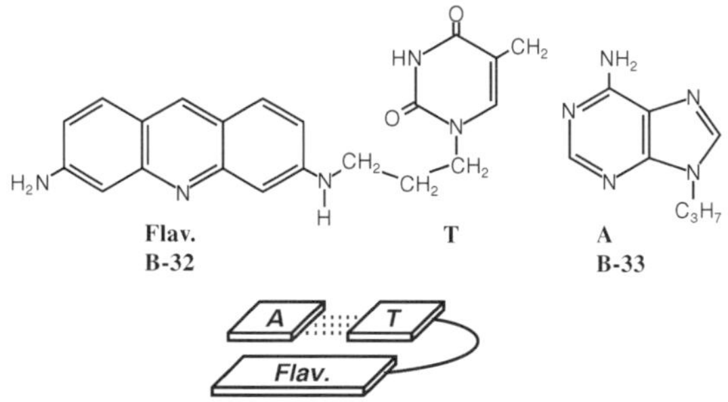

Flav.
B-32

T

A
B-33

Scheme B 18. Proflavine-based receptor for adenine.[100]

The very extended π-cloud of porphyrins **B-35** offer an opportunity for measuring interactions with aromatic guest ligands **B-36**, which have different size but are smaller than the host system (Scheme B19). Thus a full contact is always possible in face-to-face orientations which were evident from the observed NMR shifts.[101] The complexes were made water-soluble by attaching positively or negatively charged substitiuents R, X or Y at the porphyrin and, if necessary, also at the ligands. Measurements with 23 different ligands show after correction for the additional ion pair, if existent, $-\Delta G$ values around $7\,kJ\,mol^{-1}$ for all benzene-shaped compounds, $16\,kJ\,mol^{-1}$ for all naphthalene

shaped derivatives, and 18 kJ mol for anthracene-shaped. The relatively large energies are not due to hydrophobic forces, as evident from association with, for instance a saturated cyclohexane derivative of the same size; this shows only ion pair contributions with the solubilizing ionic substituents but no additional lipophilic stabilization, in contrast to the benzene derivatives.

Heteroatoms within or at the aromatic ring lead to permanent charge redistributions which favor stacked orientations, of particular importance with nucleobases. Experimental studies with modified DNA-type duplexes have shown, that formation of the double helix is as much due to nucleobase stacking as to Watson–Crick hydrogen bonds.[102] Exchange of the aminoacid tryptophane 120 in streptavidin by site-directed mutagenesis has shown that also the strong binding of biotin (Chapter A) is partially due to aromatic interactions with the indole ring.[103]

Intercalation in DNA and RNA is the biologically most important stacking phenomenon.[104] In spite of the correspondingly many investigations there is not yet a clear picture of the relation between ligand affinity and structure. One of the reasons for this is the problem often involved even with the unambiguous characterization of binding modes and affinities with the biopolymers.[105] Until now it is not clear whether unequal charge distribution, in particular positive charges within the ligand π-system will enhance binding as suggested earlier.[106]

The importance of at least partial charges within the intercalating rings, or of multipolar fragments has also been stressed in *ab initio* molecular orbital calculations, describing stacking mostly by electron correlation.[107]

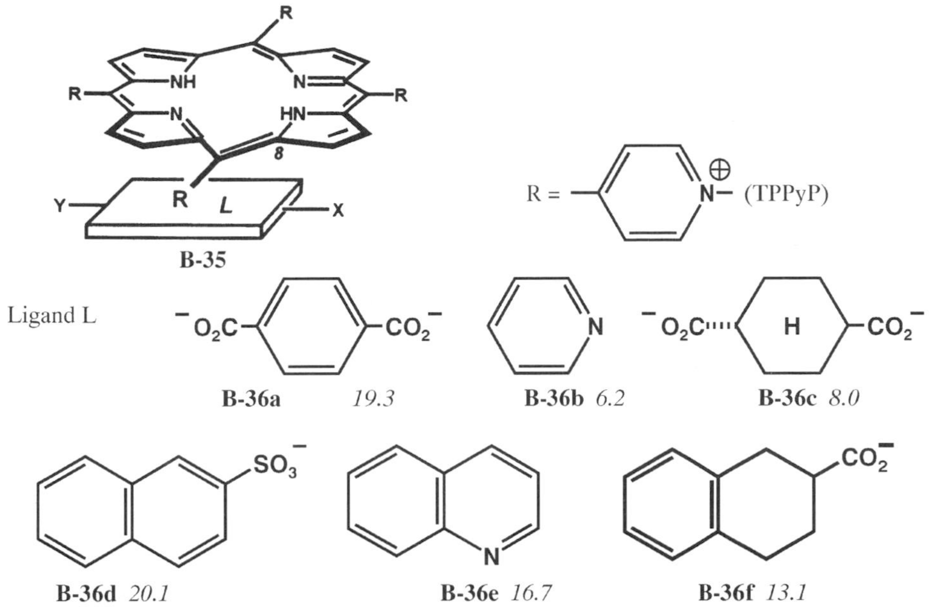

Scheme B 19. Association free energies (kJ mol^{-1}) of porphyrins in water with selected ligands L.[101]

Corresponding charge distributions are also believed to be essential for a π-charge sandwich model of nucleobase stacking.[108] This more qualitative model is contradicted by recent *ab initio* calculations favoring maximum overlap of the surfaces.[107] NMR studies with a number of mostly naphthalene-shaped intercalators also do not indicate a dependence on the number and kind of heteroatoms within the ligand.[106b] Prediction of intercalation strength is hampered by rather extreme changes of partially compensating enthalpy and entropy contributions.[109]

Interactions between aromatic systems occur in many cyclophane complexes,[110] and in most cases one observes the expected increase of binding affinity with the increased size of the π-surfaces, even if the guest molecule extends over the rim of the host cavity.[111] Interactions between the electron-poor tetracationic cyclophane **B-37** and electron-rich guest molecules play a major role in many supramolecular systems including self-assembly processes[112] (Section I2). With aceto-nitrile as solvent molecular orbital calcula-

tions indicate the guest polarizability to be the most important factor, in line with the highest association constant with the particularly polarizable benzidine as guest.[113] Association of nucleotides and nucleosides in the azonia-cyclophane **B-3** in water is dominated again by stacking with the nucleobases, with a higher affinity towards adenine derivatives, which possess a higher polarizability.[114]

Cyclodextrins (CDs, Scheme B20) are a class of electroneutral macrocyclic hosts, for which the hydroxy groups outside provide for water solubility, and the axial C–H bonds inside the cavity provide for binding of lipophilic guest molecules.[115] Associations with CDs depending on their diameter (Scheme B 21, see also website) are shape selective with respect to the guest size, but show at least as much for differences dependent on the chemical nature of the guest. Cyclodextrin complexes are by far the most widely used supramolecular systems; at the same time their binding mechanisms are the most difficult to rationalize and to predict.[116] We will illustrate this with few of the many cases, where titration

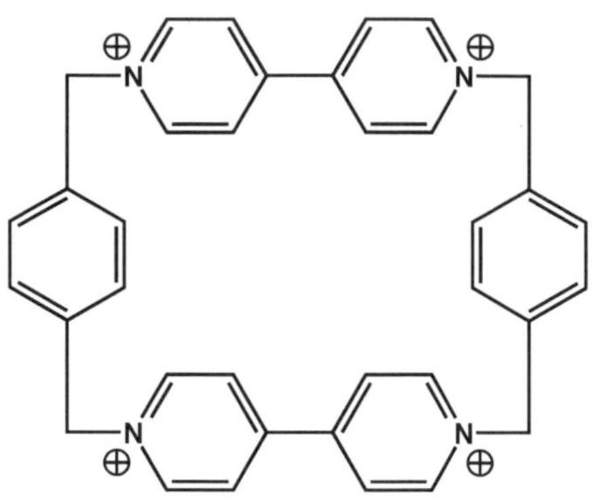

B-37a

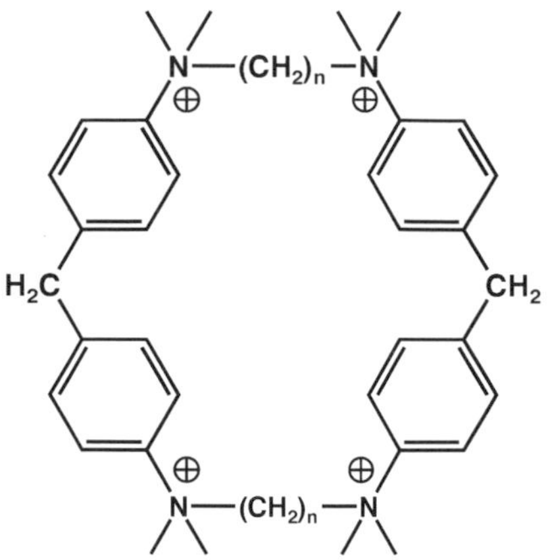

B-37b

calorimetry (Section D 5) has helped considerably to disentangle the relevant binding contributions.[117] For a series of CD complexes for which the NMR evidence for the intracavity inclusion will be discussed in Section E4,

the thermodynamic parameters in Scheme B 21 show mostly negative $T\Delta S$ contributions; they are counteracted by a ΔH advantage which reaches its optimum with phenolates, not phenoles, particularly if they fit tightly as

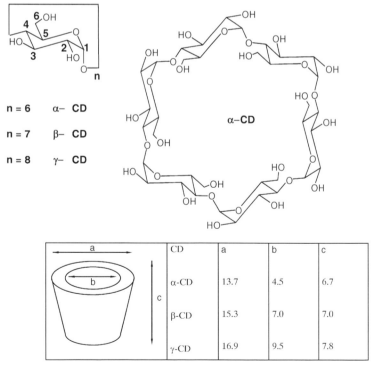

	$n = 6$	$\alpha-$ **CD**
	$n = 7$	$\beta-$ **CD**
	$n = 8$	$\gamma-$ **CD**

CD	a	b	c
α-CD	13.7	4.5	6.7
β-CD	15.3	7.0	7.0
γ-CD	16.9	9.5	7.8

Scheme B 20. α-,β-,γ-Cyclodextrins with cavity dimensions (Å).[115]

guest:	O⁻ ... NO₂	OH ... NO₂	COO⁻ (adamantane)
α-CD			
-ΔG	18.7	11.5	11.6
-ΔH	42.8	23.0	14.3
TΔS	-24.1	-11.5	-2.7
β-CD			
-ΔG	15.0	14.2	24.5
-ΔH	16.1	10.2	21.6
TΔS	-1.1	3.9	2.9

Scheme B 21. Thermodynamic data [kJ mol^{-1}] for selected cyclodextrin complexes.

with the small α-CD cavity, pointing to dominating dispersive interactions. On the other side, the $T\Delta S$ disadvantage is close to zero in complexes with loose fit, such as with β-CD and the phenylderivatives, but not so with the adamantyl derivative which fills also the wider β-CD cavity. These observations are in line with the NMR-spectroscopic evidence of a preferred loose fit in CD complexes with guest molecules offering alternatively small and large parts for inclusion in the CD cavity (Section E 4). That the entropic term is not positive, or at best close to zero, does not rule out hydrophobic interactions as driving force, as will be discussed in Section B 6. In some cases similar or even higher association free energies ΔG were observed in organic solvents than in water;[118,119] thus for m-(p-hydroxyphenylazo)benzoate and even the wide-cavity γ-CD ΔG in water is $-22.5\,\text{kJ}\,\text{mol}^{-1}$, in DMF $-23.1\,\text{kJ}\,\text{mol}^{-1}$.[118] Clearly this speaks for the absence of hydrophobic forces in these cases.

B 5.3. CHARGE TRANSFER COMPLEXES

Electron transfer from high energy occupied molecular orbitals of electron-rich compounds (donor D) into low lying unoccupied orbitals of electron-poor systems (acceptor A) are recognized by distinct charge transfer (CT) bands in the long wavelength part of UV spectra.[120] Mulliken has described CT complexes from A and D with linear combinations of wavefunctions for states resembling A–D and A^+D^-.[121] It has been shown that the interaction free energy ΔG in equilibria between the soft Lewis acid iodine I_2 as acceptor and a range of hydroxy-, thio-, and amino compounds correlates linearly with the basicity of these donors, indicating the importance of the polar contribution from the A^+D^- state.[122] Nevertheless, these complexes should not be confused with associations between hard Lewis acids and bases, and the often used term electron donor–acceptor (EDA) complexes is usually too general. The electron transfer between molecular orbitals plays a usually very small role in the energetics of host–guest equilibria.

Calixarenes like **B-38a,b** (Scheme B22) can form semi-open cavities, with conformations stabilized by circular hydrogen bonds between the phenolic groups at the lower rim of the vase-like macrocycle.[123] Fullerenes are electron deficient compounds and therefore show typical CT bands around 440 nm if mixed with electron rich compounds. Of practical interest is the complex formed, for example with a sulfonated calix[8]arene like **B-38b** ($R = SO_3H$) and the fullerene C_{60} in water as solvent.[124] (Scheme B 22, see website). Similar structures have been observed in the solid state, such as by ^{13}C NMR CP-MAS (cross polarization magnetic angle spinning) techniques, which show upfield shifts of the fullerene signals by 1.4 ppm.[125] and symmetrization of the calixarene ligand upon complexation. Related solid state complexes with, for instance *tert*-butyl instead of alkylsulfonate substituents in the calixarene hold considerable promise[126] for improved separation of fullerenes, which are obtained as multicomponent mixture from the sood of carbon plasma discharges.

Even in lipophilic solvents the calixarenes like **B-38a** can form quite stable complexes

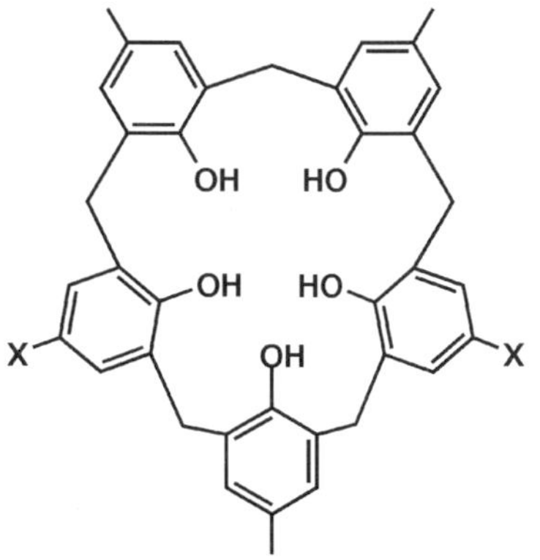

B-38a

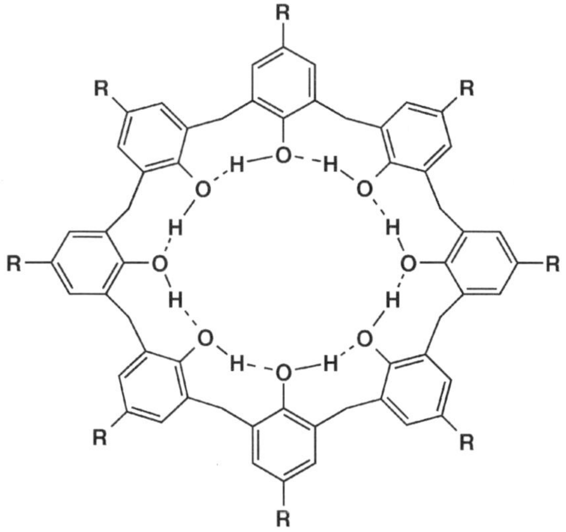

B-38b

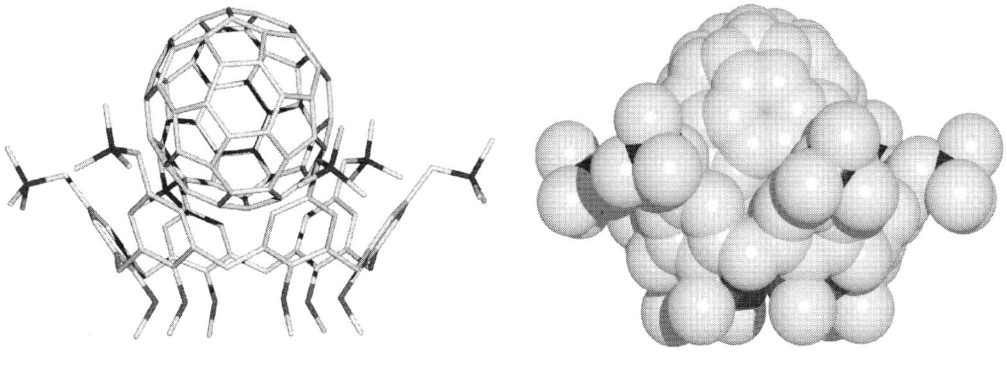

Scheme B 22. Calixarenes and their complexes with fullerene C_{60}.[124]

with fullerenes. One notes a significant affinity increase due to the large dispersive interactions with iodine as substituent at the calixarene upper rim (cf. the well known iodine–carbohydrate complexes). The stabi-

lity constants (Table B 7) as expected become lower with increasingly polarizable groups in the solvent molecules.[127]

B 6. HYDROPHOBIC (LIPOPHILIC) INTERACTIONS

Hydrophobic interactions dominate many important processes, such as aggregation of surfactants to micelles, bilayers, and other supramolecular structures, folding of proteins, protein–ligand (e.g. enzyme–substrate) and protein–protein association, solubilization of non-polar substances by surfactant

Table B 7. Stability constants (M^{-1}) of fullerene C_{60} complexes with calix[5]arenes **B-38a** in different solvents (at 298 K).[127]

Host **38a**	Benzene	CS_2	o-Dichlorobenzene
X= I	1840	660	310
X= Me	1510	600	280
X= H	460	290	210

aggregates, and supramolecular complexation of guests with non-polar parts. These interactions to some extent compensate the inefficiency of polar interactions (including hydrogen bonding) in water, which results from the high dielectric constant and strong proton donor–acceptor capacity of this solvent. Water possesses also a large internal cohesion energy density which is manifested in large vaporization enthalpy and high surface tension. As a result the energy required for formation of a cavity to accomodate the solute becomes very large and is not compensated by non-polar molecules solute-solvent weak interactions. The origin of hydrophobic interactions lies in the fact that non-polar molecules tend to avoid aqueous surrounding, as is evident from very low solubility of non-polar substances, in particular hydrocarbons, in water, and in the positive transfer free energies of such substances from organic solvents to water. Thermodynamic parameters of transfer of nonpolar molecules or moieties from a non-aqueous environment or the gas phase to water characterize quantitatively this tendency known as the *hydrophobic effect*.[85,128–131]

Table B8 shows some typical data which illustrate the general thermodynamics of the hydrophobic effect: at room temperature the enthalpy of dissolution of gaseous hydrocarbons in water is negative and is proportional to the surface area of the solute; the entropy of dissolution is large and negative; the dissolution of gaseous hydrocarbons is characterized by large positive change of heat capacity which is proportional to the solute surface area.

These facts were generalized by Frank and Evans in their classical theory of the hydrophobic effect.[131] Its main feature is formation of short-lived highly structured hydration shells (called 'icebergs', 'flickering clusters' or 'clathration shells') at the surfaces of non-polar solutes. The negative enthalpy of hydrocarbon hydration results from van der Waals interactions between hydrocarbon and water

as well as from stronger intermolecular hydrogen bonding between water molecules constituting the hydration shell (the latter is supported by downfield shifts of water H-NMR signals induced by hydrocarbon groups of different solutes, e.g. alcohols[132]). Structuring of water in the hydration shell leads to large negative entropy of hydration, and 'melting' of this ice-like shell induced by increased temperature explains the large positive heat capacity. Transfer of hydrocarbons from organic solvents to water shows generally the same thermodynamic features which agree with the Frank–Evans theory. As a result of van der Waals interactions of the solute with the organic solvent phase, the enthalpy of transfer is less negative and approaches zero for larger molecules, but the entropy is still large and negative, being the dominant component of total positive transfer free energy, and the heat capacity change also is still large and positive. The transfer free energy from a given solvent is proportional to the solute surface area.

The unfavorable hydration free energy of non-polar molecules can be partially compensated if such molecules will associate in aqueous solution, thus reducing the surface area accessible for water. Such association represents a special type of interactions called 'hydrophobic interactions' or 'hydrophobic forces'.[128] The hydrophobic interaction is thus considered as a partial reverse of the transfer process 'solute in organic solvent → solute in water' in the sense that the nearest surrounding of the solute inside the associate is partially organic-like. Comparison of the transfer free energies from different organic solvents to water shows that they are generally smaller for more polar solvents, as one can see from Table B8 comparing data for *n*-hexane and methanol. Therefore, transfer of a hydrocarbon from hexane to methanol is accompanied by positive ΔG values (4.35, 5.69 and 8.2 kJ mol^{-1} for methane, propane and *n*-hexane), which are smaller than for the transfer of the same solutes from hexane to

Table B 8. Thermodynamic parameters (kJ mol^{-1}) of transfer of some hydrocarbon solutes from gas phase and from organic solvents to water at 298 K (standard state 1 atm gas and unit mole fraction solution); compiled from published data.[130]

Solute	Transfer from	ΔG	ΔH	$T\Delta S$	ΔC_P [a]
CH$_4$	gas	26.28	-13.81	-40.09	217
	n-hexane	13.14	-11.55	-25.69	
	methanol	8.79	-5.86	-14.65	
C$_3$H$_8$	gas	26.07	-22.51	-48.58	319
	n-hexane	20.63	-8.41	-29.04	
	methanol	14.94	-6.15	-21.09	
n-C$_6$H$_{14}$	gas	28.53	-31.38	-59.91	
	n-hexane	32.55	0.17	-32.38	440
	methanol	24.35	-0.46	-24.81	

^{a} in J mol^{-1} K^{-1}

water, but are considerable enough to create a 'driving force' for at least weak association of hydrocarbons in polar organic solvents. Thus, we see that the hydrophobic interactions are an extreme case of more general phenomenon of solvophobic interactions.

Instead of saying that nonpolar molecules 'dislike' being in contact with water, one can say that they like to be in contact with other non-polar molecules. Therefore, hydrophobic interactions are often called also 'lipophilic interactions'. The difference, however, is not purely semantic. The van der Waals interactions between hydrocarbons can be stronger than those between hydrocarbons and water and this can be the actual 'driving force' for hydrocarbon association in aqueous solutions. These interactions are dealt with in Section B 5 with some typical supramolecular structures. Analysis of the macroscopic adhesion work of two liquids leads to the conclusion that attractive forces between alkanes, per unit area, are approximately equal to those between alkanes and water and, therefore, the van der Waals interactions do not contribute significantly to hydrophobic association.[133] This point needs, however, further investigation.

Considering hydrophobic interactions in terms of the classical model of hydrophobic hydration one should expect the following thermodynamic characteristics, related to partial destruction of the hydration shells of interacting nonpolar species: the association enthalpy should be positive or close to zero; the $T\Delta S$ term should be positive and dominating (at room temperature) in the association ΔG; the value of ΔC_P should be large and negative. The examples in Section B 5, Scheme B 21 show that cyclodextrin complexes are obviously not always dominated by hydrophobic interactions, in contrast to frequent assumptions. Also one should expect the association free energy to be proportional to the surface area of the mutual contact of the interacting species (ΔA) which upon association becomes inaccessible for the solvent. In addition, one should expect the association constant to decrease on going from pure water to water–organic mixtures, and to become very small and finally undetectable in pure organic solvents.

The contribution of hydrophobic interactions (ΔG_{hp}) to the total binding free energy can be estimated empirically from the equation,[133,134–136]

$$\Delta G_{hp} = \gamma^{ow} \Delta A \qquad (B 6\text{-}1)$$

where ΔA is the accessible interaction surface area mentioned above and γ^{ow} is the microscopic surface free energy related to the interfacial oil/water surface tension. A value of γ^{ow} about $200\,J\,mol^{-1}\,Å^{-2}$ was calculated from alkane solubility data.[133] This is considerably less than one would expect from the macroscopic oil/water surface tension of about $300\,J\,mol^{-1}\,Å^{-2}$, possibly due to differences in curvature between molecular and macroscopic planar interfaces and other factors.[133,135] Other evaluations of γ^{ow} are $230\,J\,mol^{-1}\,Å^{-2}$ from folding free energy of barnase mutants,[136] $160\,J\,mol^{-1}\,Å^{-2}$ from calculated surface area of cavity created in water during dissolution of gaseous hydrocarbons,[134a] and $105\,J\,mol^{-1}\,Å^{-2}$ from transfer ΔG of hydrophobic side chains of amino acids.[134b] Experimental data from associations of about 50 different aromatic compounds, which noticeably are rather independent of heteroatoms and charges within the π-systems, lead to a somewhat high value of $430\,J\,mol^{-1}\,Å^{-2}$.[137] Application of equation (B6-1) of course needs estimation of ΔA; the structure of the associate and ΔA may be evaluated with the help of modern computer aided molecular modeling methods Section E 7.

Alternatively the contribution of hydrophobic interactions can be analyzed in terms of linear free energy relationships by introducing respective hydrophobic substituent parameters.[138,139] Such parameters can be calculated from partition constants (P) of different solutes between water and an organic solvent. The largest set of P values has been obtained for *n*-octanol/water system.[139] The P values, or the corresponding substituent values[139] $\pi_X = \log(P_{R-X}/P_{R-H})$ allow one also to estimate the hydrophobicity of non-hydrocarbon moieties, such as halogen, cyano, mercapto, groups etc., which cannot be obtained easily by other methods. Directly from transfer thermodynamic parameters of linear alkanes from hexane to water one can obtain increments for the methylene group: $\Delta\Delta G = 3.85$, $\Delta\Delta H = 2.75$ and $T\Delta\Delta S = -1.1\,kJ\,mol^{-1}$.[140] Interestingly, the dominating increment is enthalpic, while the respective values for methyl group (8.4, -5.4 and

$-13.8\,kJ\,mol^{-1}$) indicate a totally entropic effect.

As discussed already in the preceding Section, complexes with cyclodextrins and cyclophanes in water are formed to a substantial degree by hydrophobic forces.

The data in Table B 9 show that thermodynamic parameters for complexation with β-cyclodextrin are remarkably similar to those for the transfer process. Complexation with α-cyclodextrin is more exothermic for all guests, in line with results in Section B 5, Scheme B 21, for compounds which fit well into the cavity and are dominated by other than hydrophobic forces. In all cases one observes large negative ΔC_P values, which were found to be strongly dependent on the temperature;[141] this is typical also for the transfer heat capacity.[130] Large ΔC_P values mean that ΔH strongly depends on temperature. For example, binding of *n*-heptane to β-cyclodextrin is endothermic at $273\,K$ ($\Delta H = 6.3\,kJ\,mol^{-1}$), but exothermic at $323\,K$ ($\Delta H = -8.8\,kJ\,mol^{-1}$).[142] Therefore, as will be discussed also later in this chapter, a better diagnostic criterion for hydrophobic interactions seems to be the observation of large negative ΔC_P rather than analysis of ΔH and ΔS values at single temperature. Note also that ΔS can give either a 'favorable' or an 'unfavorable' contribution to ΔG depending on the chosen standard state, Section A 4.

Thermodynamics of complexation of aliphatic alcohols by cyclodextrins was studied extensively by several groups.[143] For straight chain alcohols and α-cyclodextrin a very good correlation between logarithms of stability constants and the number of carbon atoms, n_C, was found[143b]

$$\ln K = -0.375 + 1.19\,n_C \qquad \text{(B 6-2)}$$

Also the complexation enthalpy and entropy are linear functions of n_C:[143b]

$$\Delta H(kJ\,mol^{-1}) = 5.13 - 3.83\,n_C \qquad \text{(B 6-3)}$$

$$\Delta S(J\,mol^{-1}K^{-1}) = 14.1 - 3.0\,n_C \qquad \text{(B 6-4)}$$

Table B 9. Thermodynamic parameters for the complexation by cyclodextrins and transfer from water to hydrocarbon media of heptane, cyclohexane and benzene at 298 K. Units: kJ mol^{-1} for ΔG and ΔH and J mol^{-1}K^{-1} for ΔS and ΔC_P. Values of ΔG and ΔS are given for two standard states: 1 M solution (column M) and unit mole fraction solution (column m.f.).[141,142]

Solute	ΔG		ΔH	ΔS		ΔC_P	Ref.
	M	m.f.		M	m.f.		
Complexation with α-cyclodextrin							
cyclo-hexane	− 15.1	− 25.1	− 19.2	− 13.8	19.8	− 303	141
n-heptane	− 21.6	− 31.6	− 21.3	1.0	34.6	− 492	141
benzene	− 8.6	− 18.6	− 13.1	− 15.1	18.5	− 272	142
Complexation with β-cyclodextrin							
n-heptane	− 19.7	− 29.7	− 0.6	64	97.6	− 305	141
cyclo-hexane	− 21.4	− 31.4	− 5.4	53.6	87.2	− 316	141
benzene	− 12.8	− 22.8	− 1.9	36.6	70.2	− 498	142
Transfer from water to hydrocarbon							
n-heptane		− 28.6	− 0.7		94		140
cyclo-hexane		− 28.2	0.1		95	− 360	130
benzene		− 19.4	− 2.1		58	− 225	130

Slopes of these correlations give the binding increments for a methylene group $\Delta\Delta G = -3.0$ and $\Delta\Delta H = -3.83$ kJ mol^{-1}, not far from the negative values of the increments given above[140] for the transfer of a methylene group from hexane to water. The entropic contribution is small, $T\Delta\Delta S = -0.83$, also in qualitative agreement with that found for hexane to water transfer. However, the comparison of complexation thermodynamics of a whole molecule, such as 1-hexanol, with transfer thermodynamics with n-hexane (Table B 8) shows that complexation is totally 'enthalpy-driven', but the transfer is totally 'entropy-driven' at room temperature. Inclusion of alcohols into the α-cyclodextrin cavity rather than external binding implied in the above discussion was not proved, but there are arguments in favor of intracavity binding.[144]

Similar trends were found for the complexation of alicyclic carboxylic acids, such as admantanecarboxylic acid, bicyclo[2.2.2]octanecarboxylic acid, etc. with cyclodextrins.[145] The increment in ΔG per methylene group for binding to β-cyclodextrin is -3.3 kJ mol^{-1} for neutral acids and -2.5 kJ mol^{-1} for their anions.[145a] The binding is 'enthalpy-driven' and is characterized by large negative ΔC_P values (-440 and -397 J mol^{-1}K^{-1} for adamantanecarboxylate binding to α- and β-cyclodextrins).[145b] We see that the complexation of aliphatic guests with polar substituents shows some clear characteristics expected for hydrophobic interactions: linear dependence of thermodynamic parameters on the number of methylene groups with increments reasonably close to those of water to hydrocarbon solvent transfer, large and negative ΔC_P values, also the inhibitory effect of organic cosolvents.[145b] On the other hand, the complexation is 'enthalpy-driven' while on the basis of the Frank–Evans theory it is expected to be 'entropy-driven'.

Negative complexation enthalpy is common for binding of non-polar molecules of different structures to cyclodextrins (see for example a compilation of thermodynamic data[117]). The situation with cyclophanes is exemplified by the study of complexation

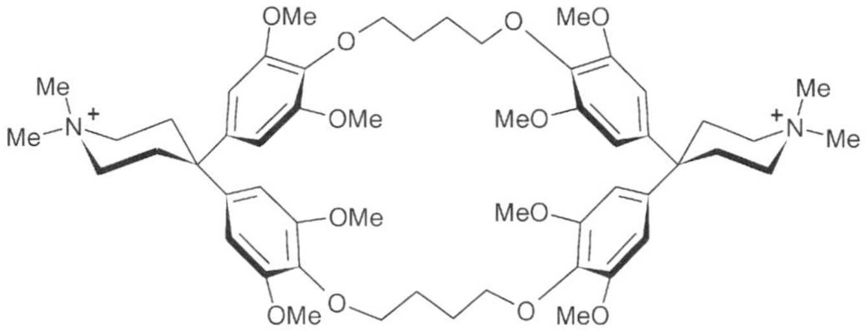

B-39

thermodynamics of a series of 1,4-disubstituted benzenes with macrocycle **B-39** (Table B 10).[146]

Complexation of all guests is 'enthalpically driven' and possess negative ΔC_P values. There is a rough correlation between complex stability and guest hydrophobicity expressed as $\log P$, but electronic and/or steric factors also are important. Stability of the complexes strongly decreases on going to methanol solvent.[146] The exothermic complexation in this and related systems was explained by tight binding of apolar solutes inside the host cavity in contrast to rather loose contacts between the solute and surrounding molecules of a hydrocarbon solvent, transfer to which models the hydrophobic interactions. This tight binding also reduces considerably the mobility of the guest and, therefore, produces a negative entropic contribution which is compensated by the interaction enthalpy (Scheme B 21). Thus, the expected gain in entropy due to dehydration of apolar moieties of guest and host molecules becomes masked by this $\Delta H/\Delta S$ compensations.

Discussed above exothermic complexation, which shows some typical features of hydrophobic interactions often is referred to as 'non-classical hydrophobic interaction'.[1a] However, taking into account large ΔC_P values and always possible contributions from other interaction types and solvation changes, which often lead to compensating changes in ΔH and ΔS, the magnitude of reaction ΔH at a single temperature cannot be considered as good diagnostic criterion. More satisfactory seems the study of structure-binding correlations, solvent effects (discussed in Section C 2) and determination of reaction ΔC_P. For complexes with cyclophanes, cyclodextrins etc. one can also view the enthalpic contribution to the hydrophobic interaction as consequence of the presence of water molecules inside or close to the cavity, which can built up a smaller number of hydrogen bonds to other water molecules;

Table B 10. Selected thermodynamic parameters (kJ mol^{-1}) for the complex formation between cyclophane **B-39** and 1,4-disubstituted benzene guests (1,4-X–C$_6$H$_4$–X) in water at 293 K.[146]

X	ΔG	ΔH	$T\Delta S$	ΔC_P [a]	$\log P$ [b]
COOMe	-28.5	-49.4	-20.9	-251	2.6
MeO	-22.5	-41.8	-19.3	-84	2.2
Me	-22.3	-30.1	-7.8	-84	3.15
CN	-21.9	-43.1	-21.2	-126	1.0
NO$_2$	-21.8	-41.0	-19.2	-167	1.5
OH	-15.4	-43.1	-27.7	-251	0.6

[a] J (mol^{-1} K^{-1}) [b] Logarithms of partition constants of guest molecules between octanol and water.

Table B 11.   Selected data for the complex formation between β-cyclodextrin and benzenes with polar substituents **R** in water, from published data.[142,147]

C_6H_5R-	R =	H	OH	O$^-$	NH$_2$	NH$_3^+$
	log K =	2.23	2.11	1.18	1.75	0.36
	R =	CH$_2$CH(NH$_2$)CONH$_2$				CH$_2$CH(NH$_3^+$)COO$^-$
	log K =	2.03				0.48

upon complexation these water molecules can then gain enthalpy by additional hydrogen bonds. The presence of high enthalpy water in cavities has been invoked, for instance for cyclodextrins complexes.[116]

Hydrophobic interactions often cooperate with polar interactions in host–guest complexes. Several examples were considered in previous chapters. An analysis of binding contributions in such polyfunctional complexes should take into account effects of polar substituents on the hydrophobicity of the hydrocarbon portion of host or guest molecules. As a rule uncharged polar groups give modest if any effects, but charged groups strongly reduce the hydrophobicity. Table B 11 illustrates the effects of polar substituents on the binding of benzene derivatives to β-cyclodextrin.[142,147]

It seems that apolar moieties of guests with neutral polar groups can penetrate into the macrocycle cavity without considerable dehydration of polar groups, while strongly hydrated charged groups do not allow complete penetration of the guest into the cavity. As a consequence of this binding differences, complexation with apolar hosts usually displaces acid–base dissociation equilibria of guests bearing ionogenic groups to their neutral forms, that is it enhances pK$_a$ values of carboxylic acids and phenols and reduces pK$_a$ values of ammonium cations (see Section D 2.2). There are, however, several interesting and still not completely understood exclusions from this general rule, among which the most widely known is the considerable reduction of pK$_a$ of *para*-nitrophenol induced by complexation with cyclodextrins.[143,147,148]

Effects of apolar substituents are different and are related to different geometric fit requirements by solvophobic interactions. All supramolecular complexes discussed in the previous chapters require geometric fit, which must be very close in the case of dispersive interactions, and relatively loose for electrostatic forces. In contrast, solvophobic interactions do in principle not rely on contacts between host and guest molecule, except for the need to replace the solvation of host and guest surface at least to some degree by any kind of intermolecular interactions discussed in the preceding chapter. This leads to another limitation of the lock-and-key principle. The consequences for the design of supramolecular complexes are illustrated with some examples, which also by other diagnostic evidence like measurements of the affinity dependence on the solvent (Section C 2) show a predominantly hydrophobic driving force.

A consequence of solvophobic effects is, that association between molecules can be significantly enhanced also by parts of the host or of the guest which are not in any contact. Thus, the cyclophane **B-37b**($n = 4$) shows binding constants with aromatic substrates, which are eight times larger for biphenyl than for benzene, although only the latter can be in full contact with the cavity.[111] Again, the observed complexation induced NMR shifts are larger for the weaker bound substrate, here due to the fast exchange between the guest parts exposed to the ring current of the cavity.

Another important consequence is that in aqueous media loose binding is preferred

a) β-cyclodextrin   b) γ-cyclodextrin

Figure B 10. Complexation induced shifts by cyclodextrins on the guest molecule ANS (in [ppm]; in D₂O).[149]

over tight fit. This is illustrated by data given in Fig. B10.

In agreement with the intermolecular NOEs (nuclear Overhauser effects) shown in Section E 4 the complexation induced NMR shifts (Fig. B 10) show larger effects, meaning preferential inclusion with the *smaller* guest molecule parts in the cavities; only with the much larger cavity of γ-cyclodextrin the CIS values indicate encapsulation of the naphthalene part. Computer aided molecular modeling shows, however, a better fit of the larger part in the cavity. Similar conclusions emerge from the binding of substrates to the water soluble cryptophane discussed in Chapter H.

The classical theory of the hydrophobic effect and hydrophobic interactions underwent considerable re-examination on the basis of results accumulated in last decades.[150] Studies of the solution thermodynamics of gaseous and liquid hydrocarbons in water over a wide temperature interval of 0–150°C show a rapid increase in transfer ΔH and ΔS with temperature (due to large and temperature-dependent transfer ΔC_P), but a relatively weak temperature dependence of transfer ΔG.[130] If at room temperature positive transfer ΔG of a hydrocarbon molecule from pure liquid to water is principally due to negative ΔS, around 100°C transfer ΔS is close to zero and the hydrophobic effect is purely enthalpic. It seems that the 'entropic nature' of the hydrophobic effect, to which so much attention was paid in the classical theory of hydrophobic hydration, may be less impor-

tant since the low solubility of hydrocarbons in water can be of both entropic and enthalpic nature, depending on temperature. It seems that among all transfer thermodynamic parameters, ΔC_P most directly reflects the unique solvating properties of water.[151]

Application of the scaled particle theory to solvation of non-polar solutes suggests that the low solubility of hydrocarbons in water is mainly due to the small volume of water molecule, that is a large number of water molecules are involved in creation a cavity of sufficient volume to accommodate the solute.[150] In view of this, formation of a structured hydration shell at the surface of non-polar solute may have nothing to do with low solubility of hydrocarbons; in contrary, the 'clathration shell' may even stabilize the solute making hydrocarbons at least slightly soluble in water.[150]

It was pointed out that a distinction must be made between 'bulk' and 'pairwise' hydrophobic interactions.[150,152] The former represent association of a large number of non-polar molecules or moieties, as it occurs in micelle formation or protein folding. Each non-polar molecule in such process changes its completely aqueous surrounding to completely (or nearly completely) non-aqueous, just as it happens during the transfer of non-polar solute from water to organic solvent. The transfer thermodynamics is, therefore, a good model for hydrophobic interactions of this type. The pairwise interaction between two single molecules (see above) can be considered as a partial reverse of the transfer process, but the validity of this assumption is not obvious. Hydrophobic interactions in host–guest complexes probably involve both bulk and pairwise interactions depending on the complex geometry.

Experimental measurements of very small association free energies between single molecules are very difficult. However, the relevant information can be obtained from measurements of solution osmotic coefficients (or other solution properties such as solubility, heat of solution, etc.) as a function of solute concentration. In ideal strongly diluted solution the osmotic pressure Π is a linear function of the density ρ, but at higher solute concentrations a non-ideality appears which is accounted for in terms of virial (polynomial) expansion as

$$\Pi/kT = \rho + B_2\rho^2 + B_3\rho^3 + \cdots$$

The second virial osmotic coefficient B_2 is related to the free energy of pairwise solute–solute interactions; it is negative when the interaction is attractive and positive if it is repulsive. Comparison of B_2 values for cyclohexane and benzene in water and in the gas phase reveal that the gas phase coeffi-

cients in both cases are considerably more negative than those in aqueous solution.[152] This means that water does not promote hydrocarbon pairwise association. This observation rises the question whether pairwise interactions between hydrocabons are solvent-driven or whether the only 'driving force' is the van der Waals interactions (Section B5) between the solutes (in other words, whether this interaction is hydrophobic or lipophilic). In conclusion, we note that after a long time of confidence in hydrophobic interactions dominating for protein folding, there is recent evidence for the primary importance of polar interactions in these processes.[153]

B 7.   EXERCISES AND ANSWERS

B 7.1.   EXERCISES

1. (a) In Scheme B 10 recalculate the ΔG values with the increments for primary and secondary interactions ($P = 7.9$ kJ mol^{-1}, $S = 2.9$ kJ mol^{-1}), also with the simple increment of 5 kJ mol^{-1}; see text.

(b) Calculate the *would-be* ΔG for complex **V** in chloroform, and calculate from that the attenuation effect of the DMSO/CDCl$_3$ (50 : 50) on the increments used here.[154]

2. Devise receptors which could make complete use of all hydrogen bonds of the nucleobases A,T,C,G (see Scheme B 11). Use any spacer between the binding sites, neglecting in the first approximation any strain in the spacers, but try to use donor D and acceptor A with orientations convergent to the nucleobase sites (see Scheme B 11 for help).

3. (a) Design complementary DAAD and ADDA hydrogen bonded systems; use e.g. pyridins, naphthyridins (1,8-diazanaphthalens) and acetamide substituents.

(b) what is the advantage of these heterodimer systems compared to homodimers compounds like ADAD etc.?

4. For the complex with double-stranded DNA shown in Scheme B 12[63] estimate the *would-be* ΔG in chloroform. From this and the reported dissociation constant of 10^{-9} M estimate the dampening effect (in ΔG per hydrogen bond) of water competing for binding in the

minor groove (assuming no other binding contributions except hydrogen bonds and ion pairing) in comparison to the values in chloroform.

5. Download from the website structures of cyclodextrin derivatives and of their complexes. Localize intramolecular and intermolecular hydrogen bonds, *van der Waals* and ion pair contacts, and try to estimate association energies. Make a list of the interactions you find with distances and assignment of the groups.

6. Download from the website structures of several supramolecular complexes; locate the important noncovalent interactions; determine the relevant intermolecular distances between ligand and receptor or metal atoms, where applicable, and try to estimate the total ΔG from increments (make a list, you may also try to sketch a scheme like in Fig A1. Simple viewing can be done with many viewing programs; for other tasks and easier selection/location you need more recent versions of CHIME and/or RASMOL or a molecular mechanics program (see guideline and access in the website for the problems).

7. Logarithms of the stability constants of ternary Cu(**A**)L complexes where **A** = α,α-bipyridyl or phenanthroline are compared in Scheme B 23 to those of the binary complexes (**A** = (H$_2$O)$_n$) for **L** = glycerol 1-phosphate (**L$_1$**) and flavin mononucleotide (**L$_2$**).[155] Discuss the possible reasons of the noticeable increase in log$K$ for **L$_2$**, but not for **L$_1$** on going to ternary complexes with aromatic ligands **A**. Compare the additional stability

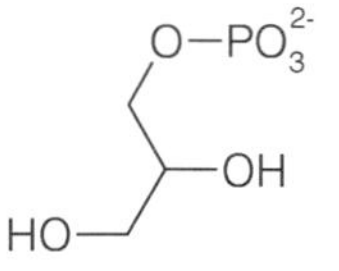

L1 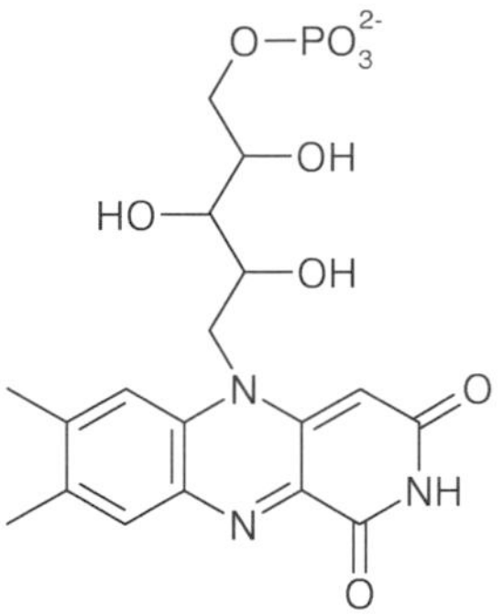   L2

$Cu(A) + L \rightleftharpoons Cu(A)L$

A	L_1	L_2
$(H_2O)_n$	2.83	3.07
Bpy	2.90	3.56
Phen	2.92	3.89

Scheme B 23. Logarithms of the stability constants of binary $(A = H_2O)$ and ternary $(A = \alpha,\alpha\text{-bipyridyl}$ or phenanthroline) $Cu(A)L$ complexes with ligands L_1 and L_2.[155]

(calculate $\Delta\Delta G$) with numbers given in the text for related interactions.

8. Incorporation of Zn^{2+} into 5,10,15,20-tetrakis(4-sulfonatophenyl)porphin (TPPS$_4$) in water is accelerated in the presence of amino acids. Second-order rate constants k_{ZnA} of metalation of TPPS$_4$ by amino acid complexes of Zn^{2+} are given below:

Amino acid	none	Gly	Ala	Val	Phe	Tyr	Trp
k_{ZnA} ($M^{-1}s^{-1}$)	1.59	3.60	12.0	19.9	43.0	52.7	82.2

Propose an explanation of the observed trend and try to analyze the effect of amino acid side chain quantitatively.[156]

B 7.2. ANSWERS

1. (b) With the inclusion of four secondary interactions with the two pyridins, which do not contribute to pri-mary bonding, one would expect with chloroform as solvent $\Delta G = -47.5$ kJ mol^{-1}; the experimental value of -29.4 kJ mol[1] reflects the attenuation effect by DMSO.

2 and 3 See structures on next page and in the Literature.[157]

3. (b) disadvantages of homodimers:

i) less stable: more repulsive secondary interactions;

ii) With heterodimers self-association can compete stronger, depending on contribution of secondary interactions.

4. (a) Calculated from increments (Section B 3) with primary interactions: $12 \times 7.9 = 94.8$ kJ mol^{-1} in CDCl$_3$ with 1 or 2 salt bridges (in water 5–10 kJ mol^{-1}); secondary repulsions diminish these values but are difficult to estimate in view of the often less pronounced vicinity of the A–A and D–D combinations. Experimentally found in water: 51 kJ mol^{-1}. The difference between calculation and experiment is mainly due to the competition of water for Hoogsteen base pairing in the DNA groove; still the resulting increment per hydrogen bond of about 2.5 kJ mol^{-1} in water is rather large, in agreement with lipophilic contributions in the deep groove.

5. See website and model answer for **6** below.

6. See website.

Use Fig. A1 as guideline if you have no experience; you also may consult the references given on the website for the different structures.

A **model answer**, for instance for the complex between streptavidin and biotin (compare Fig. A1):

hydrogen bonds from biotin to the following receptor aminoacids, with distances in [Å]:

ser88–2.68

asp128–2.78

asn23–2.99

tyr43–2.58

ser27–2.77

ser45–3.04

asn49–2.47

In hydrophobic environment like chloroform these seven interactions alone would be worth about 35 kJ mol^{-1}.

7. Increased binding for L_2 reflects intracomplex stacking interactions between flavin and aromatic ligands **A**. Using data for L_1 as correction for changes in the binding free energy through phosphate one obtains from the data for L_2 $\Delta\Delta G$ values -2.4 and -4.2 kJ mol^{-1} for Bpy and Phen, respectively. These numbers can be compared to the data shown in Schemes B 17 and B 18.

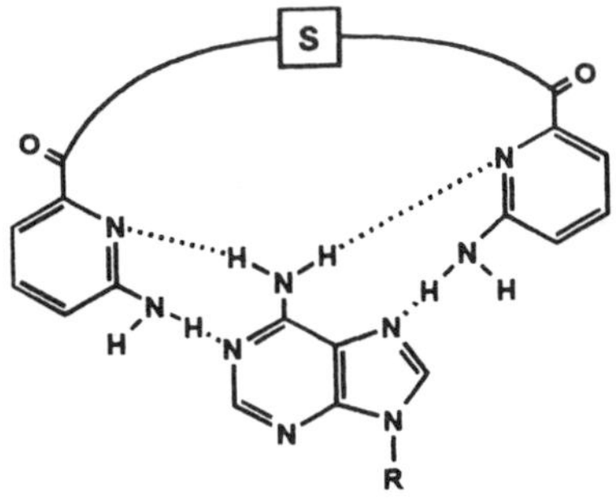

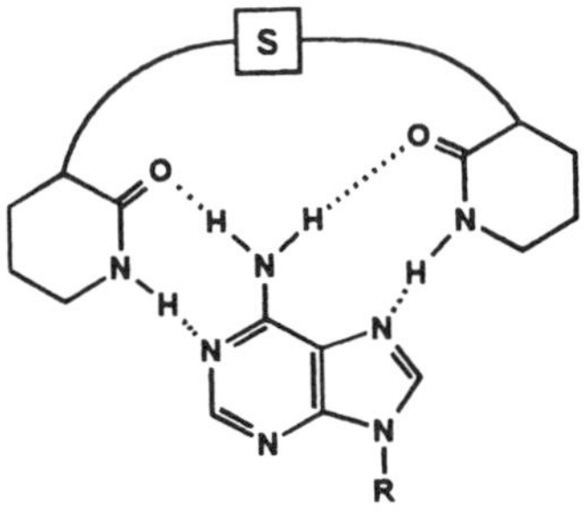

For **A**.

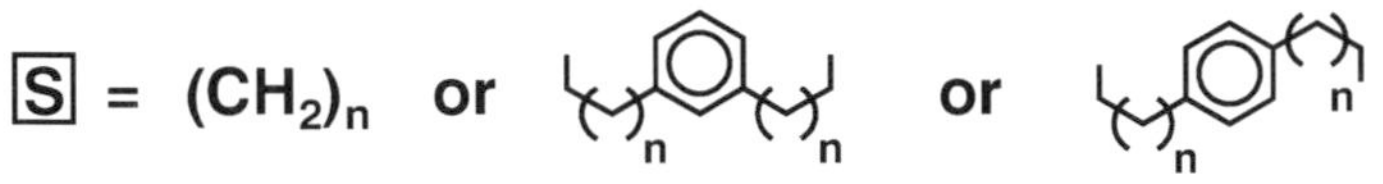

$$\boxed{S} = (CH_2)_n \quad \text{or} \quad \text{...} \quad \text{or} \quad \text{...}$$

For   **T**:

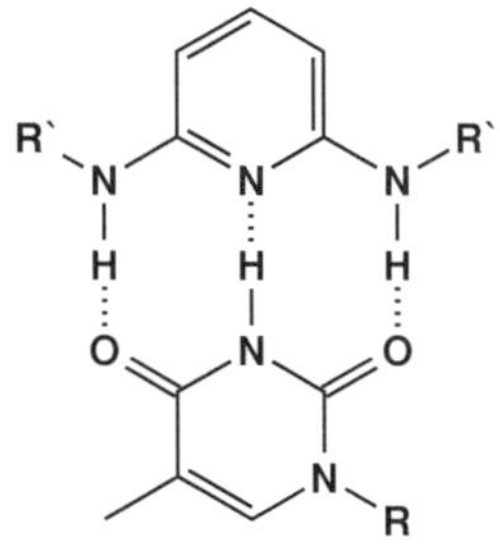

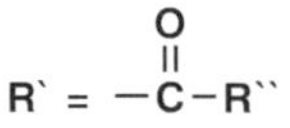

For   **G**:

$$R` = -C-R``$$

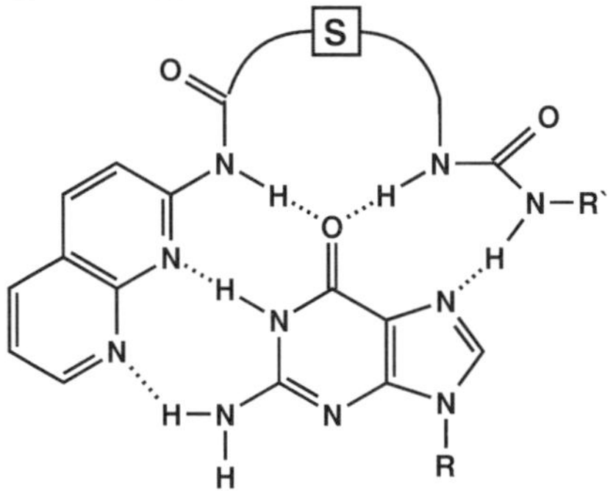

For   **C**:

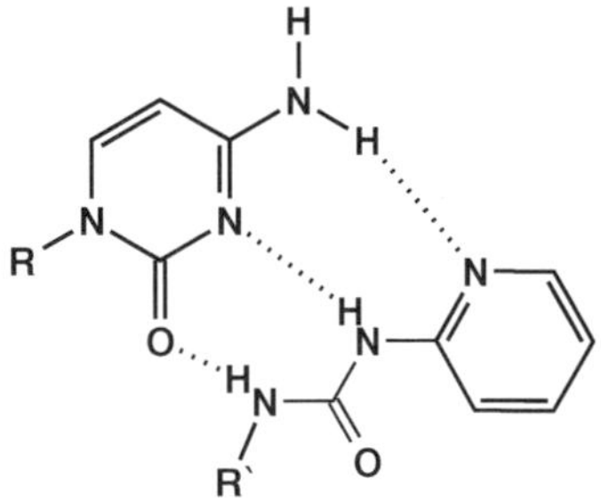

a) DAAD 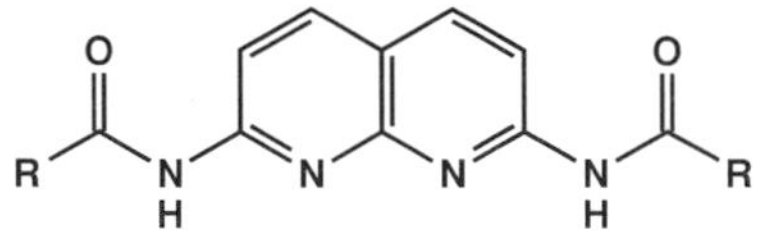   ADDA

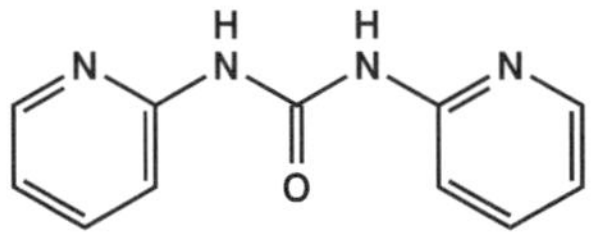

8. Non-polar side groups of amino acids produce only minor effects on their coordination properties. However, they can participate in van der Waals or hydrophobic interactions with the porphyrin plane and accelerate the reaction due to increased affinity of the metal complex to the porphyrin. If so, a correlation should exist between $\log k$ and the surface area A of the non-polar part of amino acids. The van der Waals surface areas of the non-polar parts of amino acids are available[158] and Figure B 11 shows that indeed a good correlation with the slope 0.00943 ± 0.00091 is observed. In terms of free energy this slope is equivalent to $54\,J\,mol^{-1}\,Å^{-2}$), being close to increments observed for hydrophobic interactions (Section B 6) provided that about 20% of the total surface is in the contact with porphyrin. Also good, but more empiric correlation is observed between $\log k$ and transfer free energies from water to ethanol of the amino acids employed (see the original paper).

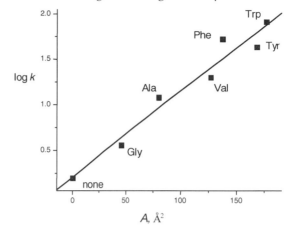

Figure B 11. Correlation of $\log k_{ZnA}$ of metalation of $TPPS_4$ by amino acid complexes of Zn^{2+} with van der Waals surface areas of the non-polar parts of amino acids.

REFERENCES

1 (a) Jencks, W.P. (1989) *Catalysis in Chemistry and Enzymology*, McGraw Hill, New York; (b) Rigby, M.; Smith, E.B.; Wakeham, W.A.; Maitland, G.C. (1986) *The Forces between Molecules*, Clarendon, Oxford, 1986; (c) Israelachvili, J.N. (1985) *Intermolecular and Surface Forces*, Academic Press, London; (d) Huyskens, P.C.; Luck, W.A.; Zeegers-Huyskens, T. (1991) *Intermolecular Forces* Springer, Berlin.

2 See Raevsky, O.A. (1997) *J. Phys. Org. Chem.*, **10**, 405.

3 Abraham, M.H. (1993) *Chem. Soc. Rev.*, **22**, 73.

4 Allen, F.H.; Kennard, O.; Taylor, R. (1983) *Acc. Chem. Res.*, **16**, 146.

5 Jeffrey, G.A.; Saenger, W. (1994) *Hydrogen Bonding in Biological Structures*, Springer Verlag, Berlin; Jeffrey, G.A.; Maluszynska, H. (1982) *Int. J. Biol. Macromol.*, **4**, 173.

6 (a) Warshel, A.; Russle, S.T. Q. (1984) *Rev. Biophys.*, **17**, 283; (b) Tam, S.-C.; Williams, R.J.P. (1985) *Struct. Bonding (Berlin)*, **63**, 103.

7 See Schneider, H.-J.; Schiestel, T.; Zimmermann, P. (1992) *J. Am. Chem. Soc.*, **114**, 7698.

8 (a) Robinson, R.A.; Stokes, R.H. (1968) *Electrolyte Solutions*, Butterworths, London; (b) Bockris, J.O'M.; Reddy, A.K.N. (1970) *Modern Electrochemistry*, Plenum, New York; (c) Kor-

tüm, G. (1965) *Treatise on Electrochemistry*, 2nd edn., Elsevier, Amsterdam.

9 Conway, B.E. (1983) in *Comprehencive Treatise on Electrochemistry*, *Vol.5*, Eds. Conway, B.E.; Bockris, J.O'M.; Yeager, E., Plenum Press, New York, p.111.

10 Yokoyama, H.; Yamatera, H. (1975) *Bull. Chem. Soc. Jpn.*, **48**, 1770.

11 Hasted, J.B. (1972) in Water, *A Comprehensive Treatise*, Franks, F. (Ed.), Plenum Press, New York, vol. **1**, p. 255.

12 (a) Yokoyama, H.; Kon, H. (1991) *J. Phys. Chem.*, **95**, 8956; (b) Yokoyama, H.; Hiramoto, T.; Shinozaki, K. (1994) *Bull. Chem. Soc. Jpn.*, **67**, 2086.

13 (a) Honig, B.; Sharp, K.; Yang, A. (1993) *J. Phys. Chem.*, **97**, 1101; (b) Honig, B.; Nicholls, A. (1995) *Science*, **268**, 1144.

14 Pethybridge, A.D.; Spiers, D.J. (1976) *J. Chem. Soc., Farad. Trans.1*, **9**, 64.

15 (a) Meyer, U. (1976) *Coord. Chem. Rev.*, **21**, 159; (b) Marcus, Y. (1988) *Chem. Rev.*, **88**, 1475.

16 (a) Schneider, H.-J. (1994) *Chem. Soc. Rev.*, **22**, 227; (b) Schneider, H.-J. (1991) *Angew. Chem., Int. Ed. Engl.*, **30**, 1417.

17 Burley, S.K.; Petsko, G.A. (1988) *Adv.Protein Chem.*, **39**, 125.

18 (a) De Stefano, C.; Foti, C.; Giuffré, O. (1996) *J. Solut. Chem.*, **25**, 155; (b) Daniele, P.G.; Prenesti, E.; De Stefano, C.; Sammartano, S. (1995) *J. Solut. Chem.*, **24**, 325; (c) De Stefano, C.; Foti, C.; Giuffré, O.; Sammartano, S. (1996) *Talanta*, **43**, 707; (d) Tam, S.-C.; Williams, R.J.P. (1984) *J. Chem. Soc., Farday Trans. 1*, **80**, 2255.

19 Martell, A.E.; Smith, R.M. Vol. 4 (1976); Vol. 5 (1982) *Critical Stability Constants*, Plenum Press, New York.

20 Hosseini, M.W.; Lehn, J.-M. (1982) *J.Am. Chem.Soc.*, **104**, 3525.

21 (a) Shinoda, S.; Tadokoro, M.; Tsukube, H.; Arakawa, R. (1998) *J.Chem. Soc. Chem. Commun.*, 181; (b) Metzger, A.; Lynch, V.M.; Anslyn, E.V. (1997) *Angew. Chem., Int. Ed. Engl.*, **36**, 862.

22 Bencini, A.; Bianchi, A.; Burguete, M.I.; Dapporto, P.; Doménech, A.; García-España, E.; Luis, S.V.; Paoli, P.; Ramírez, J.A. (1994) *J. Chem. Soc. Perkin Trans 2*, 569.

23 Schneider, H.-J.; Blatter, T.; Palm, B.; Pfingstag, U.; Rüdiger, V.; Theis, I. (1992) *J.Am. Chem. Soc.*, **114**, 7704.

24 (a) Stewart, K.D.; Gray, T.A. (1992) *J. Phys. Org. Chem.*, **5**, 461; (b) Schneider, H.-J.; Blatter, T. (1992) *Angew. Chem., Int. Ed. Engl.*, **31**, 1207; (c) Palm, B.; Schneider, H.-J., unpublished results.

25 Nation, D.A.; Reibenspies, J.; Martell, A.E. (1996) *Inorg. Chem.*, **35**, 4597.

26 Saigal, S.; Pranata, J. (1997) *Bioorg.Chem.*, **25**, 11.

27 Kim, D.H.; Park, J. (1996) *Bioorg. Med. Chem. Lett.*, **6**, 2967 (and refs.).

28 Barthel, J.; Gores, H.-J.; Kraml, L. (1996) *J. Phys. Chem.*, **100**, 1283.

29 Kato, Y.; Conn, M.M.; Rebek, Jr., J. (1994) *J. Am. Chem. Soc.*, **116**, 3279.

30 Diederich, F. (1991) *Cyclophanes*, Royal Society of Chemistry, Cambridge.

31 Hinzen, B.; Seiler, P.; Diedrich, F. (1996) *Helv. Chim. Acta.*, **79**, 942.

32 Eliseev, A.V.; Schneider, H.-J. (1994) *J. Am. Chem. Soc.*, **116**, 6081.

33 Schneider, H.-J.; Güttes, D.; Schneider, U. (1988) *J. Am. Chem. Soc.*, **110**, 6449.

34 Dill, K.A. (1990) *Biochemistry*, **29**, 7133.

35 Xu, D.; Lin, S.L.; Nussinov, R. (1997) *J. Mol. Biol.*, **265**, 68.

36 Gao, J.; Mammen, M.; Whitesides, G.M. (1996) *Science*, **272**, 535.

37 Gelb, R.I.; Schwartz, L.M.; Radeos, M.; Laufer, D.A. (1983) *J. Phys. Chem.*, **87**, 3349.

38 (a) *The Hydrogen Bond - Recent Developments in Theory and Experiments, Vol. I-III*, Schuster, P.; Ed., (1976) North-Holland, Amsterdam; (b) Jeffrey, G.A.; Saenger, W. (1994) *Hydrogen Bonding in Biological Structures*; Springer Verlag, Berlin; (c) Jeffrey, G.A. (1997) *An Introduction to Hydrogen Bonding*, Oxford University Press, New York and Oxford; (d) Hibbert, F.; Emsley, J. (1990) *Adv. Phys. Org. Chem.*, **26**, 255; (e) Perrin, C.L.; Nielson, J.B. (1997) *Annu. Rev. Phys. Chem.*, **48**, 511.

39 Pauling, L. C. (1960) *The Nature of the Chemical Bond*, 3rd edn., Cornell University Press, Ithaca.

40 (a) Vanquickenborne, L.G. (1991) in : Huyskens, P.C.; Luck, W.A.; Zeegers-Huyskens, T., Eds. *Intermolecular Forces*, Springer, Berlin, p. 31 ff; (b) Umeyama, H.; Morokuma, K. (1977) *J. Am. Chem. Soc.*, **99**, 1316; (c): for liquid formamide with an average of 2.5 hydrogen bonds $27\,kJ\,mol^{-1}$ free association energy were calculated with only 3% non-Coulombic contribution: Jorgensen, W.L.; Swenson, C.J. (1985) *J. Am. Chem. Soc.*, **107**, 569.

41 (a) Allen, F.H.; Kennard, O.; Taylor, R. (1983) *Acc. Chem. Res.*, **16**, 146; (b) Taylor, R.; Kennard, O. (1984) Acc. Chem. Res., **17**, 320.

42 Adalsteinsson, H.; Maulitz, A.H.; Bruice, T.C. (1996) *J. Am. Chem. Soc.*, **118**, 7689; a similar soft angular dependence was observed with AM1 calculations: Novoa, J.J.; Whangbo, M.-H. (1991) *ibid.* **113**, 9017.

43 Boobeyer, D.N.; Goodford, P.J.; McWinnie, P.M.; Wade, R.C. (1989) *J.Med.Chem.*, **32**, 1083.

44 Raevsky, O.A. (1990) *Russ. Chem. Rev. (Engl. Translation)*, **59**, 219.

45 Abraham, M.H. (1993) *Chem. Soc. Rev.*, **22**, 73.

46 Gutmann, V. (1978) *The Donor–Acceptor Approach to Molecular Interactions*, Plenum Press, New York.

47 Drago, R.S. (1973) in: *Structure and Bonding, Vol. 15*, Dunitz, J.D., Henmerich, P., Ibers, J.A., Jorgensen, C.K., Neilands, J.B, Reinen, D., Williams, R.J.P., Eds, Springer, Heidelberg, 73.

48 Shan, S.; Loh, S.; Herschlag, D. (1996) *Science*, **272**, 97.

49 Shan, S.; Herschlag, D. (1996) *J. Am. Chem. Soc.*, **118**, 5515.

50 (a) Kato, Y.; Toledo, L.M.; Rebek, J. (1996) *J. Am. Chem. Soc.*, **118**, 8575, and references cited therein; (b) Frey, P.A.; Whitt, S.A.; Tobin, J.B. (1994) *Science*, **264**, 1927; (c) Tobin, J.B.; Whit,

S.A.; Cassidy, C.S.; Frey, P.A. (1995) *Biochemistry*, **34**, 6919.

51 Ash, E.L.; Sudmeier, J.L.; De Fabo, E.C.; Bachovchin, W.W. (1997) *Science*, **278**, 1128; Scheiner, S.; Kar, T. (1995) *J. Am. Chem. Soc.*, **117**, 6970.

52 Cleland, W.W.; Kreevoy, M.M. (1994) *Science*, **264**, 1887.

53 Hirst, S.C.; Tecilla, P.; Geib, S.J.; Fan, E.; Hamilton, A.D. (1992) *Isr. J. Chem.*, **32**, 105.

54 Hannon, C.L.; Bell, D.A.; Kelly-Rowley, A.M.; Cabell, L.A.; Anslyn, E.V. (1997) *J. Phys. Org. Chem.*, **10**, 396.

55 Solov'ev, V.P.; Kazachenko, V.P.;. Raevsky, O.A. ; Schneider, H.-J.; Rüdiger, V. (1998) *Eur. J. Org. Chem.*, (at press).

56 (a) Fersht, A.R. Shi, J.-P.; Knill-Jones, J.; Lowe, D.M.; Wilkionson, A.J.; Blow, D.M.; Brick, R.; Carter, P.; Wyne, M.M.Y.; Winter, G. (1985) *Nature*, **314**, 235; (b) Williams, D.H. (1993) *Proc. Nat. Acad. Sci. USA*, **90**, 1172 and references cited therein.

57 Jorgensen, W.L. (1989) *Acc. Chem. Res.*, **22**, 184.

58 (a) Zimmerman, S.C. (1993) *Top.Curr.Chem.*, **165**, 71; (b) Hamilton, A.D. (1991) *Adv. Supramol. Chem.*, **1**, 1; (c) Rebek, J., Jr., (1990) *Acc. Chem. Res.*, **23**, 399.

59 Schneider, H.-J.; Juneja, R.K.; Simova, S. (1989) *Chem. Ber.*, **112**, 1211.

60 Pranata, J.; Wierschke, S.G.; Jorgensen, W.L. (1991) *J. Am. Chem. Soc.*, **113**, 2810.

61 Sartorius, J.; Schneider, H.-J. (1996) *Chemistry Eur. J.*, **2**, 1446.

62 Thong, N.T.; Helene, C. (1993) *Angew. Chem., Int. Ed. Engl.*, **32**, 697.

63 White, S.; Szewczyk, J. W.; Turner, J.M.; Baird, E. E.;Dervan, P. B. (1998) *Nature (London)*, **391**, 468.

64 Lemieux, R.U. (1996) *Acc. Chem. Res.*, **29**, 373.

65 (a) Das, G.; Hamilton, A.D., (1994) *J. Am. Chem. Soc.*, **116**, 11139; (b) Coterón, J.M.; Hacket, F.; Schneider, H.-J.; (1996) *J. Org. Chem.*, **61**, 1429.

66 Fujimoto, Takako; Shimizu, Chikao; Hayashida, Osamu; Aoyama, Yasuhiro, (1997) *J. Am. Chem. Soc.*, **119**, 6676; Aoyama, Yasuhiro (1993) *Trends Anal. Chem.*, **12(1)**, 23.

67 Desiraju, G.R. (1996) *Acc. Chem. Res.*, **29**, 441.

68 Guo, H.; Karplus, M. (1994) *J. Phys. Chem.*, **98**, 7104.

69 Hughes, M.P.; Smith, B.D. (1997) *J. Org. Chem.*, **62**, 4492.

70 (a) Bowers, M.T. Ed. Vol., 1+2, (1979), Vol 3, (1984) *Gas Phase Ion Chemistry*, Academic Press, New York; (b) Adams, N.G.; Babcock, L.M., Eds.; Vol.1, (1992); Vol. 2, (1996) *Advances in Gas Phase Ion Chemistry*, JAI Press, Greenwich, CT.

71 Taft, R.W.; Anvia, F.; Gal, J.-F.; Walsh, S.; Capon. M.; Holmes, N.C.; Hosn, K.; Oloumi, G.; Cvasamwala, R.; Yazdani, S. (1990) *Pure Appl. Chem.*, **62**, 17.

72 Meot-Ner, M.; Deakyne, C.A. (1985) *J. Am. Chem. Soc.*, **107**, 469.

73 Sunner, J.; Nishizawa, K.; Kebarle, P. (1981) *J. Phys. Chem.*, **85**, 1814.

74 Buschman, H.-J.; Cleve, E.; Schollmeyer, E. (1992) *Inorg. Chim. Acta*, **193**, 93.

75 Mock, W.L.; Shin, N.-Y. (1986) *J. Org. Chem.*, **51**, 4440.

76 (a) Ma, J.C.; Dougherty, D.A. (1997) *Chem. Rev.*, **97**, 1303; (b) Dougherty, D.A.; Stauffer, D.A. (1990) *Science*, **250**, 1558; (c) Dougherty, D.A. (1996) *Science*, **271**, 163.

77 (a) Pullman, A.; Berthier, G.; Savinelli, R. (1997) *J. Comp. Chem.*, **18**, 2012; (b) Kim, K.K.; Lee, J.Y.; Lee, S.J.; Ha, T.-K.; Kim. D.H. (1994) *J. Am. Chem. Soc.*, **116**, 7399.

78 Inokuchi, F.; Miyahara, Y.; Inazu, T.; Shinkai, S. (1995) *Angew. Chem., Int. Ed. Engl.*, **34**, 1364; Review: Lhotak, P.; Shinkai, S. (1997) *J. Phys. Org. Chem.*, **10**, 273.

79 Arneke, F.; Böhmer, V.; Cacciapaglia,R.; DallaCort, A.; Mandolini,L. (1997) *Tetrahedron*, **53**, 4901.

80 Dhaenens, M.; Lacombe, L.; Lehn., J.-M. Vigneron, J.P. (1984) *J. Chem. Soc., Chem. Commun.*, 1097.

81 see Ordentlich, A.; Barak, D.; Cronman, C.; Ariel, N.; Segall, Y.; Velan, B.; Shafferman, A. (1995) *J. Biol. Chem.*, **270**, 2082 and references cited therein.

82 Stauffer D.A.; Karlin, A. (1994) *Biochemistry*, **33**, 6840.

83 (a) Schneider, H.-J. (1991) *Angew. Chem., Int. Ed. Engl.*, **30**, 1417; (b) Schneider, H.-J.; Werner, F.; Blatter, T. (1993) *J. Phys. Org. Chem.*, **6**, 590; (c) Schneider, H.-J.; Schiestel, T.; Zimmermann, P. (1992) *J. Am. Chem. Soc.*, **114**, 7698.

84 Schwabacher, A.W.; Zhang. S.; Davy, W. (1993) *J. Am. Chem. Soc.*, **115**, 6995.

85 Ben-Naim, A. (1980) *Hydrophobic Interactions* Plenum Press, New York.

86 Neusser, H.J.; Krause, H. (1994) *Chem. Rev.*, **94**, 1829.

87 Hobza, P.; Spirko, V.; Selzle, H.L.; Schlag, E.W. (1998) *J. Phys. Chem. A*, **102**, 2501.

88 Newcomb, L.F; Haque, T.S.; Gellman, S.H. (1995) *J. Am. Chem. Soc.*, **117**, 6509.

89 Nishio, M.; Hirota, M.; Umezawa, Y. (1998) *The CH-? Interaction*, Wiley, New York.

90 Karatsu, M.; Suezawa, H.; Abe, K.; Hirota, M.; Nishio, M.; Osawa, E., (1983) *Tetrahedron*, **39**, 3091.

91 Paliwal, S.; Geib, S.; Wilcox, C.S. (1994) *J. Am. Chem. Soc.*, **116**, 4497.

92 Ren, T.; Jin, Y.; Kim, K.S.; Kim, D.H. (1997) *J. Biomol. Struct. Dynamics*, **15**, 401.

93 Cloninger, M.J.; Whitlock, H.W. (1998) *J. Org. Chem.*, **63**, 6153.

94 (a) Linse, P. (1993) *J Am. Chem. Soc.*, **115**, 8793; (b) Jorgensen, W.L.; Severance, D.L. (1990) *ibid.* **112**, 4768; (c) Chipot, C.; Jaffe, R.; Maigret, B.; Pearlman, D.A.; Kollman, P.A (1996) *ibid.* **118**, 11217 and references cited therein

95 Laatikainen, R.; Ratilainen, J.; Sebastian, R.; Santa, H. (1995) *J. Am. Chem. Soc.*, **117**, 11006.

96 Cozzi, F.; Ponzini, F.; Annunziata, R-; Cinquini, M.; Siegel, J.S. (1995) *Angew. Chem., Int. Ed. Engl.*, **34**, 1019; and references cited therein.

97 Rotello, V. M.; Viani, E. A.; Deslongchamps, G.; Murray, B. A.;Rebek, J, Jr. (1993) *J. Am. Chem. Soc.*, **115(2)**, 797; Rebek, J., Jr. (1990) *Angew. Chem.*, **102**, 261.

98 Hamilton, A. D.; Van Engen, D. (1987) *J. Am. Chem. Soc.*, **109**, 5053.

99 Zimmerman, S. C.; Wu, W.; Zeng, Z. (1991) *J. Am. Chem. Soc.*, **113**, 196; Zimmerman, S. C. (1993) *Top. Curr. Chem.*, **165**, 71.

100 Constant, J. F.; Fahy, J.; Lhomme, J.; Anderson, J.E. (1987) *Tetrahedron Lett.*, **28**, 1777.

101 Schneider, H.-J.; Wang, M. (1994) *J. Org. Chem.*, **59**, 7464.

102 Schweitzer, B.A.; Kool, E.T. (1995) *J. Am. Chem. Soc.*, **117**, 1863.

103 Sano, T.; Cantor, C.R. (1995) *Proc. Nat. Acad. Sci. USA*, **92**, 3180.

104 (a) Wilson, W.D. (1989) in *Nucleic Acids in Chemistry and Biology*, Blackburn, M.; Gait, M. Eds.; IRL Press Oxford, Chapter 8; (b) Wilson, W.D. (1998) in *DNA and Aspects of Molecular Biology*, Kool, E. Ed. Vol. 7 of *Comprehensive Natural Products Chemistry* (D. Barton and K. Nakanishi, Eds.).

105 Long, E.C.; Barton, J.K. (1990) *Acc. Chem. Res.*, **23**, 271.

106 (a) Strekowski, L.; Mokrosz, J.L.; Wilson, W.D.; Mokrosz, M.J.; Strekowski, V. (1992) *Biochemistry*, **31**, 10802; (b) Schneider, H.-J.; Sartorius, J. (1997) *J. Chem. Soc., Perkin Trans.* 2, 2319.

107 (a) Sponer, J.; Lesczynski, J.; Hobza, P. (1996) *J. Phys. Chem.*, **100**, 5590; (b) Sponer, J.; Lesczynski, J.; Hobza, P. (1996) *J. Biomol. Struct. Dynamics*, **14**, 117.

108 Hunter, C.A. (1993) *J. Mol. Biol.*, **230**, 1025.

109 Haq, I.; Ladbury, J.E.; Chowdhry, B.Z.; Jenkins, T.C. (1996) *J. Am. Chem. Soc.*, **118**, 10693; Hopfinger, A.J.; Cardozo, M.G.; Kawakami, Y. (1995) *J. Chem. Soc., Farad Trans.*, **91**, 2315 and references cited therein.

110 Diederich, F. (1988) *Angew. Chem. Int. Ed. Engl.*, **27**, 362; Diederich, F. (1991) *Cyclophanes*, Royal Society of Chemistry, Cambridge.

111 Schneider, H.-J.; Kramer, R.; Theis, I.; Zhou; M.-Q. (1990) *J. Chem. Soc., Chem. Commun.*, 276.

112 Amabilino, D.B.; Stoddart, J.F. (1995) *Chem.-Rev.*, **95**, 2725; Philp, D.; Stoddart, J.F. (1996) *Angew. Chem., Int. Ed. Engl.*, **35**, 1155; Fyfe, M.C.T.; Stoddart, J.F. (1997) *Acc. Chem. Res.*, **30**, 393.

113 Castro, R.; Berardi, M.J.; Cordova, E.; de Olza, M.O.; Kaifer, A.E.; Evanseck, J.E (1996) *J. Am. Chem. Soc.*, **118**, 10257.

114 Schneider, H.-J.; Blatter, T.; Palm, B.; Pfingstag, U.; Rüdiger, V.; Theis, I.: (1992) *J. Am. Chem. Soc.*, **114**, 7704.

115 Review: Szeijtli, J. (1998) *Chem. Rev.*, **98**, 1741 and references cited therein.

116 Review: Connors, K. A. (1997) *Chem. Rev.*, **97**, 1325.

117 Rekharsky, M.V.; Inoue, Y. (1998) *Chem. Rev.*, **98**, 1875.

118 Hamai, S. (1992) *Bull. Chem. Soc. Jpn.*, **65**, 2323.

119 Danil de Namor, A.F; Traboulssi, R.; Lewis, D.F. (1992) *J. Am. Chem. Soc.*, **112**, 8842.

120 Foster, R. (1969) *Organic Charge-Transfer Complexes*, Academic, New York.

121 Mulliken, R.S.; Person, W.B. (1969) *Molecular Complexes- A Lecture and Reprint Volume* Wiley Interscience, New York.

122 Bouab, W.; Esseffar, M.; Abboud, J.-L. M. (1997) *J. Phys. Org. Chem.*, **10**, 343.

123 Gutsche, C.D. (1989) *Calixarenes*, Royal Society of Chemistry, Cambridge; Vicent, J.; Böhmer, V. (Eds) (1990) *Calixarenes* Kluwer, Dordrecht .

124 Williams, R.M.; Verhoeven, J.W. (1992) *Recl. Trav. Chim. Pay-Bas*, **111**, 531.

125 Williams, R.M.; Zwier, J.M.; Verhoeven, J.W. (1994) *J. Am. Chem. Soc.*, **116**, 6965.

126 Atwood, J.L.; Koutsantonis, G.A.; Raston, C.L. (1994) *Nature*, **368**, 229.

127 Haino, T.; Yanase, M.; Fukazawa, Y. (1997) *Angew. Chem., Int. Ed. Engl.*, **36**, 259.

128 Tanford, C. (1980) *The Hydrophobic Effect*, 2nd ed., Wiley, New York.

129 Kauzmann, W. (1959) *Adv. ProteinChem.*, **14**, 1.

130 Privalov, P.L.; Gill, S.J. (1988) *Adv. Protein Chem.*, **39**, 191.

131 (a) Frank, H.S.; Evans, M.W. (1945) *J. Chem. Phys.*, **13**, 507; (b) Franks, F. (1973) in *Water, A Comprehensive Treatise*, Franks, F. ed., Vol.4, Chapter 1, Plenum Press, New York.

132 Zeidler, M.D. (1973) in *Water, A Comprehensive Treatise*, Franks, F. ed., Vol.2, Chapter 10, Plenum Press, New York.

133 Nicholls, A.; Sharp, K.A.; Honig, H. (1991) *Proteins*, **11**, 281.

134 (a) Hermann, R.B. (1971) *J. Phys. Chem.*, **76**, 2754; (b) Chotia, C. (1976) *J. Mol. Biol.*, **105**, 1.

135 Honig, B.; Sharp, K.; Yang, A. (1993) *J. Phys. Chem.*, **97**, 1101.

136 Serrano, L.; Neira, J.-L.; Sancho, J. Fersht, A.R. (1992) *Nature*, **356**, 453.

137 Cohen, J.L.; Connors, K.A. (1970) *J. Pharm. Sci.*, **59**, 1271.

138 Matsui, Y.; Nishioka, T.; Fujita, T. (1985) *Top. Curr. Chem.*, **128**, 61.

139 (a) Hansch, C.; Leo, A. (1979) *Substituent Constants for Correlation Analysis in Chemistry and Biology*, Wiley, New York; (b) Leo, A.; Hansch, C.; Elkins, D. (1971) *Chem. Rev.*, **71**, 575.

140 Abraham, M.H. (1982) *J. Am. Chem. Soc.*, **104**, 2085.

141 Wishnia, A.; Lappi, S.J. (1974) *J. Mol. Biol.*, **82**, 77.

142 Tucker, E.E.; Christian, S.D. (1984) *J. Am. Chem. Soc.*, **106**, 1942.

143 (a) Rekharsky, M.V.; Schwarz, F.P.; Tewari, Y.B.; Goldberg, R.N. (1994) *J. Phys. Chem.*, **98**, 10282; (b) Spencer, J.N.; DeGarmo, J.; Paul, I.M.; He, Q.; Ke, X.; Wu, Z.; Yoder, C.H.; Chen, S.; Mihalick, J.E. (1995) *J. Solution Chem.*, **24**, 601 and references therein.

144 Tee, O.S.; Gadosy, T.A.; Giorgi, J.B. (1996) *Can. J. Chem.*, **74**, 736.

145 (a) Eftink, M.R.; Andy, M.L.; Bystrom, K.; Perlmutter, H.D.; Kristol, D.S. (1989) *J. Am. Chem. Soc.*, **111**, 6765; (b) Harrison, J.C.; Eftink, M.R. (1982) *Biopolymers*, **21**, 1153.

146 Smithrud, D.B.; Wyman, T.B.; Diederich, F. (1991) *J. Am. Chem. Soc.*, **113**, 5420.

147 (a) Bertrand, G.L.; Faulkner, J.R.; Han, S.M.; Armstrong, D.W. (1989) *J. Phys. Chem.*, **93**, 6863; (b) Buvári, A.; Barcza, L. (1988) *J. Chem. Soc.,Perkin Trans.2*, 543.

148 Bergeron, R.J. (1984) in *Inclusion Compounds*, Eds. Atwood, J.L.; Davies, J.E.; MacNicol, D.D., Academic Press, London, Vol. 3, p.391.

149 Schneider, H.-J.; Blatter, T.; Simova. S. (1991) *J. Am. Chem. Soc.*, **113**, 1996.

150 Blokzijl, W.; Engberts, J.B.F.N. (1993) *Angew. Chem., Int. Ed. Engl.*, **32**, 1545.

151 Muller, N. (1990) *Acc. Chem. Res.*, **23**, 23.

152 Wood, R.H.; Thompson, P.T. (1990) *Proc. Natl. Acad. Sci. USA*, **87**, 946.

153 Ben-Naim, A. (1990) *J. Phys. Chem.*, **94**, 6893.

154 Bell, T.W.; Hou, Z. (1997) *Angew. Chem. Int. Ed. Engl.* **36**, 1536.

155 Bastian, M.; Sigel, H. (1997) *Inorg. Chem.*, **36**, 1619.

156 Tabata, M.; Tanaka, M. (1988) *Inorg. Chem.*, **27**, 203.

157 Lüning, U.; Kühl, C. (1998) *Tetrahedron Lett.*, **39**, 5735.

158 Livingstone, J.R.; Spolar, R.S.; Record, M.T. (1991) *Biochemistry*, **30**, 4237.

C1. GENERAL CONSIDERATIONS

The study of medium effects on supramolecular equilibria is important for several reasons: as diagnostic approach for the discrimination of different types of interaction mechanisms, for the search of optimum conditions for binding and/or selectivity, and for finding suitable conditions in case of solubility problems. In addition, a comparison of equilibrium constants of related reactions studied under different conditions often needs a correction for medium effects.

A general approach to the quantification of medium effects on chemical equilibrium is to express them as activity coefficient variation of components. Thus, for a 1:1 host–guest complexation reaction of the type

$$H + G \rightleftharpoons HG$$

one can write an expression for the thermodynamic equilibrium constant K_0; assuming that the activity of each component can be expressed as its concentration multiplied by the activity coefficient (γ) one obtains

$$K_0 = [HG]\gamma_{HG}/([H]\gamma_H[G]\gamma_G) = ([HG]/[H][G])$$
$$\times (\gamma_{HG}/\gamma_H\gamma_G) = K(\gamma_{HG}/\gamma_H\gamma_G)$$

where K is the experimental stoichiometric equilibrium constant. Since by definition K_0 is independent of reaction medium, salt and solvent effects on K can be treated as changes in activity coefficients in accordance with equation

$$K = K_0(\gamma_H\gamma_G/\gamma_{HG}) \qquad \text{(C1-1)}$$

It follows from equation (C1-1) that K_0 equals the stoichiometric equilibrium constant K under conditions when all activity coefficients equal 1, that is in an ideal solution. Since even very much diluted real solutions are not ideal, it is practical to assume unity values for activity coefficients in a suitably chosen reference medium in which the concentration dependence of chemical potentials of all components at least has the same mathematical expression as in an ideal solution. Commonly this is a diluted solution in a given solvent (often water) at zero ionic strength.

Although activity coefficients are known today for many important systems like numerous salt, acid and base solutions, various solvent mixtures, etc., for the practical use of equation (C1-1) one needs to know γ_{HG} which usually cannot be found experimentally. Moreover, experimental measurements of activity coefficients of often very complicated host and guest molecules may be too difficult or even impossible to perform, for example when the activity coefficient of a single ion is needed. Due to these circumstances theoretical equations for activity coefficients are of great importance.

The activity coefficient for a given species in a given medium is related to the change of standard free energy (ΔG_{tr}) upon transfer of this species from a reference medium to the given medium

$$\ln\gamma = \Delta G_{tr}/RT \qquad \text{(C1-2)}$$

The transfer free energy can be calculated from an appropriate theoretical model, e.g. with the Debye–Hückel model for electrolyte solutions, affording the respective theoretical equation for the activity coefficient. Another approach consists of application of empirical linear free energies relationships to ΔG_{tr}, which is a common practice in treating solvent effects (see Section C 2).

Traditionally equation (C 1-1) is employed for treatment of salt effects, and its equivalent with ΔG_{tr} is used for treatment of solvent effects. Combining equations (C 1-1) and (C1-2) one obtains

$$\ln K = \ln K_0 + \{\Delta G_{tr}(H) + \Delta G_{tr}(G) - \Delta G_{tr}(HG)\}/RT \qquad (C\,1\text{-}3)$$

and for complexation free energies

$$\Delta G = \Delta G_0 - \Delta G_{tr}(H) - \Delta G_{tr}(G) + \Delta G_{tr}(HG) \qquad (C\,1\text{-}4)$$

where K_0 is the equilibrium constant and ΔG_0 the complexation free energy in a reference medium. If the gas phase is chosen as a reference medium, equation (C 1-4) coincides with equation (A 1-12), Section A 1, for total complexation free energy (in this case transfer free energies are solvation free energies, and their negative values are desolvation free energies).

In the following sections we shall discuss some currently employed theoretical and empirical expressions for activity coefficients and transfer free energies of different types of species, and their use in interpretation of medium effects in supramolecular systems.

C2.   SOLVENT EFFECTS

The classical description of solvent effects takes into account only electrostatic forces and considers the solvent as an inert continuous medium characterized by the dielectric constant, ε. In terms of this approach the transfer free energy (ΔG_{tr}) of an ion of valence z_i and radius r_i can be evaluated using the Born equation for the solvation free energy which gives

$$\Delta G_{tr} = (z_i^2 e^2/2r_i)(1/\varepsilon - 1/\varepsilon_{ref}) \qquad (C\,2\text{-}1)$$

where ε_{ref} is the dielectric constant of the reference medium. The transfer free energy of

a dipolar molecule is given by the Kirkwood equation

$$\Delta G_{tr} = -(\mu^2/a^3)\{(\varepsilon - 1)/(2\varepsilon + 1)\} \qquad (C\,2\text{-}2)$$

where μ is the dipole moment, a is an effective molecular radius, and the reference medium is vacuum. These equations take into account only one solvent property, namely the dielectric constant, and in accordance with them the complexation free energy should be a linear function of $1/\varepsilon$ for ionic equilibria, or the so-called polarization $P = (\varepsilon - 1)/(2\varepsilon + 1)$ for complexation between dipolar molecules. Although correlations of free energies of ion pairing reactions with $1/\varepsilon$ are sometimes observed (Section B 1) and several examples of correlations of activation free energies with P were reported for reactions of dipolar molecules,[1] generally correlations with the dielectric constant as a single solvent parameter are poor or absent. Other important physical solvent properties are the Hildebrand solubility parameter δ_H and its square δ_H^2, the cohesive energy density related to the energy required to make a cavity in the solvent

$$\delta_H^2 = (\Delta H_{VAP} - RT)/V \qquad (C\,2\text{-}3)$$

where ΔH_{VAP} is the vaporization enthalpy and V is the molar volume of the solvent, as well as the polarizability R which is a function of the solvent refractive index n

$$R = (n^2 - 1)/(2n^2 + 1) \qquad (C\,2\text{-}4)$$

Also the solvent macroscopic surface tension was proposed as a measure of cavity formation energy in the following equation for the solvation free energy (ΔG_{solv}) of a non-ionic solute[2]

$$\Delta G_{solv} \cong a - b\mu_2^2/V_2 + cV_2^{2/3}\gamma_1 + RT\ln kT/V_1 \qquad (C\,2\text{-}5)$$

where μ_2 is the solute dipole moment, V_2 the molar volume, γ_1 the solvent surface tension, V_1 the molar volume, whereas a, b and c are constants which can be estimated.

It has long been realized, however, that the use of macroscopic (bulk) physical solvent properties as parameters for quantitative interpretation of solvent effects on reaction rates and equilibria is generally unsatisfactory. The problem is that solute–solvent interactions involve numerous contributions which range from non-specific electrostatic interactions (ion–ion, ion–dipole, dipole–dipole, induced dipoles, etc.) to specific electron or proton donor–acceptor interactions and solvophobic interactions. In the last decades, however, significant progress in theoretical calculations of solvation free energies has been achieved. This progress owes its success to development of new computational methods for free energy calculations such as free energy perturbation, thermodynamic integrations and slow growth calculations[3] (Section E 7). The approach to the quantification of solvation free energies is the calculation of relative solvation energies of pairs of molecules, one of which is considered as a mutant of another one. The principle is illustrated by calculation of relative solvation free energy of methanol and ethane.[4] The following mutation cycle is considered:

$$
\begin{array}{ccc}
CH_3OH_{(gas)} & \xrightarrow{\Delta G_1} & CH_3CH_{3(gas)} \\
\Delta G_{solv}(CH_3OH) \downarrow & & \downarrow \Delta G_{solv}(CH_3CH_3) \\
CH_3OH_{(solv)} & \xrightarrow[\Delta G_2]{} & CH_3CH_{3(solv)}
\end{array}
$$

Evidently, $\Delta\Delta G = \Delta G_{solv}(CH_3OH) - \Delta G_{solv}(CH_3CH_3) = \Delta G_1 - \Delta G_2$. The values of ΔG_1 and ΔG_2 were calculated by a free energy perturbation method which considers transformation of methanol to ethane in a given medium as mutation that goes through several distinct hybrid molecules CH_3X with properties (bond length CH_3–X and partial charges) intermediate between CH_3–OH and CH_3–CH_3. The $\Delta\Delta G$ value of hydration, calculated with precision ± 0.8 kJ mol^{-1}, was only 0.6 kJ mol^{-1} smaller than the experimental value. Absolute solvation free energies can be calculated by this approach considering a mutation of a molecule to nothing.[3a]

As discussed in Section E 7, another approach is to parametrize force field potentials specifically for interactions in a given solvent, leading to predictions of solvation free energies in water or for instance in chloroform with an error of only 3 kJ mol^{-1}.[5] In a more empirical approach, conceptually related to the treatment of hydrophobic interactions on the basis of equation (B 6-1) discussed in Section B 6, the free energy of a solute (ΔG_S) is given by equation[6,7]

$$\Delta G_S = \Sigma \sigma_k A_k(\{r_k\})$$

where σ_k is a proportionality constant called an atomic surface tension of atom k, A_k is the solvent-accessible or the exposed van der Waals surface area (depending on particular model employed) of atom k, which depends on geometry and atomic radii r_k (Section E 7). Most recent calculations by a new version, parametrized by with a set of 235 neutral solutes predict hydration free energies with a mean error of only 2 kJ mol^{-1}.[7c]

Transfer free energies of neutral species can be determined experimentally from such data as partition coefficients or solubilities. When two immiscible liquids containing a substance X are equilibrated, the ratio of concentrations of X in both liquid phases, $[X]_1$ and $[X]_2$, is given by a partition coefficient P directly related to ΔG_{tr} of X between liquids 1 and 2

$$P = [X]_1/[X]_2 = \exp(-\Delta G_{tr}/RT)$$

The value of P for two miscible liquids can be calculated from partition coefficients of a given substance between these two liquids and a third liquid, which is immiscible with both of them

$$P_{1/2} = P_{1/3}/P_{3/2}$$

Alternatively one can measure the solubilities S of a substance X in two liquids, S_1 and S_2, and calculate P from their ratio $P_{1/2} = S_1/S_2$, because the saturated solutions in each solvent are in equilibrium with the same pure substance.

Table C1.   Transfer free energies (kJ mol^{-1}) of some cations and anions from water as reference solvent.[a,9] Reprinted with permission from *Pure. Appl. Chem.*, Gritzner, G. 1988, **60**, 1743. Copyright 1988, Wiley.

Solvent	Na$^+$	K$^+$	Rb$^+$	Cs$^+$	Ag$^+$	Tl$^+$	Cl$^-$	Br$^-$	I$^-$	SCN$^-$	ClO$_4^-$
water	0	0	0	0	0	0	0	0	0	0	0
acetone	4	2	0	1	8	2	57	43	31		
acetonitrile	14	8	7	5	-22	9	42	32	19	13	4
DMF	-11	-15	-10	-9	-17	-11	46	30	19	17	0
DMSO	-14	-12	-11	-13	-34	-21	40	26	9	9	2
ethanol	9	12	11	8	7	6	21	20	19		
methanol	8	10	10	9	8	4	13	11	7	6	5
PC	15	6	3	12	16	8	38	30	18	10	16

[a] DMF, dimethylformamide; DMSO, dimethylsulfoxide; PC, propylene carbonate.

Solvation of single ions cannot be measured directly. Several extrathermodynamic assumptions were proposed to calculate single ion transfer free energies from experimental transfer functions of neutral electrolytes (see literature reviews[8,9]). All methods employed involve several assumptions, and the results of different methods not always agree satisfactorily. Nevertheless, there are extensive data on ΔG_{tr} for numerous ions and solvents which are generally consistent (see Table C 1).

Analysis of relations between values of ΔG_{tr} for ions of different nature reveals some interesting trends.[9] There is no correlation between ΔG_{tr} for cations and anions, for example, between $\Delta G_{tr}(Cl^-)$ and $\Delta G_{tr}(Na^+)$, which means that solvation of cations and anions depends on different solvent properties. Values of ΔG_{tr} for cations of similar nature, for instance Na$^+$ and K$^+$, show very good mutual correlation, and the same is true for ΔG_{tr} of similar anions, such as Cl$^-$ and Br$^-$. At the same time, correlation between ΔG_{tr} values for Na$^+$ and Ag$^+$ holds only for solvents containing oxygen donor atoms, while solvents with nitrogen or sulfur donor atoms show strong deviations. This is interpreted in terms of the hard–soft principle for solvation free energies, since Ag$^+$ is a typical soft Lewis acid while Na$^+$ is a typical hard acid. Solvent characteristics which showed the best correlation with ΔG_{tr} values are donor and acceptor numbers (see below) for cations (in 'hard' donor solvents) and anions respec-

tively. Correlations of ΔG_{tr} for cations in 'soft' donor solvents need some additional parameters.

The presently most successful quantitative treatment of solvent effects is based on the use of empirical parameters (descriptors) which give a measure of the free energy of one or another type of interaction in the frameworks of linear free energy relationships, termed here as linear solvation energy relationships (LSER). Extensive compilations of solvatochromic parameters like α, β, π^*, $E_T(30)$ and Z, as well as parameters based on solvent-dependent equilibria and rates, can be found in the literature.[10-13] Of special interest with respect to host–guest complexes driven by solvophobic interactions is the empirical solvophobicity parameter S_P, calculated from free energies of transfer of non-polar molecules (hydrocarbons, argon, etc.) from water to other solvents.[12] These free energies for different solutes, but for a given solvent can be linearly correlated to a solute parameter R_T (related to solute size) with the slope M which varies from zero for water to its maximum absolute value of 4.2024 for hexadecane, with the parameter $S_P = 1 - M/4.2024$ as a measure of solvophobic properties of the medium.

In practical applications of the solvent parameters given above (and numerous others) it is assumed that the transfer free energies (ΔG_{tr}) which appear in equations (C 1-3) and (C 1-4) can be expressed as linear combinations of different parameters, each reflecting a contribution of a certain type of intermolecular interaction

$$\log K = \log K_0 + aX + bY + cZ + \cdots$$
$$\Delta G = \Delta G_0 + a'X + b'Y + c'Z + \cdots$$

where X, Y, Z, etc. are solvent parameters and a, b, c, etc. (or a', b', c', etc.) are the coefficients of proportionality, or sensitivity parameters. These indicate the relative contributions of the respective solvent properties, which can be found by a multiparameter regression analysis. Evidently this approach implies that the chosen parameters are as far as possible independent of each other; one finds, however, that many of them are interrelated[10b,11] (α, $E_T(30)$, Z, Y, AN and A; β and DN; π^* and B), while there are no interrelations between α, β and π^*; DN and AN; A and B; S_P correlates with $E_T(30)$ in water–methanol mixtures,[14] but not for a large number of solvents.[12]

The above mentioned contributions to the solvation free energy (see equation (C 2-6)) may be more conveniently expressed in the following terms: (*i*) formation of a cavity in the bulk solvent, (*ii*) polar/dipolar solute–solvent interactions, (*iii*) hydrogen-bonding or electron donor–acceptor solute–solvent interactions, (*iv*) solvophobic effect. A logical combination of parameters involves, therefore, the Hildebrand parameter δ_H^2, which accounts for the energy of separation of solvent molecules in the cavity formation process, a polarity parameter, suitable parameter(s) for hydrogen bond or electron donation/acceptance ability, which can be chosen in accordance with the chemical nature of given solvent and solute molecules, and finally solvophobicity parameter. One of successfully employed multiparameter correlations is the following equation.[11]

$$\log K = \log K_0 + a\alpha + b\beta + s\pi^* + h\delta_H^2 \quad \text{(C 2-6)}$$

Other combinations of solvent parameters are also employed.[10-13] However, the interpretation particularly of multi-parameter LSER encounters the same general problems as the interpretation of other linear free energy relationships. Besides the problem of parameter intercorrelation there is the unresolvable controversy whether they represent a fundamental law or merely reflect empirical interrelations between experimentally observable parameters.[15] Second, the empirical character of solvent parameters makes it difficult to establish which parameter exactly measures at the molecular level. Therefore, although LSER are very much useful for the quantitative description of solvent effects, mechanistic conclusions based on them should be regarded with caution.

In order to use equation (C 2-6) and other similar equations one needs a large set of experimental data obtained in solvents which give maximum possible variations in each independent parameter. Fortunately, however, quite good correlations of equilibrium (and rate) constants are often obtained with a single solvent parameter. This may be due to the fact that the investigated process is dominated by interactions of only one type, and such a conclusion is often drawn. One must check, however, whether the set of solvents employed provides considerable variations of other parameters, and whether other parameters give statistically significantly worse correlations.

Satisfactory single-parameter correlations are found for several host–guest equilibria. Thus, formation free energies of cyclophane (**C-1**)-pyrene complex **C-2**, Scheme C1, obtained in 17 organic solvents and water shows a good linear correlation with $E_T(30)$ (Fig. C 1) according to equation[16]

$$-\Delta G(\text{kcal mol}^{-1}) = 0.25E_T(30) - 7.1$$

This is surprising in view of the expected change of binding mechanisms (large hydrophobic contributions in pure water, essentially dispersive interactions in solvents like carbondisulfide, see Chapter B), and also reflects the interrelation discussed above between solvent parameters, which is particularly seen with $E_T(30)$ values. If one wants

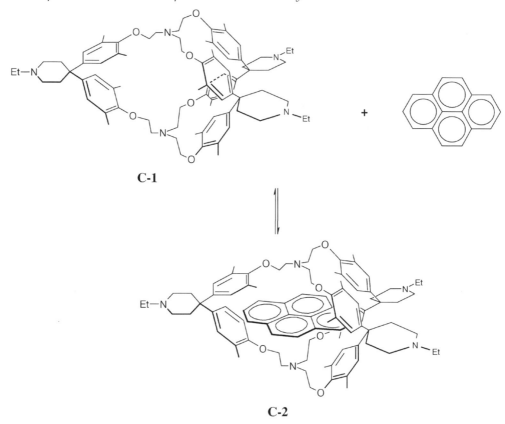

C-1

C-2

Scheme C1. Complexation equilibrium between cyclophane **C-1** and pyrene.[16] Reprinted with permission from *J. Am. Chem. Soc.*, Smithrud *et al.*, 1990, **112**, 339. Copyright 1990 American Chemical Society.

to describe all solvent effects with one single parameter the solvatochromic $E_T(30)$ value often seems to be the best compromise. Interestingly, water does not show any special behaviour which one might expect due to the hydrophobic effect (Section B6). The solvent effects can be interpreted in terms of competition of solvent molecules for the surfaces of the host and the guest (more polarizable solvent molecules are more tightly bound to these surfaces due to larger dispersion forces) as well as with the cavity formation term (release of bound solvent molecules from the interacting surfaces produces more energy for the solvents of higher density with larger cohesive energy).

Calorimetric studies show that in all solvents the complexation in Scheme C1 is predominantly or entirely enthalpy-driven.[17] Figure C2 shows the plots of ΔG, ΔH and $T\Delta S$ vs. $E_T(30)$. Evidently, the same tendency exists for all three thermodynamic parameters. However, if ΔG shows a good linear correlation, the other two parameters show strong mutually compensating deviations from linearity. This is a quite typical situation in LSER.

Solvent effects in aqueous organic mixtures on the binding of guests **C-3**–**C-8** and Et_4N^+ to the charged macrocycles **C-9**, **C-10** and to α-cyclodextrin can be described by the single solvophobicity parameter S_p, but less well with polarity parameters like $E_T(30)$, according to the equation[18]

$$\log K = a\,S_P + \log K_0 \qquad (C\,2\text{-}7)$$

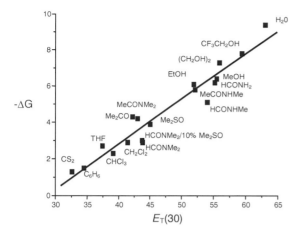

Figure C 1. Dependence of ΔG of formation of complex **C-2** on $E_T(30)$.[16] Reprinted with permission from *J. Am. Chem. Soc.*, Smithrud *et al.*, 1990, **112**, 339. Copyright 1990 American Chemical Society.

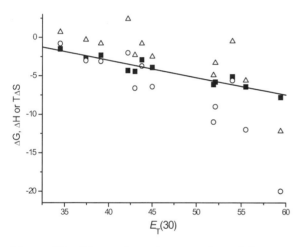

Figure C 2. Thermodynamic parameters of formation of complex **C-2** *vs.* $E_T(30)$ (from published data):[17] squares $= \Delta G$, circles $= \Delta H$, triangles $= T\Delta S$.

Table C 2 shows the values of a and $\log K_0$ for different pairs of hosts and guests.

The greatest slope a, that is, the highest sensitivity of the equilibrium constant to S_P is observed for a neutral host α-cyclodextrin which utilizes hydrophobic interactions as the principal 'driving force' for complexation (Sections B5 and B6). The smallest a value is observed for the pair **C-10**/Et_4N^+ which in

line with other data represents an essentially electrostatically bound ion pair. Combinations of other differently charged host–guest pairs give intermediate values of a. One can conclude, therefore, that the slope of correlation like (C 2-7) can serve as a measure of contribution of hydrophobic interactions to complexation free energy.

C-3

C-4

C-5

C-6

C-7

C-8

C-9

Complex formation between theophylline and methyl *trans*-cinnamate shows in several aqueous–organic mixtures and aqueous salt solutions[19] good linear correlation between

Table C2. Correlations of association constants with solvophobicity parameter in accordance with equation (C2-7).[18] Reprinted with permission from *J. Am. Chem. Soc.*, Schneider *et al.*, 1988, **110**, 6442. Copyright 1988 American Chemical Society.

System	a	$\log K_0$
α-cyclodextrin/**C-8**	7.0	− 3.9
C-9/**C-4**	5.9	− 2.2
C-9/**C-5**	4.0	− 0.5
C-9/**C-7**	3.5	− 1.3
C-9/**C-6**	2.95	1.35
C-9/**C-3**	2.72	2.78
C-10/Et_4N^+	1.25	2.3

C-10

complexation free energy and macroscopic surface tension in accordance with equation (C2-5) for both solvent and salt effects.

Correlations in mixed solvents need some additional comments. Empirical solvent parameters (mostly $E_T(30)$) were measured in mixed solvents and several equations have been proposed to calculate their values for mixtures of different compositions, such as

$$E_T(30) = E_D \ln(c_P/c^* + 1) + E_T^0(30)$$

where c_P is the molar concentration of the more polar component, $E_T^0(30)$ is the $E_T(30)$ value of the pure less polar component, and E_D and c^* are empirical constants.[20a] Other equations can be found in the literature.[20b,c] Also the solvophobicity parameter S_P was calculated for different mixed solvents.[12] However, the phenomenon of preferential solvation of solutes by one of the components of a mixed solvent poses a serious limitation

on their use, because probe molecules employed for determination of the empirical parameters also undergo preferential solvation which is not necessarily of the same type as for another solute. A successful approach to solvation in mixed solvents provides the coordination model,[21] which treats the solvation changes as being due to successive replacement of the molecules of one solvent (S_1) by those of a second solvent (S_2) in the first coordination sphere of the solute (M) in accordance with the equilibria

$$M(S_1)_n + iS_2 \rightleftharpoons M(S_1)_{n-i}(S_2)_i + iS_1 \quad (C2\text{-}8)$$

with respective equilibrium constants β_i written in terms of mole fractions x of both solvents. Then the transfer free energy from pure S_1 ($x_{S1} = 1$, $x_{S2} = 0$) to the mixed solvent is given by equation

$$-\Delta G_{tr} = RT \ln x_{S1} + RT \ln\{1 - \Sigma\beta_i(x_{S2}/x_{S1})\}$$

A recently proposed 'phenomenological model' treats solvation in mixed solvents (mainly aqueous–organic) as a sum of three contributions: solvent–solvent interactions (cavity effect, ΔG_{cav}), solvent–solute interactions (ΔG_{solv}) and solute–solute interactions (independent of solvent composition constant term).[22] The solvation is treated in terms of equation (C2-8) with $n \le 2$, and ΔG_{solv} is calculated as a weighted average of contributions by the three solvated species $M(S_1)_2$, $M(S_1)(S_2)$ and $M(S_2)_2$. The cavity term is expressed as $\Delta G_{cav} = gA\gamma$, where g is an empirical factor, A is the surface area of the cavity and γ is the solvent surface tension. In practical applications of this model to association equilibria several approximations are made in order to reduce the number of adjustable parameters.

Solvent effects on ion–molecular equilibria, such as metal ion complexation by crown ethers and related ligands, are expected to be dominated by solvation of the charged component.[23] In agreement with this, it is found that logarithms of stability constants of potassium complexes of 18-crown-6 measured in 8 organic solvents as well as that of some other ionophore complexes correlate well with ΔG_{tr} of K^+ from water to given solvent (equation C 2-9), although the point for water itself strongly deviates from the correlation.[24]

$$\log K = 5.13 + 0.115 \Delta G_{tr}(K^+) \qquad (C\,2\text{-}9)$$

Although the transfer free energies of non-electrolytes, including macrocyclic ligands, between organic solvents are usually small, they can be significant for the transfer from water to organic solvents, for instance ΔG_{tr} of 18-crown-6 from water to methanol equals 9.5 kJ mol^{-1}.[23]

Complexation energy differences between H_2O and D_2O as solvent are usually difficult to detect, as is the rule with such solvent isotope effects. Small, but reliable solvent isotope effects $K(H)/K(D) \approx 0.8$ were reported for the binding constants of different dye guests to cyclodextrins.[25] They can be significant in protolytic equilibria, such as acid dissociation constants: protic acids are stronger in H_2O than in D_2O and the isotope effect in K_a^H/K_a^D is larger for more weak acids approaching about 5.[26]

As an example of an often applied method to explore the medium polarity *inside* of receptor cavities, we cite studies with the cubic polycycle **C-11**,[27] which has a hydrophobic cavity of up to 9 Å inside diameter and binds, for example the fluorscence dye **C-3** with equilibrium constants K of up to 6×10^6 M^{-1} depending on the applied pH value (see Fig. C 3). The decrease in K at low pH can be explained by a decrease of hydrophobicity due to protonation of the host nitrogen atoms, the decrease above pH 3.7 possibly by the need to retain some cation–π interaction with the guest molecule. As a measure of polarity one used here the intrinsic emission I increase from a fluorescence dye such as **C-3**, which by comparison of I in different solvents one can translate into $E_T(30)$ values; Fig. C 3 shows a monotonous increase of polarity with decreasing pH. A $E_T(30)$ value of 61 (not far

from that of water) is reported[27] for the much more open cyclophane **C-12**.[28] Although increased emission with fluorescence dyes is usually taken as evidence of hydrophobic

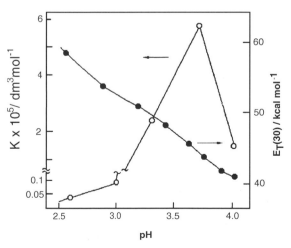

C-11

C-12

Figure C 3. pH dependent variations in association constants K (open circles) and in polarity parameters $E_T(30)$ (filled circles) derived from **C-3** fluorescence measurements with host **C-11**.[27]

pockets also in proteins, due to removal of quenching by water, there is no physical law connecting hydrophobicity with the increase of intensity in fluorescence of such dyes.[29] (Section D 3.2).

C3. SALT EFFECTS

Additions of neutral salts alter activity coefficients of both ionic and non-ionic species. Let us consider first activity coefficients of electrolytes.[30] The well-known Debye–Hückel equation for the activity coefficient of an ion of valence z_i has in water at 298 K the form

$$\log \gamma_i = -0.51 z_i^2 I^{1/2}/(1 + 0.33 a_i I^{1/2}) \quad \text{(C3-1)}$$

where a_i (Å) is the distance of the closest approach between ions, and I is the ionic strength. Estimation of a_i encounters the same problems as in the case of ion pairing (Section B1), but fortunately equations with a constant value of $0.33a_i$ for all ion types equaling for example 1.0 or 1.5 were found to be applicable for many simple electrolytes.[31,32] In very diluted electrolyte solutions ($I < 0.01$ M) the term $0.33a_i I^{1/2}$ in the denominator may be neglected, and equation (C3-1) can be simplified to an equation linear with respect to $I^{1/2}$

$$\log \gamma_i = -0.51 z_i^2 I^{1/2} \quad \text{(C3-2)}$$

which is known as the limiting Debye–Hückel law. Note that equation (C3-2) does not involve a ionic size parameter and that γ_i in this concentration range depends only on ionic charge; however, of course the deviation of γ_i from unity in this range is rather small. Equation (C3-1) in all its modifications works well at ionic strengths below and equal to 0.1 M. In more concentrated electrolyte solutions additional terms must be added. As a rule γ_i passes through a minimum at ionic strengths between 0.2–0.5 M instead of a monotonic decrease to a limiting value, in accordance with equation (C3-1). In the simplest case one can write the equation with

a term linear in ionic strength

$$\log \gamma_i = -0.51 z_i^2 I^{1/2}/(1 + 0.33 a_i I^{1/2}) + \beta I$$
$$\text{(C3-3)}$$

A widely and successfully used version of equation (C3-3) assumes a constant value of $0.33a_i = 1$ and lets β as adjustable parameter specific for a given ion.[33] Several simplified versions with constant values of both $0.33a_i$ and β for all electrolytes were proposed,[34] such as the Davies equation (C3-4).[35]

$$\log \gamma_i = -0.51 z_i^2 (I^{1/2}/(1 + I^{1/2}) - 0.2I)$$
$$\text{(C3-4)}$$

All these equations work at ionic strengths up to 0.5 M. The widely used Pitzer equation[36] for γ_i, applicable without restrictions to the highest electrolyte concentrations is semiempirical and is based on improved analysis of the Debye–Hückel model and on recent Monte Carlo calculations of interionic radial distribution functions. Its discussion is outside the scope of this book, however. Other modern equations for activity coefficients can be found in the literature.[30]

An alternative model proposed for electrolyte activity coefficients is based on *quasi-lattice theories*.[37] These consider the central ion to be surrounded by an arrangement of discrete charges rather than a smoothed-out cloud of charge as in the Debye–Hückel theory. The electrolyte solution is viewed as a crystalline salt, the lattice of which is expanded by penetrating water molecules. Since the Coulomb energy is inversely proportional to the interionic distance which in its turn is inversely proportional to a cubic root of the salt concentration, one can expect $\log \gamma$ to be a linear function of $I^{1/3}$. An often employed equation, which follows from this model has the form

$$\log \gamma_{\pm} = aI^{1/3} + bI \quad \text{(C3-5)}$$

where 'a' is a negative parameter related to the lattice Madelung constant and 'b' is a positive constant. Equation (C3-5) describes a logγ *vs. I* profile remarkably similar to that given by equation (C3-3), in particular, it predicts a minimum in γ at intermediate values of *I*.

Salt effects have been studied most thoroughly in aqueous solutions where the majority of neutral salts are completely dissociated. Main problems in studying the salt effects in organic solvents are generally low solubility of inorganic salts and the high degree of ionic association. The latter makes it virtually impossible to apply the theoretical equations given above which are derived for completely dissociated electrolytes. At the same time, salt effects in non-polar organic solvents, reported mostly for reaction kinetics,[38] are quite impressive and deserve more attention.

As mentioned above, activity coefficients of non-electrolytes in aqueous solutions also depend on addition of neutral salts. The oldest and still popular method of experimental determination of activity coefficients of non-electrolytes is based on solubility measurements in accordance with equation (C3-6)[39]

$$\gamma = S_0/S \qquad (C3\text{-}6)$$

where S_0 and S are the solute solubilities in pure water and in the presence of salt. Added salts often decrease the solubility of organic solutes (salting-out), but in some cases they produce the opposite effect (salting-in).

In the simplest case the relationship between solubility and added salt concentration is given by the Setchenow equation[39]

$$\log(S_0/S) = \log\gamma = k_S[\text{salt}] \qquad (C3\text{-}7)$$

where k_S is the salting-out (negative value) or salting-in (positive value) constant. Table C3 gives k_S values for different salts with benzoic acid as a solute and for KCl with several other organic solutes in water. The magnitude and direction of the salt effect strongly depends on the nature of the salt: salts composed of smaller cations and anions produce higher salting-out effects whereas salts of large cations and anions produce salting-in effects. It was proposed that the observed k_S value can be presented as the sum of intrinsic k_S values for cations and anions,[40] but no generally applicable intrinsic ion salting-out constants seem to have been found. Qualitatively ions may be classified as typically salting-out (cations Mg^{2+}, Li^+, Na^+, K^+, anions SO_4^{2-}, OH^-, Cl^-), typically salting-in (cations R_4N^+, anions ClO_4^-, $RCOO^-$, RSO_3^- (R=aryl)) and intermediate, which produce very small effects (cations Rb^+, Cs^+, H^+, anions Br^-, I^-, NO_3^-). By their absolute values salting-out/in constants typically are not large and activity coefficients of non-electrolytes approach those for monovalent ions at salt concentrations ranging from 0.1 to 1 M.

Table C3.   Salting-out constants for selected salts and solutes in water (from published data[39a,b]).

Salt	Solute	k_S (M)$^{-1}$	Salt	Solute	k_S (M)$^{-1}$
LiCl	Benzoic acid	0.189	KCl	Benzene	0.166
NaCl		0.182		Phenol	0.133
KCl		0.144		Nitrobenzene	0.166
KBr		0.109		Aniline	0.130
KI		0.049		*p*-Phenylenediamine	0.018
KNO$_3$		0.040		4-Nitrophenol	0.050
HCl		0.082		4-Nitroaniline	0.049
NaClO$_4$		0.052		3-Nitroaniline	0.077
NH$_4$I		0.021		H$_2$	0.102
Me$_4$NI		-0.256			
Et$_4$NI		-0.633			
(*n*-Bu)$_4$NI		-1.32			

Theories proposed for the treatment of salting-out/in of non-electrolytes[39,41] are not as elaborated as the theories of electrolyte activity coefficients. The main problem here is the necessity to account for specific ion-molecular interactions, while activity coefficients of ionic species are determined primarily by long-range non-specific electrostatic forces.

A noticeable increase in the formation constants of organic host–guest complexes upon addition of inorganic salts was reported with cyclodextrins as host molecules.[42] As an example Figure C4 shows the salt effects on the formation constants of inclusion complexes of guest **C-13** with γ-cyclodextrin. All plots are approximately linear in accordance with equation (C3-7). The effect is most pronounced with ammonium sulfate as an electrolyte and can be attributed to the salting-out of the guest molecule. Similar, but smaller effects were observed for the complexation of **C-13** with β-cyclodextrin.[42]

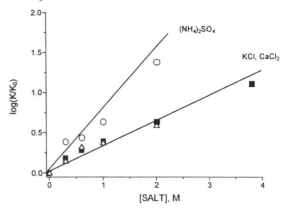

Figure C4. Binding constants of guest **C-13** to γ-cyclodextrin as a function of salt concentration. 42 Circules = $(NH_4)_2SO_4$, squares = KCl, triangles = $CaCl_2$. Reprinted with permission from *J. Am. Chem. Soc.*, Katake *et al.*, 1989, **111**, 7319. Copyright 1989 American Chemical Society.

C-13

Salt effects in ionic reactions, especially with highly charged species, are particularly significant. Application of equation (C3-1) to the experimental formation constant (equation (C1-1)) gives the following general expression

$$\log K = \log K_0 + 0.51 I^{1/2}\{z_H^2/(1 + 0.33a_H I^{1/2})$$
$$+ z_G^2/(1 + 0.33a_G I^{1/2}) - z_{HG}^2/$$
$$(1 + 0.33a_{HG} I^{1/2})\} \qquad (C3\text{-}8)$$

where z_H, z_G and z_{HG} are charges and a_H, a_G and a_{HG} are size parameters of host, guest and complex, respectively. Assuming a constant

value for $0.33a_i$, for example $0.33a_i = 1$, and taking into account that $z_{HG} = z_H + z_G$ one obtains a simplified equation

$$\log K = \log K_0 + 1.02 z_G z_H I^{1/2}/(1 + I^{1/2})$$
$$(C3\text{-}9)$$

An obvious problem in applying these equations to host–guest complexation equilibria is that charged macrocyclic hosts definitely cannot be considered as hard spherical ions for which the Debye–Hückel model is valid. Moreover, highly charged guests, such as protonated polyammonium macrocyles, can create electrostatic potentials considerably larger than kT/e, while the Debye–Hückel model has been derived for potentials much smaller than this value. We saw, however, in Section B2 that association between multiply charged hosts and guests of different structures can be satisfactorily treated in terms of simple Coulombic contributions of pairwise single-site interactions and, moreover, that in terms of the association free energies such hosts and guests often behave as normal ions bearing corresponding full charges. One may expect in such cases at least semiquantitative applicability of equations (C3-8), (C3-9) and

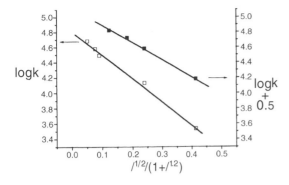

Figure C5. Salt effects of stability constants of complexes between **C-10** and Et$_4$NBr (solid squares, NaOD as added electrolyte) and between **C-9** and **C-6** (open squares, NaCl as added electrolyte) correlated in terms of equation (C3-9).[18] Reprinted with permission from *J. Am. Chem. Soc.*, Schneider *et al.*, 1988, **110**, 6442. Copyright 1988 American Chemical Society.

their forms extended to high salt concentrations.

Formation constants of supramolecular complexes of Et$_4$N$^+$ with anionic host **C-10** and of anionic guest **C-6** with cationic host **C-9** determined in aqueous solutions at different ionic strengths[18] are plotted vs. $I^{1/2}/(1+I^{1/2})$ in Figure C5. Evidently equation (C3-9) holds surprisingly well even with large anisotropic host–guest complexes, for which it was not derived. The slopes of plots equal -3.33 and -2.25 for **C-10**/Et$_4$N$^+$ and **C-9**/**C-6** complexes, respectively, and are not far from the theoretically expected value of -4.08 which corresponds to $z_G = 1$ and $z_H = 4$. In the case of **C-10**/Et$_4$N$^+$ complexes the results at ionic strengths below 0.01 M are available, which allows one to apply also the limiting Debye–Hückel law, equation (C3-2). In this range the slope of the plot of log$K$ *vs.* $I^{1/2}$ equals -4.13 in perfect agreement with the theoretical value. Equation similar to (C3-9) was successfully applied also for the treatment of salt effects on the kinetics of incorporation of Cu^{2+} and Zn^{2+} cations into peripherically charged porphyrins (equation (C3-8) is applicable both for equilibrium and rate

constants with the only difference that in the latter case the activity coefficient of the complex must be substituted by the activity coefficient of the transition state).[43] Just as in the cases discussed above, good linear correlations were observed between log$k$ and $I^{1/2}/(1+I^{1/2})$ with the slopes of expected sign and order of magnitude, but lower by their absolute values than those calculated from the products of the charges of metal ions and porphyrins. Incidentally, even salt effects on kinetics of ion-protein reactions[44] often follow the Debye–Hückel model.

An important aspect of the study of salt effects on ionic reactions is the specificity of the observed effect, that is its dependence on the nature of the electrolyte ions employed for the ionic strength variation. Although some degree of specificity is expected in concentrated electrolyte solutions, at concentrations below about 0.1 M the Debye–Hückel salt effect should be non-specific. Therefore, any study of salt effects should involve tests with different salts. Observation of considerable specificity means usually that the salt effect is due to complexation of reaction components with either cations or anions of salts employed. Even more care is necessary with the use of many organic electrolytes, which are often added as buffers. Lipophilic or other residues in these can lead to *specific* binding to host molecules, with the consequence of lowering association constants with the then competing guest ligands. Thus, the presence of only 0.1 M glycine lowers the apparent binding constant for the association between the macrocycle **C-9** and the anionic fluorescence dye **C-3** by a factor of 70.[18]

If one of the components of a host–guest equilibrium is an ion, but another one is a neutral species ($z_i = 0$) equation (C3-9) predicts the absence of any salt effect (this is the situation with e.g. cation–crown ether complexation reactions). Nevertheless, salt effects are observed in such cases also. For example, stability constants of complexes of Na$^+$ with 18-crown-6 in methanol decrease about two times upon addition of tetrabutylammonium hydroxide (0.005–0.5 M), which was interpreted as being due to an increase in the cation size on passing from free to complexed sodium cation.[45] Considerable positive salt effects on the stability of cation–neutral macrocycle complexes were attributed to ion-pairing between macrocyclic complexes and salt anions. For example, stability constants of some copper(II) complexes with cyclic polythia ethers in water increase about 10 times upon addition of $NaClO_4$ (0.01–1.0 M).[46a] Ion-pairing between aqua-cation Cu^{2+} and anion ClO_4^- is negligible, but it increases strongly for cationic Cu(II) complexes in accordance with the tendency of higher stability of ion pairs beween ions of larger sizes discussed in Section B 2. Other examples of anion effects on stability of supramolecular complex can be found in the literature.[46b,c]

Salt effects in aqueous solutions are often discussed in terms of ion effects on the structure of water. Strong intermolecular hydrogen bonding between water molecules makes liquid water to a fairly ordered medium. The introduction of an ion affects this order in several ways: small ions possessing high charge densities form tightly bound hydration shells from adjacent water molecules and this can be regarded as a structure-making effect (although the structure of hydration water of course is quite different from that of bulk water); larger ions possessing weaker electric fields do not orient considerably water molecules, but their electric fields compete with polar water–water interactions producing a structure-breaking effect (the same effect should be observed in the second hydration shell of small ions); finally, large organic ions like tetraalkyammonium cations possessing weak, strongly delocalized electric fields are expected to behave like organic co-solvents, which increase the order of water molecules adjacent to their surfaces (see Section B6).

There are several parameters which are considered as measure of salt effects on the water structure: the viscosity B coefficient in the Jones–Dole equation ($\eta = \eta_0(1 + Ac^{1/2} + Bc)$), where c is the electrolyte concentration), hydration entropy, salt effects on the rate of self-diffusion and on the dielectric relaxation time of water molecules. Also the structure of water can be defined by means of the avarage number of hydrogen bonds per water molecule, which can be evaluated from the free energy of transfer of the solute form H_2O to D_2O.[47a] There is some disagreement between conclusions from different methods, especially in borderline cases, but undoubtedly a general classification of ions by their effects on the structure of water can be established. Table C4 shows the order of the structure-breaking/structure-making effects for some common ions. Note that the arrangement of anions in the Hofmeister series (Section B2) follows the decreasing structure-breaking effect.

Table C4. Arrangement of ions by their effects on the structure of water[47b] with kind permission from Kluwer Academic Publishers from the *J. Solut. Chem.*, 1994, **23**, 831. (Determination of equilibrium constants by gel chromatography: binding of small molecules to cyclodextrins) Marcus, Y. Table II.

Structure-breaking ions	Borderline ions	Structure-making ions
I^-, ClO_4^-, $Fe(CN)_6^{3-}$, $Fe(CN)_6^{4-}$, >	Na^+, Ag^+, Et_4N^{+a}, Ba^{2+},	$(n-Bu)_4N^+$, UO_2^{2+} >
Br^-, SCN^-, BF_4^- >	F^-, HCO_3^-, $H_2PO_4^-$	Mg^{2+}, Zn^{2+}, La^{3+}, PO_4^{3-} >
K^+, Rb^+, Cs^+, Tl^+, Cl^-, CN^-, N_3^-,		$(n-Pr)_4N^+$, Cu^{2+} >
NO_3^-, S^{2-} >		Ca^{2+}, HPO_4^{2-} >
$MeNH_3^+$; Me_4N^+, BrO_3^-, $P_2O_7^{4-}$ >		Li^+, Sr^{2+}, OH^-, CH_3COO^-,
NH_4^+, $B(OH)_4^-$, SO_4^{2-}, $C_2O_4^{2-}$		Ph_4B^-, CO_3^{2-}, SO_3^{2-}

^{a} may be classified by different criteria as both structure-breaking and structure-making ion.

C4. EXERCISES AND ANSWERS

C4.1. EXERCISES

1. (a) For the determination of salt effects in the complexation of dication **C-14** and dianion **C-15** the association constants K (given below) were measured in water. The problem here is that during titration the concentration of the added ligand **C-15** and therefore the total ionic strength I also changes. One can approximate the average I by using different starting concentrations of **C-14**. Calculate the ionic strength I at 50% of the NMR shift titration.

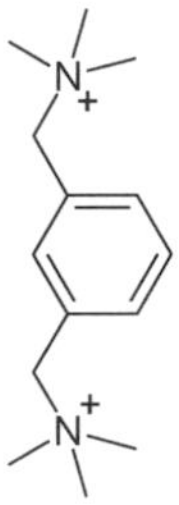

C-14 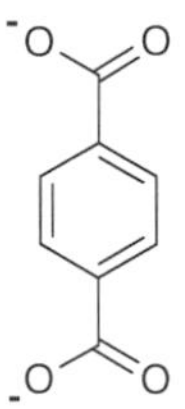   **C-15**

Observed with

$$[\text{C-14}]_0(M) = \quad 0.002 \quad 0.01 \quad 0.02$$
$$K(M^{-1}) = \qquad\quad 185 \qquad 58 \qquad 24$$

(b) for the same complex one observes with $[\text{C-14}]_0 = 0.02$ M in presence of added NaCl the following K values:

$$[\text{NaCl}](M) = \quad 0.083 \quad 0.167 \quad 0.333$$
$$K(M^{-1}) = \qquad\quad 16 \qquad 12.5 \qquad 8.5$$

Calculate the average ionic strength I for each K determination.

Table C5. Association constants between cationic cyclophane **C-9** and the naphthalene-β-carboxylate anion at different ionic strengths.

[NaCl] (M)	$K_A \times 10^{-3} \ (M^{-1})$
–	5.97
2.60×10^{-2}	4.71
5.20×10^{-2}	4.13
7.34×10^{-2}	3.80
1.06×10^{-1}	3.45
2.40×10^{-1}	2.65
4.78×10^{-1}	2.06
8.56×10^{-1}	1.63

Analyze the salt effect on the equilibrium with all six K values in terms of equation (C 3-9). Compare the calculated product of charges of host and guest ions with the theoretically expected.

2. (a) Try similar salt effect correlations for the complex between the azoniacyclophane **C-9** and naphthalene-β-carboxylate, Table C 5.

(b) Interpret the observed slope in terms of equation (C3-9) from this correlation.

(c) Which slopes (sensitivities) would you expect for other equilibria such as:

(i) cyclodextrin + ANS; (ii) tetraphenolate host **C-10** + choline; (iii) **C-9** + naphthalene (order in sequence)

3. Analyse solvent effects on stability constants (log K) of cations with the macrocycle **C–16**, Table C 6.[48]

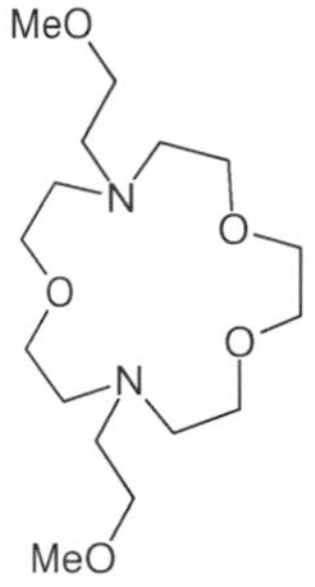

C-16

4. (a) Try to correlate the association constants K between cyclophane **C-9** and anion **C-3** obtained in different aqueous–organic mixtures with various solvent parameters, Table C 7; interpret the results.

(b) Calculate the K value expected in:

(i) ethylene glycol ($S_p = 0.376$); (ii) 50% dimethylformamide (DMF, $S_p = 0.470$); (iii) a hydrocarbon like n-hexane ($S_p = 0.00$).

Table C6. Logarithms of stability constants of macrocycle **C-16** with cations in different solvents.

Solvent	Li^+	Na^+	Ag^+
MeCN	9.13	8.17	7.08
propylene carbonate	7.0	7.1	12.2
MeOH	3.01	4.89	9.86
DMF	2.23	3.5	8.37
H_2O	<2	<2	7.57
pyridine	5.08	6.71	1.8

Table C 7.   Association constants for the complex between **C-9** and **C-3**

no.	Solvent	%	E_T	Y	S_p	$K \times 10^{-5} (M^{-1})$
1	H$_2$O	100	63.1	3.49	1.000	3.54
2	MeOH	10	62.2	3.28	0.942	2.20
3	MeOH	20	61.0	3.02	0.881	1.53
4	MeOH	30	60.0	2.75	0.808	1.50
5	MeOH	50	58.3	1.97	0.631	0.31
6	MeOH	80	56.6	0.38	0.354	0.017
7	EtOH	20	60.0	3.05	0.820	0.55
8	EtOH	40	56.6	2.20	0.585	0.35
9	EtOH	60	55.0	1.12	0.345	0.19
10	dioxane	20	58.6	2.88	0.846	1.30
11	dioxane	40	55.6	1.94	0.646	0.23

C 4.2 ANSWERS

1. The driving force for these complexes between dianions and dications is only ion pairing; this is evident from the large observed slope ($1.02z_G z_H = m = -4.6$, theoretically expected: -4.1) in the Debye–Hückel correlation, equation (C 3-9).[49]

2. The slope of the Debye–Hückel correlation equaiton (C 3-9) is only $m = -1.45$; for a complex stabilized only by all possible ion pairs one would expect $m = -4.1$. Other investigations show that indeed these complexes are dominated by van der Waals interactions.[50]

3. Results for Li and Na cations show similar tendencies, but with Ag$^+$ two solvents, MeCN and Py, show strong negative deviations, probably due to soft charac-

ter of this cation. Data for other solvents show a good correlaiton with solvent donor numbers (DN). There is a rough correlation also between $\log K$ and cation transfer free energies (see Table C1).

4. **(a)** The best linear correlation ($r = 0.935$) is observed with S_p values (equation C 2-7) with the slope a and $\log K_0$ given in Table C 2; with for example E_T parameters the correlation coefficient is 0.898. This is in line with predominant lipophilic/hydrophobic binding mechanisms.

(b) (i) ethylene glycol ($S_p = 0.376$):
$$K = 6.35 \times 10^3 \, M^{-1}$$
(ii) 50% dimethylformamide (DMF, $S_p = 0.470$)
$$K = 1.14 \times 10^4 \, M^{-1}$$
(iii) hydrocarbon like n-hexane ($S_p = 0.00$)
$$K = 6.02 \times 10^2 \, M^{-1}$$

REFERENCES

1 Amis, E.S.; Hunton, J.E. (1973) *Solvent Effects on Chemical Phenomena, Vol. 1*, Acadamic Press. New York.

2 Sinanoglu, O. (1968) in *Molecular Associations in Biology*, Ed. Pullman, B.; Academic Press, New York.

3 (a) Kollman, P. (1993) *Chem.Rev.*, **93**, 2395; (b) Jorgensen, W.L. (1989) *Acc. Chem. Res.*, **22**, 184.

4 Jorgensen, W.; Ravimohan, C. (1985) *J. Chem. Phys.*, **83**, 3050.

5 McDonald, N.A.; Carlsson, H.A.; Jorgensen, W.L. (1997) *J. Phys. Org. Chem.*, **10**, 563.

6 Still, W.C.; Tempczyk, A.; Hawley, R.; Hendrickson, T. (1990) *J. Am. Chem. Soc.*, **112**, 6127.

7 (a) Cramer, C.J.; Truhlar, D.G. (1991) *J. Am. Chem. Soc.*, **113**, 8305; (b) Cramer, C.J.; Truhlar,

D.G. (1992) *Science*, *256*, 213; (c) Hawkins, G.D.; Cramer, C.J.; Truhlar, D.G. (1997) *J. Phys. Chem. B*, **101**, 7147 (and references therein).

8 Marcus, Y. (1985) *Ion Solvation*, Wiley New York, Chichester.

9 Gritzner, G. (1988) *Pure. Appl. Chem.*, **60**, 1743.

10 (a) Reichardt, C. (1988) *Solvents and Solvent Effects in Organic Chemistry*, 2nd ed.; VCH, Weinheim, (b) Reichardt, C. (1994) *Chem. Rev.*, **94**, 2319.

11 Marcus, Y. *Chem.Soc.Rev.*, (1993), **22**, 409.

12 Abraham, M.H.; Grellier, P.L.; McGill, R.A. (1988) *J. Chem. Soc. Perkin Trans.*, 339.

13 Gutmann, V. (1978) *The Donor-Acceptor Approach to Molecular Interactions*, Plenum Press, New York.

14 Schneider, H.-J.; Wang, M. (1994) *J. Org. Chem.*, **59**, 7464.

15 Wold, S.; Sjöström, M. (1986) *Acta Chem. Scand. B*, **40**, 335.

16 Smithrud, D.B.; Diederich, F. (1990) *J. Am. Chem. Soc.*, **112**, 339.

17 Smithrud, D.B.; Wyman, T.B.; Diederich, F. (1991) *J. Am. Chem. Soc.*, **113**, 5420.

18 Schneider, H.-J.; Kramer, R.; Simova, S.; Schneider, U. (1988) *J. Am. Chem. Soc.*, **110**, 6442.

19 Connors, K.A.; Sun, S. (1971) *J. Am. Chem. Soc.*, **93**, 7239.

20 (a) Langhals, H.; (1982) *Angew. Chem. Int. Ed. Engl.*, **21**, 724; (b) Bosch, E.; Rosés, M. (1992) *J. Chem. Soc. Faraday Trans.*, **88**, 3541; (c) Skwierczynski, R.D.; Connors, K.A. (1994) *J. Chem. Soc. Perkin Trans 2*, 467.

21 Waghorne, W.E. (1993) *Chem. Soc. Rev.*, **22**, 285.

22 Connors, K.A.; Khossravi, D. (1993) *J. Solut. Chem.*, **22**, 677; Connors, K. A. (1997) *Chem. Rev.* **97**, 1325, and references therein.

23 Cox, B.G.; Schneider, H. (1992) *Coordination and Transport Properties of Macrocyclic Compounds in Solution*, Elsevier, Amsterdam, Chapter 2.8.

24 Solov'ev, V.P.; Strakhova, N.N.; Raevsky, O.A.; Rüdiger, V.; Schneider, H.-J. (1996) *J. Org. Chem.*, **61**, 5221.

25 Wang, A.S.; Matsui, Y. (1994) *Bull. Chem. Soc. Jpn.*, **67**, 2917.

26 Delgado, R.; Fraústo Da Silva, J.J.R.; Amorim, M.T.S.; Cabral, M.F.; Chaves, S.; Costa, J. (1991) *Analyt. Chim. Acta*, **245**, 271.

27 Murakami, Y.; Kikuchi, J.; Hiseada, Y.; Ohno, T. (1991) in *Frontiers in Supramolecular Chemistry and Photochemistry*, Schneider, H.-J.; Dürr, H., Eds; VCH Weinheim, p. 145 ff.

28 Tabushi, I.; Kuroda, Y.; Kimura, Y. (1976) *Tetrahedron Lett.* 3327.

29 Kosower, E.M. (1976) *Acc. Chem. Res.* **15**, 259.

30 Horvath, A.L. (1985) *Handbook of Aqueous Electrolyte Solutions*, Ellis Horwood Ltd., Chichester.

31 Guntelberg, E. (1926) *Z. Phys. Chem.*, **123**, 199.

32 Scatchard, G. (1959) in *The Structure of Electrolyte Solutions*, Ed. Hamer, W.J., Wiley, New York, p. 9.

33 (a) Guggenheim, E.A. (1935) *Phil. Mag.*, **19**, 588; (b) Guggenheim, E.A.; Turgeon, J.C. (1955) *Trans. Farad. Soc.*, **51**, 747.

34 Sun, M.S.; Harriss, D.K.; Magnuson, V.R. (1980) *Can. J. Chem.*, **58**, 1253.

35 Davies, C.W. (1938) *J. Chem. Soc.*, 2093.

36 (a) Pitzer, K.S. (1973) *J. Phys. Chem.*, **77**, 268; (b) Pitzer, K.S.; Mayorga, G. (1973) *J. Phys. Chem.*, **77**, 2300; (c) Pitzer, K.S. (1977) *Acc. Chem. Res.*, **10**, 371; (d) Pitzer, K.S. in *Activity Coefficients in Electrolyte Solutions*, 2nd. edn., Pitzer, K.S., ed., CRC Press, Boca Raton, 1991, 75.

37 (a) Chapter B1, ref. b, p. 267; (b) J.-F. Chen; Choppin, G.R.; (1995) *J. Solut. Chem*, **24**, 465; (c) Vilariño, T.; Fiol, S.; Armesto, X.L.; Brandariz, I.; Sastre de Vicente, M.E. (1997) *Z. Phys. Chem.*, **199**, 69.

38 Pocker, Y.; Buchholz, R.F. (1970) *J. Am. Chem. Soc.*, **92**, 4033.

39 (a) Long, F.; McDevit, W. (1952) *Chem. Rev.*, **51**, 119; (b) Sergeeva, V.F. (1965) *Russ. Chem. Rev.*, **34**, 309.

40 Gross, P. (1933) *Chem. Rev.*, **13**, 91.

41 Conway, B.E. (1985) *Pure & Appl. Chem.*, **57**, 263.

42 Katake, Y.; Jansen, E.G. (1989) *J. Am. Chem. Soc.*, **111**, 7319.

43 Nwaeme, J.; Hambright, P. (1984) *Inorg. Chem.*, **23**, 1990.

44 Cummins, D.; Gray, H.B. (1977) *J. Am. Chem. Soc.*, **99**, 5158.

45 Smetana, A.J.; Popov, A.I. (1979) *J. Chem. Thermodynamics*, **11**, 1145.

46 (a) Young, I.R.; Ochrymowyez, L.A.; Rorabacher, D.B. (1986) *Inorg. Chem.*, **25**, 2576; (b) Batinichaberel, I.; Crumbliss, A.L. (1995) *Inorg. Chem.*, **34**, 928; (c) He, G.X.; Kurita, M.; Ishi, I.; Wada, F.; Matsuda, T. (1992) *J. Membrane Sci.*, **69**, 61.

47 (a) Marcus, Y.; Ben-Naim, A. (1985) *J. Chem. Phys.*, **83**, 4744; (b) Marcus, Y. (1994) *J. Solut. Chem.*, **23**, 831.

48 Lucas, J.B.; Lincoln, S.F. (1994) *J. Chem. Soc., Dalton Trans.*, 423.

49 Hossain M.A.; Schneider, H.-J. (1999) *Chem. Eur. J.* in print.

50 Schneider, H.-J.; Theis, I. (1992) *J. Org. Chem.* **57**, 3066.

D ENERGETICS OF SUPRAMOLECULAR COMPLEXES: EXPERIMENTAL METHODS

D1. STUDY OF CHEMICAL EQUILIBRIA

The evaluation of association free energies and the respective enthalpic and entropic contributions in solution is of central significance for the functional description of host–guest systems, even if they are to be applied not in homogeneous solution but to aggregates on surfaces, or in films etc. Knowledge of corresponding values for non-covalent interactions is furthermore a decisive prerequisite for the understanding of natural and synthetic supramolecular complexes, and therefore for the design of new hosts or of new ligands with biological activity. Several textbooks and reviews are available on the subject,[1] which are, however, often restricted to specific systems such as UV/vis-spectroscopic[1c,f] or potentiometric[1e] methods, or primarily cover fields like classical coordination chemistry.[1b,d] Our aim here is to give a concise introduction to such methods as are available in most modern research laboratories, and are broadly applicable to systems containing two or more organic molecules. We restrict our discussion first to simple 1:1 complexes; many of these considerations also apply to complexes of different stoichiometry which are dealt with later.

The equilibrium constants discussed below are the stoichiometric constants defined by application of the mass action law to the concentrations of all components in equilibrium under given conditions. This can be done rigorously only if the activity coefficients of all components remain constant. To assure this constancy, which usually is only approximate, one must use sufficiently diluted solutions of the interacting species, as well as maintain constant solvent and salt composition of the reaction medium in addition to constant temperature and pressure.

When an experimental method exists that allows one to determine the concentrations of all components under equilibrium, the calculation of the equilibrium constant is straightforward. As a rule this is impossible, however. Fortunately, if the concentration of at least one component can be determined, or if merely a solution property proportional to the concentration of at least one component can be measured, the equilibrium constant can be calculated from the changes of this property upon variation of component concentrations. The following sections describe how to do this in different cases.

D1.1. GENERAL AND PRACTICAL CONSIDERATIONS WITH 1:1 EQUILIBRIA

Let us consider an equilibrium between host and guest species H and G forming a complex H–G. The most common procedure for determination of the equilibrium constants involves measurements at variable concentration of one component and at fixed concentration of another one. Both host and guest compounds can be either the variable or constant component; usually the variable component is called 'ligand' (L) and the constant (often minor) component 'substrate' (S). For the formal description there is, of course, no difference which one component shall be host and which one guest. In this terminology the simplest 1 : 1 equilibrium is

$$S + L \rightleftharpoons SL \qquad (D1\text{-}1)$$

with the respective equilibrium constant

$$K = [SL]/([S][L]) \qquad (D1\text{-}2)$$

The experiment consists of mixing S and L in known total concentrations $[S]_t$ and $[L]_t$ and subsequent determination of their concentrations when the system is under equilibrium. In this particular case is sufficient to determine the equilibrium concentration of only one component, S, L or SL, because the other two can be calculated from the mass balance equations

$$[S]_t = [S] + [SL] \qquad (D1\text{-}3)$$
$$[L]_t = [L] + [SL] \qquad (D1\text{-}4)$$

D 1.1.1.   Direct determinations with slowly equilibrating complexes

If formation and dissociation rates of the complex SL are slow enough on the NMR scale, the signals of free and complexed species appear in the NMR spectrum separately, and all concentrations in equation (D1-2) can be determined from their intensities.

With very slowly equilibrating, strong complexes, which occur more often with biopolymers, one can use centrifugation or filtration techniques or even simple washing in order to separate complexed from non-associated material, and then measure their quantities. This is the basis of techniques like the radioimmunoassay (RIA)[2a] or enzyme immunoassay (e.g. ELISA),[2b] Section D4.

D 1.1.2.   Spectrometric and other titrations

As mentioned above, the equilibrium constant can be calculated from the concentration dependence of any property proportional to the concentration of one of the components. Such a property (X) can be absorbance,

fluorescence, position of a NMR signal, reaction rate, conductance, etc., and the experiment performed to obtain its concentration dependence is referred to as titration.

The proportionality between X and the concentration of a given substance S is expressed as

$$X = x_S[S] \qquad (D1\text{-}5)$$

where x_S is an intrinsic molar property of S, like the molar absorptivity ε, rate constant k, etc. The observed value of X is considered to be the sum of contributions from all components (the additivity principle). Let us consider first the situation when the observed X value (X_{obs}) is due to the contribution of only one complexation partner S in free and complexed forms, while $x_L = 0$

$$X_{obs} = x_S[S] + x_{SL}[SL] \qquad (D1\text{-}6)$$

D 1.1.2a.   Determination of K with known intrinsic spectroscopic property of the complex

With strong enough complexes one can at sufficiently high concentration of L directly measure the intrinsic property x_{SL} of the complex SL at nearly 100% complexation. The intrinsic value may also be found by extrapolation from measurements with high concentration of L. It is convenient to do this by using the plot of X_{obs} vs. $1/[L]$ and extrapolate it to $1/[L] = 0$. If one now measures the spectroscopic parameter X_{obs} at lower concentrations of L, one obtains directly the SL concentration from x_{SL} and equations (D1-3) and (D1-6)

$$[SL] = (X_{obs} - X_0)/(x_{SL} - x_S) = \Delta X/\Delta x \qquad (D1\text{-}7)$$

where $X_0 = x_S[S]_t$ is the value of X in the absence of L

With the value of x_{SL}, the degree of complexation p can be calculated

$$p(\%) = 100[SL]/[S]_t = 100\Delta X/(\Delta x[S]_t) \qquad (D1\text{-}8)$$

From this and the known total concentrations $[S]_t$ and $[L]_t$ one calculates K

$$K = p/\{(100-p)[L]\} \\ = p/\{(100-p)([L]_t - p[S]_t/100)\} \qquad \text{(D1-9)}$$

The procedure is repeated preferably in several experiments with variable total concentrations, if possible evenly distributed between about 10 and about 90% complexation. If these experiments give the same K value the assumed x_{SL} value and the calculational model of a 1:1 complex are likely correct, in particular if several signals or wavelengths are used (see below).

This approach can also be used to estimate K if solubility limitations and/or small complexation constants make it impossible to determine x_{SL} from titration. The intrinsic value may also be approximated by comparison to chemically related complexes. The procedure is particularly useful for the estimation of lower and upper limits of association constants. For this one inserts conservative, cautiously chosen lower limits of the spectroscopic intrinsic property x_{SL} into equation (D1-8), and can thus obtain upper and lower limits of K.

Instead of performing the explicit calculations one may also use nomograms (Fig. D1), which are also provided as help to set up titration experiments in the desirable concentration ranges (Section D1.1.2c). The nomograms also illustrate that the concentration of the minor partner (typically S) (the signal of which is usually the observed one during the titrations) plays almost no role, in contrast to the ligand used in excess. The simple experiments described here are also recommended before one starts a full titration in order to find suitable concentration ranges.

D1.1.2b. *The general case and the fitting procedures*

Most often the intrinsic property x_{SL} cannot be found so easily. Then one needs to determine two unknowns: K and x_{SL} in the situation

presented by equation (D1-6), which is quite common in such experiments. Often the ligand can be added at concentrations much higher than the substrate, which makes it possible to equate the concentration of L at the equilibrium to its total concentration: $[L] = [L]_t$. With this simplification the degree of complexation p does not depend on the concentration of the minor component S and is given by equation (D1-10), which follows from (D1-2)–(D1-4)

$$p(\%) = 100K[L]_t/(1+K[L]_t) \qquad \text{(D1-10)}$$

The expression for X_{obs} takes the form

$$X_{obs} = (x_S + x_{SL}K[L]_t)[S]_t/(1+K[L]_t) \qquad \text{(D1-11)}$$

It is convenient to introduce the observed value of the intrinsic property of the substrate (x_{obs}) given by

$$X_{obs}/[S]_t = x_{obs} = (x_S + x_{SL}K[L]_t)/(1 + K[L]_t) \qquad \text{(D1-12)}$$

Equation (D1-12) allows to use variable substrate concentration for the titration curve provided the condition $[S]_t \ll [L]_t$ is always fulfilled. If $[S]_t$ is really maintained constant, one can use directly X_{obs}, e.g. the observed absorbance, and equation (D1-11) which can be rewritten in this case as

$$X_{obs} = (x_S[S]_t + x_{SL}[S]_tK[L]_t)/(1 + K[L]_t) \\ = (X_0 + X_\infty K[L]_t)/(1 + K[L]_t) \qquad \text{(D1-13)}$$

where $X_0 = x_S[S]_t$ is the value of X for the substrate alone and $X_\infty = x_{SL}[S]_t$ is the value of X for the complex.

The known value of x_S can be subtracted from x_{obs} and after that equation (D1-12) takes another useful form

$$x_{obs} - x_S = \Delta x_{obs} = \Delta x K[L]_t/(1 + K[L]_t) \qquad \text{(D1-14)}$$

where $\Delta x = x_{SL} - x_S$.

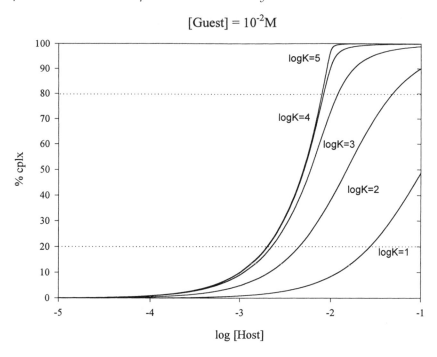

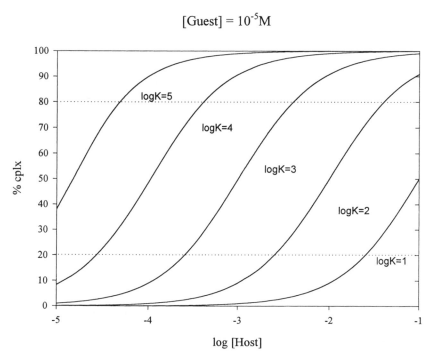

Figure D1. Nomogram for 1:1 association between host and guest; the complexation degree [% cplx] as function of host concentration for different association constants.

Equations (D1-11)–(D1-14) which describe the titration curves (isotherms) are hyperbolic functions of the variable $[L]_t$. They tend to 'saturate' at high ligand concentrations when $K[L]_t \gg 1$ and are approximately linear at low ligand concentrations when $K[L]_t \ll 1$.

In the pre-computer époque the fitting of experimental results to a theoretical equation was most often done by means of a linear regression. To this end equation (D1-14) can be linearized using one of the following three transformations:

Benesi–Hildebrand double-reciprocal plot (known as the Lineweaver–Burk plot when X is the rate of an enzymatic reaction)[3a,4]

$$1/\Delta x_{obs} = 1/\Delta x + (1/K\Delta x)(1/[L]_t) \quad (D1\text{-}15)$$

Scatchard plot (Eadie–Hofstee plot in enzyme kinetics)[3b,4]

$$\Delta x_{obs}/[L]_t = K\Delta x - K\Delta x_{obs} \quad (D1\text{-}16)$$

Scott plot[3c,4]

$$[L]_t/\Delta x_{obs} = [L]_t/\Delta x + 1/(K\Delta x) \quad (D1\text{-}17)$$

The use of linearization methods, including equations (D1-15)–(D1-17), is discussed in details in several biochemical textbooks.[4]

Linear equations, of course, are very simple. However, it should be noted that the fitting of transformed results of the measurements statistically is not equivalent to the fitting of primary results because any transformation changes the statistical weights of points. As a result, the weighting (usually ignored) becomes very important in the use of equations (D1-15)–(D1-17) and the values of K found from the same experimental data by using different linear unweighted fits do not coincide.

Instead of traditional linearization it is preferable to carry out a non-linear least squares curve fitting. This has become an easy matter with many commercially available programs (Section D1.4), in which one only has to insert the original hyperbolic (or any other, see below) equation, and which not only calculate the values of parameters, but

also visualize the results of calculation graphically by fitting curves.

Many available programs provide the possibility to introduce different weights to the data points, if (as it is frequently the case) some points for technical reasons are of lower accuracy or precision. The weight w is defined as a reciprocal function of the variance s^2 (the square of the standard deviation): $w = k/s^2$, where k is a proportionality constant, which cancels out in the final least-squares equations. Usually the independent variable ($[L]_t$ in equations (D1-11)–(D1-14)) is considered to be known exactly and its variance is assumed to be zero. The variance of dependent variable (X_{obs} or x_{obs}) can be calculated from repetitive experiments or estimated from known instrumental errors. If s^2 is set constant for all measurements, its value will cancel out from the least-squares equations and this will be equivalent to unweighted regression. Therefore the weighting is meaningful only if different weights are given to the different experimental points. If the experimental measurements are used directly for non-linear regression, the weights can be assigned simply as the inverse values of the variances. If a linearized plot is used, s^2 for a transformed function (e.g. the inverse value) must be calculated from the variances of the directly measured variables by applying the standard equation of the propagation of mean square error to the function employed to generate transformed values. Evidently, non-linear fitting has the advantage of more simple weighting the experimental points. The expressions for the weighting factors for equations (D1-15)–(D1-17) are given in the literature.[1a]

Other obvious advantages of the non-linear fitting is the possibility of finding more than two parameters from a given fit and its applicability to the cases where linearization is difficult or impossible. For example, equation (D1-12) can be linearized only if x_S is set constant, but this parameter has its own error which can affect the calculation of x_{SL} and K.

This error can be minimized by using the non-linear fit with x_S as an adjustable parameter.

An obvious disadvantage of the equations discussed above is that the condition $[S]_t \ll [L]_t$ must always be maintained. The condition of a high excess of one of the equilibrium partners is often fulfilled in biological systems where extremely low concentrations of the minor component, such as an enzyme or labeled antibody, can be detected. Synthetic complexes often do not allow one to stay within the above-mentioned concentration range for application of such methods. When the total concentrations of S and L are close to each other, one must take into account the second mass balance equation (D 1-4). Moreover, there is often interference of the optical or other signals from two or even three participating species S, L and/or SL. If, for instance in a spectrophotometric titration one adds L to S, and L itself has an appreciable absorption one must correct for this by subtraction or inclusion of the respective term into the fitting equation, requiring known molar absorptivity ε_L. The resulting equation for X_{obs} takes the form

$$X_{obs} = x_S[S]_T + x_L[L]_t + 0.5\Delta x([S]_t + [L]_t$$
$$+ K_D - \{([S]_t + [L]_t + K_D)^2$$
$$- 4[S]_t[L]_t\}^{1/2}) \qquad \text{(D 1-18)}$$

or by analogy with equation (D 1-12)

$$x_{obs} = (X_{obs} - x_L[L]_t)/[S]_t = x_S + 0.5\Delta x([S]_t$$
$$+ [L]_t + K_D - \{([S]_t + [L]_t + K_D)^2$$
$$- 4[S]_t[L]_t\}^{1/2})/[S]_t \qquad \text{(D 1-19)}$$

where $\Delta x = (x_{SL} - x_S - X_L)$ and $K_D = 1/K$ is the dissociation constant of complex SL.

Equations (D 1-18) and (D 1-19) can hardly be linearized, but the fitting of the experimental results to them by using a non-linear regression with any commercially available program is quite simple. A most popular treatment of a complexation equilibrium with close to each other concentrations of S and L by using a linear regression is the method of

Rose and Drago.[5] In this method one uses K and x_{SL} as variables and constructs a series of lines in the co-ordinates $1/K$ *vs.* $x_{SL} - x_S$ (or $x_{SL} - X_L$), the point of intersection of which gives the true values of K and x_{SL}. One then can avoid this rather laborious method and fit the experimental data directly to the appropriate equation.

If one solves equation (D 1-6) together with equations (D 1-2) and (D 1-3) leaving the concentration of free L as a variable, the resulting equation (D 1-20)

$$X_{obs}/[S]_t = x_{obs} = (x_S + x_{SL}K[L])/(1 + K[L])$$
$$\text{(D 1-20)}$$

is formally similar to (D 1-12). It finds applications when a suitable experimental technique is used, which measures the free ligand concentration (e.g. potentiometry for the study of protonation equilibria).

In practical realization of a titration experiment, one preferably uses stock solutions of the added compound L, which, however, is not always possible due to solubility limitations. In such cases a time-consuming alternative is to prepare each solution separately, instead of adding L into a cuvet or NMR tube. After corrections for self-absorbance, if necessary (in other words, if the $x_L[L]_t$ term in equation (D 1-19) is not zero), one fits the observed spectroscopic data like apparent molar absorptivity, or NMR shifts with non-linear least-squares techniques to equation (D 1-19), or (D 1-12) when $[L]_t \gg [S]_t$. If evaluation of 7–10 data points between about 20 and 80% complexation shows no systematic deviation from the least-squares fit curve (Section D.1.1.2c) the resulting K and x_{SL} values are fairly trustworthy, although additional checks are recommended in order to eliminate a contribution from complexes of other stoichiometries which might not show up in a simple fit (Section D.1.2).

Another possibility, useful in particular for weak complexes which require high concentrations, consists in preparing solutions at the highest possible concentration, and then to dilute them stepwise. In this case it is convenient to take S and L in equal total concentrations: $[S]_t = [L]_t = C_t$. Under such conditions equation (D1-19) takes a more simple form

$$x_{obs} = x_S + \Delta x\{1 + 0.5(K_d/C_t)$$
$$(1 - (1 + 4C_t/K_d)^{1/2})\} \qquad \text{(D1-21)}$$

Equation (D1-21) is useful also for the other extreme, when a very stable complex is formed and any excess of L leads to practically complete complexation of S even in diluted solutions.

D1.1.2c. Experiment planning: choice of method

The information provided by the titration curve (often called the binding isotherm) for the calculation of unknown parameters is a very important factor which determines the precision of the formation constant. When one needs to determine two unknowns (K and x_{SL}) linear, or close to linear titration curves (observed at $K[L]_t \ll 1$) are useless: they will always give excellent fit to many arbitrary combinations of K and x_{SL}, which give the same product $x_{SL}K$. Also of little use are the curves obtained under conditions close to saturation: in this case one can determine quite precise value of x_{SL} but error in K is very large. Information theory predicts that the minimum error for K is observed from measurements between 20 and 80% complexation of the minor component, whereas maximum information for x_{SL} is obtained from the later points of the titration curve (see Figs. D2a and D2b).[6] When a titration experiment is performed with comparable concentrations of L and S one often starts at $[L]_t < [S]_t$ and finishes at $[L]_t > [S]_t$ which means that initially the minor component is L but later S. Of course, in this case the criterion of minimum error in K must be applied to the respective minor component at both parts of the titration curve.

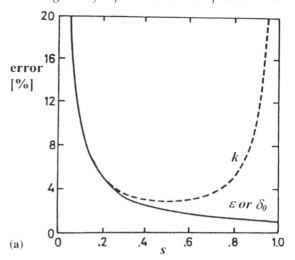

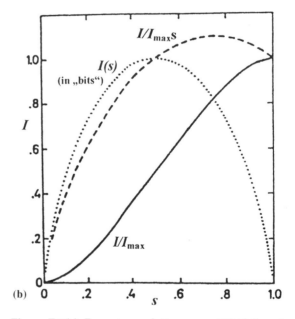

Figure D2(a). Percentage relative errors $100\Delta k/k$ and $100\Delta\varepsilon/\varepsilon = 100\Delta\delta_0/\delta_0$ as a function of the saturation fraction s; the curves are given for $\Delta s = 0.01$, which is equivalent to plotting the reduced relative errors $\Delta k/k$ Δs (k = equilibrium constant) and $\Delta\varepsilon/\varepsilon\Delta s = \Delta\delta_0/\delta_0\Delta s$.[6] **(b)** Information $I(s)$ as a function of s (dotted curve); accumulated information relative to total possible information (I/I_{max}) as s varies from $0 \rightarrow 1$ (solid curve); and rate of accumulation of information ($I/I_{max}s$) as s varies from $0 \rightarrow 1$ (dashed curve).[6] Reprinted with permission from *J. Am. Chem. Soc.*, Deranleau, D.A., 1969, **91**, 4044. Copyright 1969 American Chemical Society.

All equilibrium determinations require the simultaneous presence of associated and free host and guest molecules in measurable concentrations. The nomograms (Fig. D 1) can help to select suitable concentration ranges from expected K values. Alternatively, a simple program (available from the website) can be used to calculate optimal concentrations for complexation degrees between 20 and 80% with an optimal distribution of usually at least 8–10 points over this range; the program also allows one to overwrite the planned stock solution concentrations by those actually used and gives the exact volumes to be added.

The choice of a suitable measuring technique is dictated in the first place by the concentrations which must be chosen according to the equilibrium constant K (Fig. D 1). It should be remembered that it is primarily the minor partner which dictates the choice of the method, although the excess partner which is added during the titration may of course interfere with the measurement. The upper limit of K which can be determined reliably by using a given titration technique is dictated by necessity to obtain at least several points below 80% of complexation. For example, if the minor component can be measured only at concentrations above 0.01 M and K is larger than 10^3 M^{-1}, addition of an equimolar amount of the titrant will give already more than 75% of complexation. In consequence, 13-C NMR-spectroscopy as a still quite insensitive method can only be applied for titrations if $K < 10^2$ M^{-1}, unless one wants to spend hours of signal accumulation time for one single point. The limit for 1H-NMR is usually $K < 10^4$ M^{-1}. For UV/vis spectroscopy with intermediate molar absorptivities around 10^4 $M^{-1} cm^{-1}$ values of $K < 10^5$ M^{-1} can be measured. Fluorescence methods can be used for constants up to 10^8 M^{-1}, radioactivity measurements for even larger constants. A useful approach for very large formation constants is the use of competition titration (see below). The lower limit of

stability constants which can be determined by a titration experiment is dictated principally by problems with solution non-ideality. Thus, if one wants to determine $K \approx M^{-1}$ the ligand must be added up to 1M concentration in order to get at least 50% complexation and usually at such high concentrations the activity coefficients, reflecting the medium polarity etc., undergo substantial changes.

D 1.1.2d.   Association constants from competition experiments

Often it is convenient, or even necessary to use an easily monitored additional species, such as a dye with favorable spectroscopic properties as a competitor, towards a ligand the association constant of which one wants to determine. This procedure suits particularly well equilibria with ligands which show no or very small spectral changes upon complexation, such as cyclodextrins. The binding constant of the competing indicator dye I with the ligand should not be too far from the K value of the substrate to be analyzed and must be determined independently by the procedures described above. Also, the molar absorptivities of free and complexed indicator (ε_I and ε_{IL}, respectively) must be determined. Usually one adds S to a mixture of L and I taken at constant concentrations.

In this experiment one must take into account two expressions for the equilibrium constants

$$K = [SL]/([S][L])$$
$$K_I = [IL]/([I][L])$$

and three equations of mass balance

$$[S]_t = [S] + [SL] = [S](1 + K[L])$$
$$[L]_t = [L] + [SL] + [IL] = [L](1 + K[S] + K_I[I])$$
$$[I]_t = [I] + [IL] = [I](1 + K_I[L])$$

Assuming that only I and IL absorb light at the given wavelength one can calculate the ratio of concentrations of free and bound indicator (the 'indicator ratio') $Q = [I]/[IL]$ from the observed molar absorptivity (this can easily be shown combining equations (D1-6) and (D1-7))

$$Q = (\varepsilon_{obs} - \varepsilon_{IL})/(\varepsilon_I - \varepsilon_{obs}) \qquad \text{(D1-22)}$$

From the value of Q one can calculate the concentration of free ligand at any concentration of added substrate:

$$[L] = 1/(K_I Q)$$

Now the mass balance equation for L can be written as

$$\begin{aligned}
[L]_t &= [L](1 + K[S] + K_I[I]) \\
&= (1 + K[S]_t/(1 + K/(K_I Q)) + K_I[I]_t/ \\
&\quad (1 + 1/Q))/(K_I Q) = 1/(K_I Q) + K[S]_t/ \\
&\quad (K + K_I Q) + [I]_t/(1 + Q) \qquad \text{(D1-23)}
\end{aligned}$$

and introducing parameter P defined as[1a]

$$P = [L]_t - 1/(K_I Q) - [I]_t/(1 + Q) \quad \text{(D1-24)}$$

one obtains

$$P = K[S]_t/(K + K_I Q)$$

and

$$[S]_t/P = (K_I/K)Q + 1 \qquad \text{(D1-25)}$$

A plot according to equation (D1-25) then yields the ratio K_I/K from the slope and K can be calculated on the basis of the known equilibrium constant K_I. The intercept may deviate substantially from the theoretical value of 1.00 as consequence of very small changes in P. Alternatively, one can calculate K values directly from equation (D1-23) with known total concentrations of I and S as well as known K_I and Q and then average the obtained values.

Another important application of competition experiments is the determination of very large formation constants when the direct titration gives immediately a too high degree of complexation. In this case the binding process is monitored as usually by measurements of a suitable physical property of S and the addition of a third component I serves to reduce the concentration of free ligand and, consequently, the degree of complexation.

The competition of L with another ligand L' with known formation constant K' for the same substrate S also can be used. In particular, if S is a minor component for both L and L', titration of S in the presence of L' will follow the same equations (D1-11)–(D1-13) as in the absence of L' but with apparent formation constant $K_{app} = K/(1 + K'[L'])$.

D1.1.2e. Formation of isomeric complexes

The stochiometric 1:1 complexation can actually involve formation of different isomeric complexes. This happens when guest and host molecules can associate through several different binding sites (for example, a guest can possess different groups capable for binding) or can be included into the host in different orientations, etc. Equilibrium constants which correspond to formation of each isomer are referred to as *microscopic* binding constants, which are related to the stochiometric binding constant. We shall analyze this relation first for the simplest case of formation of only two isomeric complexes.

Let us suppose that the complexation of a substrate S with a ligand L affords two isomers SL and LS with two respective microscopic binding constants

$$K_{SL} = [SL]/([S][L])$$

and

$$K_{LS} = [LS]/([S][L])$$

Let us suppose also that each complex has its own value of an intrinsic property x: x_{SL} and x_{LS}. Assuming for simplicity that $[L]_t \gg [S]_t$ (this assumption does not affect the general result), one obtains the mass balance equations in the form

$$\begin{aligned}
[S]_t &= [S] + [SL] + [LS] \\
&= [S](1 + (K_{SL} + K_{LS})[L]) \\
[L] &= [L]_t
\end{aligned}$$

and the expression for x_{obs}

$$\begin{aligned}
x_{obs} = (x_S + (x_{SL}K_{SL} + x_{LS}K_{LS})[L]_t)/ \\
(1 + (K_{SL} + K_{LS})[L]_t) \qquad \text{(D1-26)}
\end{aligned}$$

Comparison of equation (D 1-26) with (D 1-12) shows that it describes exactly the same hyperbolic dependence of the observed intrinsic parameter on the ligand concentration as for the case of formation of a single complex and does not provide any information about the existence of isomers. It follows from this comparison also that the stochiometric binding constant is the sum of the microscopic constants

$$K = K_{SL} + K_{LS}$$

and instead of individual values of the intrinsic properties x_{SL} and x_{LS} one obtains as x_{obs} at saturation (x_∞) an apparent parameter equal to

$$x_\infty = (x_{SL}K_{SL} + x_{LS}K_{LS})/(K_{SL} + K_{LS})$$

It is important that the microscopic constants cannot be determined even in a case when both complexes give separate non-overlapping signals. Indeed, in this case one of the intrinsic values in a given spectral range equals zero, $x_{LS} = 0$, and equation (D 1-26) takes the form

$$x_{obs} = (x_S + x_{SL}K_{SL}[L]_t)/(1 + (K_{SL} + K_{LS})[L]_t)$$

Evidently, the stochiometric binding constant again is the sum of microscopic constants and the parameter $x_\infty = x_{SL}K_{SL}/(K_{SL} + K_{LS})$ again does not give the expected x_{SL} value. Generally, the microscopic binding constants never can be calculated without additional assumptions from the stochiometric constants determined experimentally by titration.[1c] We shall analyze this situation in more detail for a particular case of protolytic microequilibria (Section D.2.2). The situation, of course, becomes different when a method exists for determination of concentrations of isomeric complexes *without* the knowledge of intrinsic properties x, that is when the concentrations rather than solution properties proportional to them are measured. This can be accomplished by using NMR or ESR (electron spin resonance) techniques under conditions of slow (in the time scale of given method) exchange between free and bound states of the monitored species, when relative concentrations of all complexes can be determined by integration of the respective signals (Section D 8).

In a general case when the complexation affords N isomeric complexes SL_i (i from 1 to N) with microscopic binding constants K_i and intrinsic properties x_{SLi}, the expression for x_{obs} takes the form

$$x_{obs} = (x_S + (x_{SL1}K_1 + x_{LS2}K_2 + \cdots x_{LSN}K_N) [L]_t)/(1 + (K_1 + K_2 + \cdots K_N)[L]_t)$$

Thus, the experimentally determined stochiometric constant is the sum of all microscopic binding constants

$$K = K_1 + K_2 + \cdots K_N$$

and x_{obs} at saturation is a weighted mean of intrinsic parameters of all isomers

$$x_\infty = (x_{SL1}K_1 + x_{SL2}K_2 + \cdots x_{LSN}K_N)/ (K_1 + K_2 + \cdots K_N)$$

Note that an increased number of possible isomers enhances the apparent complex stability.

D 1.2.   SELF-ASSOCIATION

D 1.2.1.   Preliminary checks

Self-associations of a ligand can be easily overlooked and should always be checked by independent dilution experiments. For this one measures, for instance, NMR shifts of a pure substrate at the highest concentration used in the titrations, and compares it at the lowest concentration one can timewise afford to measure. With substances bearing suitable chromophores one can check for UV/vis absorbance linearity over a large range of concentrations, or use fluorimetry in a similar way. However, these optical methods often are not sensitive towards submicellar or higher aggregations. Alternative methods employ particle size or average molecular weight measurements, such as freezing point changes, osmotic pressure etc. (Section D 8).

Self-association in water with some more lipophilic cyclophanes of the type **D-1** occurs[7] at concentrations as low as 10^{-5} M when R = Me (1.6×10^{-4} M when R = H) which must be taken into account in the determination of association constants with guest molecules.

monomeric units while a hetero-associate SL from two different species. A nomogram is provided, showing the degree of self-association as a function of concentration for different association constants (Fig. D 3).

Analysis of the systems which involve higher associates is difficult. Approaches proposed for treatment of stepwise self-association equilibria can be found in ref. 1a.

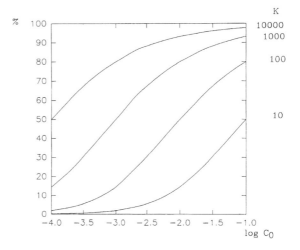

D-1

D 1.2.2. Measuring self-association constants

The simplest but most important case of self-association is the dimerization of the substrate:

$$2S \rightleftharpoons S_2$$

In this case the mass balance and equilibrium constant expressions take the forms

$$[S]_t = [S] + 2[S_2] \qquad K = [S_2]/[S]^2$$

and the expression for the observed intrinsic property is

$$x_{obs} = x_S + \Delta x \{1 + 0.25(K_d/[S]_t)$$
$$(1 - (1 + 8[S]_t/K_d)^{1/2})\} \qquad (D\,1\text{-}27)$$

where x_S and x_{S2} are the intrinsic properties of monomer and dimer respectively and $\Delta x = x_S - 0.5x_{S2}$. Equation (D 1-27) is, of course, similar to (D 1-21) but contains different coefficients due to statistical factors: a dimer is constructed from two equivalent

Sometimes self-association is highly cooperative as in a case of micelle formation. This process resembles a phase transition (like precipitation) and usually can be adequately described in terms of a critical concentration instead of the mass action law (Section G2).

Figure D 3. Self-association degree as a function of concentration for different association constants.

D1.3. COMPLEXES WITH STOICHIOMETRIES OTHER THAN 1:1

The presence of higher complexes such as SL_2 etc is often overlooked, in particular if one settles with measurements of only one NMR signal, absorbance at one wavelength, or if one stays outside the desirable measuring range of 20–80% complexation. Complexes with small binding constants, and/or with a weaker signal contribution (smaller CIS or, in general, Δx value) are particularly likely to be missed. They can, however, be quite important in terms of some supramolecular function. Thus, the minor complex of lower thermodynamic stability can be the catalytically most active one, or can give a stronger signal in sensor applications. Therefore, a reasonable effort should be paid to secure the number, strength and type of complexes present in solution. The special techniques to detect different partners in a chemical equilibrium which are used in optical spectroscopy (like isosbestic points, absorbance-difference diagrams etc.) will be discussed in the appropriate section below. Here we shall discuss some general methods and types of binding isotherms.

Only if data accumulated between about 20% to at least 80% complexation show no systematic deviation from a fit to the 1:1 model, calculational models with higher stochiometries make no sense, as they can give only arbitrary choices of the then at least few variable parameters. One also should be aware of systematic deviations which might be due not to more complicated stochiometries, but to systematic changes during titration, such as that of apparent extinction coefficients, medium properties, temperature, activity coefficients, etc. (these factors can lead, of course, also to apparent 1:1 or any other stoichiometry, when actually no complexation occurs at all and observed nonlinearity of the signal is entirely due to external factors). Before one applies other equations than the 1:1 model, it is therefore advisable to check for these external changes during titration and disregard artifacts also by asking whether non-trivial complex stochiometries used are chemically (structurally) reasonable.

D1.3.1. Methods for stoichiometry determination

The oldest way to infer the number of ligands involved in complex formation was to isolate if possible crystalline material from a solution made up from S and L, and then to determine the relative amounts of S and L in the complex. Formation of such complexes in the solid state does of course not imply that they are also present in solution, and other methods are necessary to establish the stoichiometry of solution equilibria.

A traditional way to distinguish between different stochiometries is the method of continuous variations (the so-called Job method). In this method one prepares a series of mixed S and L solutions of a constant total concentration $[S]_t + [L]_t$ but with variable ratios $[L]_t / [S]_t$, and measures absorbancies or other properties of these solutions, which are plotted as a function of a 'mole fraction' $f_L = [L]_t / ([S]_t + [L]_t)$ (Fig. D4). The presence of an extreme (a maximum or a minimum) is indicative of an interaction between the components affording a complex of composition S_nL_m and the value of f_L at the extreme (f_{extr}) is related to the complexation stoichiometry according to equation

$$m/n = f_{extr}/(1 - f_{extr}) \qquad (D1\text{-}28)$$

For example, a maximum at $f_{extr} = 0.5$ indicates a complex for which $m/n = 1$. Note that the method gives only the ratio of the stoichiometric coefficients and that such a complex can be either of 1:1 or 2:2 or of higher stoichiometry. Often both free ligand and substrate possess a non-zero absorbance at the working wavelength. In such cases one must calculate the absorbances expected at each value of f_L without interaction (i.e. on the

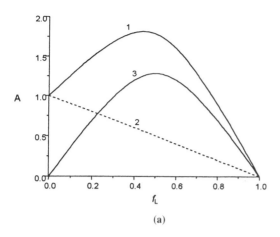 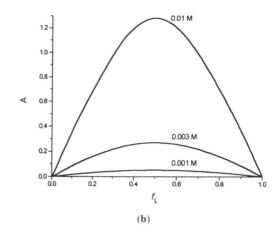

(a) (b)

Figure D 4. Simulated Job plots at different total concentrations of components. (A) Curve 1 = absorbance *vs.* f_L plot for the mixtures of S and L possessing $\varepsilon_S = 100$ $M^{-1}cm^{-1}$, $\varepsilon_L = 0$ and $\varepsilon_{SL} = 1000$ $M^{-1}cm^{-1}$ at $[S]_t + [L]_t = 0.01$ M; formation constant of the complex SL equals 1000 M^{-1}. Line 2 = absorbance calculated on assumption of no interaction. Curve 3 = corrected absorbance *vs.* f_L plot (difference between curve 1 and line 2). (B) Corrected absorbance *vs.* f_L plots for the same system at total $[S]_t + [L]_t$ concentrations 0.01; 0.003 and 0.001 M, indicated on the curves.

basis of simple additivity) and subtract the calculated values from the observed ones.

The sharpness of the extreme, and therefore the precision of the method, depends on the complexation strength. Actually, the criterion is a high degree of complexation under conditions of the experiment which depends both on K and component concentrations. Therefore one needs to use a sufficiently high total concentration to observe a well defined peak, as is illustrated in Fig. D 4b. It is useful to construct Job plots at several wavelengths because if only a single complex is formed f_{extr} must be independent of the wavelength while the observation of such dependence is indicative of formation of several complexes under the conditions employed. If several complexes of different stoichiometric ratios are present over the measured range, the Job method usually gives poorly interpretable results.

Another method known as the mole ratio method is based on analysis of a usual titration curve with fixed concentration of S and varied L, but under conditions where the degree of complexation is always very high. If

the complex stability is high enough, the concentration of the complex will increase linearly until $[L]_t / [S]_t$ reaches m/n, and than will remain constant since all substrate will be already consumed. In consequence, there will be a break in the titration curve at this point. Evidently, the method needs a sufficiently high concentration of S or large stability constant to assure the necessary high complexation degree.

Both methods described above need experimental conditions under which the complexation is maximally strong. In contrast, conditions required for reliable determination of equilibrium constants are those under which the degree of complexation is not too high, ranging from 20 to 80%. Therefore, one should use different concentration ranges for determination of stoichiometry and of equilibrium constants. This may lead in some cases to a disagreement between the stoichiometry determined by one of the methods discussed above, and that observed by fitting the binding isotherm. For example, if a given pair of substrate and ligand produces complexes of different stoichiometries, the dominating species may be different at high and low concentration ranges.

D 1.3.2.   Titration curves for stoichiometries higher than 1:1

With respect to the numerous possible combinations of n and m in S_nL_m complexes we shall restrict ourselves to the cases where $n = 1$ which represent the majority of practically important supramolecular systems. The stepwise stoichiometric formation constants for equilibria

$$S + L \rightleftharpoons SL, SL + L \rightleftharpoons SL_2, SL_2 + L$$
$$\rightleftharpoons SL_3, \ldots SL_{N-1} + L \rightleftharpoons SL_N$$

are defined as

$$K_i = [SL_i]/([SL_{i-1}][L])$$

The overall formation constants (known as Adair constants in the biochemical literature) for equilibria

$$S + L \rightleftharpoons SL, S + 2L \rightleftharpoons SL_2, S + 3L$$
$$\rightleftharpoons SL_3, \ldots S + NL \rightleftharpoons SL_N$$

are defined as

$$\beta_i = K_1 K_2 \ldots K_{i-1} K_i$$

In addition one can define a single site microscopic constant Q as

$Q = $ [occupied single site]/([free single site][L])

A useful function for the analysis of binding is the Bjerrum complex formation function $\bar{n}$

$$\bar{n} = \left(\sum_{i=1}^{N} i\beta_i[L]^i \right) \Bigg/ \left(1 + \sum_{i=1}^{N} \beta_i[L]^i \right)$$

(D 1-29)

The quantity $\bar{n}$ gives an average number of ligand molecules bound per substrate and ranges from 0 to the total number of binding sites N. Let us consider first the case where the substrate possesses N identical and independent sites. In this case

$$K_i = Q(N - i + 1)/i$$

and the ratio of two successive stability constants is

$$K_{i+1}/K_i = i(N - i)/\{(i + 1)(N - i + 1)\}$$

(D 1-30)

Evidently each next stability constant is smaller than the preceding one due to only statistical factors. Any deviation from the statistical ratio K_{i+1}/K_i implies a non-equivalence of binding sites or some sort of interaction between sites when the occupation of one site leads to a modification of the affinity of the other site(s) due, for example, to a conformational change (Section A9). In the latter case deviations from statistical binding are commonly referred as cooperativity, which can be positive or negative depending on whether the ratio of successive stability constants is higher or lower than the statistically expected value.[8]

The Bjerrum equation (D 1-29) for statistical binding takes a very simple form

$$\bar{n} = NQ[L]/(1 + Q[L])$$

(D 1-31)

The linearized form

$$\bar{n}/[L] = NQ - \bar{n}Q$$

is known as the Scatchard equation.[8] Equation (D 1-31) describes a simple hyperbolic binding isotherm like in a case of a 1:1 complexation. The cooperative binding, however, modifies to a smaller or greater extent the shape of the isotherm. The oldest and still popular way of diagnostics of cooperativity by the analysis of the shape of the binding isotherm (Section A9) is the calculation of the so-called Hill coefficient h, which is the slope of the plot of $\log(Y/(1 - Y))$ vs. $\log[L]$ (see Fig. A21):

$$h = d\log(Y/(1 - Y))/d\log[L]$$

(D 1-32)

where Y is the degree of complexation (also called degree of saturation), expressed in a general case as

$$Y = \bar{n}/N$$

(D 1-33)

The Hill equation was derived originally[9] for a complexation process which involves only one equilibrium between S and h molecules of L affording the single complex SL_h. The degree of saturation is given in this case by equation

$$Y = [SL_h]/[S]_t = \beta_h[L]^h/(1 + \beta_h[L]^h)$$

and it can be easily shown that

$$\log(Y/(1 - Y)) = h\log[L] + \log\beta_h$$

Evidently this equation implies an infinite (never observed) cooperativity when the binding constants for all complexation processes with $i < h = N$ equal zero. In real cases the value of h defined as in equation (D 1-32) is approximately constant only in a restricted range of concentrations of L; it approaches N when the cooperativity is high and equals unity for statistical binding when $\bar{n}$ is given by equation (D 1-33). A more detailed consideration of the Hill plot is given in Section A9.

In the following consideration we restrict ourselves with a case of a substrate with only two binding sites which allows one to analyze in a simplest form the essential features of different binding models. In order to discriminate the binding sites we shall enumerate them and mark the ligand bound to the given site with a respective number. The general scheme involving single site binding constants will have the following form

$$S + L \rightleftharpoons SL^1 \quad (Q_1) \tag{D 1-34}$$
$$S + L \rightleftharpoons SL^2 \quad (Q_2) \tag{D 1-35}$$
$$SL^1 + L \rightleftharpoons SL_2 \quad (\alpha Q2) \tag{D 1-36}$$
$$SL^2 + L \rightleftharpoons SL_2 \quad (\alpha Q1) \tag{D 1-37}$$

where α is the interaction factor which shows how the binding constant for a given site is changed upon occupation of the other site. Since the four equilibria form a cycle, the product of the binding constants of steps (D 1-34) and (D 1-36) must be equal to the product of the binding constants of steps (D 1-35) and (D 1-37). Therefore, the interaction factor α must be the same for both single binding sites, that is occupation of site 1 induces the same effect on the affinity of site 2 as the occupation

of site 2 induces for site 1. The expressions for the stoichiometric binding constants take the form

$$K_1 = Q_1 + Q_2 \tag{D 1-38}$$
$$K_2 = \alpha Q_1 Q_2/(Q_1 + Q_2) \tag{D 1-39}$$

In the case of statistical binding $Q_1 = Q_2 = Q$ and $\alpha = 1$; therefore $K_1 = 2Q$, $K_2 = Q/2$ and the ratio $K_2/K_1 = 1/4$ in accordance with equation (D 1-30). If the binding sites are equivalent ($Q_1 = Q_2 = Q$), but not independent ($\alpha \neq 1$) one observes positive ($K_2/K_1 > 1/4$) or negative ($K_2/K_1 < 1/4$) cooperativity. Alternatively, if the binding sites are non-equivalent ($Q_1 \neq Q_2$), but independent ($\alpha = 1$), the ratio K_2/K_1 equals

$$K_2/K_1 = Q_1 Q_2/(Q_1 + Q_2)^2 = (Q_2/Q_1)/ \\ (1 + Q_2/Q_1)^2 \tag{D 1-40}$$

Equation (D 1-40) predicts that K_2/K_1 has the maximum value of $1/4$ at $Q_2/Q_1 = 1$ and, therefore at all non-equal values of Q_1 and Q_2 the ratio K_2/K_1 is less than $1/4$ and one observes in this case a negative cooperativity without any negative interaction between the binding sites.

Different types of the binding isotherms which illustrate the plots of Y (calculated in accordance with equation (D 1-33)) *vs.* [L] for statistic (for positively and for negatively cooperative complexation) are shown in Fig. D 5. Note that a 'sigmoid' curve which sometimes is considered as a criterion of the positive cooperativity is clearly seen (Fig. D 5B) only when α is as large as 10.

The equation for the titration curve has the form ($x_L = 0$)

$$x_{obs} = (x_S + x_{SL}K_1[L] + x_{SL2}K_1K_2[L]^2)/ \\ (1 + K_1[L] + K_1K_2[L]^2) \tag{D 1-41}$$

where K_1 and K_2 are related to the single site binding constants by equations (D 1-38) and (D 1-39) and x_{SL} is an apparent parameter equal to a weighted mean of intrinsic parameters of SL^1 and SL^2 complexes (Section D.1.1.2e). Under conditions of a high excess

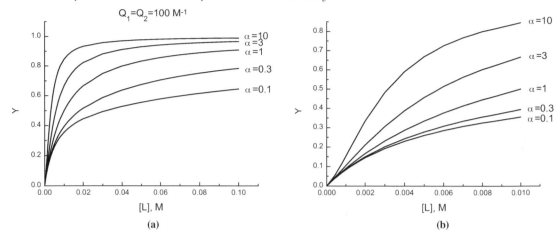

Figure D 5.   Binding isotherms, saturation (Y) *vs.* ligand concentration profiles, for different types of two-site complexation. The curves are calculated for $Q_1 = Q_2 = 100\,\mathrm{M}^{-1}$ with different values of the interactions factor α indicated on the curves. (A) Ligand concentration range 0–0.1 M. (B) Ligand concentration range 0–0.01 M.

of the ligand over substrate equation (D 1-41) describes the real x_{obs} *vs.* $[\mathrm{L}]_t$ profiles. Fig. D 6 illustrates different types of titration curves.

The case when K_1 and K_2 are similar is illustrated in Figs D 6a,b. Four situations are shown: (*i*) the intrinsic properties of both 1:1 and 1:2 complexes are similar and larger than that of the substrate ($x_S < x_{\mathrm{SL}} = x_{\mathrm{SL2}}$), (*ii*) the intrinsic properties of the substrate and 1:1 complex are similar and smaller than that of 1:2 complex ($x_S = x_{\mathrm{SL}} < x_{\mathrm{SL2}}$), (*iii*) all three species have different increasing intrinsic properties ($x_S < x_{\mathrm{SL}} < x_{\mathrm{SL2}}$), (*iv*) all three species have different intrinsic properties, that of 1:1 complex being the largest ($x_S < x_{\mathrm{SL}} < x_{\mathrm{SL2}}$). Figure D 6A shows the titration curves obtained in a wide concentration range of the ligand up to nearly complete saturation of the substrate. In Figure D 6B a shorter range is shown where the shape of the profiles below saturation is better seen. The simulated profiles are shown here as the points which are fitted to a simple hyperbolic isotherm for a 1:1 complexation, equation (D 1-12), and the fitting results are shown as curves. Although in all cases the points deviate from the curves systematically, the deviations in the cases (*i*)

and, especially, (*iii*) are rather small and easily can be overlooked in the limits of experimental errors. It is very important, therefore, to fit the titration data in a whole accessible spectral range with different relative values of the intrinsic properties of the components of a complexation equilibrium in order to avoid an occasional situation when a 1:2 complex is missing. Thus, titration curves at different wavelengths must be analyzed in the spectrophotometric titration, different chemical shifts must be used in the NMR titration, etc.

The case of strong positive cooperativity, $K_1 \ll K_2$, is illustrated in Fig. D 6C. The same four situations are shown and in all of them the curves have a 'sigmoid' shape clearly distinguished from a simple hyperbola. Finally the case of strong negative cooperativity, $K_1 \gg K_2$, is illustrated in Fig. D 6D and in a shorter concentration range in Fig. D 6E. As is seen from the last figure the plots in situations (*i*) and (*ii*) perfectly fit to equation (D 1-12) and the presence of two complexes can be detected only in situations (*iii*) and (*iv*), that is only if all three components, the substrate and both complexes, possess different values of the intrinsic properties. The importance of testing different signals for the construction and fitting the titration curves is again clearly seen.

A useful way of analysis of the titration curves is a presentation with semilogarithmic coordinates. In the case of 1:1 complexation a plot of x_{obs} *vs.* log[L] has the shape of a wave with an inflection point which occurs at log[L] = $-$log$K$. When two successive 1:1 and 1:2 complexes are formed with substantially different binding constants, $K_1 \gg K_2$, and different values of the intrinsic properties for both complexes, two distinct waves are observed, Fig. D6, and the respective inflection points give the approximate values of the binding constants. If the constants are strongly different, but the binding shows a positive cooperativity, $K_1 \ll K_2$, the semilogarithmic plot shows only one wave which is more steep, however, than the wave for a 1:1 complexation.

In the case of homotropic binding discussed above the single site binding constants and the site interaction factor cannot be calculated from the experimentally determined constants K_1 and K_2 without additional assumptions concerning possible equivalence of mutual dependence of the binding sites. The situation becomes more definite for the heterotropic binding when each site is occupied by a different ligand, L or M, in accordance with the scheme

$$S + L \rightleftharpoons SL \quad (Q_1) \qquad (D1\text{-}42)$$
$$S + M \rightleftharpoons SM \quad (Q_2) \qquad (D1\text{-}43)$$
$$SL + M \rightleftharpoons SLM \quad (\alpha Q_2) \qquad (D1\text{-}44)$$
$$SM + L \rightleftharpoons SLM \quad (\alpha Q_1) \qquad (D1\text{-}45)$$

In this case the expression for the titration curve takes the form

$$x_{obs} = (x_S + x_{SL}Q_1[L] + x_{SM}Q_2[M]$$
$$+ x_{SLM}\alpha Q_1 Q_2[L][M])/(1 + Q_1[L]$$
$$+ Q_2[M] + \alpha Q_1 Q_2[L][M]) \qquad (D1\text{-}46)$$

Equation (D1-46) predicts a hyperbolic dependence of x_{obs} on concentration of each ligand at fixed concentration of another ligand

and the fitting of these concentration dependencies allows one to determine all single site binding constants and the interaction factor.

Returning to the homotropic complexation it is worth to note that not only the negative cooperativity can be observed in the absence of any negative site interaction, but also the positive cooperativity can be due to a mechanism which does not involve any positive site interactions. Such mechanism, originally proposed by Monod, Wyman and Changeux (MWC) for interpretation of the cooperative binding of molecular oxygen to hemoglobin,[10] is based on complexation-induced shift of pre-equilibrium between two isomers of the substrate which exist in different conformations possessing different affinity to the ligand (Section A9).

In a more general case when the substrate possesses N>2 binding sites the titration curve is given by the following equation:

$$x_{obs} = (x_S + x_{SL}\beta_1[L] + x_{SL2}\beta_2[L]^2$$
$$+ \cdots x_{SLN}\beta_N[L]^N)/(1 + \beta_1[L]$$
$$+ \beta_2[L]^2 + \cdots \beta_N[L]^N) \qquad (D1\text{-}47)$$

which can be used for the fitting under conditions of a high excess of the ligand over the substrate. If the concentration of free ligand is unknown special methods developed principally for metal complexation equilibria[1b,d,f] or computer programs (see below) must be used. The most complicated systems are those which involve simultaneous complexation and self-association, affording complexes like S_mL_n with $m \geq 2$. Even in a simplest case when $m = 2$ and $n = 1$ the algebraic equation for X_{obs} (or x_{obs}) is a complicated function of total concentrations of S and L which results from solution of a quadratic equation when a high excess of the ligand is used and of a cubic equations when total concentrations of S and L are similar. In all such cases one can try to use an experimental technique, such as potentiometry, which allows one to measure the concentration of free ligand (or substrate) under equili-

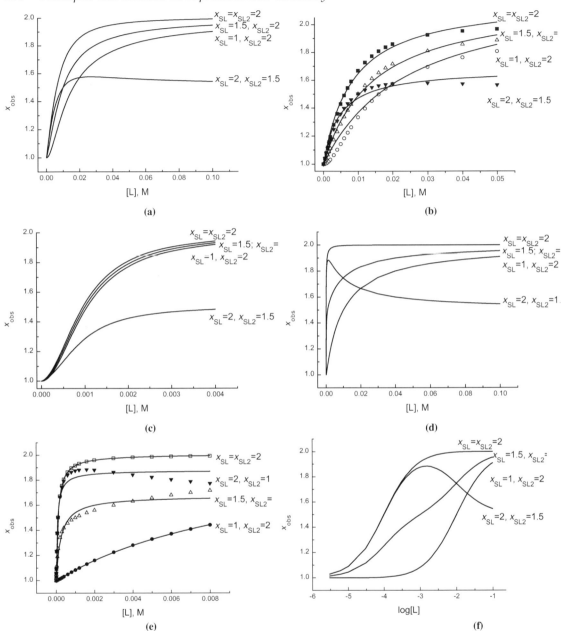

Figure D 6. Simulated titration curves for different combinations of the binding constants and intrinsic properties x in equation (D 1-41). In all cases $x_S = 1$, $x_L = 0$ and values of x_{SL} and x_{SL2} are given on the curves. (a) $K_1 = K_2 = 100$ M^{-1}, ligand concentration range 0–0.1 M. (b) $K_1 = K_2 = 100$ M^{-1}, ligand concentration range 0–0.05 M, points = simulated titration data, curves = calculated profiles obtained by fitting of the points to equation (D 1-12). (c) $K_1 = 100$ M^{-1}, $K_2 = 10000$ M^{-1}. (d) $K_1 = 10000$ M^{-1}, $K_2 = 100$ M^{-1}, ligand concentration range 0–0.1 M. (e) $K_1 = 10000$ M^{-1}, $K_2 = 100$ M^{-1}, ligand concentration range 0–0.008 M, points = simulated titration data, curves = calculated profiles obtained by fitting of the points to equation (D 1-12). (f) $K_1 = 100$ M^{-1}, $K_2 = 10000$ M^{-1}, semilogarithmic plot.

brium, or to use special computer programs for the calculation of the formation constants.

D1.4. COMPUTER PROGRAMS FOR EQUILIBRIUM CONSTANT CALCULATION

Calculation of the binding constant for a 1:1 complexation reaction by fitting of a single titration curve obtained either under conditions of a large excess of the ligand over the substrate or at similar concentrations of both components can be easily performed by using a personal computer with any commercially available program for non-linear least-squares minimization like SigmaPlot, Origin, etc. Titration curves for the formation of higher complexes obtained under conditions of a large excess of the ligand also can be treated with these programs. These situations are very common in practice, but, of course, there is a multitude of more complicated cases which cannot be treated so easily. The main problem with the application of the aforementioned programs is the necessity to have an analytical expression for X_{obs} or x_{obs} in form of an algebraic equation which is used in the fitting procedure. In order to obtain such a function one has to solve analytically the expressions for concentrations of all components. However, already in the case of 1:2 stoichiometry with similar S and L concentrations these expressions are derived from the roots of a cubic equation with a very complicated form. Such cases can be solved numerically by an appropriate computer program.

At present there are numerous computer programs suitable for calculation of equilibrium constants from titration data.[11–13] With a few exceptions all these programs use an iterative procedure based on a least-squares minimization by means of the Gauss–Newton–Marquardt (GNM) algorithm. The free concentrations of components are calculated by solving the mass balance equations using most often the Newton–Raphson method. The data points are weighted by formulas which correspond to the experimental method employed, or (often) an unweighted regression analysis is performed. Sometimes activity coefficient corrections are introduced by, for example the Debye–Hückel equation. Several programs, like SIRKO[14a] and MICMAC[14b] are claimed to be applicable to the analysis of any kind of experimental data, but the majority of the programs are designed to handle the data obtained by only one or two methods. They will be discussed, therefore, more appropriately in the respective sections devoted to use of different experimental techniques.

Generally, first of all, a proper set of reaction equilibria (protonation, complex formation, self-association, etc.) (the 'model') must be chosen, which is thought to represent the system under investigation. Then, before the beginning of the iterative fitting procedure, starting values of equilibrium constants and, if necessary as in a spectroscopic titration, of other parameters (molar absorptivities, chemical shifts, etc. of reaction products) must be introduced. This is the important step as badly chosen starting values may lead to false minima and incorrect final parameter evaluations. As was mentioned above, in some cases preliminary values of parameters can be found from linearized plots. Often some parameters or their combinations can be evaluated from analysis of the external ranges of the binding isotherm (at very low and/or very high ligand concentrations often only one or two species can dominate). Non-iterative procedures do not need starting values of calculated parameters, but they are restricted to linear and exponential functions only, while the iterative procedure can be applied to any mathematical function. If for a chosen model no satisfactory fit can be obtained, other improved models must be tested. Usually, one begins with the simplest possible model with a minimum of equilibria; improvement consists then in introduction of additional equilibria. This should be done with special care because additional parameters always improve the fit, but then the system can become rapidly underdetermined, and many different sets of constants which equally well fit the experimental data are the result. In such cases the use of additional experimental techniques is strongly recommended.

Testing goodness of fit to a given stochiometric model consists in the analysis of distribution of deviations of experimental points from the theoretical fitting curve. The deviations are reasonably small and randomly distributed if the applied model is correct. The most simple way of such testing is to plot the residuals in observed values, and those calculated from the regression line of the dependent variable *vs.* the independent variable: if the model is adequate, the residuals should be randomly distributed about the abscissa. Also, the dispersion of the experimental points about the regression line should be in the range of the standard deviation in a series of repetitive measurements of the dependent variable. In practice one often observes systematical deviations which may reflect inadequacy of the model. However, as mentioned at the beginning of Section D1.3, problems with possible external effects can also lead to non-randomly distributed deviations and must be carefully checked.

D2.   ELECTROCHEMICAL METHODS

When at least one component of an equilibrium is electroactive, that is it forms a redox couple or has a charge, its complexation can be monitored by one of the numerous electrochemical methods. The most important and most widely used method of this group is potentiometry, which allows one to determine the concentration of a free ion in a mixture of free and complexed species using the concentration dependence of the cell potential. The method is especially valuable for determinations of dissociation/protonation constants of ligand acid–base groups and of stability constants of metal complexes. Classi-

cal polarography and cyclic voltammetry are useful methods for monitoring the complexation of redox active species due to complexation induced shifts of half-wave potentials. Ionic association can be conveniently monitored by conductometry, but this method has more restrictions (no supporting electrolyte for maintaining constant ionic strength can be added, buffers usually cannot be used, etc.).

D2.1.   POTENTIOMETRY

Potentiometric determination of equilibrium constants has been extensively reviewed.[1b,e;11,12,15,16] Generally the method is based on a titration procedure (which can be acid–base, complexation, precipitation, etc., depending on the type of equilibrium studied) carried out in a galvanic cell (D2-1)

reference electrode ‖ titration solution |

   indicator electrode   (D2-1)

The potential E of the cell (D2-1) is given by

$$E = E_{\text{ref}} + E_{\text{j}} - E_{\text{i}}   \text{(D2-2)}$$

where E_{ref} is the potential of the reference half-cell, E_{j} is the liquid junction potential, and E_{i} is the potential of the sample half-cell. An indicator (or working) electrode immersed in the sample half-cell should respond reversibly to the electroactive ion to be determined (X); its potential is then the function of the concentration of X. In the ideal case E_{i} is given by the Nernst equation, which is equally applicable to the cases where the potential is created by a redox process and where it is due to the gradient of ion concentration (ion-selective electrodes)

$$E_{\text{i}} = E_{\text{i}}^0 + (RT/nF) \ln a_{\text{X}}   \text{(D2-3)}$$

where E_{i}^0 is the standard potential and a_{X} is the activity of X. In practice the electrode can show only the 'Nernstian behavior', that is a proportionality between the potential and the logarithm of the concentration of X under given conditions

$$E_{\text{i}} = E_{\text{i}}^{0'} + S \log [\text{X}]   \text{(D2-4)}$$

where parameters $E_i^{0'}$ and S are calculated from a calibration experiment.

D 2.1.1. Potentiometric titration with a glass electrode

For detailed information regarding the construction, characteristics and use of the glass electrode the literature can be consulted.[17a–c] The potential of cell (D2-1) with a glass electrode as a working electrode is given by

$$E = E^0 + E_{ref} + E_j - (2.3\,RT/F)\log([H^+]\gamma_H)$$
$$(D\,2\text{-}5)$$

where E^0 is the standard potential of the glass electrode and γ_H is the activity coefficient of hydrogen ions. At higher concentrations hydrogen and hydroxide ions begin to contribute considerably to the solution conductance due to very high equivalent conductactivity of these ions as compared to common inorganic ions of a supporting electrolyte. As a result the junction potential changes more than allowed by the limits of experimental errors. Therefore, the working limits of a glass electrode with respect to the pH is from 2 to 12.

Under experimental conditions, ensuring the constancy of E_j and γ_H equation (D2-5) takes the form

$$E = E^{0'} + Sp[H] \qquad (D\,2\text{-}6)$$

where
$E^{0'} = E^0 + E_{ref} + E_j - (2.3\,RT/F)\log\gamma_H, S = (2.3RT/F)$ and p[H] is defined by equation

$$p[H] = -\log[H^+] \qquad (D\,2\text{-}7)$$

Evidently p[H] does not coincide with the commonly used scale of pH defined as

$$pH = -\log a_H = -\log([H^+]\gamma_H) \qquad (D\,2\text{-}8)$$

where a_H is the activity of hydrogen ions. The use of the hydrogen ion concentration in potentiometry is reasonable first of all because it is the concentration rather than activity of hydrogen ions which appears in mass balance and electroneutrality equations solved with the potentiometric titration data. It should be noted that the commonly used practice of calibration of pH meters with for instance National Bureau of Standards recommended buffer solutions,[17b,18,19] or commercially available buffers, creates a scale which measures neither activity nor concentration of the hydrogen ions.[20,21] Such a 'practical scale' gives pH meter readings which are between these two values, surprisingly closer to p[H] then to pH, with an interval between these two scales usually of about 0.1–0.15 logarithmic units.[20]

On the basis of the above arguments the calibration of the cell with a glass electrode should be performed with a series of solutions of known concentrations of hydrogen ions and of the same ionic strength (created with the same supporting electrolyte) as in the titration solution. It is convenient to calibrate the cell by titration of a known concentration (about 0.005 M) of a strong acid in the solution of a supporting electrolyte by the standard base solution of the same concentration and the same ionic background as the supporting electrolyte maintaining a constant ionic strength (e.g. 0.1 M KOH should be used if 0.1 M KCl is the supporting electrolyte). Usually the results of titration of a strong acid with a strong base are analyzed using Gran plots.[22a–c] which allow one to check the concentration of the employed acid, to calculate K_W under conditions of the given experiment, as well as to evaluate the concentration of carbonate in base solution. At each titration point (10 or more approximately equally spaced points should be used) p[H] is calculated from the volume of the standard base added; then a plot of pH meter reading *vs.* p[H] is constructed and used as a calibration graph. The calibration with buffer solutions is also possible, provided the buffer is made from the components, such as acetic acid/sodium acetate or hydrogen phosphate/dihydrogen phosphate for which acid dissociation constants are known with high precision at different values of ionic strength, and the concentration of hydrogen ions can be calculated under the desired conditions.

An important application of potentiometric titrations is the determination of proton dissociation constants of molecules and ions in solution. The thermodynamic dissociation constant of an acid LH (here and subsequently the charges will be omitted for simplification) is defined by

$$K_a^T = a_H a_L / a_{LH} = [H][L]\gamma_H \gamma_L / [LH]\gamma_{LH}$$
$$(D2\text{-}16)$$

where a_H, a_L and a_{LH} are the activities and γ_H, γ_L and γ_{LH} are the activity coefficients of the respective species. On the other hand, the stochiometric dissociation constant is defined as

$$K_a^C = [H][L]/[LH] \qquad (D2\text{-}17)$$

Evidently the constants are related by

$$K_a^C = K_a^T \gamma_{LH} / \gamma_H \gamma_L \qquad (D2\text{-}18)$$

and under the condition that all activity coefficients are constant, the stochiometric acid dissociation constant is unchanged, meaning that it can be treated as a 'normal' equilibrium constant under these conditions. When the pH meter is calibrated with hydrogen ion concentrations, as recommended, the value of K_a^C is calculated from the titration data. The thermodynamic constant can be then calculated from equation (D2-18), either by using known (usually theoretically calculated) activity coefficients or by repeating determinations of K_a^C at different experimental conditions, and extrapolation to zero ionic strength if the supporting electrolyte is the only added component.

Another term used in the literature is the 'mixed' or Bronsted constant defined as

$$K_a^B = a_H[L]/[LH] \qquad (D2\text{-}19)$$

This definition is convenient when the ratio [L]/[LH] can be measured independently, e.g. spectrophotometrically, and the hydrogen ion activity is measured with a glass electrode, of course with the uncertainty discussed above.

Although acid dissociation constants are conventionally used parameters, their inverse values, protonation constants (K_H, note that $pK_a = \log K_H$), are more convenient for the treatment of potentiometric titration data. Thus for a polyacid LH_n the protonation constants are given by

$$K_{H1} = [LH]/[L][H]$$
$$K_{H2} = [LH_2]/[LH][H]$$
$$\cdots\cdots\cdots\cdots\cdots\cdots\cdots\cdots\cdots\cdots$$
$$K_{H_n} = [LH_n]/[LH_{n-1}][H$$

or generally

$$K_{H_i} = [LH_i]/[LH_{i-1}][H], \ i \text{ from 1 to } n$$
$$(D2\text{-}20)$$

where L is a completely deprotonated form of the acid.

Using a set of equations (D2-20) one can write two following mass balance equations valid at any point of the titration curve

$$[L]_t = [L] + [LH] + [LH_2] + \cdots [LH_n]$$
$$= [L](1 + K_{H1}[H] + K_{H1}K_{H2}[H]^2$$
$$+ \cdots K_{H1}K_{H2} \cdots K_{Hn}[H]^n) \qquad (D2\text{-}21)$$

and

$$[H]_t = [H] - [OH^-] + [B] + [HL] + 2[LH_2]$$
$$+ \cdots n[LH_n] = [H] - K_W/[H] + [B]$$
$$+ [L](K_{H1}[H] + 2K_{H1}K_{H2}[H]^2$$
$$+ \cdots nK_{H1}K_{H2} \cdots K_{Hn}[H]^n) \qquad (D2\text{-}22)$$

where $[L]_t$ is the total concentration of the acid (ligand), $[H]_t = n[L]_t$ is the total concentration of ionizable hydrogen, [B] is the concentration of titrant added and K_W is the ionic product of water ($pK_W = 13.891$ at 25°C and ionic strength 0.1M). Equation (D2-22) can be obtained also from the electroneutrality (charge balance) equation, which sometimes is used directly together with the mass balance equation for the ligand.[23]

From equations (D2-21) and (D2-22) the Bjerrum function (Section D1) can be calculated in the form

$$\bar{n} = \frac{[H]_t - [H] + K_w/[H] - [B]}{[L]_t}$$

$$= \frac{K_{H1}[H] + 2K_{H1}K_{H2}[H]^2 + \cdots nK_{H1}K_{H2}\cdots K_{Hn}[H]^n}{1 + K_{H1}[H] + K_{H1}K_{H2}[H]^2 + \cdots K_{H1}K_{H2}\cdots K_{Hn}[H]^n}$$

(D2-23)

where [H] is calculated from pH meter reading and $[H]_t$, $[L]_t$ and [B] are calculated from known initial concentrations and the dilution factor at each point of the titration curve, which is a plot of pH meter reading *vs.* added volume of the titrant or the equivalents of base per ligand equivalent. A typical titration curve is shown in Fig. D7, taken from the literature[24] for a pyridine-containing tetraaza macrocyclic ligand **D-2** used as a dihydrochloride (**D-2 · 2HCl**).

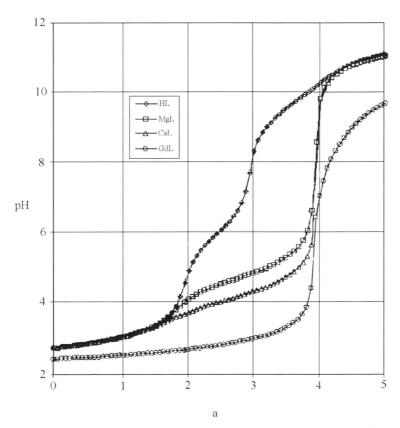

D-2

As was mentioned before (Section D1) the fitting of experimental results to the Bjerrum function can be done with any commercial non-linear least-square regression program, but for this particular case numerous special programs are available and the use of them is, of course, more convenient. The majority of them are, however, more general programs

Figure D7. Potentiometric titration curves of **D-2 · 2HCl** with and without metal (Mg^{2+}, Ca^{2+} and Gd^{3+}) present. '**a**' is the equivalent of OH^- per ligand equivalent.[24] Reprinted with permission from *Inorg. Chem.*, Kim *et al.*, 1995, **34**, 2225. Copyright 1995 American Chemical Society.

developed for the calculation of metal–ligand complex formation constants, which will be discussed later.

The determination of ligand protonation constants is important by itself, as it will be decisive for instance for the possible use of a host for complexation by salt bridges or hydrogen bonds at a desired pH. In addition, if the determination of the cation–ligand complex formation constants is the final goal, the protonation constants should be determined first and then used as fixed parameters in the refinement of the formation constants. In such a case for a polyacid LH_n, which serves as a ligand for a cation M, a set of the complex formation equilibria with the respective equilibrium constants should be added to the set of protonation equilibria represented by equation (D 2-20). Depending on the nature of a particular system these equilibria can involve formation of different species.

In a simplest case only stepwise formation of mononuclear complexes of different stochiometries with L is observed

$$M + L = ML, M + 2L = ML_2, M + 3L$$
$$= ML_3, \text{etc.}$$
$$\beta_1 = K_1 = [ML]/[M][L], \beta_2 = K_1 K_2$$
$$= [ML_2]/[M][L]^2, \beta_3 = K_1 K_2 K_3$$
$$= [ML_3]/[M][L]^3, \text{etc.}$$

where β_i and K_i are the overall and stepwise stochiometric formation constants. With polydentate ligands complexes with partially protonated forms LH_j (j from 1 to $n-1$) can exist, for instance

$$M + L + H = MLH, \beta_{MLH} = [MLH]/[M][L][H]$$

In basic solutions metal hydroxo complexes are formed frequently, for example

$$M + L + OH = MLOH, \beta_{MLOH} = [MLOH]/$$
$$[M][L][OH] = [MLOH][H]/[M][L]K_W$$

Finally formation of a variety of polynuclear complexes, which can involve partially protonated ligands or OH-anions, can complicate the system, for instance

$$2M + L + OH = M_2LOH, \beta = [M_2LOH]/$$
$$[M]^2[L][OH] = [M_2LOH][H]/[M]^2[L]K_W$$

For some ligands the last protonation constant is so large (e.g. for a ligand with alcohol or amide donor group) that in absence of a metal ion the completely deprotonated form will never be present in aqueous solution. In such cases the complex formation involves displacement of one or more protons as shown below

$$LH + M = ML + H, \quad K = [ML][H]/[M][LH]$$

In all these equilibria cations compete with protons for ligand donor groups, acidifying the solution. Therefore the titration curve of the ligand will change in the presence of cations in such a way that for a given volume of added base the pH value will be lower. This is clearly seen in the titration curves in Fig. D 7. The curve with Gd^{3+}, which forms the most stable complex with **D-2**, approaches the titration curve of a strong acid with a strong base as a result of nearly complete expulsion of ligand protons due to strong complexation with the deprotonated ligand under these conditions.

If the ligand is positively charged in the protonated form, as in a case of polyamines, it can form complexes also with anions due to electrostatic attraction or/and hydrogen bonding (Section A 6). The equilibrium constants of anion complexation also can be calculated from the titration data (Section B 2; ref. 18 a–c). Anion complexation stabilizes the protonated forms of the ligand; it therefore restrains the dissociation of protons, and shifts the titration curve to higher pH values, as exemplified in Fig. D 8, for the case of fluoride binding to the azacryptand **D-3**.

The mathematical analysis of the titration curves obtained in the presence of complexing agents (metal ions or anions) is based on the solution of equations (D 2-21) and (D 2-22), which involve now the concentrations of

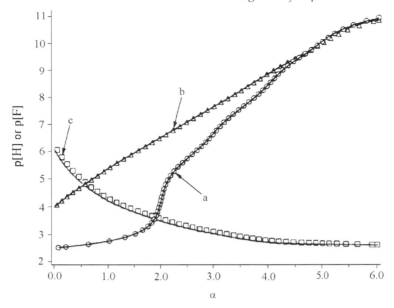

Figure D 8. Potentiometric p[H] titration of **D-3** in the absence (curve a) and presence of fluoride ion (curve b). Ion-selective-electrode p[F] measurements in the presence of fluoride ion are shown as curve c. α is defined as mmol of base per mmol of **D-3**.[25] Reprinted with permission from *Inorg. Chem.*, Relly *et al.*., 1995, **34**, 569. Copyright 1995 American Chemical Society.

D-3

complexed ligand and complexed partially protonated ligand, together with a mass balance equation for total complexing agent concentration $[M]_t$. For example, Ca^{2+} forms with EDTA (a tetracarboxylic acid, LH_4) two complexes, one with a completely deprotonated and another with a monoprotonated ligand. Three respective mass balance equations take in this case the following form

$$[L]_t = [L] + [LH] + [LH_2] + [LH_3] + [LH_4]$$
$$+ [CaL] + [CaLH]$$

$$[H]_t = [H] - [OH^-]$$
$$+ [B] + [HL] + 2[LH_2] + 3[LH_3] + 4[LH_4]$$
$$+ 3[CaLH] \qquad (D2\text{-}24)$$

$$[M]_t = [Ca] + [CaL] + [CaLH]$$

Then the concentrations of all protonated and complexed species are expressed as functions of free metal, deprotonated ligand and proton concentrations by using the respective expressions for the formation and protonation constants. This procedure converts equations (D 2-24) into three polynomial equations with three variables, which must be calculated for each point of the titration curve. This is a difficult mathematical problem, solved with the use of computer programs.

There are numerous programs developed for the determination of metal–ligand (anion–protonated ligand) formation constants from potentiometric data (reviews[11–13,16,26], comparisons between computation programs[27–29]). The more widely known versions are LETAGROP,[30] SCOGS,[31] MINIQUAD,[32] TITFIT,[33] SUPERQUAD,[34] and BEST,[35] amongst which the last two programs are the most powerful. Recently reported suite of 10 programs HYPERQUAD (recent version

HYPERQUAD 2000 available via Internet http://www.chim1.unifi.it/group/vacsab/hq2000.htm) can determine the equilibrium constants from both potentiometric and spectrophotometric titration data.[13] There are also several programs specifically designed for protonation equilibria, amongst which ACBA,[36] MUPROT,[37] and PKAS[38] are the most popular. Most of these programs are based on the method of least-squares; they differ in the choice of parameter x utilized for minimization of the residual sum of squares $U = \Sigma w_i (x_{exp,i} - x_{calc,i})^2$, where w_i is the statistical weight, and the available minimization algorithms (Section D1). One can choose pH meter or potential readings as the independent variable and minimize U for the added volumes V of the titrant ($x = V$). Alternatively one can take as the independent variable V and minimize U for $x = $ pH or E. Most frequently the unweighted regression analysis is performed, but some programs, such as SUPERQUAD, use w_i expressed as inverse variance of the measurement at the given point

$$1/w_i = s_i^2 = s_E^2 + (dE/dV)_i^2 s_V^2 \quad (D2\text{-}25)$$

where s_i^2 is calculated by using the standard error propagation formula from variances of the electrode and volume readings, s_E^2 and s_V^2 respectively, and $(dE/dV)_i$ is the slope of the titration curve at the given i-th point. The principal advantage of this approach is that it gives lower weights to experimental points in the region of an equivalence point, where dE/dV is high and errors are large.

It was shown that errors in reactant concentrations, initial volume, previously fixed values of protonation constants, electrode calibration or the presence of unknown impurities can lead to systematic deviations of experimental points from the theoretical fitting curve.[39] The fitting of pH changes instead of absolute pH values was proposed to minimize the influence of such errors.[39]

In some special cases high concentrations of supporting electrolyte cannot be used. For example, determinations of small complex formation constants of the order of 1–10 M^{-1} need very high concentrations of components, which then lead to variations of the ionic strength during titration. Correction factors for activity coefficients, junction potential and glass electrode selectivity, which should be applied in such cases, are discussed in the literature[40]. Special computer programs (BSTAC, ESAB, ES4EC) which take into account variable ionic strength are available.[41]

An obvious limitation of the discussed technique is the working range $2 < \text{pH} < 12$ of the glass electrode. Ligand protonation constants lower than about 100 M^{-1}, and higher than about 10^{12} M^{-1}, as well as complex formation constants with such ligands, cannot be measured because too acidic ligands always remain unprotonated, and too basic always remain protonated in the accessible pH range. Nevertheless, few other methods allow to measure binding constants over such a wide range.

D 2.1.2. Potentiometry with ion-selective electrodes

For comprehensive reading on theory and applications of ion-selective electrodes (ISEs) several books,[42–44] and review articles,[45] should be consulted, in particular Koryta's repeatedly updated reviews.[46] A semipermeable electrode membrane, either solid or liquid, can be made specific to different cations and anions. Many ISEs are produced commercially. The electrode potential is created, as in the case of a glass electrode, by diffusion of a given ion through the membrane, which behaves as an ion exchanger. The membrane material can be an ionic insoluble solid phase, such as a glass for cations, a hydrophobic salt (e.g. tetraphenylarsonium tetraphenylborate), a polymeric ion exchanger (e.g. a polyamine), or a neutral ion carrier, such as crown compounds or natural ionophores (for application of macrocycles to

ISEs[47]), incorporated into a polymeric membrane.

All problems associated with the liquid junction potential and activity coefficients discussed above for the glass electrode persist for ISEs, and are solved in the same way by calibration of the ISE with known concentrations of a given ion under conditions of the titration and subsequent calculation of stochiometric equilibrium constants. It is worth noting that the selectivity of ISEs to a given ion, as a rule, is lower than that of a glass electrode to hydrogen ions. The selectivity coefficient K_{XY} of an electrode specific for an ion X in the presence of another ion Y is defined by the following semiempirical equation

$$E = (RT/nF)\ln([X] + K_{XY}[Y]) + \text{constant}$$

The use of ISEs for determination of complex formation constants of alkali cations and NH_4^+ with macrocycles has been reported.[48–50] In this method a solution of the salt (e.g. NaCl) of a given concentration $[M]_t$ is titrated by a solution which contains the same salt concentration and an excess of the neutral macrocyclic ligand (e.g. 18-crown-6) in the cell with ISE for a given cation and Ag/AgCl reference electrode. The calibration plot usually shows the ideal Nernst slope equal to $RT/F = 59.16$ mV, and therefore the concentration of free cation in each point of the titration curve can be calculated by using the following equation

$$[M^+] = [M]_t 10^{-\Delta E/59.16} \qquad (D\,2\text{-}26)$$

where ΔE is the difference between meter readings in pure salt solution and in the presence of the ligand. The complex formation constants can be calculated then as described in Section D 1. ISE measurements with weak complexes, where the ionic strength cannot be kept constant, are considered in the literature.[51]

An example of ISE measurements with a fluoride-specific electrode is shown in Fig. D 8 (curve c, open squares) for fluoride binding to protonated D-3. As should be expected, the concentration of free F^- increases upon neutralization of the cationic polyamino ligand. The solid line is calculated from potentiometric titration results, and shows very good agreement with ISE data.

In some cases an ISE specific to a desired ion can be prepared by relatively simple modification of a commercial ISE. Thus, the binding of cations (D-4a,b) and some other drugs containing ternary or secondary protonated amino groups to α- and β-cyclodextrins was studied by using a series of ISEs in which new plasticised poly(vinyl chloride) membranes specific for the respective drug were installed.[52] Each membrane was prepared by exchange of protons of $-SO_3H$ end-groups of sulfonated poly(vinyl chloride) with drug cations in THF and subsequent precipitation of conditioned polymer by dilution with water.

D-4a   **D-4b**

D 2.1.3. Potentiometry with metal–metal ion electrodes

If the indicator electrode itself represents a metal, usually in a form of a wire, which is immersed into a solution of a salt of the same metal, like an Ag electrode in solution of a salt of Ag^+ or Cu electrode in solution of a salt of Cu^{2+}, etc., the potential of the resulting half-cell is given by equation (D 2-3) where a_X is the activity of a respective cation. Such electrodes, known also as electrodes of the first order, would be very convenient for the measurement of free metal ion concentrations; unfortunately, however, they are reversible only for few metals, such as Ag, Cu, Pb, Cd and Hg.

Many complex formation constants of macrocyclic ligands with Ag^+ and other monovalent cations, the latter by competition with Ag^+, were determined by the potentiometric titration with a silver electrode in water and organic solvents.[15,53] In this method a solution of $AgNO_3$ in a given solvent containing a supporting electrolyte (tetraethylammonium perchlorate (TEAP)) is titrated with a ligand solution in the same solvent containing the same concentration of TEAP and, in a competition experiment, a salt of another metal. The titration solution is connected with a reference half-cell, which is filled with the same silver nitrate solution as the titration half-cell, by a salt bridge containing TEAP in the same solvent. Both indicator and reference electrodes are silver electrodes immersed in titration and reference solutions, respectively. The complex formation constants can be calculated from the results obtained in this way by the competition method discussed in Section D1.

D2.2. MACROSCOPIC AND MICROSCOPIC PROTONATION CONSTANTS AND ACID–BASE BEHAVIOR OF MACROCYCLES

The experimentally determined protonation constants defined as in equation (D2-20) are macroscopic parameters (Section D1) in the sense that for polyacids or polybases each of them corresponds actually to a set of protonation equilibria between all isomeric forms of a form LH_{i-1} and all isomers of LH_i. For example, in a case of an α-amino acid with a neutral side group R there are four protonation 'microequilibria' shown in Scheme D-1, but only two experimentally determined protonation constants

$$K_{H1} = ([AH] + [AH^\pm])/[A^-][H^+] = K_1 + K_2$$
$$(D2\text{-}27)$$

$$K_{H2} = [AH_2^+]/([AH] + [AH^\pm])[H^+]$$
$$= K_{12}K_{21}/(K_{12} + K_{21}) \qquad (D2\text{-}28)$$

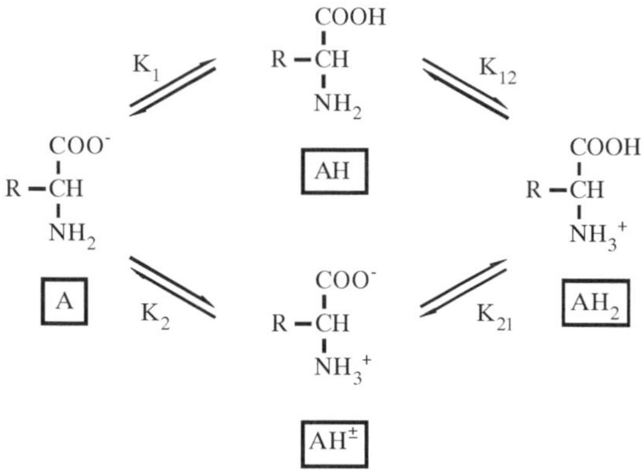

Scheme D1. Protonation microequilibria for an amino acid.

The microequilibria in Scheme D1 form a cycle and, therefore, an additional relationship exists between microconstants K_i

$$K_1 K_{12} = K_2 K_{21} \qquad (D2\text{-}29)$$

Evidently four microconstants cannot be calculated from only three equations, and if the problem is the determination of microconstants at least one of them must be evaluated independently. In this particular case the protonation constant of a methyl ester of an amino acid can be taken for the evaluation of K_{12} as long as the inductive effects of COOH and COOMe groups are considered to be similar. For example, in the case of glycine (R = H) the logarithms of protonation macroconstants are $\log K_{H1} = 9.8$ and $\log K_{H2} = 2.35$, while the logarithm of protonation constant of H_2NCH_2COOMe is $\log K_H = 7.6 = \log K_{12}$. With the latter value equation (D2-28) gives K_{21}, practically coinciding with K_{H2}. Then equations (D2-27) and (D2-29) give $K_1 = 3.5 10^4$ M^{-1} and K_2 practically coinciding with K_{H1}. Thus, $K_2 \gg K_1$ and $K_{12} \gg K_{21}$ in good agreement with what should be expected from the known fact that aliphatic carboxylate anions are much less basic than aliphatic amines. The ratio of equilibrium concentrations of zwitterionic and neutral monoprotonated forms equals $[AH^\pm]/[AH] = K_2/K_1 = 1.8 \times 10^5$. Evidently the dominant form is zwitterionic, and if the neutral form were excluded from equations (D2-27) and (D2-28) these would give $K_{H1} = K_2$ and $K_{H2} = K_{21}$ with fairly high precision. Of course, exclusion of a microscopic form renders impossible the determination of all microscopic constants, but it strongly simplifies the treatment, and allows one to assign immediately macroscopic protonation constants.

The number of microequilibria increases rapidly with the number of ionogenic groups. Scheme D2 shows the equilibria existing for a diaza macrocycle **D-5** with two pendent carboxyl groups. In this case there are sixteen microconstants, but only four experimentally determined macroscopic protonation constants. These are defined by equations analogous to (D2-27) and (D2-28), but involve more intermediate microscopic forms: two

monoprotonated, four diprotonated and two triprotonated. The scheme involves ten cycles and, consequently, there are ten additional relationships between microconstants. As a result, there are fourteen equations to calculate sixteen microscopic constants and, therefore, two of them should be evaluated independently.

When some ionogenic groups of a given ligand definitely differ strongly in their basicities, several microscopic forms can be excluded, as was discussed above for Scheme D1. When two or more groups possess more or less similar basicities the study of the ligand protonation by other experimental techniques is usually very helpful. For example, in a case of histidine the protonation of the imidazole ring can be followed by ^{1}H or ^{13}C NMR even when histidine forms part of a protein.[54] Spectrophotometry is also useful

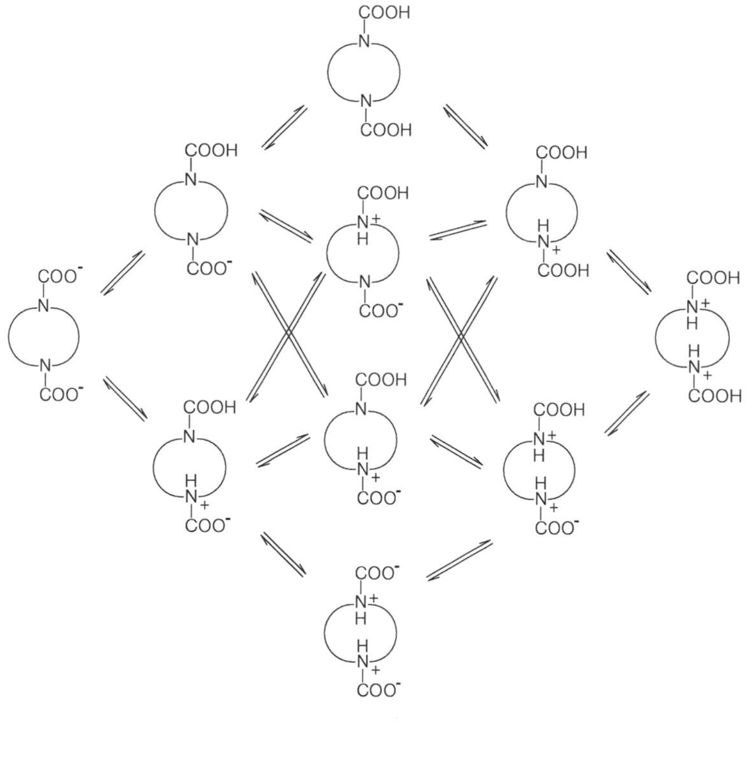

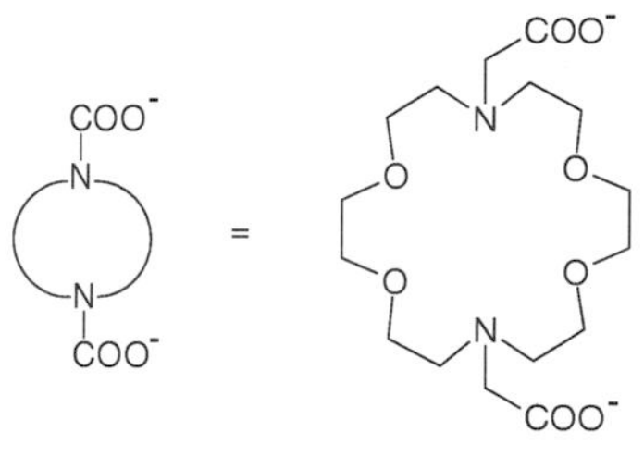

Scheme D2.   Protonation microequilibria for aza macrocyle **D-5**.

when the ionogenic group is a chromophore, like a phenol, or in flavonoles, some siderophores etc. Another method based on Hammet-type relations was applied for the calculation of dissociation microconstants of isomeric nicotinic acids.[55]

An important special case is when all base or acid groups of a ligand are chemically equivalent, such as in polycarboxylic acids or polyamines. A set of microscopic equilibria must in principle be written for the protonation of such a ligand also, but the equivalence of the groups often allows one to simplify the treatment by introducing statistical factors only. Let us consider protonation of an aliphatic diamine of the general structure $H_2N(CH_2)_nNH_2$ (or a dicarboxylic acid dianion $^-OOC(CH_2)_nCOO^-$). Since the amino groups are equivalent we can assign to each of them an intrinsic protonation constant K_0 and take into account that the concentrations of neutral and diprotonated diamines should be duplicated in order to give the concentrations of neutral and protonated amino groups respectively. Now the first macroscopic protonation constant can be expressed as

$$K_{H1} = [H_2N(CH_2)_nNH_3{}^+]/$$
$$[H_2N(CH_2)_nNH_2][H^+] = 2K_0$$

and the second constant as

$$K_{H2} = [H_3N(CH_2)_nNH_3{}^{2+}]/$$
$$[H_2N(CH_2)_nNH_3{}^+][H^+] = K_0/2.$$

These expressions are correct if protonation of the first amino group does not influence the pK_a of the second group. In such a case, therefore, $K_{H1}/K_{H2} = 4$ (see equation (D1-30)) or $\Delta \log K_H = 0.6$.

Table D1 shows some representative data for protonation of diamines and dianions of diacids. Evidently, $\Delta \log K_H$ approaches a purely statistical value 0.6 only for ligands with large distances between the groups, indicating that usually protonation of the first group is thermodynamically more favorable than protonation of the second group. This

Table D1. Logarithms of protonation constants of some aliphatic primary diamines and anions of dicarboxylic acids.

Diamine or diacid	$\log K_{H1}$	$\log K_{H2}$	$\Delta \log K_H$
H_2NNH_2	7.9	−1.05	8.95
$H_2N(CH_2)_2NH_2$	9.93	6.85	3.08
$H_2N(CH_2)_4NH_2$	10.8	9.35	1.45
$H_2N(CH_2)_8NH_2$	11.0	10.1	0.9
$^-OOC-COO^-$	4.3	1.3	3.0
$^-OOC(CH_2)_2COO^-$	5.6	4.2	1.4
$^-OOC(CH_2)_5COO^-$	5.4	4.5	0.9
$^-OOC(CH_2)_7COO^-$	5.4	4.55	0.85

effect is explicable in terms of electrostatic destabilization of doubly charged forms due to a mutual repulsion of similarly charged groups. In particular, for 1,2-diamines and 1,2-diacids a stabilization of the monoprotonated species by intramolecular hydrogen bonding should be taken into account.

The acid–base behavior of macrocycles and other concave structures has been reviewed.[56] Ionogenic functional groups of macrocyclic ligands can be part of a macrocycle (nitrogen atoms of azamacrocyles), or can be attached more or less flexibly to a macrocycle (lariat ethers etc.). The functional groups located outside the macrocycle behave generally like the same groups in acyclic molecules, but the groups inside the macrocycle can be strongly influenced by their microenvironment.

A macrocycle is conformationally less flexible than a chemically similar open chain molecule. As a result the functional groups are forced to remain at only slightly variable distances and the above discussed electrostatic repulsion effect can be more pronounced for macrocycles. For example, the logarithms of two successive protonation constants of non-macrocyclic diamine **D-6a** differ by $\Delta \log K_H = 0.8$, but for the macrocyclic diamine **D-6b** $\Delta \log K_H$ increases to 3.65; for **D-6c**, where nitrogen atoms are more separated, it equals 1.05. Trends in stepwise protonation constants of polyazamacrocycles also show the importance of the distance between ring amino groups.[56,57a] In Fig. D9 the protonation constants for macrocycles **D-7 a–d** are plotted against the number of protonated nitrogen atoms. In each case a sharp decrease in $\log K_{Hi}$ is observed when all distant nitrogens of a given macrocycle are already protonated.

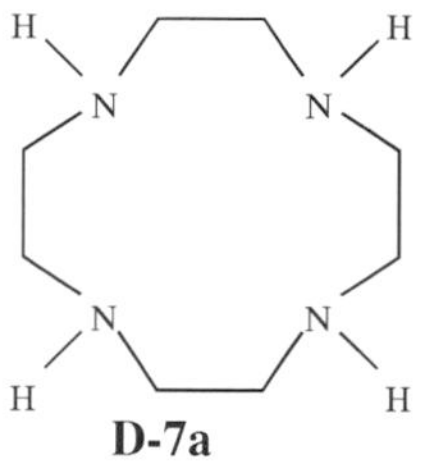

D-6a

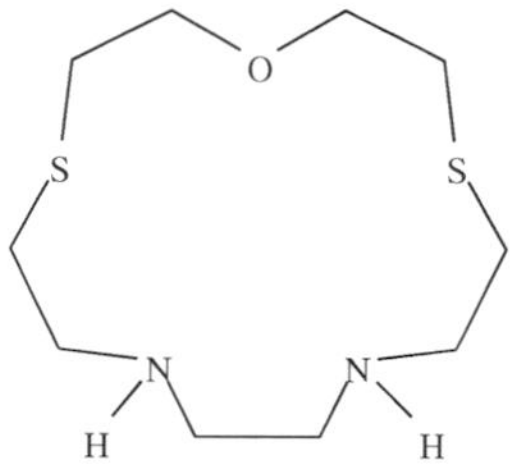

D-6b

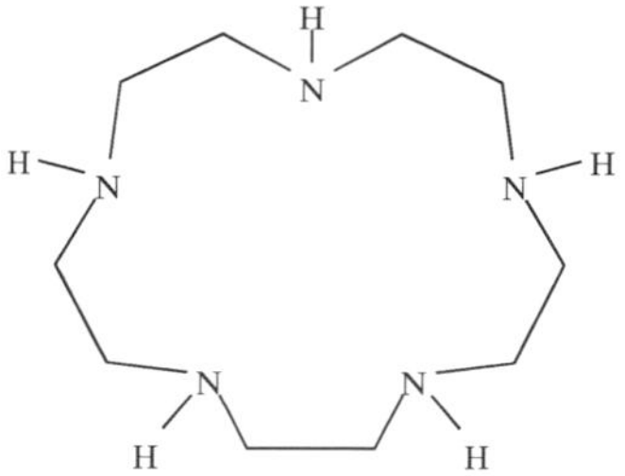

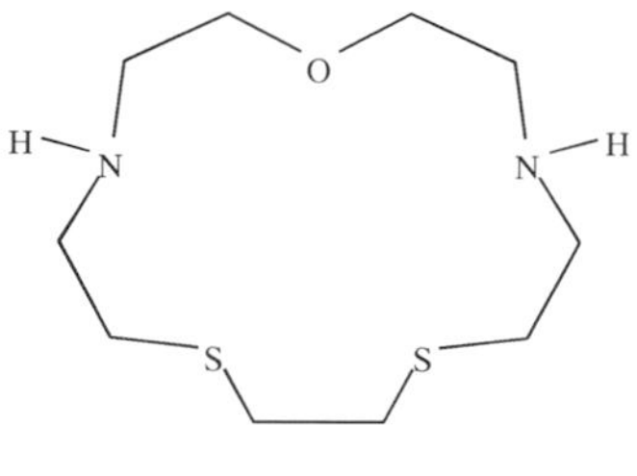

D-6c

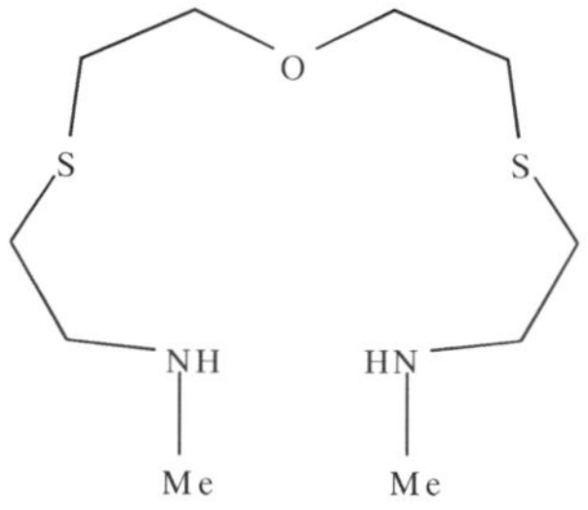

D-7a

D-7b

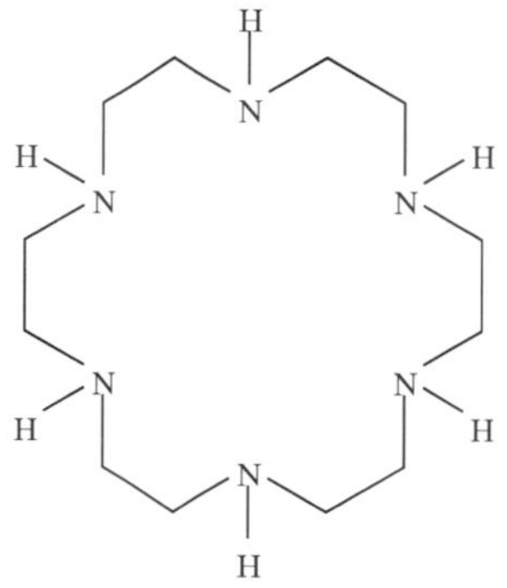

D-7c

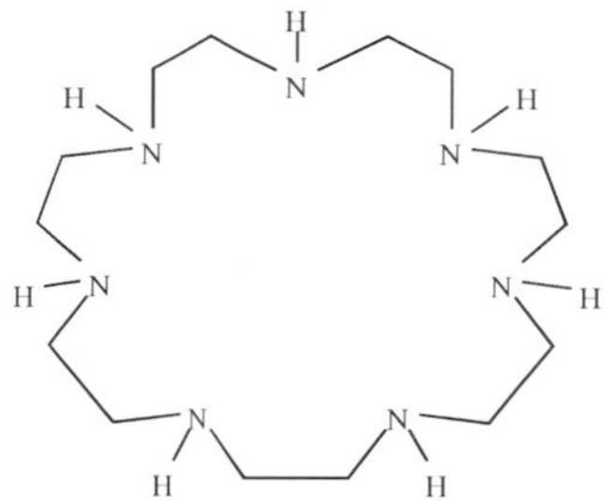

D-7d

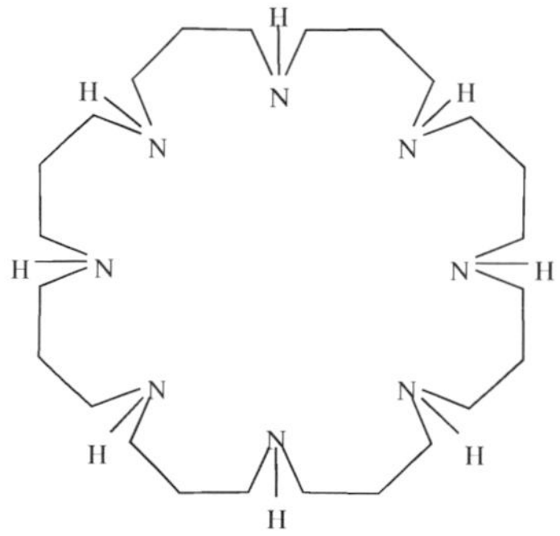

D-7e

Sites of protonation in partially protonated azamacrocycles can be identified in principle by NMR spectroscopy. As an example let us consider the results of ^{1}H and ^{13}C NMR study of protonation of macrocycle **D-8** shown in Fig. D 10.[57b] Fig. D 10a is the distribution diagram of the species formed by protonation of **D-8** as a function of pH calculated with the protonation constants determined by potentiometric titrations. Figure D 10b shows the NMR

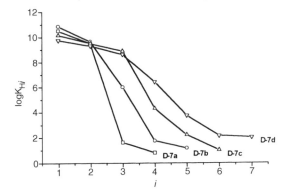

Figure D 9. Stepwise protonation constants for aza macrocycles.

chemical shifts of different hydrogen atoms of methyl and methylene groups of the macrocycle at various pH measured in D_2O. In the pH range 8–11, as the mono- and diprotonated species are formed the resonances of *all* protons undergo the downfield shifts. This indicates that two first protons are distributed among all nitrogens and since the protonation equilibrium is fast in the NMR scale one observes an averaged shift of all signals. Addition of the third and forth protons induce further downfield shifts of the ^1H signals of hydrogen atoms of both C2 and C3, located in

the α positions with respect to secondary more basic nitrogens, but signals of hydrogens in the α positions with respect to tertiary nitrogen (C1 and C4) tend to return to their initial values observed in the neutral ligand. This means that in the tetracation of **D-8** protonation sites are secondary nitrogens while the tertiary nitrogens remain neutral. Protonation-induced changes in the ^{13}C NMR spectrum of **D-8** also agree with this conclusion.

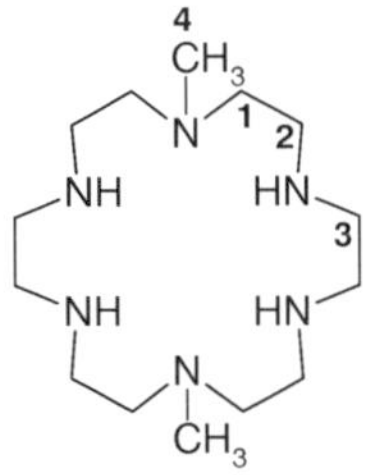

D-8

The above mentioned intramolecular hydrogen bonding, which can persist even in aqueous solutions, causes specific effects on protonation constants of macrocycles. Thus, the anomaleously low (compared to the respective phenols) first pK_a value of calixarenes, Scheme D 3, is attributed to hydrogen bond stabilization of the appearing phenolate

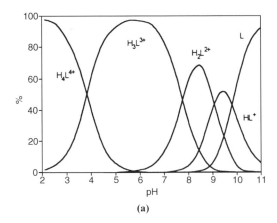

(a)

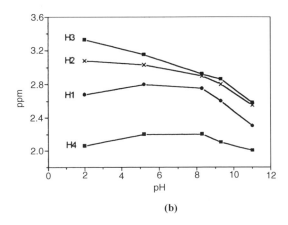

(b)

Figure D 10. (a) The distribution diagram of the species formed by protonation of **D-8** as a function of pH; (b) ^1H NMR chemical shifts as function of pH. [57b] Reproduced by permission of the Royal Society of Chemistry from *J. Chem. Soc., Perkin Trans 2*, Andres *et al.*, 1994, 2367.

D-12

anion.[58] With the octaphenol **D-9** the pK_a for the abstraction of the first four protons is two units lower than that of the parent resorcinol, whereas the pK of the remaining four protons involved in the especially strong hydrogen bonds is above even methanol.[59] Another manifestation of this effect was found in the basicities of macrocyclic pyridines **D-10** in water, methanol and water–methanol mixtures.[60] Logarithms of the protonation constants of **D-10** (R = H) in water are 4.88, 4.95, 4.16, 3.95, 3.70, 3.53 and 3.36 for $n = 2$–8, while for non-macrocyclic pyridine **D-11** $\log K_H =$ 3.36. Enhanced basicity of these molecules is due to hydrogen bond stabilization of the protonated pyridine through an included water molecule, as shown in **D-12**, a maximum being observed for $n = 3$ which provides the optimal complementarity of interactions.

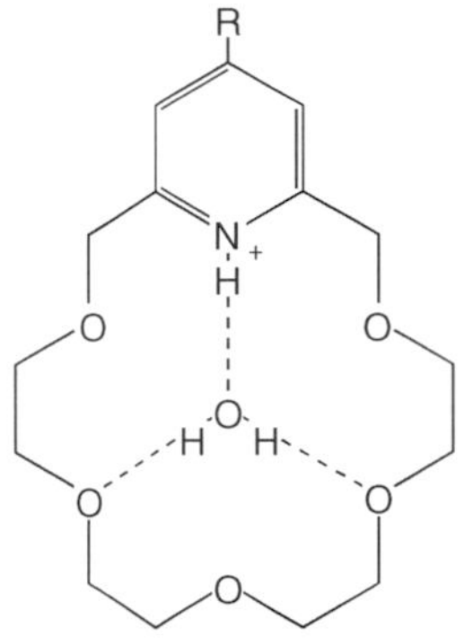

$pK_{a1} = 3.3$ 　　　 $pK_a = 8.9$

$pK_{a1} = 2.9$ 　　　 $pK_a = 7.0$

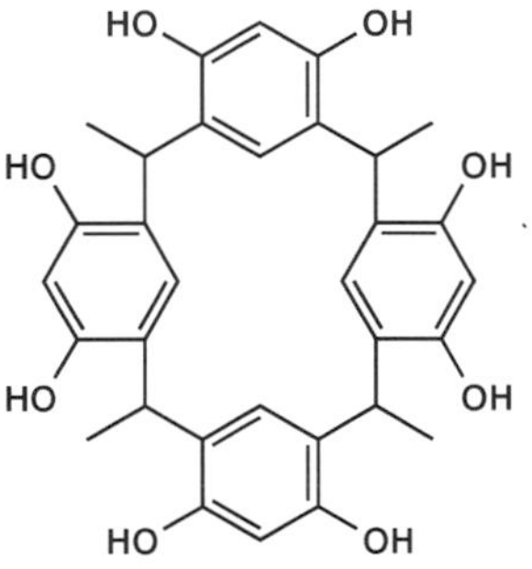

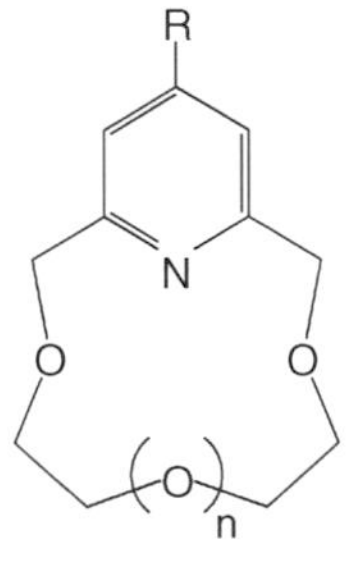

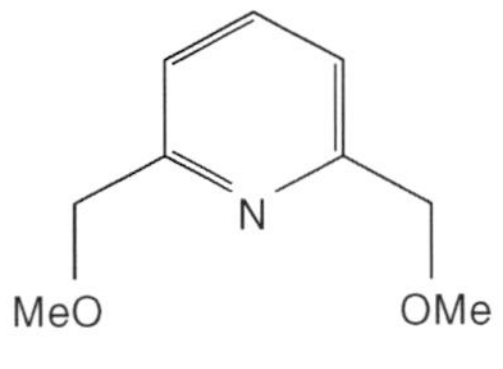

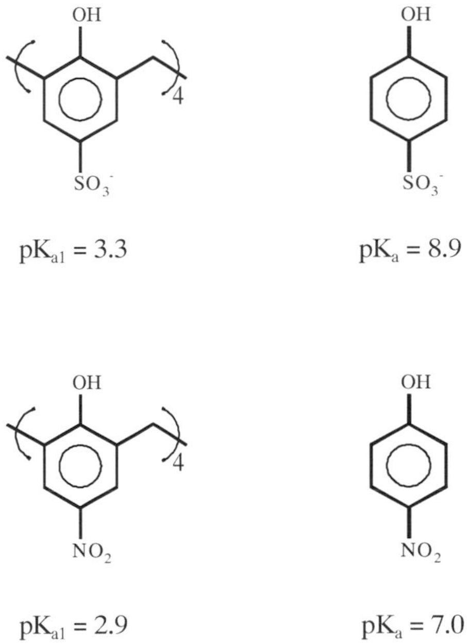

Scheme D 3. Acidities of calixarenes and the respective phenoles.[58]

D-9

D-10 　　　 **D-11**

The most unusual effects were found for some macrobicyclic molecules, however. Besides anomalous protonation constants they show also anomalously slow kinetics of

protonation/deprotonation reactions.[61,62] The basicity of cryptands **D-13**, which possess first $\log K_H$ values between 9.7 and 11.2, is quite typical for a tertiary amine, but the second-order rate constants of proton transfer from protonated cryptand to outside OH-anions are three to seven orders of magnitude lower than those observed for deprotonation of aliphatic ammonium ions by OH-anions.[63] In a structure like **D-13** the nitrogens can be protonated outside (*o* or exo conformation), or inside (*i* or endo conformation) the cavity. Relatively large macrocycles **D-13** allow rapid interconversion of conformations, which in this case do not differ considerably in their energies. With the smaller cryptand **D-14** this interconversion becomes slow, and the processes of outside and inside protonation can be discriminated kinetically. Perhaps, the most impressive fact is the extremely slow deprotonation of internally double protonated **D-14**, which remains in double protonated form for weeks in 5 M KOH. Scheme D 4 shows all rate and equilibrium constants determined for protonation of this cryptand. Note, first, the low pK_a for a tertiary amine for outside protonation of **D-14** (reflecting the unfavorable exo conformation in this case), and, second, the extremely small rate constants of second protonation of the internally monoprotonated cryptand, as well as of the deprotonation of both mono and internally double protonated forms. The H/D exchange rate between protons inside the protonated bicycle D-12 and outside D$_2$O is spectacularly slow.

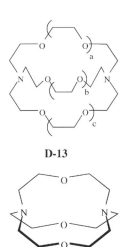

D-13

D-14

Host–guest complexation can considerably affect the dissociation constants of host and guest molecules. In the simplest case of interaction between an ionizable substrate and non-ionizable ligand the following equilibria must be considered

$$SH \rightleftharpoons S + H^+ \qquad (K_a^{SH})$$
$$S + L \rightleftharpoons SL \qquad (K_1)$$
$$SH + L \rightleftharpoons SHL \qquad (K_2)$$

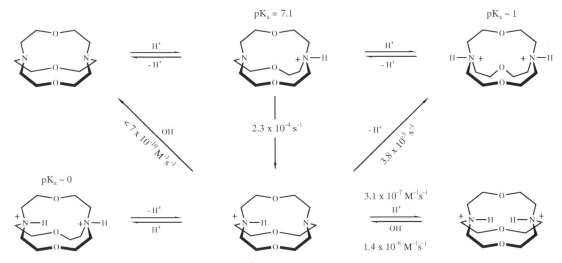

Scheme D 4. Protonation equilibria with **D-14**.[62] Reprinted with permission from *J. Am. Chem. Soc.*, Smith *et al.*, 1981, **103**, 6044. Copyright 1981 American Chemical Society.

The apparent formation constant (K_{app}) is pH-dependent according to equation

$$K_{app} = ([SL] + [SHL])/([S] + [SH])[L]$$
$$= (K_1 + K_2[H^+]/K_a^{SH})/(1 + [H^+]/K_a^{SH})$$
$$(D\,2\text{-}30)$$

In accordance with equation (D 2-30) K_{app} equals K_1 at extremely high pH and K_2 at extremely low pH. Just as K_{app} depends on pH, an apparent dissociation constant of the substrate $K_{a,app}^{SH}$ is a function of the ligand concentration

$$K_{a,app}^{SH} = ([S] + [SL])[H^+]/([SH] + [SHL])$$
$$= K_a^{SH}(1 + K_1[L])/(1 + K_2[L]) \quad (D\,2\text{-}31)$$

Note that only the dissociation constant of free substrate is needed for the description of the system. The dissociation constant of bound substrate is given by equation

$$K_a^{SHL} = [SL][H^+]/[SHL] = K_a^{SH}K_1/K_2 \quad (D\,2\text{-}32)$$

or

$$pK_a^{SHL} = pK_a^{SH} - \log(K_1/K_2) \quad (D\,2\text{-}33)$$

The complexation-induced shifts in the acidity constants of ionizable substrates (or ligands) produce, of course, corresponding shifts in pH, allowing potentiometric titrations. In the case of ligand–metal complexation the metal ion usually competes with a proton for the same donor atom and complexation induces large shifts in solution acidity. In the case of organic host–guest complexation, like, the extensively studied binding of organic acids to cyclodextrins,[64] complexation-induced shifts in pK_a are rather small, typically in the range 1–2 units or less. Nevertheless, measurements of shifts in pH (typically less than 1 unit) upon addition of cyclodextrins to solutions of organic acids can be used for determination of binding constants.[64a–d] In terms of equation (D 2-32) these shifts are due to different affinities of the host to neutral and deprotonated forms of an acid substrate. As was mentioned in Section B6,

ionic substituents considerably decrease binding of organic guests. Therefore, usually $K_2 > K_1$ and complexation leads to an increase in pK_a and proton uptake. Some exclusions, however, were reported, among which the most known is much stronger binding of *para*-nitrophenolate anion than *para*-nitrophenol to α-cyclodextrin.[64b]

D 2.3. POLAROGRAPHY AND CYCLIC VOLTAMMETRY

There are many techniques based on current–voltage measurements,[65] amongst which we will discuss here only classical polarography and cyclic voltammetry (CV). These two methods have been most widely used for the study of complex formation equilibria and for characterization of redox properties of supramolecular systems. For general reading on theory and applications of these methods several works[66–69] can be recommended. In addition, introductory texts on CV, which is a relatively new technique, are available.[70,71]

Classical polarography deals with current–voltage curves obtained with a dropping mercury electrode (DME). If a reversible redox process, like reduction of a metal cation, takes place at the electrode surface the equation for experimentally obtained current(i)–potential (E) profile takes the form

$$E = E_{1/2} - (2.3RT/nF)\log(i/(i_d - i))$$
$$(D\,2\text{-}34)$$

where $E_{1/2}$ is the half-wave potential ($E = E_{1/2}$ when $i = i_d/2$) given by equation (D 2-35)

$$E_{1/2} = E^{0\prime} - (2.3RT/nF)\log(I_{ox}/I_{red})$$
$$(D\,2\text{-}35)$$

and i_d is the diffusion current, which is the limiting value of the current under conditions when the surface concentration of the oxidized form is zero (infinitely large negative potential). In equation (D 2-35) $E^{0\prime}$ is the formal potential of the given redox couple and I is the diffusion current constant.

Polarographic studies of complex formation equilibria are based on complexation-induced shifts in half-wave potentials $E_{1/2}$. For the determination of stepwise formation constants of the complexes with redox active metal ions the equation of DeFord and Hume (D2-36)[72] is used

$$\text{antilog}\{(0.4343nF/RT)[(E_{1/2})_f - (E_{1/2})_c]$$
$$+ \log(I_f/I_c)\} = 1 + \beta_1[L] + \beta_2[L]^2$$
$$+ \beta_3[L]^3 + \cdots \qquad \text{(D2-36)}$$

where the subscripts 'f' and 'c' indicate the parameters obtained for the free and the complexed with L metal ion, respectively. The ratio I_f/I_c is calculated as the ratio of diffusion currents in the absence and in the presence of the ligand:

$$I_f/I_c = (i_d)_f/(i_d)_c$$

The electrode process considered for the derivation of equation (D2-36) is the reversible reduction of metal ions and complexes at DME affording the metal amalgam.

Polarography is a very sensitive and selective method, which allows one to work with metal concentrations of the order 10^{-6} M, and to use a high excess of the ligand. Under such conditions the concentration of free ligand equals approximately its total concentration, and the fitting of experimental results to equation (D2-36) can be done with any non-linear least-squares regression program. The case of low ligand to metal ratios was analyzed for both linear and cyclic voltammetric data.[73]

Complex formation constants obtained by polarography are less precise than those calculated from potentiometric measurements, principally due to poor precision in the determination of $E_{1/2}$.[74] In addition, adsorption of electroactive species on the electrode strongly affects the determination of the formation constants.[75]

Polarography of redox active organic ligands was used for determination of the constants of ion-pair association between protonated macrocyclic polyamines like **D-7** and anions of di- and tricarboxylic acids.[76] Anodic waves attributed to the process

$$\text{Hg} + \text{LH}_i = \text{HgL} + i\text{H}^+ + 2e^-$$

were observed for these macrocyles at DME.

While in classic polarography the voltage is varied linearly only in one direction, in CV the voltage applied to the working electrode is scanned from a starting value to a predetermined limit (switching potential); then the direction of the scan is reversed. A hanging mercury drop electrode (HMDE) and various solid electrodes such as platinum, gold, glassy carbon are commonly used as working electrodes in CV. A typical cell for CV contains three electrodes: working, reference and auxiliary. The latter is placed very close to the working electrode and provides the necessary current. Thus, practically no current flows through the reference electrode and its potential remains constant.

Figure D11 (solid line) shows a typical CV experiment with a cyclic voltammogram of $[\text{Fe(CN)}_6]^{4-}$. Initially the potential is scanned to more positive values, as indicated by the arrow. When it becomes sufficiently positive to oxidize Fe(II), an anodic current appears due to the electrode reaction

$$[\text{Fe(CN)}_6]^{4-} = [\text{Fe(CN)}_6]^{3-} + e^-$$

and increases until the concentration of $[\text{Fe(CN)}_6]^{4-}$ at the electrode surface is considerably decreased due to its conversion into the oxidized form. The switching potential is 0.7 V; during the subsequent negative scan a cathodic current is observed due to reduction of $[\text{Fe(CN)}_6]^{3-}$ accumulated during the first scan. It is essential that the particles of both the oxidized and the reduced form of the redox-active substance move in the vicinity of the working electrode only by diffusion. Convection, which can rapidly remove the product of the electrode reaction formed

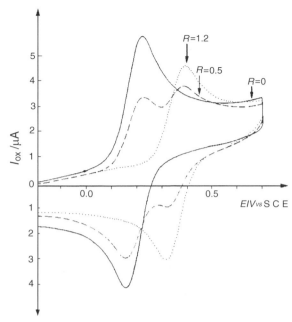

Figure D 11. Cyclic voltammetry (50 mV s^{-1}) on Pt disc, in 0.1 M KCl, pH 5.5; R = [**D-7e** 8H$^+$]/ [Fe(CN)$_6$]$^{4-}$.[77]

during the first scan, should be minimized (no stirring, no vibrations, etc.).

The important parameters of a cyclic voltammogram are: anodic peak current (i_{pa}) and anodic peak potential (E_{pa}) taken from the peak position of anodic current and the respective parameters i_{pc} and E_{pc} from the peak position of cathodic current. If the redox couple is reversible the half-wave potential is centered between E_{pa} and E_{pc}

$$E_{1/2} = (E_{pa} + E_{pc})/2 \qquad (D\,2\text{-}37)$$

Frequently $E_{1/2}$ calculated in this way is considered simply as the formal potential because the diffusion coefficients have a small effect. The difference between peak potentials is related with the number of electrons transferred in the electrode reaction

$$\Delta E_p = E_{pa} - E_{pc} = 0.059/n \qquad (D\,2\text{-}38)$$

For the voltammogram in Fig. D2-5 $E_{pa} = 0.22$ V, $E_{pc} = 0.16$ V (*vs.* SCE, standard calomel electrode) and $\Delta E_p = 0.06$ V are in a good agreement with what is expected for a one-electron reversible redox reaction.

Electrochemical 'reversibility' means that the electrode reaction is fast enough to maintain the surface concentrations of the oxidized and reduced forms at equilibrium with each other, which is an indispensable condition for the validity of the Nernst equation. Evidently one should use conditions which provide the most possible approximation to reversible behavior. An important factor is the working electrode preparation because the electron transfer kinetics depends on the type and condition of the electrode surface. Solid electrodes should be polished, usually with Al$_2$O$_3$, and cleaned, for example by sonification in water. Some authors recommend 'activation' of electrodes by application of a sequence of extreme positive and negative potentials. Another important factor is the potential scan rate, which can be as high as 1000 V s^{-1}. With too fast scans irreversible behavior can be observed because the electrode does not spend enough time for equilibration at each potential value. This leads to an increase in ΔE_p above the value predicted by equation (D2-34). In such a case the scan rate should be slowed down until ΔE_p decreases to its normal value. There is a lower limit for the scan rate, however: the whole cycle should be scanned in less than about 30 s in order to avoid interference from convection. Thus, scan rates of the order 0.05–0.1 V s^{-1} in the potential range 0.5–1 V are often used.

Redox couples whose peak separation increases with increasing scan rate are classified as 'quasi-reversible'. Estimation of $E_{1/2}$ with equation (D2-37) from such voltammograms is rather unreliable. It should be noted that even for reversible couples estimation of $E_{1/2}$ in CV is less precise than in classical polarography. In particular, $E_{1/2}$ depends on the choice of the switching potential, which should be as far as possible from the middle of peak potentials.

The peak current for a reversible reaction is given by equation

$$i_p = (2.69 \times 10^5) n^{3/2} A D^{1/2} C v^{1/2} \quad \text{(D2-39)}$$

where i_p is given in Amps, n is the number of electrons transferred in the electrode reaction, A is the electrode surface (cm^2), D is diffusion coefficient $(\text{cm}^2 \text{s}^{-1})$, C is the concentration (mol cm^{-3}) and v is the scan rate (V s^{-1}). The most important feature of equation (D2-39) is the proportionality of the peak current to concentration and to the square root from the scan rate. Such proportionality exists also for completely irreversible couples, but for quasi-reversible couples the current is not proportional to the square root from the scan rate. Observation of such proportionality together with a correct distance between peak potentials in accordance with equation (D2-38) are usual experimental criteria of reversibility of the system under investigation.

An important application of CV is the evaluation of formal potentials from equation (D2-37). This has been done for many redox-active macrocycles.[78] In spite of being rather imprecise the values of $E^{0'}$ obtained by using CV are quite suitable for comparative studies.

Complexation-induced shifts of the peak currents and potentials can be used for the evaluation of formation constants, although this method is less accurate than others. In a case when the formal potentials of both free $(E_f^{0'})$ and complexed $(E_c^{0'})$ forms of a redox-active component can be obtained, the ratio of complex formation constants for reduced (K_{red}) and oxidized (K_{ox}) states of this component can be calculated from the difference between the formal potentials. Indeed, from the cycle shown in Scheme D5, where M represents the redox-active component and L is the ligand, it follows that

$$RT \ln K_{ox} + nFE_c^{0'} = nFE_f^{0'} + RT \ln K_{red}$$

and

$$\Delta E^{0'} = E_c^{0'} - E_f^{0'} = (RT/nF) \ln(K_{red}/K_{ox})$$
$$\text{(D2-40)}$$

Such treatment is very useful when the formation constant for one of the oxidation states is known from other data, and when the complexation is strong enough to ensure a complete formation of the complex under experimental conditions. This is the case, for example, in the interactions of highly charged redox-active anions like $[\text{Fe(CN)}_6]^{4-}$ or $[\text{Ru(CN)}_6]^{4-}$ with cationic azamacrocycles like protonated **D-7**.[77,79] Cyclic voltammograms of $[\text{Fe(CN)}_6]^{4-}$ in the presence of macrocycle **D-7e** are shown in Fig. D11. When the ratio [macrocycle]/$[\text{Fe(CN)}_6]^{4-}$ is less than unity two redox couples attributed to free and complexed $[\text{Fe(CN)}_6]^{4-}$ are observed. With an excess of the macrocycle,

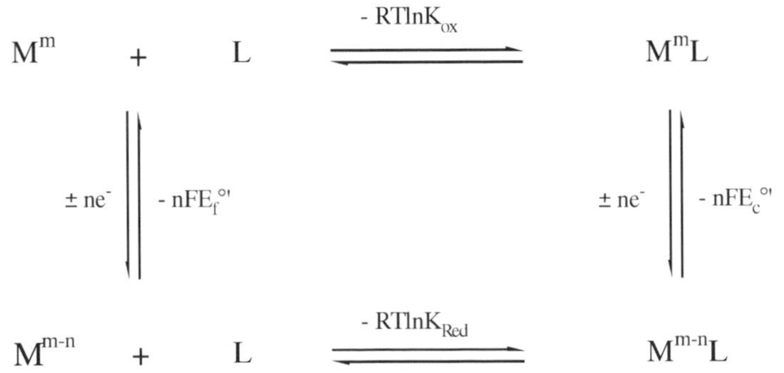

$$
\begin{array}{ccccc}
M^m & + & L & \xrightleftharpoons{\;-RT\ln K_{ox}\;} & M^m L \\[2em]
\pm ne^- \big\updownarrow \; -nFE_f^{0'} & & & & \pm ne^- \big\updownarrow \; -nFE_c^{0'} \\[2em]
M^{m-n} & + & L & \xrightleftharpoons{\;-RT\ln K_{Red}\;} & M^{m-n} L
\end{array}
$$

Scheme D5. Redox and complexation equilibria with a redox-active component M.

however, again only one couple with a shifted formal potential is detected. This new voltammogram remains unchanged on further increase in ligand concentration and therefore can be attributed to the complex. The formation constant for $[Fe(CN)_6]^{4-}$ was determined potentiometrically, and the observed shift in the formal potential was used for the calculation of the formation constant for $[Fe(CN)_6]^{3-}$.[77] A generalized method of calculation of the formation constants from CV data for such systems, which takes into account ligand protonation and allows the determination of the interaction stochiometry, has been developed.[79b]

Weak complexations can be studied using ligand-induced shifts both in $E_{1/2}$ and i_p. In the first case the same treatment based on equation (D2-36) as for polarography is applicable. In the second case i_p (e.g. i_{pc}) is presented as a sum of two currents: one due to the reduction of the free redox-active component, and the other one due to the reduction of its complexed form, each being expressed as in equation (D2-39) with different diffusion coefficients and concentrations of the respective forms, related by equations of mass balance and the formation constant. Both methods were used in the study of complexation of β-cyclodextrin with ferrocenecarboxylic acid, which undergoes a reversible one-electron oxidation.[80] Many other cyclodextrin inclusion complexes were studied by polarography and CV; the subject has been reviewed.[81]

A related method based on complexation-induced changes in diffusion coefficients obtained by using rotating disk electrode voltammetry was developed, and applied to the study of cyclodextrin complexation of ferrocene derivatives.[82]

D2.4. CONDUCTOMETRY

The electrical conductivity of an electrolyte solution depends on both the nature and the concentration of ions presenting in the solution. This makes it possible to use conductometry for the determination of formation constants. The literature[83–85] should be consulted for the theory of the method. A review on conductometric determination of cation–macrocycle complex formation constants is available.[86]

By definition, the conductance κ is the reciprocal of the solution resistance, and the molar conductance (or conductivity) Λ of a given electrolyte MX is defined as

$$\Lambda = \kappa/[MX]_t \qquad (D2\text{-}41)$$

where $[MX]_t$ is the total salt concentration. Let us suppose that the cation M and the anion X can associate with the equilibrium constant K_{MX}

$$K_{MX} = [MX]/[M][X]$$

Evidently the neutral associate MX will not conduct the electricity and the measured conductance will be proportional to the concentration of unassociated salt

$$\kappa = \Lambda \, \alpha [MX]_t$$

where α is the fraction of completely dissociated electrolyte. Because α depends on $[MX]_t$, the observed molar conductance, defined as in equation (D2-41), will depend on the electrolyte concentration:

$$\Lambda_{obs} = \kappa/[MX]_t = \alpha\Lambda \qquad (D2\text{-}42)$$

Now if Λ can be measured under sufficiently high dilution when $\alpha = 1$, equation (D2-42) will allow one to calculate α at any total concentration of the electrolyte, and then the equilibrium constant can be calculated with equation (D2-43)

$$K_{MX} = (1 - \alpha)/(\alpha^2 [MX]_t) \qquad (D2\text{-}43)$$

The problem is, however, that the molar conductance depends on the electrolyte concentration even in the absence of ion pairing, due to simultaneous action of the interionic attraction effect (retardation of the ion movement caused by the attraction of a given ion to the shell of counterions), and the electrophoretic effect (retardation caused by a counterflux of water immobilized in the hydration shells of counterions). For electrolyte concentrations less than approximately 0.001 M this dependence is given by the limited form of the Onsager equation

$$\Lambda = \Lambda^0 - SI^{1/2} \qquad \text{(D2-44)}$$

where Λ^0 is the limited molar conductance at infinite dilution, S is a constant for a given electrolyte, which follows from the theory, and I is the ionic strength. For a 1:1 electrolyte $I = \alpha[MX]_t$ and equation (D2-42) together with (D2-44) takes the form

$$\Lambda_{obs} = \alpha(\Lambda^0 - S(\alpha[MX]_t)^{1/2}) \qquad \text{(D2-45)}$$

Values of α calculated by using equation (D2-45) from the experimental dependence of the conductance on electrolyte concentration obtained at sufficiently high dilution are very precise and give equally precise values of K_{MX} from equation (D2-43). The only problem is that such experiment must be performed with variable ionic strength, which affects not only the conductance, but also the activity coefficients of associating ions. A necessary correction (actually very small) can be made with the Debye–Hückel equation.

Since equation (D2-45) is valid only at very high dilution, it can be applied only for determination of sufficiently large association constants greater than 100–1000 M^{-1}. With smaller association constants one needs to go at higher electrolyte concentrations and to use more complicated expressions for conductivity. Different theoretical treatments lead to a general expression for the conductivity of a non-associated electrolyte of the form[87]

$$\Lambda = \Lambda^0 - SI^{1/2} + EI \ln I + J_1 I - J_2 I^{3/2} \qquad \text{(D2-46)}$$

where S and E depend on the charge type of the electrolyte, mobility of the ions, dielectric constant, viscosity and temperature, while J_1 and J_2 depend on the theoretical treatment employed and on the distance of closest

approach between ions: recommended values of parameters for different types of electrolytes are available.[87]

Another method, widely applied for the determination of cation–macrocycle complex formation constants,[86] is based on the conductance changes induced by the binding of the cation to a neutral ligand, such as a crown ether, at constant total electrolyte concentration. The electrolyte solution, like an alkali metal perchlorate in propylene carbonate, is prepared sufficiently diluted to neglect the ionic association, and is titrated with the ligand solution. The observed molar conductance decreases or increases depending on whether the mobility of the complex is lower or higher than of the free cation. The titration is extended to 'saturation' and the limiting value of the observed molar conductance is taken as the molar conductance of the complex salt (Λ_{MLX}), while the initial value of the molar conductance in the absence of the ligand equals the molar conductance of the pure salt Λ_{MX}). At intermediate points the conductance represents the sum of conductivities of two salts: simple salt MX, consisting of ions M$^+$ and X$^-$, and complex salt MLX, consisting of ions ML$^+$ and X$^-$

$$\kappa = \Lambda_{obs}[M]_t = \Lambda_{MX}[MX] + \Lambda_{MLX}[MLX] \qquad \text{(D2-47)}$$

where $[M]_t$ stands for total cation concentration, which equals the initial concentration of MX corrected for dilution if it takes place during the titration.

Equation (D2-47) represents one of variants of the general equation (D1-6) and may be solved by any of the methods discussed in Section D1. In particular, some[86] use an equation of type (D1-9) improved by a procedure of successive approximations. Also an equation of the form (D1-19) was applied for 1:1 complexation of cyclodextrins with ionic surfactants.[88]

Some restrictions of the method should be mentioned. First, complexation not always leads to a noticeable change in the ion mobility. Second, only neutral ligands can be studied. Third, both MX and MLX must be completely dissociated. A general treatment of the results of titration experiment, which takes into account the self-association of both MX and MLX, based on application of equation (D 2-46) was developed.[89]

D 3. SPECTROSCOPIC METHODS

D 3.1. UV-VISIBLE AND INFRARED SPECTROSCOPY

Ultraviolet and visible spectrophotometry operates in a spectral range corresponding to electronic transitions which are observed with many organic molecules and with inorganic transition metal ions. This is a relatively old method, which nevertheless still undergoes considerable improvements in precision, rapidity and other characteristics.[1c,90]

The quantification of spectrophotometric data is based on the Beer–Lambert–Bouger law (commonly referred as Beer's law) for the absorbance A of a given substance S

$$A = \varepsilon l[S] \qquad (D\,3\text{-}1)$$

where ε is the molar absorptivity usually expressed in $M^{-1} cm^{-1}$, and the optical pathlength l is expressed in cm and concentration [S] in M. The range of linearity between A and [S] depends on the instrument employed and is principally restricted by the effect of stray light.[90a] Typically the linearity is maintained up to absorbances 2 or 3, however even in the linear range the precision in A is lower at both small and large absorbances and reaches its maximum at intermediate values of A around 0.4.

The proportionality between the absorbance and concentration makes the spectrophotometry especially useful for determination of the solution composition and, consequently, the reaction stochiometry. Important information about the number of species present can be obtained from analysis

of the spectra of solutions containing different amounts of S and L. The superposition of such spectra (usually recorded in a whole set of experimentally accessible range of wavelengths and corrected if necessary for the absorbance of L) in the coordinates ε *vs.* wavelength (λ) (or directly A *vs.* λ if the total concentration of S is maintained constant) often shows one or more isosbestic points, Fig. D 12. The observation of isosbestic point(s) is considered to be an indication of the presence of only two species, such as S and SL, in an equilibrium under given conditions. Indeed, if the spectra of S and SL intercross, there is a point where ε values for both free and complexed substrate coincide and, therefore, all spectra obtained at different degrees of complexation must pass through this point. Of course, more than two species may in rare cases have identical molar absorptivities at a given wavelength and the observation of an isosbestic point does not exclude the presence of more than one equilibrium.

A more sophisticated treatment based on the matrix analysis is possible, which allows in principle the rigorous determination of total number of absorbing species existing in solution in a given concentration range.[1c] Let us consider a set of absorption spectra of a number N of solutions containing different S and L concentrations which we enumerate from 1 to N. Each spectrum can be represented by the absorbances measured at a number M of different wavelengths which we enumerate from 1 to M. Thus we obtain a number $N \times M$ of absorbances A_{mn} which can be arranged in a form of a rectangular matrix

$$
\begin{array}{ccccccc}
A_{11} & A_{21} & A_{31} & \cdots & A_{m1} & \cdots & A_{M1} \\
A_{12} & A_{22} & A_{32} & \cdots & A_{m2} & \cdots & A_{M2} \\
A_{13} & A_{23} & A_{33} & \cdots & A_{m3} & \cdots & A_{M3} \\
\cdots & \cdots & \cdots & \cdots & \cdots & \cdots & \cdots \\
A_{1n} & A_{2n} & A_{3n} & \cdots & A_{mn} & \cdots & A_{Mn} \\
\cdots & \cdots & \cdots & \cdots & \cdots & \cdots & \cdots \\
A_{1N} & A_{2N} & A_{3N} & \cdots & A_{mN} & \cdots & A_{MN}
\end{array}
$$

If there is no interaction between S and L and the only absorbing species is S each element of

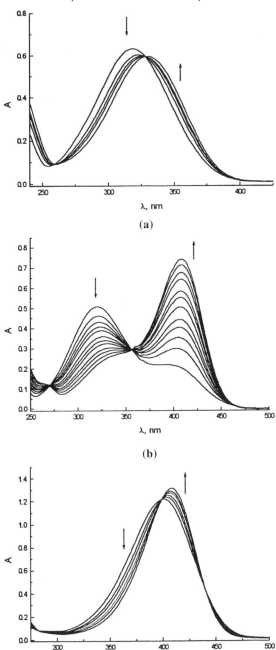

(a)

(b)

(c)

Figure D 12. Absorption spectra of 6.6×10^{-5} M 4-nitrophenol (S) at different concentrations of α-cyclodextrin (L) at pH values 2.6 (a), 6.5 (b) and 10.1 (c).

the matrix A_{mn} is simply the product of the molar absorbtivity of S at the wavelength 'm' and the total analytical concentration of S in the solution number 'n' (assuming $l = 1$ cm)

$$A_{mn} = \varepsilon_m [S]_n$$

Therefore, the ratio of absorbances at a given wavelength m in two solutions, 1 and 2, equals the ratio of the substrate concentrations in these solutions

$$A_{m1}/A_{m2} = [S]_1/[S]_2$$

and the ratio of absorbances at two wavelengths, 1 and 2, but in the same solution number 'n' equals the ratio of the molar absorbances at these wavelengths

$$A_{1n}/A_{2n} = \varepsilon_1/\varepsilon_2 \qquad \text{(D 3-2)}$$

Since these relations are correct for any combination of the numbers 'm' and 'n', the elements of each row can be obtained from the elements of another row by multiplication of the latter by the ratio of the substrate concentrations in the respective solutions (calculated from known initial S concentration and dilution factor) as well as each column can be obtained from another column by multiplication of the latter by the ratio of the molar absorptivities at the respective wavelengths. The rows or columns which can be obtained by multiplication of another row or column by a constant are called linearly dependent. Therefore, in this case we have only one linearly independent row or column. The number of linearly independent rows or columns in a given matrix is called its rank and the matrix rank equals the number of absorbing species in a system consisting of a set of solutions studied.

Mathematically the matrix rank equals the order of the largest non-zero determinant that can be obtained from the elements of the matrix. For a large number of absorbing species calculation of the matrix rank is rather difficult. Several programs such as TRIANG[1b] allow such a so-called multicomponent analysis and are available often from the instrument makers. The main problem in such calculations is that determinants calculated from experimentally obtained matrix elements never equal exactly zero due to inevitable uncertainties in measured absorbances. Therefore the method needs highly precise data and a reasonable statistical criterion to decide which values of determinants must be considered zero. Another way of determination of the number of absorbing species is based on the mathematical procedure of factor analysis,[1c] which needs, however, the same problem of statistical criteria to be solved.[1c]

If the number of species is less than, or equal to three the matrix elements can be analyzed by graphical methods. For the single species this analysis can be done by using equation (D 3-2). It follows from this equation that the absorbances in all solutions at a given wavelength are directly proportional to the absorbances of the same solutions at another wavelength.

$$A_{1n} = (\varepsilon_1/\varepsilon_2)A_{2n}$$

Therefore a plot of A_{1n} values vs. A_{2n} taken at different n must be a straight line passing through the origin. Each other combination of the wavelengths will give also a straight line

$$A_{1n} = (\varepsilon_1/\varepsilon_3)A_{3n}, A_{2n} = (\varepsilon_2/\varepsilon_3)A_{3n},$$

etc. Thus, a series of straight lines is produced which intersect in the origin.

With two species, S and SL, a similar series of straight lines is obtained if the ratios of absorbances at two wavelengths, 1 and 2, are plotted vs. the ratios of absorbances at another wavelength, 3, and the wavelength 2, that is A_{1n}/A_{2n} vs. A_{3n}/A_{2n} for different n. In the case when the total concentration of absorbing species is held constant, $[S]_t$ is constant while $[L]_t$ is variable, a series of straight lines passing through the origin must be observed on plotting the differences of absorbances at one wavelength, like 1 and different solutions n and i, $A_{1n} - A_{1i}$, vs. similar differences at

another wavelength, like 2: $A_{2n} - A_{2i}$; the numbers 1, 2, etc. refer to the wavelengths used and the letters n, i, etc. to the solutions used. In a case of three species with constant $[S]_t$ the analysis involves data at three wavelengths, like 1, 2 and 3, in two solutions, n and i, and the plotting functions are $(A_{1n} - A_{1i})/(A_{2n} - A_{2i})$ vs. $(A_{3n} - A_{3i})/(A_{2n} - A_{2i})$.[1b] It should be mentioned also that spectrophotometry perfectly suits the method of continuous variations (Section D 1).

A problem in application of UV-vis spectrophotometry for the study of supramolecular complexation is that often there are small complexation-induced changes in the absorbance of chromogenic substrates. Coordination to transition metal cations leads to a substantial perturbation of metal and ligand electronic levels and as a rule produces large spectral changes. Not surprisingly, therefore, electronic spectroscopy finds considerably wider applications for the study of coordination than organic supramolecular equilibria.

In some cases, however, the complexation-induced spectral changes are very large and the respective guests can be used as indicator dyes for the competitive method (Section D 1). This happens when the absorbing species undergoes an additional reversible change between two forms which strongly differ in their absorbances and have different affinity to the ligand. For example, the dianion of phenolphthalein completely loses its strong absorbance at 520 nm upon binding to β-cyclodextrin due to the complexation-induced lactonization which destroys the quinoid structure of the dye, Scheme D 6,[91a,b] (there is an alternative explanation of the complexation-induced decolorization of the dye[91c]).

A more general case is the complexation-induced shift of acid–base equilibria of organic (in particular) dye molecules which usually possess very much different absorptivities in different protonation forms. Fig. D 12a–c illustrates spectral changes

purple   colorless

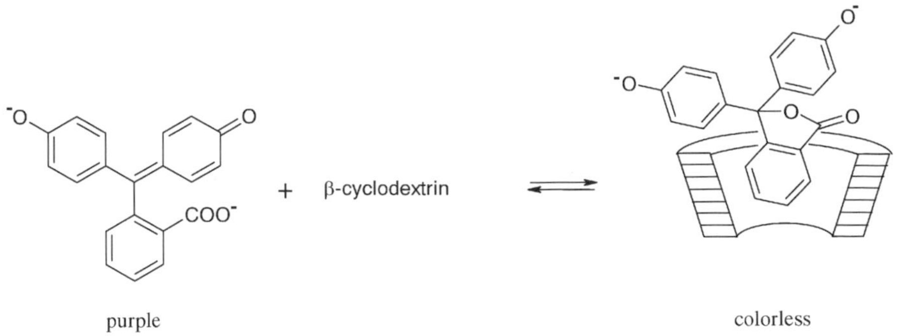

Scheme D 6.   Complexation-induced lactonization of the phenolphthalein dianion.[91a,b]

produced by addition of α-cyclodextrin to solutions of 4-nitrophenol at pH values below, around and above the pK_a of the substrate. The observed molar absorptivity is given by equation

$$
\varepsilon_{obs}
$$
$$
= \frac{\varepsilon_S + \varepsilon_{SH}[H]/K_a^{SH} + (\varepsilon_{SL}K_1 + \varepsilon_{SHL}K_2[H]/K_a^{SH})[L]}{1 + [H]/K_a^{SH} + (K_1 + K_2[H]/K_a^{SH})[L]}
$$
$$
(D3\text{-}3)
$$

where ε_S, ε_{SH}, ε_{SL} and ε_{SHL} are the molar absorptivities of S, SH, SL and SHL respectively and [H] is the concentration of hydronium ions in solution; K_1, K_2 are binding constants of the deprotonated and the protonated substrate. Equation (D3-3) predicts a hyperbolic plot of ε_{obs} *vs.* ligand concentration (see equation (D1-12), at any pH value, which is typical for a 1:1 complexation

$$
\varepsilon_{obs} = (\varepsilon_{S,app} + \varepsilon_{SL,app}K_{app}[L])/(1 + K_{app}[L])
$$
$$
(D3\text{-}4)
$$

where $\varepsilon_{S,app}$ and $\varepsilon_{SL,app}$ are the pH-dependent apparent molar absorptivities of free and complexed substrate (observed absorptivities in the absence of L and at saturation) and K_{app} is the apparent pH-dependent binding constant (Section D2.2) given by equations

$$
\varepsilon_{S,app} = (\varepsilon_S + \varepsilon_{SH}[H]/K_a^{SH})/(1 + [H]/K_a^{SH})
$$
$$
(D3\text{-}5)
$$

$$
\varepsilon_{SL,app} = (\varepsilon_{SL}K_1 + \varepsilon_{SHL}K_2[H]/K_a^{SH})/
$$
$$
(K_1 + K_2[H]/K_a^{SH})
$$
$$
(D3\text{-}6)
$$
$$
K_{app} = (K_1 + K_2[H]/K_a^{SH})/(1 + [H]/K_a^{SH})
$$
$$
(D3\text{-}7)
$$

Equation (D3-5) describes the pH-dependence of the molar absorptivity of the substrate and is a standard equation employed for determination of pK_a values by spectrophotometric titration.[1c] It can be rewritten in a form which allows to calculate pK_a^{SH} from the observed absorptivity at a given pH when the absorptivities of both protonated and deprotonated forms are known

$$
pK_a^{SH} = pH - \log\{(\varepsilon_{S,app} - \varepsilon_{SH})/(\varepsilon_S - \varepsilon_{S,app})\}
$$
$$
(D3\text{-}8)
$$

Equations (D3-5) and (D3-6) show that at pH values substantially lower than pK_a^{SH} (when the ratio $[H]/K_a^{SH}$ is very large), $\varepsilon_{S,app}$ and $\varepsilon_{SL,app}$ equal ε_{SH} and ε_{SHL} respectively, and at pH values substantially higher than pK_a^{SH} (when the ratio $[H]/K_a^{SH}$ is very small), $\varepsilon_{S,app}$ and $\varepsilon_{SL,app}$ equal ε_S and ε_{SL} respectively. These situations are illustrated in Figs D 12a,c where small complexation-induced changes typical of inclusion processes are observed. However, when pH is lower than $pK_a^{SH}=7.1$ but not too much, Fig. D12b, addition of the host shows a much greater effect. Under these conditions the spectrum of free substrate corresponds to predominantly neutral form while the spectrum of the complex corresponds to the predominantly deprotonated form. This is due to the fact that α-cyclodextrin forms a much more stable complex with the 4-nitrophenolate anion ($K_1 = 3700$ M^{-1}) than with neutral 4-nitrophenol ($K_2 = 385$ M^{-1})[92]; this difference in the binding constants produces a complexation-induced shift of the protolytic equilibria towards the deprotonated highly absorbing form.

Several computer programs are available for the calculation of equilibrium constants from spectrophotometric data[1-13], such as SPECFIT[93] or SQUAD.[94] SQUAD calculates both the number of absorbing species and equilibrium constants from the titration data obtained at a series of wavelengths. The recently published program HYPERQUAD[13] can be applied to spectrophotometric and/or potentiometric data and is especially valuable when both types of data are available for a given system.

Infrared (IR) spectroscopy is of a more limited use in the determination of binding constants. The main disadvantages are often low molar absorptivities of chromophores in this spectral region, difficulties to measure in water with the conventional transmission techniques, which is changed with the advent of FT-IR (Fourier transform-IR) techniques, and high background signals from majority of common organic solvents. Traditionally IR spectroscopy has dominated the measurement of equilibrium constants of hydrogen-bonded associates in organic solvents like CCl$_4$ (see ref. 22 Section A1). The advent of modern FT-IR instruments, which have high sensitivity, in particular due to signal accumulation techniques, and allow more variety of solvents combined with computerized editing of spectra including background subtraction holds promise for wider applications. A problem with IR is that (like with UV-vis spectra) the formation of non-covalent bonds often does not lead to significant spectral changes. More suitable for equilibrium studies are near infrared (NIR), also as here these changes must be larger. NIR as well as Raman spectroscopy is also easier to use in aqueous solutions with glass cells. As an illustration we mention the NIR study of complexation of substituted phenols with cyclodextrins.[95] The bands employed for determination were those of phenols in the range 1100–1700 nm with molar absorptivities in the range 0.03–0.5 M^{-1}cm^{-1}. The experiments were performed in D$_2$O, which is more transparent in the near-IR region than H$_2$O. Addition of cyclodextrins enhanced the absorbance by 3–5 times, but still high concentrations of components should be employed. The results were in a qualitative agreement with those from other methods. Raman spectroscopy was used, for example for determination of acidity of dicarboxylic acids and their complexation with metal cations.[96]

D 3.2. FLUORESCENCE SPECTROSCOPY

Some absorbing species loose excitation energy by a radiative process generally termed luminescence. Absorption of a photon produces a singlet excited state which can return to the ground state by a fast radiative process (fluorescence) or non-radiative (internal conversion), or can be transformed into a triplet state (intersystem crossing) of lower energy. The lifetime of a singlet excited state is very short ($<10^{-7}$ s), but triplets are long-lived (10^{-6} to 0.1 s) because both radiative and non-radiative decay to the ground singlet state are spin-forbidden.[97] The (relatively slow) radiative decay of a triplet state is called phosphorescence. For low molecular weight compounds usually only fluorescence is detected in solution at room temperature while longer-living triplet states are collisionally quenched (see below).

Although it is not completely clear which molecular characteristics are responsible for manifestation of fluorogenic properties, it is generally believed that besides the presence of conjugated double bonds rigidity is an important factor; both fluorescein (**D-15**) and the chemically similar phenolphthalein (Scheme D 6) strongly absorb light, but only the former is fluorogenic.

D-15

Fluorescence intensity (I_f) is measured as number of counts per second, but depends so strongly on the instrument characteristics that often one measures simply relative units. Under certain conditions I_f is proportional to the fluorogenic solute concentration and under such conditions fluorescence spectroscopy can be applied for the binding study. Generally

$$I_f = \varphi I_{abs}$$

where φ is the fluorescence quantum yield and I_{abs} is the intensity of the absorbed light. The latter can be expressed as an exponential function of the absorbance $A = \log\{I_0/(I_0 - I_{abs})\}$, where I_0 is the intensity of the excitation light

$$I_{abs} = I_0(1 - 10^{-A})$$

Therefore

$$I_f = \varphi I_0(1 - 10^{-A})$$

At low absorbance $A \leq 0.05$ the exponential term can be approximated by a linear function $10^{-A} = 1 - 2.3A$ and together with the Beer's law for the absorbance one obtains the following equation

$$I_f = 2.3\varphi I_0 \varepsilon l[S] = k[S] \qquad \text{(D 3-9)}$$

where k is a coefficient of proportionality between I_f and solute concentration. Thus the range of linearity of the fluorescence extends only to very diluted solutions. At higher concentrations I_f tends to 'saturate' and than even decreases due to the effect of inner filter when the excitation light does not fully reach the center of the cell.[90a] On the other

hand the sensitivity of the method is very high and good fluorophores like derivatives of naphthalene, pyrene, anthracene, indole or iso-indole, can be detected at concentrations in the nano- to micromolar range.

The fluorescence intensity is much more sensitive to the micro-environment of the solute than the absorbance. As a rule inclusion of fluorogenic species into macrocycles, such as cyclodextrins, enhances the fluorescence due to protection of the excited state from the quenching by solvent molecules.[98] The opposite complexation-induced decrease of fluorescence is observed if, for example lone pairs in the guest or the host molecule lead to quenching. In both cases fluorescence can be a valuable method for the study of supramolecular association. It should be noted, however, that the precision of fluorimetric measurements is usually lower than that of absorbance measurements due to often observed irreproducible quenching effects of various impurities and light scattering (e.g. by dust particles) and stronger dependence of the signal on instrumental parameters. Due to quenching effects etc. temperature constancy is much more important than with UV/vis techniques. Equation (D 3-9) is similar to Beer's law for the absorbance and the mathematical treatment of fluorimetric titration experiments is similar to that of spectrophotometric titration.[99]

A simplified kinetic scheme of the various processes involved in formation and decay of excited molecules (A^*) is the following

light absorption $\quad A + h\nu \rightarrow A^* \qquad \text{(D 3-10)}$

fluorescence $\quad A^* \rightarrow A + h\nu'(k_F) \qquad \text{(D 3-11)}$

non-radiative decay (internal conversion and intersystem crossing)

$$A^* \rightarrow A(k_{NR}) \qquad \text{(D 3-12)}$$

bimolecular quenching

$$A^* + Q \rightarrow A + Q \ (k_Q) \qquad \text{(D 3-13)}$$

From this scheme for the steady-state fluorescence one obtains the expression known as the

Stern–Volmer equation[97]

$$I_f = I_0/(1 + K_{SV}[Q]) \qquad (D3\text{-}14)$$

where I_0 is the fluorescence intensity in the absence of quencher and K_{SV} is the Stern–Volmer constant

$$K_{SV} = k_Q/(k_F + k_{NR}) \qquad (D3\text{-}15)$$

A more familiar linear form of the Stern–Volmer equation has the form

$$I_0/I_f = 1 + K_{SV}[Q] \qquad (D3\text{-}16)$$

In the absence of quencher the decay of fluorescence involves two parallel steps (D3-11) and (D3-12). The decay kinetics are therefore pseudo-first-order processes that can be characterized by the life time τ given by equation

$$\tau_0 = 1/(k_F + k_{NR})$$

This gives another useful expression for K_{SV} of the form

$$K_{SV} = k_Q\tau_0 \qquad (D3\text{-}17)$$

The rate constant k_Q is close to the diffusion-controlled limit of the second-order rate constant, but can be considerably smaller if the fluorescent group is protected from collisions with the quencher by its surrounding, like inside the host or a protein molecule. This creates the possibility of using quenching for the study of accessibility of, for example fluorescent amino acids (tryptophane, tyrosine) or special fluorescent markers in proteins[97,100] as well as of fluorescent guests in their inclusion complexes.[101] The approach consists of determination of Stern–Volmer constants (usually from steady-state quenching data) for free and bound guests, measurements of τ_0 also for free and bound guests in the absence of the quencher and comparison of k_Q values.

Life-time measurements can be accomplished by phase modulation or by single counting techniques (the latter is more reliable especially for multi-step processes, but needs more sophisticated and more expensive equipment).[97,102] Life-times are sensitive to the micro-environment of the fluorophore and can be used for estimation of local polarity.[103] In the presence of quencher the life-time is given in accordance with the scheme (D3-11)–(D3-13) by the expression

$$\tau = 1/(k_F + k_{NR} + k_Q[Q]$$

Therefore for the dynamic (bimolecular) quenching the ratio of life-times in the absence and in the presence of the quencher depends on the quencher concentration in accordance to equation similar to (D3-16)

$$\tau_0/\tau = 1 + K_{SV}[Q] = I_0/I_f \qquad (D3\text{-}18)$$

Note that equation (D3-14) describes a hyperbolic dependence of the fluorescence intensity on the quencher concentration of exactly the same form as (D1-13) with $X_\infty = 0$. As a result, the dynamic quenching, which does not involve any complexation of the quencher with fluorophore, can be confused with the situation when the quencher does form a non-fluorescent complex with the fluorophore. The latter is known as the static quenching.[97] Static and dynamic quenching can be discriminated by measurement of life-times: equation (D3-18) is correct only for dynamic quenching, in the case of static quenching only free molecules of the fluorophore produce fluorescence and therefore $\tau_0/\tau = 1$. Also dynamic quenching does not affect the absorption spectra of the fluorophore in contrast to static quenching, although the complexation-induced spectral changes can be rather small (Section D3.1).

In general static quenching is not complete and is described by equation of the type (D1-13) with $X_0 > X_\infty$, which in traditional Stern–Volmer coordinates gives a curve instead of straight line. This is illustrated in Fig. D13 which shows the results of quenching of a cyclodextrin-sandwiched porphyrin **D-16** by quinones: benzoquinone which does not form a complex with the host shows a linear dependence of I_0/I_f typical for dynamic quenching, but naphthoquinone, which forms a fairly stable complex ($K = 7400$ M^{-1}) with

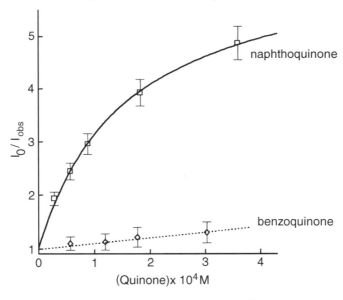

Figure D 13.   Stern–Volmer plots for quenching of **D-16** by quinones.[104] Reprinted with permission from *J. Am. Chem. Soc.*, Kuroda *et al.*, 1993, **115**, 7003. Copyright 1993 American Chemical Society.

D-16 shows a downward curvature typical for the static quenching.[104]

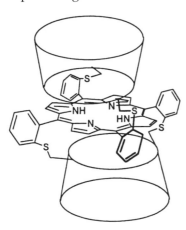

D-16

The high sensitivity of the fluorescence to the micro-environment makes fluorescent probes useful for the evaluation of micro-environment characteristics inside host cavities. Photophysical properties of many commonly used fluorescence probes have been collected.[102]

Time-resolved fluorescence measurements are important of course for the study of dynamic properties of supramolecular complexes, Section F2. They also provide valuable information for the binding studies. The fluorescence decay is usually very fast even in comparison with life-times of non-covalent complexes. Therefore, the fluorescence decay curves for systems containing both free and bound fluorescence molecules as a rule show separate exponential components for free and bound molecules. This can be used in a qualitative way to prove the presence of complexes of different stochiometry.[99] Also, the concentration dependence of the fractions of decay components characteristic for free and bound species can be used for calculation of the formation constant.[105]

D 3.3.   CHIROPTICAL METHODS

The rotation of linearly polarized light by chiral molecules is proportional to the concentration of the chiral solute, and if the molar rotations of free and bound substrate are different one can use the optical rotation for stability constant determination in the same manner as in the case of spectrophotometry. The largest change in the optical activity occurs when an achiral guest is incorporated into the cavity of chiral host, which induces the optical activity of the guest. This phenomenon is well known from studies of natural chiral hosts like cyclodextrins as well as synthetic chiral macrocyles,[106] and finds primarily applications for studies of guest orientation inside the macrocycle cavity (Section E5). One can also make use of the relation between the optical rotation, α, and the concentration of an optically active host or guest compound, provided the α value changes upon complexation. However, both the often small absolute values of α and of its change make it usually necessary to employ high concentrations, which produces difficulties with respect to changes in activity coefficients and to non-linear correlations between α and [H] or [G]. Instead of simple

polarimetric methods one therefore prefers the more sensitive measurement of changes in the optical rotatory dispersion (ORD) or its equivalent, the circular dichroism (CD). Modern CD instruments allow data collection in quite dilute solution, also with sensitivity-enhancing signal accumulation. The CD effect occurs due to different absorption of right and left circularly polarized beams by chiral chromophores and is measured by the differential dichroic absorption $\Delta\varepsilon$ defined as

$$\Delta = \varepsilon_L - \varepsilon_R$$

Another often used measure proportional to $\Delta\varepsilon$ is the ellipticity (θ) (Section E 5).

As an example Fig. D 14 shows the CD and absorption spectra of two achiral guests **D-17a,b** in the presence of β-cyclodextrin. Maxima in induced CD spectra are observed in the areas of absorption; they can be positive (Fig. D 14a) or negative (Fig. D 14b). The sign of induced CD is related to the orientation of the guest (Section E5), which in this particular

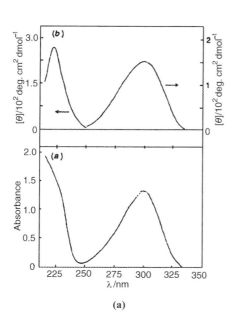

(a)

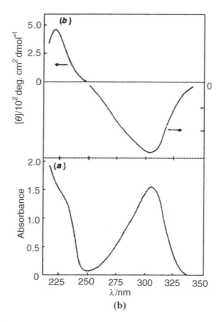

(b)

Figure D 14. Absorption (a) and CD (b) spectra of 2×10^{-4} M guests **D-17a** (A) and **D-17b** (B) in the presence of 0.015 M β-cyclodextrin in 4%(v/v) MeOH–H$_2$O.[107] Reproduced by permission of the Royal Society of Chemistry from *J. Chem. Soc., Perkin Trans. 2*, Sakurai *et al.*, 1994, 1929.

case is dictated by the presence of the *ortho*-methyl substituent.[107] Independently of the sign, the observed ellipticity can be used for the calculation of the formation constant for a 1:1 complexation equilibrium with equation of type (D1-12) where $x_S = 0$ and x_{SL} is the molar ellipticity (note, that in the case of induced CD there is no background signal). Such calculations performed for the complexes in **D-17** and a series of related guests with β-cyclodextrin as the host give the stability constants coinciding in limits of errors with those determined by spectrophotometry.[107] For other examples of application of induced CD to determination of binding constants, see the literature.[106,108]

D-17; R = H (**a**), Me (**b**)

D3.4. NUCLEAR MAGNETIC RESONANCE

D3.4.1. NMR shift titrations and experimental considerations

The substantial change of NMR shieldings which one usually observes upon formation of host–guest complexes (Scheme E2)[109] is the major reason why this technique has become by far the most important method of measuring association constants, and to a still lesser degree also to characterize conformations of such complexes (Section E4). Proton NMR is used most widely not only because of practical reasons such as availability, sensitivity etc. Firstly and simply, hydrogen in contrast to carbon atoms are positioned at the periphery of molecules and therefore are more exposed to intermolecular screening effects. Secondly, intermolecular anisotropy effects

have the same absolute magnitude on protons as on carbon: for C-13. However, they present only a small contribution to the total C-13 NMR shift range, which is about 20 times larger than for protons. C-13 NMR shieldings reflect to much larger degree than proton shifts *intra*molecular geometry variations; thus differences in torsional angles alone may lead to more than 10 p.p.m. changes. For this reason 13-C NMR shift can be a valuable tool for following conformational changes upon complexation, but not for measuring association, in spite of its inherently large spectral dispersion. Fluorine-19 NMR-spectroscopy to some degree combines advantages of H-1 and C-13 NMR, but is (as techniques with other nuclei) of course confined to corresponding compounds. A considerable number of binding studies with crown ethers and other ligands and alkali cations was performed with magnetically active nuclei of alkali metals.[110]

The great advantage of NMR-spectroscopy to provide several independent signals for the evaluation of supramolecular equilibria should always be used, as is demonstrated in Fig. D15.[111] Application of the fitting methods described in Section D1 yields titration association constants K from NMR, which often differ by less than 10%. A precision of $\pm 1\%$ often obtained in the fitting of single signal shifts may be misleading and may mask really much lower accuracy due to unproperly chosen titration conditions and/or systematic errors. Thus, a fit to a nearly linear shift *vs.* concentration profile may give $< 1\%$ statistical error, but is meaningless if both K and CIS (the shift at 100% complexation) are unknown. If deviations larger than 10% between the association constants obtained with different signals are observed, the calculational model needs to be modified.

Under properly chosen conditions (see experimental planning considerations in Section D1) one can obtain reliable association constants with high field instruments even if

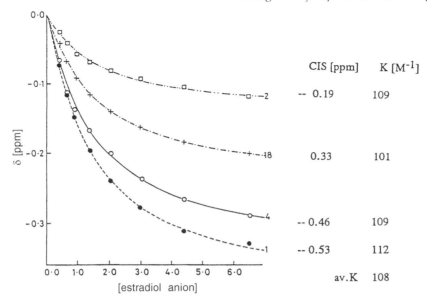

	CIS [ppm]	K [M^{-1}]
-- 2	-- 0.19	109
+-18	0.33	101
4	-- 0.46	109
-1	-- 0.53	112
	av. K	108

Figure D 15. 1-H NMR shift titration with estradiol (anion) as guest (monitored compound) and the azoniacyclophane **B-23a** (n = 6) as host, with experimental points and computer fit for a 1:1 association model; for steroid protons H2, H18, H4, and H1, with *K* values from each fit.[111] Reprinted with permission from *J. Chem. Soc., Perkin Trans. 2*, Kumar *et al.*, 1989, 245. Copyright 1989, Royal Society of Chemistry.

the maximally observed shifts changes are < 0.03 p.p.m. Control of constant temperature, of pH (pD), of solvent, salt conditions etc. then is of particular importance. In hydrogen bonding associations the largest shifts are of course occurring at the protons involved in the bridges; at the same time these shifts can be blurred by exchange with other protons (from the solvent), and/or by traces of acids or bases. Nevertheless, measurements of, N–H shifts in amides, nucleotides etc. can be a valuable and often only help; they are in principle also possible in protic solvents by suitable solvent signal suppression techniques.

Usually one adds within the desired range of 20–80% complexation (Section D 1), 8–10 accurately known amounts of stock solutions to the host or guest compound, whichever signals one wants to observe. (It may be necessary to chose smaller intervals if signals cross each other during titration.) Which way to go should be decided before one starts a full titration by a preliminary measurement, using

either host or guest in moderate excess (which will also ease signal assignment), but at the highest possible concentration of one of the partners. This routine experiment will also help to estimate, if necessary, a value for the expected association constant for planning the titration according to Section D 1, for which a suitable program can be used (see website). If solubility of one partner is much lower than the (expected) dissociation constant one can prepare solutions at the highest possible concentration and then dilute them stepwise, instead of adding stock solutions. Stock solutions are for solubility reasons often made up by dissolving the compound first in, for example DMSO with subsequent dilution with water. As long as the cosolvent concentration in the final aqueous sample stays below 5% this does usually not affect the complex significantly, in contrast, of course, to measurements in non-polar solvents. Alternatively one starts with a solution already containing the cosolvent in a concentration corresponding to the half way of the titration, or, more time consuming, one adjusts the solvent composition after each addition to the same value by adding appropriate amounts of the cosolvent.

D 3.4.2. Evaluation of association constants and complexation-induced shifts (CIS)

An important difference between NMR and any kind of optical spectroscopy is that the former operates at much lower frequencies and possesses relatively slow 'time scale', see also Section F1. In the case of fast exchange, which is most common for supramolecular complexes one observes only one sharp signal at an intermediate mean weighted position between the signals of free and complexed molecules

$$\delta_{obs} = \delta_S f_S + \delta_{SL} f_{SL}$$

where δ_S and δ_{SL} are the chemical shifts of a monitored nucleus in free and complexed substrate molecules and f_{SL} and f_S are the molar fractions of S and SL respectively. Substituting the expressions for f_{SL} and f_S in accordance with a given complexation scheme leads to equations which coincide with those for x_{obs} given in Section D1 (e.g. D1-12, D1-19, etc.); explicit equations which can be copied as input in non-linear fitting programs are made available on the website. With large excess of ligand one obtains for a 1:1 complexation:

$$\delta_{obs} = (\delta_S + \delta_{SL}K[L])/(1 + K[L])$$

In the case of fast exchange one obtains from the fitting procedure simultaneously with the association constants the complexation-induced shifts (CIS) value for the fully complexed material, sometimes also called bound shift. Comparison of this value with those expected from other data can also help to decide whether some shifts and the corresponding association constants are artifacts, due to contaminations by acids or bases.

Without shift titration the CIS can be obtained only, if the solubility allows a measurement at concentrations about ten times above the dissociation constants, leading to > 90% complexation (one can use the nomograms in Fig. D1 or a simple program from the website). With weaker complexes this method requires high concentrations which may change bulk properties of the solution and therefore influence the actual CIS and K value. On the other hand, the CIS values can be calculated from *single* measurements at (in principle) only one exactly known concentration if the association constant is known, or if the constant can be reliably estimated from other data. Mass action law then allows one to calculate the degree of complexation for each partner (host and guest) (program on the website). For obtaining the CIS values, the shifts observed simultaneously for H and G only need to be divided by the corresponding mole fraction. Even though one single measurements allows to extract in this way the CIS, it is advisable to use at least two measurements in order to check consistency of the calculational model of complex formation. The method of extracting shifts of both host and guest molecule from only one or two measurements is particularly important if the spin system is too complicated to observe all signals in the usual one-dimensional spectra, or if the signal/noise ratio is insufficient to allow a time consuming full titration. In such cases a single and then longer measurement makes C-13 CIS values accessible, and known two-dimensional techniques such as COSY will allow analysis also of many overlapping signals.

Several computer programs are available for the calculation of formation constants form NMR data under conditions of fast exchange.[112] In principle one can use any program such as SIGMAPLOT, ORIGIN etc. for the always recommended least-square fitting procedure and insert the corresponding equations.

A comment is in order with respect to the use of NMR data for determination of the stoichiometry of the fast complexation reactions. The molar ratio method can be applied for NMR data without any modification, but the Job plot for Δ_{obs} does not show an expected extreme defined by equation (D1-28). The problem is that it is the concentration of the complex and not the complexation fraction which actually reaches a maximum upon variation in f_L and produces at this point a highest deviation of the observed property form additivity. Since δ_{obs} is not proportional to the concentration it does not behave like, the absorbance and shows different pattern upon variation in f_L. The expected extreme can be observed, however, in modified coordinates, namely, when the product $(\delta_{obs} - \delta_S)f_S$ is plotted *vs.* f_L. In the case of slow exchange, of course, one can simply use the areas of the separate peaks to calculate the reaction stoichiometry.

If exchange between host and guest is slow on the NMR time scale, which is less frequently the case, both association constants and CIS values are directly accessible from the NMR spectra. As an example slow complexation of nucleotide bases to molecular tweezers in chloroform may be cited.[113] Particularly with 'container' hosts the exchange might actually be too slow to achieve sufficiently fast equilibrium, or to observe lines which are not too much exchange-broadened. In such cases one can in principle take advantage of lower magnetic field strength; or, more practically, one can increase either the measuring temperature, or the reaction time before spectra recording. Alternatively, one will use separation techniques described elsewhere (Section D4). The accuracy of the binding constant values derived from NMR signal integrals is limited by the known limitations of NMR signal area measurements, but fortunately not by the need to use calibration methods or coefficients, as it is necessary with optical techniques. Besides the usual precautions for obtaining correct signal integrals (not intensities!) advantage should be taken of the usually many available NMR signals, now even from 2D-spectra, which help to average out errors.

D3.4.3. Affinities from NMR relaxation measurements

Binding constants can also be extracted from NMR relaxation times.[1a,110] In supramolecular complexes rotational correlation times of guest compounds are often increased, which allows one to use corresponding change of relaxation rates as another spectral parameter. If the guest G is the observed minor component and the host H is in excess, the transverse relaxation time T_2, is the weighted average of the times for the free and bound guest, T_{2free} and T_{2bound}, respectively; for simplicity one may use the half-height width ($W_{1/2} = T_2^{-1}$) of a Lorentzian-shaped NMR signal

$$T_{obs} = f\,T_{2free} + f\,T_{2bound} \qquad \text{(D3-19)}$$

Here f and f are the molar fractions of free and bound guest molecule , which by application of mass law allow to derive the corresponding equilibrium constant K. Describing the change of relaxation times before and after complexation as ΔT_2, the association constant K for a 1 : 1 complex can be obtained from the slope of a linear plot described for $[H] \gg [G]$ by[1a]

$$[G]\Delta T_2^{-1} = T_{2bound}K + T_{2bound}[H] \qquad \text{(D3-20)}$$

which is a variation of the Scott plot, equation (D1-17), with $\Delta x_{obs} = \Delta T_2$, $\Delta x^{-1} = T_{2bound}$ and K being the dissociation constant (K^{-1} in equation (D1-17)).

NMR relaxation data were used for instance with ^{81}Br relaxation not only to measure K for bromide itself, but by competition also the association constants of many anions with cyclodextrins.[114] The method is of obvious value for complexes with no or too small changes of NMR shift, or of optical signals. The substantial change in correlation time which a small ligand experiences upon binding to a large host molecule leads to a large negative Nuclear Overhauser effect (NOE) instead of weak NOE signal. This so-called transfer NOE (trNOE) builds up with the fast cross-relaxation rates of macromolecules and can be distinguished from host signals with spin-lock filter techniques, thus allowing to identify bound ligands even in mixtures of possible ligands.[115] The method, which also can give structural information, Section E7, does not yet yield accurate formation constants but is particularly suited for screening many ligands which bind to large biomolecules with association constants above 10^3 M^{-1}. One may use as little as 10 nM of a protein receptor with a usually 10-fold excess of ligand. The use of dynamic information from NMR-spectroscopy in the form of T_1 and T_2 relaxation with respect to kinetics and mobility of supramolecular complexes is discussed in Section F2.

Finally we mention that differences in diffusion coefficients between free and complexed substrates can also be used for the determination of equilibrium constants K, as long as the K values are well below $10^5\,M^{-1}$ at most.[116] The underlying diffusion measurements by pulsed gradient spin echo NMR experiments are described in several reviews.[117]

D3.5.   MASS SPECTROMETRY

Mass spectrometry (MS) gains rapidly growing importance in the study of both solution and gas-phase supramolecular reactions. The common steps in the numerous MS techniques are transfer to the gas phase and ionization of the substance(s) of interest, and subsequent separation of gaseous ions by their mass-to-charge (m/z) ratios. New developments in MS involve first of all so-called soft non-destructive ionization techniques, which are capable of producing molecular ions for species of masses up to hundreds kDa and even for weak non-covalent associates. Also important improvements were made in manipulation with gaseous ions, which allow study of thermodynamics and kinetics of a wide range of gas-phase ion-molecular reactions, thus creating a possibility to compare supramolecular complexes in solution and in gas phase.

D3.5.1.   Mass spectrometry study of solution equilibria

Mass spectrometry can be used as an analytical tool for determination of solution composition under conditions of equilibrium, several reviews have been published.[118–123] The central problem here is the application of a sufficiently soft ionization technique, which does not destroy the solution host–guest complex, and allows to transfer the often non-volatile heavy host molecules to the gas phase.

First introduced for this purpose were field desorption (FD) and fast atom bombardment (FAB) techniques. In FD ions are desorbed from a coated emitter after solvent evaporation. Peak intensities in the observed mass spectra do not reflect, however, solution composition but rather secondary reactions at the gas phase or the emitter surface, as was concluded from the study of complexation equilibria between crown ethers and alkali metal cations by FD-MS.[124] At the same time, FAB-MS was found to be practically free from any contribution of secondary gas phase processes when applied to the same crown ether–cation system.[118] However, with peptide ligands complexation with cations was found to occur in the gas phase rather than in solution.[125] In the FAB technique, a fast beam of neutral atoms (usually Xe) promotes desorption of ionized species directly from the liquid matrix. Only viscous non-volatile liquids, typically glycerol or *meta*-nitrobenzyl alcohol, can be used as liquid matrixes, which of course restricts applications of FAB-MS to few solvents. FAB-MS study of crown ether cation complexation revealed good correlations with solution equilibria;[126] in particular, stability constants of crown ether–alkali cation complexes in glycerol determined by

FAB-MS agreed well with those found calorimetrically under similar conditions.[127] Many applications of this technique to supramolecular systems can be found.[118,119]

FAB-MS is often referred to as liquid-assisted secondary ion MS due to the key importance of the liquid matrix to assist the desorptive ionization.[128] Similar by its physical basis is the liquid secondary ion mass spectrometry (LSI-MS), which utilizes the same viscous liquid matrix, but fast beam of atomic ions (usually Cs^+) instead of neutral atoms. LSI-MS was also applied to determination of stability constants of crown ether–alkali cation complexes.[129] FAB/LSI-MS allows one to produce molecular ions from species of masses up to several kDa. However, besides the analyte mass and the necessity to use only certain viscous solvents mentioned above, there are several more limitations of this technique. Although molecular ions $[M^+]$ are usually dominating, appreciable fragmentation of desorbed species and formation of secondary ions (protonated $[M+H]^+$ and 'cationized' $[M+C]^+$, where C is a metal cation), persists in FAB spectra, complicating spectra interpretation. More importantly, in spite of successful detection of molecular ions from metal ion complexes, less stable organic host–guest complexes undergo dissociation under desorption conditions and are undetectable with this technique.[118]

Other desorption techniques applied to the study of host–guest complexation (mainly crown ether–cation) are ^{252}Cf plasma desorption (PD),[130] desorption chemical ionization (DCI),[131] and electrohydrodynamic (EH)[132] mass spectrometry. A promising technique is matrix-assisted laser desorption ionization (MALDI)[128] which can produce intact molecular ions from species with masses above 100 kDa. The sample is dispersed in a radiation-absorbing matrix (different non-volatile, typically aromatic, organic compounds), placed on a metallic target, dried and illuminated by short intense light pulses from a laser source, which produce molecular ions, as well as protonated and cationized molecules through still rather obscure mechanisms. The matrix selection is very important and usually several matrix materials must be tested in order to get optimum performance. One disadvantage of this technique is that it produces primarily singly charged species, which for large polymers like protein molecules possess too high, poorly resolved m/z values.

The enormous success of electrospray ionization (ESI) (and its modification ion spray (IS)) technique[120–123] has shown this technique to be the most promising MS approach to the study of host–guest complexation. In this process, a diluted solution of the analyte in water or in another polar solvent flows through a small diameter capillary to which a high voltage is applied. If the capillary is the positive electrode, some anions will move to the capillary wall and some cations will move to the liquid surface at the capillary tip. As a result, liquid is expelled from the capillary as positively charged droplets of micrometer size. The stability of charged droplets depends on charge-to-volume ratio which cannot exceed certain limit (the Rayleigh limit) when the surface charge density exceeds the liquid's surface tension. It is thought that solvent evaporation from the droplets increases the charge-to-volume ratio till the Rayleigh limit, which results in their disintegration referred to as 'Coulomb explosion' or 'droplet fission'. Further solvent evaporation again makes the droplets unstable and leads to further disintegration until small droplets of nanometer size containing single macromolecules of protein are formed. If the starting solution also contains a ligand for a given macromolecule, the complexes between them will probably survive inside small droplets since the macromolecular surrounding does not change considerably, and non-covalent interactions formed in solution can persist inside the droplet. Then macromolecules and/or their complexes gradually lose the solvent and are transformed into the desolvated molecular ions detected ultimately. Collisions between desolvated species can lead to their dissociation. The sequence of the events occurring during ESI is shown in Fig. D 16.

The IS technique differs from ESI in the mechanism of droplet formation: in this technique a turbulent flow of nitrogen gas is applied to nebulize the liquid sample and the high voltage serves principally to charge the droplets. Both techniques give very similar results.[133] The molecular ions detected are multiply charged and even for large molecules the m/z ratios turn out to be within the range where conventional mass spectrometers work well. An example of an ESI-MS spectrum of the α-chain of the human hemoglobin is shown in Fig. D 17. It represents a set (so called 'charge structure') of differently charged molecular ions ('charge states') which contain different numbers of attached protons. In turn, each charge state is a multiplet (isotopic envelope) of signals which differ by their isotope composition (mainly ^{12}C and ^{13}C for organic compounds).

More frequently produced positively charged species contain multiply protonated $[M + nH]^{n+}$ or/and cationized $[M + nC]^{n+}$ forms, depending on solution composition and affinity of the analyte to protons and cations (Na^+, K^+, NH_4^+, etc.). The absolute values of m and z can be determined both from charge structure and from a single charge state.[122a] In the first case the assumption is that each peak differs form another one only by charge, that is by number of protons in the example in Fig. D 17a. Thus, for two peaks, one of which is located at a lower m/z value $(m/z)_{LOW}$ and another one is a nth peak away from the first located at higher m/z value $(m/z)_{HIGH}$ one can write

$$z_{LOW} = z_{HIGH} + n$$

and

$$(m/z)_{HIGH} - (m/z)_{LOW} = m(1/z_{HIGH} - 1/z_{LOW})$$
$$= m(z_{LOW} - z_{HIGH})/(z_{LOW}z_{HIGH})$$
$$= (m/z)_{HIGH}(n/z_{LOW})$$

From the last equation

$$z_{LOW} = n(m/z)_{HIGH}/\{(m/z)_{HIGH} - (m/z)_{LOW}\}$$

With known charge of a single peak, charges and masses of all peaks for a given charge structure are easily calculated. In the fine structure of a single charge state peaks differ from each other by number of ^{13}C atoms incorporated. Therefore the mass difference between two adjacent peaks is simply the mass difference between ^{13}C and ^{12}C which equals 1.0034 atomic mass units. Since all peaks of a given isotopic envelope have the same z one can write for two adjacent peaks $(m/z)_i$ and $(m/z)_{i+1}$

$$z = 1/\{(m/z)_{i+1} - (m/z)_i\}$$

Besides these simple methods, a number of sophisticated algorithms have been developed for the interpretation of mass spectra containing multiple charge states of multiple components.[134]

Several reviews are available on the application of ESI-MS to the study of metal complexation with crown ethers,[118,119,123] and of association processes with biopolymers.[120,121,123] There is convincing evidence of the 'survival' of non-covalent associates under ESI conditions, although the method does not always provide a representation of the solution equilibria accurate enough for calculation of stability constants. In general, complexes formed by electrostatic interactions are expected to be strengthened in the gas phase, but binding contributions from solvophobic interactions will be absent. In some occasions equilibrium constants or at least their relative values (from a competition experiment) reasonably close to those found in solution were determined from integrated ion abundances of reaction components. Fig. D 18 exemplifies IS-MS spectra obtained for solutions of a macromolecule (human cytoplasmic receptor FKBP, a small protein of molecular mass 11 812 Da) mixed with a ligand FK506 (**D-18**) (an immunosuppresive agent).[135] Peaks of both free and bound

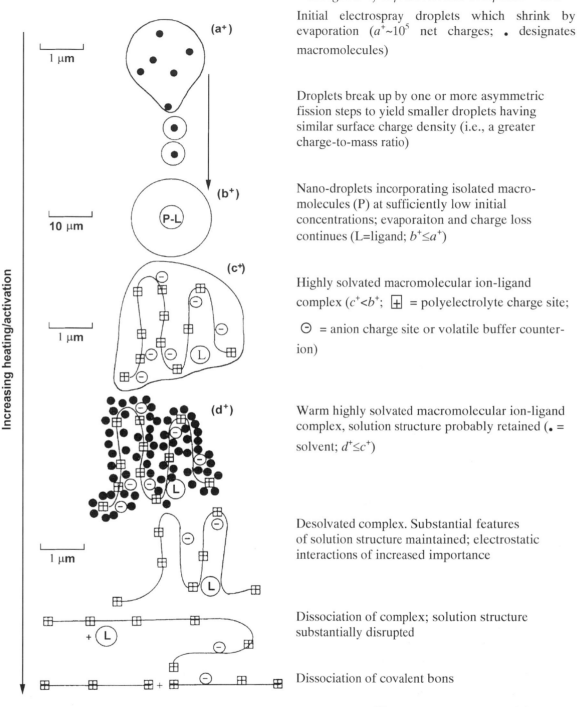

Initial electrospray droplets which shrink by evaporation ($a^+\sim10^5$ net charges; • designates macromolecules)

Droplets break up by one or more asymmetric fission steps to yield smaller droplets having similar surface charge density (i.e., a greater charge-to-mass ratio)

Nano-droplets incorporating isolated macromolecules (P) at sufficiently low initial concentrations; evaporaiton and charge loss continues (L=ligand; $b^+\leq a^+$)

Highly solvated macromolecular ion-ligand complex ($c^+<b^+$; ⊞ = polyelectrolyte charge site; ⊖ = anion charge site or volatile buffer counter-ion)

Warm highly solvated macromolecular ion-ligand complex, solution structure probably retained (• = solvent; $d^+\leq c^+$)

Desolvated complex. Substantial features of solution structure maintained; electrostatic interactions of increased importance

Dissociation of complex; solution structure substantially disrupted

Dissociation of covalent bons

Figure D 16. Model for formation of macromolecular ions during ESI.[121] Reprinted with permission from *Chem. Soc. Rev.*, Smith, R.D. *et al.*, 1997, **26**, 191. Copyright 1997, Royal Society of Chemistry.

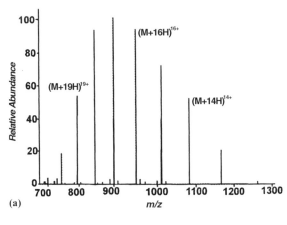

(a)

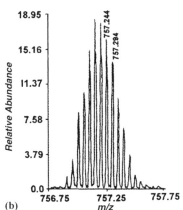

(b)

Figure D 17.   ESI-MS spectrum of the α-chain of the human hemoglobin.[122a] (a) Charge structure observed at low resolution. (b) The isotopic envelope of a single charge state at high resolution. Reprinted with permission from the *Journal of Chemical Education*, Hofstadler *et al.*, **73**, 1996, p. A82. Copyright 1996, Division of Chemical Education.

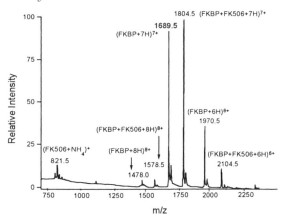

Figure D 18.   IS-MS spectra[135] of a mixture of FKBP receptor with FK506 ligand (**D-18**) at pH 7.5. Reprinted with permission from *J. Am. Chem. Soc.*, Li *et al.*, 1994, **116**, 7487. Copyright 1994. American Chemical Society.

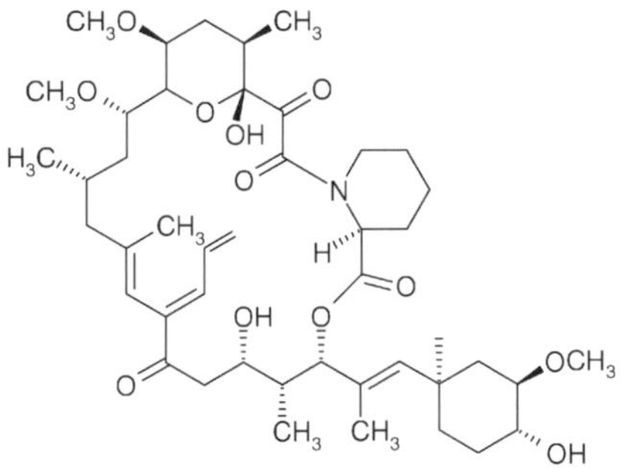

D-18

protein are observed (Fig. D 18) and complex peaks disappeared upon acidification, which induces denaturation of the protein. The latter serves as an evidence of specific protein–ligand interaction (see below). Competition experiments with FK506 and other ligands were used in this study for determination of the relative binding constants to the receptor from the relative intensities of the peaks of complexes with competing ligands.[135]

Besides the problem of missing insufficiently stable complexes, there is the serious problem of so-called non-specific associations (associates which are weak or non-dectectable in solution but are strengthened in the gas phase and become detectable by MS), and secondary gas-phase processes. Proposed approaches to distinguish specific and non-specific associations involve[120] (a) searching experimental conditions under which the complex dominates and the observation of its correct stochiometry, (b) testing effects of changes in solution conditions (temperature, solvent, pH, etc.) which can destroy specific interactions, and (c) testing selectivity to structural modification of interaction components.

Among many organic host–guest complexes characterized by ESI-MS,[118,123] cyclodextrin complexes with amino acids and peptides have received considerable attention.[136] Fig. D 19a shows the ESI spectrum of β-cyclodextrin (β-CD) from aqueous solution, which represents a set of cationized peaks. The absence of protonated peak indicates very low affinity of β-CD to protons and relatively high affinity to sodium cations: under the given conditions no sodium salt was added to the solution, and β-CD is charged by traces of sodium present as a common laboratory contaminant only. However, neither protonated nor complexes of sodium cations with cyclodextrin exist in solution. Evidently charging occurring in the gas phase and detection by MS ions does not relate to solution composition. This illustrates one of important restrictions of MS: only charged species can be detected and if the analyte is a neutral species in solution, it can be detected only if it becomes charged somehow in the gas phase; the intensity of resulting peak may be a rather imprecise measure of the analyte concentration in such a case.

Figures D 19b,c show the spectra of β-CD in the presence of two amino acids: phenylalanine, which does form a complex with β-CD in solution by inclusion of its phenyl ring into the macrocycle cavity, and more polar histidine, for which no complexation with β-CD was reported. Nevertheless, the relative intensity of the $[CD + His + H]^+$ peak is considerably larger than that of $[CD + Phe + H]^+$ peak. Tests with mixtures of β-CD with amino acids and peptides of different structures showed formation of adducts with all of them, but instead of the expected more intensive peaks with aromatic amino acids, the greatest peaks were observed with most basic (arginine, lysine) amino acids.[136d] In addition, ESI-MS spectra of β-CD from solutions containing residual ammonium ion contamination show only peaks of β-CD

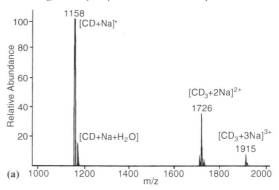

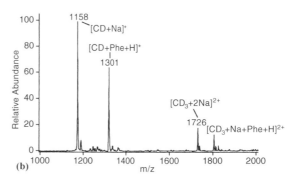

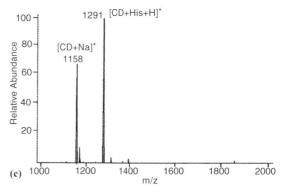

Figure D 19. ESI-MS spectra of β-CD alone and in the presence of amino acids in aqueous solution. (a) Spectrum of β-CD alone. (b) Spectrum of the mixture of β-CD with phenylalanine. (c) Spectrum of the mixture of β-CD with histidine.[136d] Reprinted by permission of Elsevier Science from (False positives and the detection of cyclodextrin inclusion complexes by electrospray mass spectrometry) by Cunniff *et al.*, *Journal of the American Society for Mass Spectrometry*, Vol. **6**, pp 437. Copyright 1995, American Society for Mass Spectrometry.

cationized by NH_4^+.[136d] Based on these observations the authors conclude that instead of expected inclusion complexes of β-CD with aromatic amino acids pre-formed in solution, ESI-MS detects electrostatic adducts of β-CD with protonated amino groups of amino acids most probably formed in the gas phase.[136d]

No complexes were detected by ESI-MS for the mixtures of β-CD with hexane and toluene, which definitely form inclusion complexes in solution.[136d] This result is, perhaps, not surprising since purely hydrophobic adducts can hardly survive ESI conditions. Alternatively, these adducts could be destroyed by cationization of β-CD with Na^+, which must occur since only charged species are detected. It is worth to noting that by themselves complexes between neutral host and neutral guest molecules can be quite stable in the gas phase. Thus, formation of benzene and other hydrocarbon complexes of neutral cavitands was observed in the gas phase by DCI-MS technique.[137]

ESI is considered to be 'soft' ionization technique in the sense that usually no fragmentation occurs and only molecular ions are detected. However, with respect to weak noncovalent adducts, especially those formed by solvophobic interactions, ESI is 'hard' enough because it leads to complete desolvation of gaseous species and strongly shifts the balance between relative contributions of noncovalent interactions of different types in favor of purely electrostatic binding. As a result, this technique may give a more accurate representation of solution equilibria when complexation is dominated by electrostatic forces and will be less reliable when other interaction types are dominating. Undoubtedly, ESI is a very powerful technique for highly sensitive detection mostly of relatively specific associations, for instance of drug–receptor or other ligand–protein complexes, and finds application, in particular, in rapid screening of combinatorial libraries.[121]

D 3.5.2.   Mass spectrometry study of gas phase equilibria

The study of host–guest complexation equilibria in the gas phase is of an obvious importance since it provides the thermodynamic parameters free of any solvent effect.[118,137–139] We already discussed the gas-phase data, such as for cation–macrocyle (Sections A2 and A3) or cation–π (Section B4) complexation. Since the 90s gas-phase interactions between macrocyles and metal cations have been studied systematically.[138,139] Two ion-trapping techniques, Fourier transform ion cyclotron resonance (FTICR)[140] and quadrupole ion trap (QIT)[141] mass spectrometry, were proven to have the highest capabilities for the study of ion–molecular reactions in the gas phase.[138] These techniques allow the storage and manipulation of ions over a period of several seconds, sufficiently long for both kinetic and equilibrium studies, as well as for testing multi-stage reaction sequences.

The sensitivity of detection in FTICR is rather low: about 100 ions are required to produce a detectable signal, while the electron multipliers used in conventional MS instruments can detect single ions. Very low pressures typically below 10^{-8} torr must be used, under which the coherent motion may last for tens of seconds, enough to produce very high resolution spectra. This long detection time allows also the equilibration of the interacting species making FTICR useful for the study of gas-phase equilibria.

An important technique in the study of gas-phase processes is the tandem MS, also referred to as MS^n.[142] The principle of this technique is the use of a sequence of steps of ion manipulation, typically collisionally activated dissociation (CAD), with subsequent mass analysis after each step. Most often only an MS/MS ($n = 2$) experiment can be performed due to several limitations.[138] Ions from an ion source initially pass through a mass analyzer, which select ions of a desired

m/z ratio, then through a collision cell (if CAD is a chosen method of ion manipulation) and again through a mass analyzer, which performs the final mass analysis and detection.

Determination of equilibrium constants is performed by monitoring the composition of the reaction mixture as a function of time. Once the system reaches the state of equilibrium, the equilibrium constant can be calculated from the abundances of ionic reactants and products and known vapor pressures of components. The main restriction of this method is the difficulty of measuring accurately the pressure of the reagents, especially with highly involatile large host molecules. As a result, the absolute complexation free energies can be rather inaccurate. For example, free energies of gas-phase transfer reactions of alkali metal cations from macrocyclic to macrobicyclic ligands (equations (A2-4) and (A2-5) measured with different macrocycles, but with the same macrobicyclic ligand, must give by subtraction the free energies of transfer of given cations between two macrocycles (equation A3-1). It follows from results given in ref. 39 of Chapter A, that, with [2.2.2] cryptand as the macrobicyclic ligand, the free energies of transfer of Li^+, Na^+, K^+, Rb^+ and Cs^+ from 18-crown-6 to 21-crown-7 must be -9.18, $+1.14$, $+2.21$, -2.08 and $-16.22 \, kJ \, mol^{-1}$. These numbers are, however, substantially, up to $10 \, kJ \, mol^{-1}$, different from those measured directly under the same conditions (Table A3-2). On the other hand, the sequence of cations arranged in order of decreasing ΔG values is generally the same, indicating, that perhaps these techniques are more suited to obtain relative rather than absolute gas-phase thermodynamic data.

When direct equilibrium measurements are impossible, at least relative binding affinities can be determined by so-called kinetic and bracketing (ligand exchange) methods.[138,139] In the kinetic method a cation-bound complex of two different ligands (e.g. a proton-bound

pair of bases or a sandwich-type cation crown ether complex) is generated in the gas phase and then is collisionally activated to induce dissociation. Schematically this can be shown as follows

$$L_1 + C^+ + L_2 \rightarrow (L_1)C(L_2)^+ \xrightarrow{CAD} L_1 + CL_1^+$$
$$+ L_2 + CL_2^+$$

Dissociation at the cation bridge produces a mixture of products formed by cleavage of either $C–L_1$ or $C–L_2$ bonds; the ratio of abundances of CL_1^+ and CL_2^+ cations then reflects the ratio of the respective bond energies. The method is inappropriate when L_1 and L_2 of very different structural types are compared. In the bracketing method the cationic complex with one ligand is allowed to react in the gas phase with another ligand and the appearance of transfer product is monitored. Usually, the cation-transfer reaction is performed in both directions to confirm the order of affinities.

The use of MS for the study of gas-phase complexation between neutral guests and neutral hosts (calixarenes, cavitands) was recently explored.[118,137,143] In most of experiments the host was vaporized from a DCI probe within the ion source in an atmosphere containing two volatile guests in known relative pressures and an excess of a chemical ionization (CI) reagent gas (usually methane). The relative affinities of the host to the guests were evaluated from the abundances of the corresponding charged complexes in the mass spectrum. It was concluded, however, that complexation occurs predominantly through interaction of charged hosts with neutral guests, although interactions between neutral species also may contribute significantly under specific conditions.

D 4. SOLUBILITY AND SEPARATION-BASED METHODS

Although the methods discussed in this chapter often do use spectroscopic techniques for concentration measurements, they do not need any complexation-induced spectral changes. Complexation-induced increase of substrate solubility in a given solvent, or extractability from water into an organic phase has been used since long for the determination of formation constants. Besides their economy these methods have the advantage to be applicable also to sparingly soluble guest molecules, but are relatively inaccurate due to various, not easy to control sources of systematic errors. Separation-based methods which use complexation-induced changes in chromatographic retention time (elution volume) or electrophoretic mobility of the substrate need more sophisticated instruments, but are more accurate, use very small amounts of the substrate, can operate at varied temperature, and can be easily automated; they gain growing importance for the study of chemical equilibria also in view of their applications in separation technology.

If dissociation of a supramolecular complex becomes slow in comparison to separation of complexed and free substrates, and if this separation is complete before re-establishing the equilibrium, one can use any determination of free or bound substrate for the evaluation of the apparent binding constant. Container molecules (Section H1) usually already meet the condition of exchange which is slow in comparison to analytical separation. Slow equilibration is the basis of many affinity determinations with proteins,[2] (Section D1.1.1) which not only can exert so strong binding that the dissociation is slow enough but due to their large mass also can be easily separated from low molecular weight substrates. In the so-called radioimmunoassay (RIA) one uses a radioactive label which is liberated upon addition of the 'cold' substrate to be evaluated, and after separation from the protein complex the activity is measured with a counter. One can avoid the use of radioactive techniques by using fluorescent or enzyme labels instead. With the EIA (enzyme immunoassay) or the ELISA test (enzyme

linked immunosorbent assay) one uses signal multiplication by coupling ligand or substrate molecule to an enzyme which catalytically develops, for instance, a colored indicator. With ELISA an antibody is linked to a insoluble carrier, which allows easy separation of the complex formed with a test ligand, to which an appropriate enzyme like peroxidase is covalently attached. These techniques are expected to gain more popularity also in synthetic host–guest chemistry with the increasing binding strength in supramolecular complexes and therefore slower dissociation rates, and the increasing size of host molecules, which make fast separations more feasible. The very high sensitivity, gained with RIA by measuring radioactivity, and with EIA and ELISA by enzymatic proliferation, also allows measurements with extremely very small quantities. On the other hand, the always to lesser or larger observed degree of non-specific sorption of the labeled component on the carrier, and the possible influence of the carrier on the ligand affinity may lead to errors with these techniques.

D4.1. SOLUBILITY CHANGES

This technique has been thoroughly described.[1a] The solubility (s, the concentration of a substrate in saturated solution) is determined in a series of solutions with variable ligand concentrations at constant temperature, ionic strength and solvent composition. The experimental procedure consists of equilibration by the use of shaking machines or ultrasonic baths (which may take hours or even days and should be checked until constant concentrations are observed) of a series of thermostatted vials or ampoules containing an excess of the substrate and increasing concentrations of the ligand with subsequent analysis of the solution phase after separation from the solid phase by filtration or centrifugation. The substrate must have a measurable solubility in the absence of ligand (s_0), but s_0 should be small enough to observe a noticeable increase in solubility on ligand additions. If only very high ligand concentrations affect the substrate solubility, this most probably reflect changes in the substrate activity coefficient rather than complex formation (in such cases both increase or decrease in solubility can be observed). If reasonably low ligand concentrations cause an increase in the solubility, this is interpreted as being due to complexation. Often the UV/vis absorption is used for determination of concentration of dissolved substrate; in this case an independent experiment must be performed to prove that the molar absorptivity of the substrate at chosen wavelength is not changed by complexation.

Depending on the complexation stoichiometry and solubility of the complex, different types of solubility *vs.* ligand concentration profiles (solubility diagrams) are observed, as illustrated schematically in Fig. D 20. In the simplest case of a single 1:1 soluble complex between S and L, the experimentally measured solubility in the presence of ligand equals the sum of concentrations of all forms of substrate existing in the solution

$$s = [S]_t = [S] + [SL] \qquad (D4\text{-}1)$$

The concentration of free S in all solutions is equal to s_0 since all solutions are in equilibrium with the same solid phase. Substituting

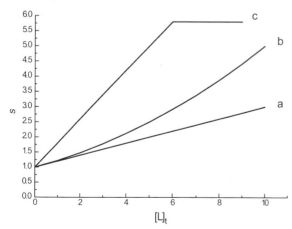

Figure D 20. Schematic presentation of typical solubility diagrams.

s_0 for [S] and using the mass balance equation for the ligand $([L]_t = [L] + [SL])$ and the expression for the formation constant $K = [SL]/([S][L])$ one obtains the expression

$$s = s_0 + \{Ks_0/(1 + Ks_0)\}[L]_t \qquad (D4\text{-}2)$$

The binding constant is calculated with this equation from the slope and intercept of the solubility diagram of the type shown in Fig. D 20, line 'a'. Note that the slope of the line cannot be larger than unity.

If complexes containing more than one substrate molecule per one ligand molecule are formed, the solubility diagram remains linear. Indeed, for a set of complexes SL, S_2L, S_3L, etc. with overall formation constants $\beta_i = [S_iL]/([S]^i[L])$ both the solubility and total ligand concentration are linear functions of free ligand concentration

$$
\begin{aligned}
s &= [S] + [SL] + [S_2L] + [S_3L] + \cdots \\
&= s_0 + [L](\beta_1 s_0 + 2\beta_2 s_0^2 + 3\beta_3 s_0^3 + \cdots)[L]_t \\
&= [L] + [SL] + [S_2L] + [S_3L] + \cdots \\
&= [L](1 + \beta_1 s_0 + \beta_2 s_0^2 + \beta_3 s_0^3 + \cdots)
\end{aligned}
$$

Therefore, s will be a linear function of $[L]_t$. The fact that solubility data do not discriminate between complexes with different number of substrate molecules is not unexpected: in this method the concentration of the substrate cannot be a variable since it always equals to s_0. The slope of the solubility diagram can serve as a qualitative indication of formation of higher complexes: if it is larger than unity such complexes must be present. Of course, slopes less than unity do not prove formation of a single 1:1 complex.

If higher complexes of the type SL_2, SL_3, etc. are formed in addition to the SL complex, the solubility diagram becomes curved upward, as shown in Fig. D 20, curve 'b'. In this case equation (D4-1) must be extended to include these higher complexes

$$
\begin{aligned}
s = [S]_t &= [S] + [SL] + [SL_2] + [SL_3] + \cdots \\
&= s_0(1 + \beta_1[L] + \beta_2[L]^2 + \beta_3[L]^3 + \cdots)
\end{aligned}
$$
$$(D4\text{-}3)$$

Also the mass balance equation for the ligand will in this case be a polynom with respect to [L]

$$[L]_t = [L] + [SL] + 2[SL)_2] + 3[SL_3] + \cdots$$
$$= [L] + s_0(\beta_1[L] + 2\beta_2[L]^2 + 3\beta_3[L]^3 + \cdots)$$
$$\text{(D4-4)}$$

If only 1:1 and 1:2 complexes are present the system of equations (D4-3) and (D4-4) can be solved analytically or graphically by a suitable linearization procedure.[1a] If higher complexes are present, an expression for s as a function of $[L]_t$ cannot be obtained and the equations should be solved numerically. When stability constants are small, and only a small fraction of added ligand is converted to the complexes the assumption $[L] = [L]_t$ may be valid and equation (D4-3) can be used directly to fit the experimental solubility diagram. A solubility diagram which reaches a plateau region (curve 'c') is observed in the cases when the ligand or the complex possess a limited solubility in the concentration range employed (this also can be a trivial result of complete dissolution of the substrate, which of course can be easily checked).

D4.2.   EXTRACTION

The equilibrium distribution of a species S between two immiscible liquid phases, most often water and a lipophilic organic solvent, is described by the partition constant P_S, assuming unity activity coefficients of S in both phases[144]

$$P_S = [S]_O/[S]_W \qquad \text{(D4-5)}$$

where subscripts 'O' and 'W' refer to organic and water phases, respectively. Addition of a ligand capable of complexation of S in one or both phases will change the total concentrations of S in these phases, which can be used for stability constant determination. Experimentally one determines the distribution ratio (an apparent partition constant) D_S, which equals the ratio of total concentrations of S in

both phases determined analytically

$$D_S = [S]_{t,O}/[S]_{t,W} \qquad \text{(D4-6)}$$

With some sacrifice in accuracy it is sufficient to determine $[S]_t$ in only one of the phases since its concentration in the other phase can be calculated from the mass balance equation. In the simplest case both S and L are neutral species and L as well as the complex SL is soluble only in one of the phases, like in water. Such a case is found, for example, in a study of extraction of carbazole in heptane from water in the presence of β-cyclodextrin.[145] In this case $[S]_{t,O} = [S]_O$, $[S]_{t,W} = [S]_W + [SL]_W$, and after introducing the complex formation constant in water $K = [SL]_W/([S]_W[L]_W)$ the expression for D_S takes the form

$$D_S = [S_O/\{[S]_W(1 + K[L]_W)\}$$
$$= P_S/(1 + K[L]_W) \qquad \text{(D4-7)}$$

If the ligand is used in a high excess over the substrate, $[L]_W = [L]_{t,W}$ and equation (D4-7) can be directly applied to fit the experimental dependence of D_S on ligand concentration. Otherwise the mass balance equation must be used to exclude the free ligand concentration.

The more general case which takes into account partition of all reaction components is shown in Scheme D7. The expression for D_S will take the form

$$D_S = P_S(1 + K_O[L]_O)/(1 + K_W[L]_W) \qquad \text{(D4-8)}$$

To use equation (D4-8) one needs to know the free ligand concentrations in both phases. They are related by a respective partition constant P_L which must be determined independently and under conditions of high excess of L over S both concentrations can be easily calculated from P_L and the mass balance equation. Since the system is heterogeneous, the mass balance equation must take into account the volumes of each phase. For example, if the ligand solution is prepared initially in one of the phases, like in water, at total concentration $[L]_i$ in the volume V_W and than a volume V_O of the organic solvent is

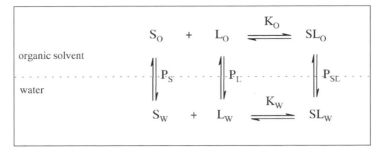

Scheme D 7. Partition and complexation equilibria for extraction of a neutral solute.

added, the mass balance equation will take the form

$$[L]_i V_W = [L]_{t,W} V_W + [L]_{t,O} V_O \qquad (D4\text{-}9)$$

Evidently, if equal volumes of two solvents are used, the volume cancels out.

Besides the usual complications (formation of complexes of different stochiometries, self-association, phase separation difficulties etc.) there is a specific problem in the extraction technique, which makes it generally inaccurate: this is the mutual solubility of two phases which furthermore can be influenced by the added solutes. Even traces of the organic solvent in water can be harmful if they compete with the guest for a hydrophobic host molecule. More harmful are, however, traces of water in the organic phase due to often observed specific hydration of polar species, especially ionic guests, discussed below. As a rule, the mutual solubility depends on usually rather high ligand con-

centration which makes the system difficult to control.

Nevertheless, the extraction technique has been used traditionally for the study of complexation of cations with crown ethers and related ligands.[146–148] A simplified scheme of an extraction process of a salt MA in the presence of a neutral ligand L is shown in Scheme D 8. It is assumed that no ion pairing between anion and free or complexed cation occurs in water and that no free ions, as well as no higher aggregates than ion pairs exist in the organic solvent. The complex formation constant in water (K_W) refers to association of free metal cation M^+ with L, the formation constant in organic solvent refers to association between the ion pair MA and L

$$K_O = [MLA]_O/([MA]_O[L]_O)$$

Evidently K_O defined in that way must depend on the counterion A which must be

MA$_O$ $+$ L$_O$ $\xrightleftharpoons{K_O}$ MLA$_O$

organic solvent
- - - - - - - - - - - - K_{MA}^{EX} - - - - - - P_L - - - - - - - - - - - - - - - - K_{MLA}^{EX} - - - -
water

M^+_W $+$ A^-_W $+$ L_W $\xrightleftharpoons{K_W}$ ML^+_W $+$ A^-_W

Scheme D 8. A simplified scheme of an extraction process of a salt MA in the presence of a neutral ligand L.

the same for different metals in order to make it possible to consider K_O as a measure of relative affinity of a given ligand to different cations. Commonly metal picrates are used for extraction experiments, as the picrate ion concentration can be easily determined spectrophotometrically.

Equilibria between charged species in water and neutral ion pairs in organic solvent are expressed by extraction constants defined as

$$K_{MA}^{EX} = [MA]_O/([M^+]_W[A^-]_W) \quad \text{(D4-10)}$$
$$K_{MLA}^{EX} = [MLA]_O/([ML^+]_W[A^-]_W) \quad \text{(D4-11)}$$

For ligands which are practically insoluble in water it is convenient to introduce the extraction constant K^{EX} which refers to formation of a complex in the organic phase from aqueous cation and anion and ligand present in the organic phase

$$K^{EX} = [MLA]_O/([M^+]_W[A^-]_W[L]_O) \quad \text{(D4-12)}$$

From Scheme D8 it is evident that

$$K^{EX} = K_O K_{MA}^{EX} \quad \text{(D4-13)}$$

The treatment of experimental results is especially simple in this case because only two equilibria must be considered: extraction of uncomplexed and complexed ion pairs represented by equations (D4-10) and (D4-12) respectively. The value of K_{MA}^{EX} is measured in the absence of L, and the extraction constant K^{EX} is easily calculated from total analytical concentrations of cations in both phases and mass balance equations. In the spectrophotometric picrate method the anion concentration is determined from the observed absorbencies of equilibrated solutions, and since both phases are electrically neutral this also gives the cation concentrations. In the 'NMR' extraction method[148] the fraction of bound ligand in the organic phase is determined by integration of signals of ligand protons. Then the complexation constant in the organic phase is calculated with equation (D4-13). If ligand solubility in water is appreciable the

treatment becomes rather difficult because in this case all equilibria shown in Scheme D8 must be considered.[149–151] However, this case seems to be less important because with water-soluble ligands many other, more direct methods can be (and should be) used.

D4.3. CHROMATOGRAPHY AND ELECTROPHORESIS

Different chromatographic techniques such as gas liquid chromatography (GLC), high-performance liquid chromatography (HPLC), thin layer chromatography (TLC), and size exclusion chromatography (SEC) have been employed for determination of association constants.[152] When determination of the formation constant in solution used as a mobile phase is the goal, the observed retention time (t_r) or elution volume (V_e) of a substrate S is altered by addition of a ligand L to the mobile phase (eluent) due to different retention of free S and its complexed form SL on the given stationary phase. Alternatively, one measures corresponding changes with stationary phases on which the host or guest are immobilized. One should be aware that contributions to binding between host and guest in solution and with immobilized host (or guest) may differ, for example due to differences in conformations, in local dielectrics etc, with subsequent variations of association constants. The method is of obvious value with gas chromatographic separations, where accurate temperature control can also via the van't Hoff method provide ΔH and ΔS values.

The physico-chemical basis of reversed-phase HPLC (the most often employed technique for the study of equilibria in aqueous solution) is in many aspects similar to that of extraction: in both cases the separation of solutes is achieved by distribution between an aqueous solution and a non-polar phase, which in the case of HPLC is an inert non-polar solid stationary phase, like a surface-modified by bulky alkyl groups on silica. The important advantages of HPLC are that both phases are completely immiscible and that equilibrium distribution of the solutes between phases is established rapidly.

In reversed-phase HPLC retention of solutes is determined principally by their polarity: it is large for hydrophobic neutral organic molecules and is absent for ordinary inorganic ions. Neutral organic hosts like cyclodextrins capable of inclusion of non-polar organic guests usually possess a hydrophobic cavity and a hydrophilic external surface. Therefore, both hosts and their inclusion complexes are retained very weakly on non-polar stationary phases while free guests show considerable retention.[153] Similarly, retention of non-polar organic hosts considerably decreases on complexation with polar ionic guests, as it occurs, for instance, on complexation of crown ethers with metal cations.[154] Alternatively, complexation of practically unretained ionic species with oppositely charged ligands affords neutral complexes with detectable retention.[154]

In SEC (most often gel-permeation (filtration) chromatography) the separation of solutes is achieved according to their size due to sieving effect or steric exclusion. Large molecules that cannot penetrate pores of the stationary phase are eluted rapidly in a void volume of the column, but smaller molecules are eluted in a larger volumes. Evidently if a small molecule can interact with a large molecule the complex will also be large and will move rapidly together with the large molecule. This makes SEC particularly useful for the study of complexation of proteins and other macromolecules or surfactant micelles with low molecular weight guests. However, even some organic hosts like cyclodextrins are large enough to be separated from small guests by gel-filtration with a suitable stationary phase.[155]

There are two principal methods of chromatographic study of association equilibria. In one of them, often called the retention method, retention of S is measured in a series of mobile phases containing increasing concentrations of the ligand. In each run the column is equilibrated with a solution containing given concentration of L and then a small amount of S is injected. For the quantitative analysis of complexation equilibrium the capacity factor k' of S is considered,[152] which is the ratio of the amount of the solute in the stationary phase to that in the mobile phase. It is calculated from the retention parameters according to equation (D4-14)

$$k' = (t_r - t_0)/t_0 = (V_e - V_m)/V_m \quad (D4\text{-}14)$$

where t_0 and V_m are dead time and the mobile phase volume which correspond to observed retention time and elution volume of an unretained solute. On the other hand, k' is given by equation (D4-15)

$$k' = ([SA]V_S)/([S]V_m) \quad (D4\text{-}15)$$

where V_S is the volume of the stationary phase, [S] is the substrate (solute) concentration in the mobile phase and [SA] is the concentration of S in the stationary phase. The latter is treated as the concentration of complexes between S and the stationary phase binding sites A. Introducing the respective equilibrium constant $K_S = [SA]/([S][A])$ and the phase ratio $\phi = V_S/V_m$ into equation (D4-15) one obtains

$$k' = K_S \phi [A] \quad (D4\text{-}16)$$

When S is injected at sufficiently low concentrations so that the amount of S is always much less than the number of the stationary phase binding sites, [A] is approximately constant and equal to the total concentration of the binding sites. Under such conditions k' is independent of [S] and represents itself an intrinsic property of the solute. For a rapidly established equilibrium the observed k' has the mean weighted value between those of free and complexed molecules[152,154]

$$k' = f_S k'_S + f_{SL} k'_{SL}$$

where K'_S and K'_{SL} are the capacity factors of free and complexed substrate molecules, and f_{SL} and f_S are the mole fractions of S and SL, respectively (the situation is formally similar to that discussed for NMR spectra under conditions of fast exchange, Section D 3.4). Substitution the expressions for f_{SL} and f_S in terms of concentrations leads to equations which coincide with those for x_{obs} given in Ch. D 1. In particular, for a 1:1 association equilibrium with the formation constant K one obtains

$$k' = (k'_S + k'_{SL}K[L])/(1 + K[L]) \qquad (D 4\text{-}17)$$

Under the condition of a high excess of L over S, which is easily achieved in a chromatographic experiment, [L] is equal to the total concentration of L in the mobile phase and the equation (D 4-17) can be used directly. Often $k'_{SL} = 0$ (e.g. in the reversed-phase HPLC with cyclodextrins or in SEC with sufficiently large hosts) and equation (D 4-17) takes an especially simple linearized form. Different versions of such linearized equations have been summarized.[9] In a general form they can be presented as follows

$$X = Z + ZK[L] \qquad (D 4\text{-}18)$$

where Z is a constant which involves such parameters as K_S, ϕ, $[A]$, V_m, and some others and X is an experimentally determined variable such as $1/k'$ or $1/(V_e - V_m)$ or $1/(t_r - t_0)$ for HPLC; $1/(V_e - V_0)$ for SEC (V_0 is the void volume), and $R_f/(1 - R_f)$ for TLC (R_f is the retention fraction).

If higher complexes of stoichiometries SL_i are formed, the expression for k' takes a form similar to equation (D 1-47)[154]

$$k' = (k'_S + k'_{SL}K_1[L] + k'_{SL2}K_1K_2[L]^2 + \cdots)/$$
$$(1 + K_1[L] + K_1K_2[L]^2 + \cdots) \qquad (D 4\text{-}19)$$

When capacity factors of all complexes equal zero, equation (D 4-19) after linearization with inverse coordinates gives a simple polynomial dependence of $1/k'$ on [L], from which all stepwise formation constants can be determined.[156]

In an alternative approach, known as Hummel and Dreyer method,[157] the column is equilibrated with a mobile phase containing a given concentration of monitored substrate S, and then a small amount of L is injected. If the complex SL has different retention times from that of free S, but similar to that of L, a positive peak of the complex SL appears at the retention time of the ligand and a negative peak is observed at the retention time of the substrate. This is illustrated in Fig. D 21 for the determination of association constant between α-cyclodextrin and the *p*-nitrophenolate anion.[158b]

The main requirement for this method is a good separation of the peaks. Positive or negative, or both peak areas can be chosen for calculation of the amount of bound S depending on the degree of interference of the signal from S with L and sharpness of the peaks. For calculation of the equilibrium

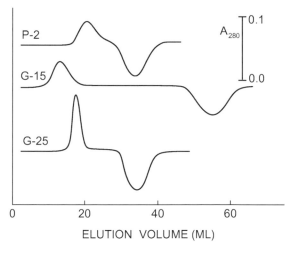

Figure D 21. Hummel–Dreyer elution profiles on Bio-Gel P-2, Sephadex G-15 and G-25; the column was equilibrated with a buffer solution pH 10, containing 0.1 mM *p*-nitrophenolate; 0.25 ml sample of 10 mM α-cyclodextrin in the same buffer was added.[158b] Reprinted from *Journal of Chromatography*, Vol. **290**, Korpela *et al.* (Determination of equilibrium constants by gel chromatography: binding of small molecules to cyclodextrins), pp 351, copyright 1984 with permission from Elsevier Science.

constant two assumptions are important:[158a] the whole amount of added L is present together and in equilibrium with the complex in the positive peak, and the concentration of free S in the positive peak is equal to the concentration of S in the eluent solution with which the column is equilibrated. With these assumptions and known reaction stochiometry the equilibrium constant can be calculated just from a single experiment as follows (for a 1:1 complexation)

$$K = [SL]/([S][L]) = Q_{SL}/([S]Q_L) = Q_{SL}/ \{[S](Q_{L,t} - Q_{SL})\} \qquad (D4\text{-}20)$$

where Q_{SL} and Q_L are molar amounts of the complex and free ligand (calculated as difference between measured Q_{SL} and known total amount of ligand $Q_{L,t}$), and [S] is the concentration of free substrate. A more precise value of K is obtained from measurements with different [S] by regression analysis in terms of the rearranged equation (D4-21)[158]

$$Q_{SL}/Q_{L,t} = K[S]/(1 + K[S]) \qquad (D4\text{-}21)$$

Both methods give consistent results when applied to the same system.[159] The advantage of the retention method is the possibility to vary known ligand concentration in wide intervals which allows to study complex equilibria. In the Hummel and Dreyer method the ligand concentration is unknown and cannot be as high as in the retention method due to unavoidable dilution of the injected ligand solution. The latter method is therefore not suitable for the study of non-trivial equilibria, but in simple cases the method is less laborious and even can give higher precision.[159]

Recent progress in instrumentation of capillary electrophoresis makes this technique increasingly more important for both analytical and physico-chemical applications. When electrophoretic mobilities of free (μ_f) and complexed (μ_c) substrate are different, the measured mobility (μ_i) in the presence of ligand L forming a 1:1 complex with S is given by equation (D4-22) analogous to (D4-17)[152,160a,b]

$$\mu_i = (\mu_f + \mu_c K[L])/(1 + K[L]) \qquad (D4\text{-}22)$$

The observed solute mobility must be corrected by subtraction of the electroosmotic flow mobility. Therefore the migration times, which are inversely proportional to the observed mobility and are used in calculation of the latter, cannot be directly used for determination of association constants.[152,160a] There are several sources of considerable systematic errors in this technique. These involve changes in the viscosity and conductivity of the solution in presence of ligand, which produce respective changes in the electrophoretic mobility not associated with complex formation; therefore some correction methods must be applied (e.g. the use of an electroosmotic flow marker which does not associate with the ligand). In addition, mobility changes can be caused by the binding of a solute to the capillary wall.

D5.   CALORIMETRY

Recent progress in construction of highly sensitive microcalorimeters which operate with small sample volumes makes this technique increasingly more popular for determination of equilibrium constants. Titration calorimetry was applied for study of many supramolecular systems discussed in previous chapters. The main advantage of titration calorimetry is that it allows simultaneous determination of both free energy and enthalpy changes at a given temperature, as well as precise determination of heat capacity changes from measurements at different temperatures. As pointed out previously (Section A4) calorimetry has an unquestionable advantage for determination of ΔH and ΔC_P values over application of the van't Hoff equation to the temperature dependence of an equilibrium constant. Calorimetry does not need any change in spectral or other physico-chemical characteristics to be induced by complexation. It requires, however, the reaction heat effect to be large enough to produce a measurable temperature change. Other than with spectroscopic methods the measured signal depends itself on the strength, specifically the ΔH of the complex. Equilibria which are essentially entropy driven, such as ion pairs, often are nearly undetectable by calorimetry, if they are even not sufficiently endothermic. All processes accompanying the titration experiment, such as stirring, mixing or dilution effects, as well as side reactions, such as in particular any proton transfer can greatly contribute to the measured signal and therefore falsify the results, and require extensive corrections. Calorimetric measurements become also difficult in non-aqueous solvents due to the much lower heat exchange rates compared to water. There is ample general reading on calorimetry and titration calorimetry.[161–165]

There are several operational modes used in calorimetry which differ in heat measurement principles. In adiabatic calorimeters there is no heat exchange between the reaction vessel and the surroundings. The heat produced or consumed during a reaction is determined from the measured temperature change and the known heat capacity of the vessel. In isoperibolic (semi-adiabatic) calorimeters there is some heat flux between the reaction vessel and the surroundings. This mode of operation is often used in titration calorimetry, especially for continuous titration when the titrant (guest or host solution) is added to the other component at a constant rate during one run (of course, this type of titration can be used only for rapid reactions when the state of equilibrium is reached immediately upon mixing the components). The reaction vessel is placed in a constant temperature environment with which it can exchange the heat through a thermal resistance of finite magnitude (in adiabatic calorimeters this resistance is infinitely large). The temperature during the titration run is continuously monitored and after corrections for

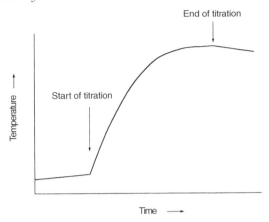

Figure D 22. Schematic view of an isoperibolic titration experiment.

the heat exchange and some other effects (see below) is converted to the reaction heat using the known heat capacity of the reaction vessel.

A typical isoperibolic titration experiment is illustrated schematically in Fig. D 22. Before the start and after the end of titration the temperature of the reaction vessel changes slowly due to the heat effect from agitation of the solution and heat exchange with the surroundings. The slopes of these portions of the total temperature–time profile are used for calculation of the respective correction terms assuming that the rate of agitation heating is constant and the rate of heat exchange is directly proportional to the temperature difference between the reaction vessel and surrounding. These corrections can be calculated accurately only for a limited period of time; therefore the isoperibolic titration is restricted to processes which are complete within less than one hour or so. Other corrections are introduced for the temperature difference between titrant and titrate solutions and for dilution of reactant solutions.

The main advantage of isoperibolic continuous calorimetric titration is the possibility to obtain a complete titration curve from a single run. An alternative way is the incremental titration in which the titrant is added by portions and the temperature of the reaction vessel returns to its initial value each time before a new portion of the titrant is added. This procedure has the advantage that slower reactions, which need considerable time for equilibration of the reactants and products, also can be studied with heat conduction calorimeters which operate in the *isothermal* mode. The reaction vessel is connected to the surrounding through an infinitesimally small thermal resistance, thus allowing the heat produced in the vessel to flow easily to the constant temperature surrounding without noticeable heating (or cooling if the reaction is endothermic) of the vessel. Of course, one should not interpret the term 'isothermal' too strictly: there must be a small temperature difference between the vessel and the surroundings otherwise no heat flow will be produced. In this operation mode the heat flux (or 'thermal power') $P = dQ/dt$ is measured after mixing of reactants and the heat Q produced (or consumed) during reaction is calculated by using the respective calibration parameter. Under steady-state conditions (e.g. in a flow calorimeter) P is directly proportional to the detector thermopile potential U: $P = \varepsilon_C U$, where ε_C is the calibration constant, and the heat produced during a time interval Δt equals $Q = \varepsilon_C U \Delta t$. If P depends on time, this can be taken into account by using Tian's equation $P = \varepsilon_C (U + \tau dU/dt)$, where τ is the time constant of the instrument, and subsequent integration of measured P over the time interval of the experiment gives the value of Q.[165a] The advantages of isothermal calorimetry are that no heat capacity measurements and no correction for the heat exchange between the reaction vessel and the surroundings are required. In addition, the baseline of isothermal calorimeters is very stable, allowing long lasting experiments. Of course, the other corrections mentioned above must also be introduced here. An isothermal incremental titration experiment is illustrated in Fig. D 23 for titration of α-cyclodextrin with heptan-1-ol.[166]

The finally obtained heat *vs.* titrant concentration profiles have the same form, and are analyzed in the same way as other, spectroscopic, binding isotherms (Section D 1). The calculational methods most often employed in calorimetric titrations for determination of

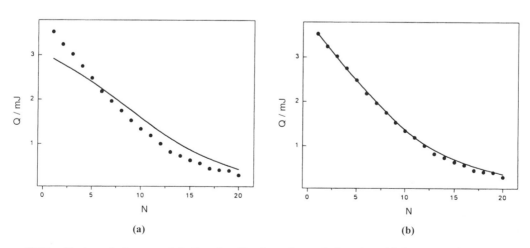

(a) (b)

Figure D 23. Heat evolution per injection for titration of α-cyclodextrin with heptan-1-ol plotted against injection number:[166] (a) fitting to a 1:1 binding model, (b) fitting to a model including both 1:1 and 2:1 host–guest complexes. Reprinted by permission from *J. Chem. Soc., Farad. Trans.* Hallén, D. *et al.*, 1992, **88**, 2859 (Figs. 1b,c). Copyright 1992, Royal Society of Chemistry.

equilibrium constants have been reviewed.[167] The corrected heat produced during an association reaction between S and L is related to the reaction enthalpy by equation

$$Q = [SL]V\Delta H$$

where V is the solution volume. Introducing an expression for [SL] from the mass balance equation for a 1:1 complexation leads to the equation for Q analogous to equation (D1-18) of the form

$$Q = 0.5V\Delta H([S]_t + [L]_t + K_d - \{([S]_t + [L]_t + K_d)^2 - 4[S]_t[L]_t\}^{1/2})$$ (D5-1)

The thorough analysis of both random and systematic errors of the method lead some authors to conclude that constants higher than 10^3 and smaller than $1\ M^{-1}$ cannot be determined accurately, in contrast to the reaction enthalpies ΔH, which of course can be measured accurately with larger K values. The upper limit of reliable determination of K was calculated by application the standard error propagation equation to the expression for K as a function of concentrations of reactants and experimental reaction heat assuming a constant relative error 1% in the reaction heat and reactant concentrations ranging above 0.01 M. For reactions with large ΔH values both the relative error in the reaction heat and reactant concentration range can be smaller and, therefore, larger K values can be determined with good precision.

D6.   KINETIC METHODS

Determination of the equilibrium constant K from the ratio of rate constants of complex formation (k_f) and dissociation (k_d) reactions according to

$$K = k_f/k_d$$ (D6-1)

can be useful for very slow complexation processes, or for unstable compounds which decompose before reaching the equilibrium. For example, this method was applied to the complexation of alkali picrates with several

spherands[168] and a calixspherand[169] in chloroform. In these systems, studied by the extraction picrate method (Section D4.2), equilibration was incomplete even after several days.

Kinetic determination of equilibrium constants from equation (D6-1) is used sometimes for fast reactions also. For example, association constants of isomeric hydroxy-2-naphthalenecarboxylate guests with cyclophane **D-19** and cyclodextrins were deter-

D-19

mined by the temperature-jump relaxation technique with spectrophotometric monitoring.[170] The measured parameter here is the relaxation time τ, which for a 1:1 association reaction between H and G is given by equation

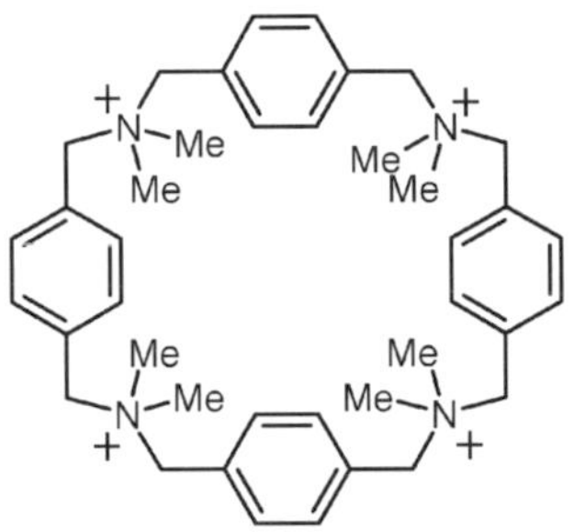

$$1/\tau = k_f([H] + [G]) + k_d$$ (D6-2)

where [H] and [G] are the concentrations of H and G under equilibrium. When one of the components is taken in large excess over another one, so that $[H]_t \gg [G]_t$, the equation can be simplified to

$$1/\tau = k_f[H]_t + k_d$$ (D6-3)

and k_f and k_d can be calculated from the slope and intercept, respectively, of the plot of $1/\tau$ *vs.* total host concentration. Note that in this approach molar absorptivities (extinction coefficients) of free and bound guest are not required.

More often reaction kinetics are employed for indirect determination of binding constants using the complexation-induced changes in substrate reactivity in a manner analogous to spectroscopic or calorimetric titrations. The simplest reaction scheme considered in this case involves two parallel routes of monomolecular (usually pseudo-monomolecular, see below) reactions of free and complexed substrate S to the product P

$$S \rightarrow P(k_0) \qquad (D6\text{-}4)$$

$$S + L \rightleftharpoons SL\,(K) \qquad (D6\text{-}5)$$

$$SL \rightarrow P(k_C) \qquad (D6\text{-}6)$$

Assuming that the complexation reaction (D6-5) is fast enough to keep always S, L and SL in equilibrium during the reaction course, one obtains for the observed first-order rate constant the expression

$$k_{obs} = (k_0 + k_C K[L])/(1 + K[L]) \qquad (D6\text{-}7)$$

analogous to equation (D1-20). Of course, when $[L]_t \gg [S]_t$ one can assume $[L] = [L]_t$, which gives an equation similar to (D1-12). If complexes with higher stochiometries (SL$_2$, SL$_3$, etc.) are formed, the expression for k_{obs} takes the form analogous to equation (D1-47)

$$k_{obs} = (k_0 + k_{C1}\beta_1[L] + k_{C2}\beta_2[L]^2 + \cdots)/$$
$$(1 + \beta_1[L] + \beta_2[L]^2 + \cdots)$$

where k_{C1}, k_{C2}, etc. are the rate constants of substrate transformation in the complexes SL, SL$_2$, etc.

Equation (D6-7) was applied for numerous reactions with reactive substrates capable to complexation with different hosts (Section I3). Most often the reactions studied are not monomolecular and involve interaction of S with another substrate, like with water or hydroxo anions in hydrolytic reactions. If the concentration of the second substrate is maintained constant (for example by using it in a high excess over the first substrate), and if it does not interact with L, the same scheme (D6-4)–(D6-6) and equation (D6-7) can be used with k_0 and k_C as apparent pseudo-first-order rate constants. When both substrates can form complexes with L the analysis in general is difficult, and requires numerical

solutions.[171] It becomes easier in extreme cases when either both substrates are used in a higher excess over the L (a situation typical for enzyme kinetics when L is an enzyme), or if the ligand is used in a high excess over both substrates (see for example the analysis of cyclodextrin effects on kinetics of the bromination of phenols).[172] The main purpose, however, of analyzing two-substrate kinetics is usually the confirmation of the reaction mechanism and calculation of rate rather than equilibrium constants, which often are measured independently by other techniques.[171,172]

One characteristic features of kinetics is that the same rate equation can follow from quite different reaction schemes. This is true also for the equation (D6-7): observation of hyperbolic 'saturation' kinetics (often referred also as the 'Michaelis–Menten type' kinetics) in accordance with this equation does not mean necessarily that the reaction involves the steps (D6-4)–(D6-6). One of possible reaction schemes involves steps (D6-4) and (D6-5), but with the complex being unreactive (so-called 'non-productive binding') and with the chemical transformation of the substrate occurring through a bimolecular step (D6-8)

$$S + L \rightarrow P(k_B) \qquad (D6\text{-}8)$$

Assuming again the complex SL to be in equilibrium with S and L in accordance with equation (D6-5) one obtains from (D6-4), (D6-5) and (D6-8) an expression for the observed pseudo-first-order constant (D6-9) which has the same form as (D6-7), with k_B instead of $k_C K$

$$k_{obs} = (k_0 + k_B[L])/(1 + K[L]) \qquad (D6\text{-}9)$$

The non-productive binding was proven, for instance, for the cyclodextrin-promoted cleavage of 4-nitrophenyl acetate (Section I3) and can be expected for many other systems since the discrimination of mechanisms represented by steps (D6-4)–(D6-6) with 'productive' binding and by steps (D6-4), (D6-5), (D6-8) with 'non-productive' binding cannot be made on the basis of quasi-equilibrium or steady-state (see below) kinetics only. In such a case the value of K is determined correctly although the increase in the observed rate constant is not due to complexation of the substrate.

Numerous reaction schemes, which lead, to rate equations of general type (D6-7) have been analyzed for enzyme kinetics.[173] These are catalytic reactions with respect to L (enzyme) usually studied under conditions of high excess of S over L. Since the catalytic activity of enzymes is typically very large, spontaneous, uncatalyzed transformation of S can be neglected ($k_0 = 0$). Chemical transformation of bound S in the non-covalent enzyme-substrate complex is often very fast, so that the quasi-equilibrium approximation used above may be not valid and must be substituted by more accurate steady-state approximation. For the complexation step (D6-5) now both formation and dissociation rate constants must be considered, and step (D6-6) now leads to formation of the product P and regeneration of the enzyme L

$$S + L \underset{k_d}{\overset{k_f}{\rightleftharpoons}} SL \overset{k_C}{\rightarrow} P + L \qquad \text{(D6-10)}$$

Assuming steady-state conditions for SL ($d[SL]/dt = k_f[S][L] - k_d[SL] - k_C[SL] = 0$) and $[S]_t \gg [L]_t$ one obtains for the reaction rate the classical Michaelis–Menten equation

$$d[P]/dt = k_C[S]_t[L]_t/(K_m + [S]_t) \qquad \text{(D6-11)}$$

where $K_m = (k_d + k_C)/k_f$. Evidently, K_m becomes equal to the equilibrium dissociation constant of the SL complex $K_S = k_d/k_f$ only if $k_C \ll k_d$. It is worthwhile to note, however, that although equation (D6-11) is formally valid for a large number of enzymatic reactions, the mechanism shown on scheme (D6-10) represents itself a great simplification.

Binding constants of a considerable number of guests to cyclodextrins were determined by kinetic competition method, using non-reactive guests as inhibitors for the cleavage of activated esters by cyclodextrins.[174–176] The

treatment discussed in Section D1.1.2d for a competition method with a light-absorbing indicator I can be applied for the kinetic competition method as well. In this case the reaction scheme involving steps (D6-4) (D6-6) together with an equilibrium (D6-12) of the binding of L to a non-reactive competing guest G is considered.

$$G + L \rightleftharpoons GL \, (K_G) \qquad \text{(D6-12)}$$

The substrate S serves here as the indicator and the 'indicator ratio' Q is given by equation

$$Q = [S]/[SL] = (k_C - k_{obs})/(k_{obs} - k_0) \qquad \text{(D6-13)}$$

instead of (D1-22). The expression for parameter P defined by equation (D1-24) takes the form

$$P = [L]_t - 1/(KQ) - [S]_t/(1 + Q)$$

and the final equation (D1-25) takes the form

$$[G]_t/P = (K/K_G)Q + 1 \qquad \text{(D6-14)}$$

In practice several useful simplifications were introduced.[174,175] Usually $[L]_t \gg [S]_t$, which allows one to simplify the expression for parameter P and to exclude S from mass balance equations for $[L]_t$ and $[G]_t$. In addition, often the condition $[G]_t \gg [L]_t - [L] = [GL]$ is fulfilled which allows to derive from equation (D6-14) the following linear expression

$$[G]_t = (K/K_G)[L]_tQ - 1/K_G \qquad \text{(D6-15)}$$

used for calculation of K_G.[174] Also calculation of K_G directly from mass balance equations at each concentration of the guest using the free ligand concentration estimated as $[L] = 1/(K_I Q)$ (Section D1.1.2d) with subsequent averaging was performed.[175] It should be noted that the application of the kinetic competition method requires the 'productive' binding of S to L; in case of 'non-productive' binding, guests expected to be the competitive inhibitors behave as inert 'spectators' or even promoters occupying the host cavity.[175]

D 7.   DIRECT MEASUREMENTS OF AFFINITIES AND SURFACE PLASMON RESONANCE

Recent developments allow to measure interactions between an immobilized receptor and a ligand from electrical or optical signals in suitable devices. One way to determine directly non-covalent forces between molecules is derived from surface imaging techniques like atomic force microscopy (AFM); the basic method is therefore described together with structural techniques in Section E3. For measuring forces the receptor is fixed on a surface and the ligand is immobilized on the tip of ultrathin lever, which is moved over the surface (Fig. E2).

In contrast to AFM the surface plasmon resonance technique[177,178] (SPR) basically measures interactions between a ligand dissolved in any media, and a receptor which is immobilized e.g in a hydrogel matrix on thin gold film, yielding not only masses or concentrations of the interacting ligand but in principle also equilibrium constants for the interaction. In the SPR device (Fig. D24) the gold surface with the receptor, mounted on a small sensor chip, is in contact with the dissolved ligand flowing over the surface. Binding of the ligand causes an increase of its concentration on the surface, and a corresponding increase of the refractive index. This increase is measured by an optical response with a polarized light beam directed to, and reflected from the opposite side of the gold surface, and can be observed with extremely high sensitivity when surface plasmon waves are excited at the metal-liquid interface. The response, expressed in resonance units 'RU' of picogram per mm^2 of sensor surface, is equivalent to the effective mass change on the surface, independent of the nature of receptor and ligand, with a sensitivity of about $RU = 1$ for 1 $pg\,mm^{-2}$.[179]

The SPR method has been mostly applied to interactions between biomolecules of high molecular weight,[181] where the mass changes upon binding are correspondingly large and the affinities typically also are high . Technical improvements, including the use of reference flow cells for correction of signals due to any changes in the bulk property of the flow medium, and of gradient techniques, allow to measure also small molecule interactions.[182] Thus, the binding constant between a nonapeptide and immobilized streptavidin was determined to be $K = 4.3 \times 10^4$ M^{-1} with a standard deviation of about 10%. Figure D25 illustrates the sensograms obtained with a ligand as small as maltose, and the signal-concentration profile, yielding a K_A value of 10^4 M^{-1} with the anti-maltose antibody used

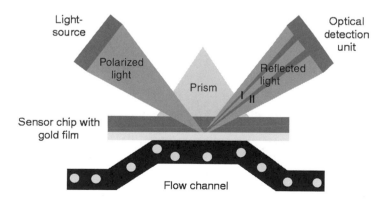

Figure D 24.   Schematic drawing of a surface plasmon resonance device.[180] After BIACORE 3000 Product Information 1998. With permission from Biacore AB, copyright 1998.

as immobilized receptor in this study.[181] With sufficiently strong binding substrates one can also deduce the kinetics of association/dissociation by measuring the time dependence of SPR signals.

The major advantage of techniques like SPR or of modified AFM (Section E3) lies in the speed of determination, enabling a high, automated throughput of many samples, and the small amount of material needed; with SPR one needs 0.1 to $100 \times K_D$ (where K_D is the dissociation constant) in, say, 100 µl; meaning for analytes of MW around 500 and with K_D around mM about 50µg substance. Both aspects make in particular SPR a promising method to complement the biological screening of combinatorial libraries by binding studies, which provide firmer grounds for rational drug design. Of course one has to take into account the necessary immobilization step prior to measurements; the methods are obviously most suitable if

one wants to evaluate affinities of many ligands (or in synthetic supramolecular chemistry also host compounds) to one target substrate, such as for sensor development. Differences of SPR to solution techniques by spectral methods are that affinity data alone do not provide information on interaction modes or structures (unlike e.g. NMR), that the binding mechanism may change due to immobilization (e.g. favoring long range electrostatic interactions). With force measurements by AFM conformational distortions due to induced fit with subsequent strain energy disadvantage will not necessarily show up for the immobilized partner.

D 8.   MISCELLANEOUS TECHNIQUES AND COMPARISON OF METHODS

The experimental methods discussed in previous Sections of this chapter are most widely used already; some, such as mass-spectro-

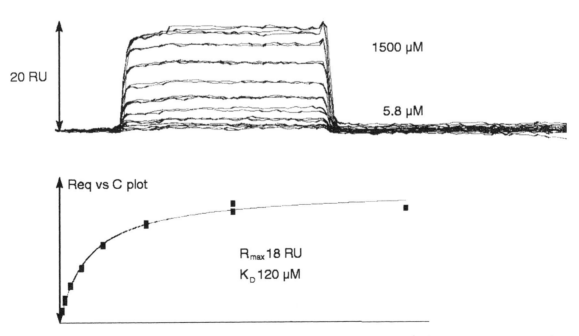

Figure D 25.   SPR Measurement with maltose as ligand and an anti-maltose antibody as receptor; concentration range 5.8–1500 µM.[180] After BIACORE 3000 Product Information 1998. With permission from Biacore AB, copyright 1998.

metry, AFM or SPR methods are expected to find more widespread application in the near future. Any experimental technique capable of discriminating and quantifying species of different nature can be used in principle for the determination of equilibrium constants. This Section discusses briefly several other methods which occasionally are applied to the study of supramolecular equilibria.

Equilibrium dialysis and ultrafiltration are separation-based methods traditionally employed for the study of binding of low molecular weight compounds to macromolecules, such as proteins and synthetic polymers.[1a] They can be applied also for sufficiently large host compounds like cyclodextrins.[183] In these methods two compartments of solutions are separated by a semipermeable membrane which allows free transfer of one component, L, between the compartments, but is impermeable for another one, S. Under equilibrium the concentration (more exactly the activity) of free L is identical in both compartments, but S together with SL and higher complexes (if they occur) are present in only one of the compartments. Determining the total analytical concentrations of L in both compartments under equilibrium one can calculate concentrations of free and bound ligand and, consequently, the complexation degree, which than is analyzed as function of [L] by procedures outlined in Section D 1 in order to obtain reaction stochiometry and formation constants. Significant complications arise for charged solutes, when the Donnan effect must be taken into account; another complication is due to the osmotic pressure created by the presence of macromolecules in only one of the compartments. Also the equilibration across the membrane is slow and takes up to 1 or 2 days.

Methods based on measurements of colligative solution properties give the number of solute particles present in solution and respective techniques, cryoscopy, ebulliometry and vapor pressure osmometry, are often employed for determination of self-association.

Electron spin resonance (ESR) spectroscopy is similar by its theoretical basis to NMR, but is restricted to paramagnetic species. Binding constants of various nitroxide radicals to cyclodextrins were determined by this technique.[184,185] In comparison to NMR, ESR has about 10^3 times shorter a time scale; therefore some supramolecular association equilibria which are fast in the NMR scale (Section F1) can be slow in the ESR time scale.[185a] This creates a possibility of discriminating different modes of inclusion leading to usually indistinguishable microequilibria (Section D 1). Such a discrimination was accomplished for inclusion of the nitroxide radical **D-20** into β-cyclodextrin.[185a] Figure D 26 shows the ESR spectrum of **D-20** in the presence of increased concentrations of β-cyclodextrin in water. Computer simulation of the spectra in the presence of the host shows that there are three species with different hyperfine splitting constants which

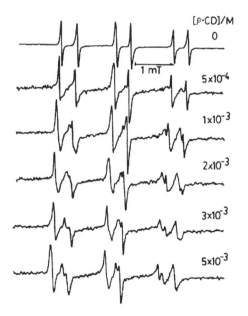

Figure D 26. ESR spectra of **D-20** in the presence of increased concentrations of β-cyclodextrin in water.[185a] Reprinted with permission from *J. Am. Chem. Soc.*, Kotake *et al.*, 1988, **110**, 3699. Copyright 1988. American Chemical Society.

Table D2. Overview of the most usual methods for binding affinity measurements.

No.	Method	Medium	Typical log K	Typical conc (10^x M)	Typical quantity of minor compound	Necessary purity L/M/S	Special problems
1	NMR 1-H	any (deuterated)	<4	> −3	0.1 to 2 mg	M	paramagnetic material small shift changes
2	NMR 13-C	any	<3	> −2	5 to 20 mg	S	paramagnetic material
3	UV/vis	H_2O; + restricted	f(ε)*	−2 to −4	10 to 100 µg	L	overlap
4	Fluorometry	H_2O; + restricted	2 to 7	−3 to −6	1 to 10 µg	LL	overlap, quenching
5	IR	few solvents	f(ε)*	f(ε)*	1–100 µg	L	overlap
6	Polarimetry	H_2O; + restricted	<2*	−1 to −2*	10–100 mg*	L*	lower accuracy
7	CD/ORD	H_2O; + restricted	1 to 6	−3 to −5	0.01 to 1 mg	L*	
8	Calorimetry	H_2O; + restricted	f(ΔH)*)	−1 to −3	1 to 10 mg	LLL	f(ΔH)*, side reactions
9	Potentiometry						
a	pH	mainly H_2O	2 to 12	−1 to −4	0.1 to 10 mg	LL*	proton-dependend
b	ISE	mainly H_2O	2 to 6	−2 to −5	0.01 to 1 mg	L	alkali cations/halides
10	polarography	mainly H_2O	2 to 7	−2 to 7	1 µg to 1 mg	LL*	redox-active substr.
11	conductometry	any	2 to 6	−2 to −6	1 µg to 1 mg	LL*	charged substrates,
12	MS	any gas			<1 ng	S	low accuracy, unspecific associations
13	solubility	any	1 to 5		0.1 to 10 mg	M	microdisperse sln.
14	extraction	water immiscib.			0.01 to 1 mg	M	mutual solubility
15	chromatography	mainly H_2O	2 to 5	−2 to −6	0.1 to 1 µg	M	substrate retention on stationary phase
16	electrophoresis	H_2O	2 to 5	−2 to −6	0.1 to 1 µg	M	needs charged cpds
17	kinetic	any	2 to 5	−2 to −6	0.1 to 1 µg	M	needs slow complex association/dissociation or reactive substrates
18	SPR	H_2O; + restricted	3 to 15				needs immobilized ligand
19	RIA	H_2O	<15*				needs radioactive test + slow complex dissociation
20	ELISA	H_2O	<15*				needs enzyme label + slow complex dissociation

Where applicable higher limits and lower for K determinations are given (by $\langle \log K$ and $\log K \rangle$); typical concentrations and necessary quantities are given for the *observed* compound, which is usually the minor one (for the major compound the K values is decisive); necessary purities are denoted L for large; LL for extra large; M for medium; S for small requirement; other explanations see text and other footnotes. Special conditions or limitations are marked by *, see footnotes. Restrictions for medium, such as optical transparency etc. Special remarks to the methods:

1), 2) (Refer to shift titrations); paramagnetic impurities, even dissolved oxygen can broaden lines, detrimental in particular for C-13 NMR signals; the use of O–H or of N–H signals (important for hydrogen bonded associations) can suffer from acid–base contaminations and exchange problems, then dry solvents are important.

3) Band overlap problems in particular if host and guest have chromophores; check for optimal wavelength by spectrum subtraction of neat and complexed compounds, perhaps use deconvolution techniques. The limits for $\log K$ measurements are dictated by the extinction coefficients ε and the false light of the instruments (in case of high ε and low K values). With small extinction coefficients, e.g. with CT complexes and bands, low K values can also be determined.

4) May involve similar problems as with 3); other limitations: quenching by external impurities (in particular paramagnetic) and by substrate–substrate interactions, inner filter effects; dust particles; temperature variations. Small constants can be measured if the complexing ligand has low UV absorptions, like with cyclodextrins and fluorescence dyes.

5) FT-IR allows in ATR mode also data collection in aqeous medium; again: problem of overlapping bands, may need deconvolution techniques.

6) Log K and quantities depend on rotations and their changes, both are often small; low sensitivity.

7) Log K and quantities depend on ellipticities; optically active impurities intefer particularly.

8) Calorimetry is limited by the complexation ΔH; if ΔH is low, side reactions from impurities generate more problems; the necessary quantities depend on ΔH and on technique.

9) Generally for electrochemical techniques only contamination are detrimental with compounds which interfer with the signal (such as acids/bases, redox materials etc); in most cases one needs the presence of electrolytes which may influence the K values.

9a) Protonation and/or protonation-dependent equilibria can be studied; requires very careful control of medium and temperature.

9b) Generally limited to alkali cations and halide anions; requires careful medium control.

10) Substrate must form a reversible redox couple; sensitive, but less accurate than potentiometry method.

11) When applied for study of ion pair formation needs extensive corrections due to necessary variable ionic strength; difficult to apply for non-symmetrical ion pairs.

12) This is almost the only technique that permits study of gas-phase equilibria involving ions; very sensitive, but usually not very accurate method, more suitable for identification and characterization of complexes than for thermodynamic studies; problems to distinguish unspecific associations form supramolecular complexes.

13) Is limited to combinations of sparingly soluble substrates and highly soluble ligands and complexes.

14) with ionic substrates allows determination of very high (10^5–10^{12} M^{-1}) formation constants in organic water-immiscible phases; has low accuracy due to mutual solubility of the phases.

15) and 16) In many aspects similar techniques; the latter is applicable only for charged substrates and requires special corrections for medium effects.

17) Limited to complexes with slow association rates, or to substrates, which undergo chemical reactions influenced by the complex formation.

are attributed to free **D-20** and its complexes with β-cyclodextrin formed by inclusion of either the phenyl or the *tert*-butyl group of the guest. Equilibrium constants for each complex were calculated by using equation $K = (f_1/f_2)([H]_t - [G]_t f_1/(f_1+f_2))^{-1}$, where $[H]_t$ and $[G]_t$ are the total concentrations of host and guest respectively; f_1 and f_2 are the area intensities of ESR spectrum for the included and free species.

D-20

In principle, any of the methods discussed in this chapter should give correct values of the equilibrium constants when properly applied. In practice, the method of choice is dictated by numerous factors related to the physical properties of reactants, the desired or necessary reaction conditions, restrictions due to solubility or instability of reactants and others. Generally, spectroscopic titrations have the advantage of providing more or less extensive structural information about the complex(es) formed, especially with NMR techniques. Calorimetric titration is the method of choice when ΔH and ΔS, and especially ΔC_P values are of interest, but is like several others more demanding with respect to the necessary equipment, instrumental expertise and measuring time. Potentiometric titrations, which measure the concentration of one of the reaction components rather than a physical property related to concentration, have the advantage of a smaller number of fitting parameters, and being applicable over a particular large range of equilibrium constants. In these methods only equilibrium constants are the adjustable parameters in the mathematical analysis of titration experimen-

tal data, while spectroscopic and other titrations need also determinations of proportionality coefficients (intrinsic properties) for all species present, that is, the number of fitting parameters increases by a factor of two, which can create serious problems in the analysis of complex equilibria.

The most difficult problem is, of course, the presence of systematic errors, which often only become evident when different methods are applied to the same system. Published equilibrium constants for a given system often differ considerably. For example, binding constants reported in the literature for the widely used fluorescent probe 1-anilino-8-naphthalenesulfonate with cyclodextrins range from 11 to 110 M^{-1} for β-cyclodextrin and from 128 to 1270 M^{-1} for γ-cyclodextrin.[176] Partially this is due to differences in reaction conditions, but it also reflects substantial systematic errors of some determinations and/or the applied calculational methods. Equilibrium measurements should always be accompanied by a error estimation; if calculation by error propagation seems to be mathematically too complicated one can always quickly check the expected errors by numerical estimates, using realistic lower and upper limits for the accuracy of the experimental data, including volumes/concentrations, signal to noise ratios in spectra, possible deviations from Beer's law, temperature, pH or medium polarity changes etc. It is advisable to carry out such an error estimation, for which one should of course allow that all errors go to the same direction yielding the highest and lowest K value possible on this basis, prior to setting up any experimental method, and to keep the experimental raw data available for later recalculations, e.g. with improved mathematical models.

In Table D2 we give an overview of the techniques applied most often for affinity determinations, which of course can only serve as a first lead with many approximations. The typical log K values, concentrations and necessary quantities refer to low molecular weight organic or inorganic compounds, as well as to normal instruments and accumulation times for data aquisition. Instrument improvements and additional signal accumulation techniques will further decrease the necessary concentrations and quantities. Of course, the first requirement always is signal change upon complex formation. Optical and electrochemical methods obviously require a corresponding specific sensing property of one of the reaction partners, a limitation which can often be overcome using competition methods. NMR shift titrations have almost no limitations in this respect, although complexation with alkali metal ions produce usually no changes in ligand signals. Calorimetry is also generally applicable, but, as said above, the signal change here depends on the complexation ΔH itself and side reactions or impurities can particularly blur the results. Techniques like SPR and ELISA as well as some chromatographic methods require immobilization of one partner prior to measurements, but then allow automated fast data collection with many ligands. Separation-based techniques like ELISA require rather high binding affinities, occurring typically with bioreceptors.

D 9.   EXERCISES AND ANSWERS

D 9.1.   EXERCISES

1. Explore some properties of binding isotherms given by equations like (D 1-11)–(D 1-14) and their linearized forms (D 1-15)–(D 1-17), specifically, when X is the absorbance (A) and x is the molar absorptivity (ε). Answer the following questions:

(a) how does the initial slope in the plot of ε_{obs} *vs.* $[L]_t$ (if $K[L]_t \ll 1$) depend on K and the molar absorptivities of free and bound substrate?

(b) at what concentrations of L will the $\Delta\varepsilon_{obs}$ values be equal to (i) 10%, (ii) half, (iii) 90%, (iv) 99% of its maximum value for a given stability constant K?

(c) find the coordinates of the intersection points of the lines given by equations (D 1-15)–(D 1-17) with the abscissa.

2. Table D 3 shows the results of a spectrophotometric titration of 0.066 mM 4-nitrophenol with α-cyclodextrin at three pH values (see Fig. D 12 for complete spectra). Calculate from these results binding constants at each pH, using non-linear least-squares fitting, and linear least-squares fitting with the equations of Benesi–Hildebrand, Scatchard and Scott. Use the values of these constants to calculate the binding constants of neutral 4-nitrophenol and its anion (pK$_a$ of 4-nitrophenol is 7.2).

3. Table D 4 shows the effects of several cyclodextrins on the fluorescence intensity I (counts per second) of 5 µM solution of the substrate **D-21** (from the condensation of malondialdehyde and thiobarbituric acid; employed for analytical determination of the former).[186] Calculate from these data the binding constants and fluorescence enhancement factors for different cyclodextrins, and discuss their magnitude (HP-β-CD is a β-cyclodextrin derivative alkylated by 2-hydroxypropyl substituents at the primary 6 hydroxy groups).

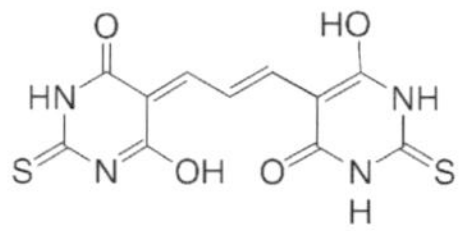

D-21

4. The complexation of β-cyclodextrin with n-butylamine was studied by competition of the guest with the fluorescent probe 1-anilino-8-naphthalenesulfonate anion (ANS) at pH 11.6.[176] Data obtained with constant concentrations of β-cyclodextrin (0.01 M) and ANS (0.075 mM) and variable guest concentration are given in Table D 5:

The equilibrium constant $K_{ANS} = 37.3$ M^{-1} for complexation between β-cyclodextrin and ANS without n-butylamine was obtained from the fluorimetric titration of ANS with the host, results of which were fitted to the equation

$$I_{rel} = (1 + I_C K_{ANS}[\text{Host}])/(1 + K_{ANS}[\text{Host}])$$

where I_{rel} is the observed relative fluorescence intensity ($I_{obs}/I_{\text{free ANS}}$) and $I_C = 46.7$ is the relative fluorescence intensity of bound ANS. Calculate from these results the binding constant of n-butylamine to β-cyclodextrin.

5. Choose an appropriate method for the determination of equilibrium constants of the following processes.

Table D 3.

pH 2.6		pH 6.5		pH 10.1	
[α-CD] (M)	A (318 nm)	[α-CD] (M)	A (409 nm)	[α-CD] (M)	A (409 nm)
0	0.633	0	0.189	0	1.167
0.00259	0.594	0.00031	0.281	0.0001	1.191
0.00505	0.578	0.00062	0.339	0.0002	1.212
0.00739	0.566	0.00093	0.390	0.0004	1.237
0.00964	0.558	0.00123	0.429	0.00059	1.251
0.0118	0.554	0.00183	0.492	0.00079	1.265
0.0138	0.550	0.00241	0.540	0.00098	1.274
0.0158	0.548	0.00299	0.572	0.00136	1.284
0.0177	0.546	0.00437	0.630	0.00262	1.306
0.0195	0.545	0.00570	0.663	0.00348	1.313
0.0212	0.545	0.00697	0.684		
0.0245	0.542	0.00818	0.701		
0.0275	0.541	0.0100	0.717		
		0.0121	0.728		

Table D 4.

Host	[H] (M)	I	Host	[H] (M)	I	Host	[H] (M)	I
none		1170	HP-β-CD	0.0043	2760	γ-CD	0.0039	1360
β-CD	0.00090	1480		0.012	3720		0.0079	1600
	0.0018	1560		0.017	4110		0.019	2050
	0.0036	1660		0.022	4210		0.040	2950
	0.0054	1750		0.027	4310		0.054	3930
	0.0072	1850		0.032	4470		0.070	4740
	0.0090	1820		0.037	4500		0.085	5050
							0.10	5590
							0.12	6030
							0.13	5860

Table D 5.

[n-butylamine] (M)	I
0	0.0309
0.04	0.0193
0.08	0.0138
0.12	0.0111
0.16	0.00914
0.18	0.00855
0.2	0.00795

(a) ion-pair association between $^+Me_3N(CH_2)_2NMe_3^+$ and $C_2O_4^{2-}$

(b) ion-pair association between $^+H_3N(CH_2)_2NH_3^+$ and $^-O_2C(CH_2)_2CO_2^-$

(c) complexation of Pb^{2+} with 1,7-dithia-12-crown-4

(d) complexation of Ni^{2+} with 1,10-diaza-18-crown-6

(e) inclusion of benzoic acid into β-cyclodextrin

(f) inclusion of hydroquinone into β-cyclodextrin

(g) complexation of Cu^{2+} with cyclen (ethylenediamine)

(h) binding of Na^+ to 4,10-diaza-15.crown-5 in methanol

(i) inclusion of flavonols in an azoniacyclophane like **CP66** (structure C-9)

6. Describe protonation and complexation equilibria for the macrocycles **D-7** (consider complexation with benzoic acid, or phthalic acid as guests); indicate values for statistical corrections and find relations between microscopic and macroscopic protonation constants. Use any available program which you might have to calculate distribution curves of differently protonated species of macrocycle **D-7b** (take approximate pK_a values from Fig. D 9).

7. Which protonation forms do you expect to dominate for the macrocycle **D-5** (Scheme D 2) at the following pH values: 2, 5, 8, 11 (use approximate protonation constants typical for ionogenic groups of the macrocycle)?

8. Dimerization of valerolactam in CDCl$_3$ was studied by NMR spectroscopy.[187,188] Chemical shifts of NH protons (at 60 MHz, at 306 K) as the function of total substrate concentration are given in Table D 6. Calculate form these data chemical shifts of monomeric and dimeric forms of valerolactam and the dimerization constant.

D 9.2. ANSWERS

1. Let us use for analysis equation (D 1-12), which takes in this particular case the form

$$\varepsilon_{obs} = (\varepsilon_S + \varepsilon_{SL}K[L]_t)/(1 + K[L]_t) \qquad (D\,9\text{-}1)$$

where ε_S are ε_{SL} are the molar absorptivities of free and bound substrate.

(a) Differentiation of (D 9-1) by $[L]_t$ gives $d(\varepsilon_{obs})/d[L]_t = (\varepsilon_{SL} - \varepsilon_S)K/(1 + K[L]_t)^2$, and at $[L]_t = 0$ the initial slope equals $(\varepsilon_{SL} - \varepsilon_S)K$. Thus, other than intuitively expected, the slope depends not only on K and molar absorptivity of the complex, but also on the absorptivity of the free substrate.

(b) (i) 0.11/K (ii) 1/K, (iii) 9/K (iv) 99/K. Note, that in order to cover the interval of complexation degrees from 10–90% one needs to vary [L] about 100 times.

(c) $-K$, Δx and $-1/K$, respectively.

2. Figures **D 27a–d** illustrate different types of fitting with the results at pH 6.5. In all cases the guest concentrations is maintained constant and [α-CD] $\gg$ [guest]. This allows us to use directly absorbances A and the simplified equations (D 1-13) and (D 1-15)–(D 1-17) with A as X_{obs} or x_{obs}.

The values of K (M^{-1}) are the following (errors are the standard deviations):

pH:	2.6	6.5	10.1
Non-linear fitting:	220 ± 11	508 ± 12	1720 ± 61
Benesi-Hildebrand equation:	212 ± 6	568 ± 29	1610 ± 84
Scatchard equaiton:	211 ± 7	522 ± 13	1700 ± 50
Scott equation:	227 ± 14	521 ± 10	1700 ± 40

In these experiments large intervals of complexation degree were covered, and all treatments give satisfactory results with K values differing from each other not more than by their errors. One can see, however, that points are distributed evenly in Fig. D 27a, but are concentrated at the begining or at the end of abscisse in other plots. This makes the linear plots more sensitive to systematic errors at larger or smaller host concentrations.

With its pK$_a$ = 7.2, 4-nitrophenol exists in solution as practically pure neutral and pure anionic forms at pH 2.6 and 10.1, respectively. Therefore the observed binding constants at these pH values, 220 and 1700 M^{-1}, can be attributed to these forms. The value of K at pH 6.5 expected in accordance with equation (D 2-30) is therefore 470 M^{-1}, reasonably close to the measured value.

3. The equation of type (D 1-13) with $X = I$ can be used for the fitting. As observed in many other cases, the fluorescence enhancement depends primarily on the cavity size (β-CD < HP-β-CD < γ-CD, calculated enhancement factors at 'saturation' are 1.7; 5 and 15.2 respectively), rather than on the complexation strength (γ-CD < HP-β-CD < β-CD, formation constants 3.5; 180 and 600 M^{-1} respectively). These observations are in a qualitative agreement with idea that the fluorescence enhancement effect is primarily due to protection of the guest from quenching by solvent.

4. Use the treatment described in Section **D 1.1.2d** assuming that the concentration of ANS can be neglected in mass balance equations for S and L; $K = 28.1$ M^{-1}.

Table D 6

$[S]_t$ (mol kg^{-1})	δ (Hz)	$[S]_t$ (mol kg^{-1})	δ (Hz)	$[S]_t$ (mol kg^{-1})	δ (Hz)
0.0839	382	0.6541	444.8	2.8432	484.6
0.1695	408.8	0.7519	453.3	3.56	484.7
0.2569	420.7	0.8316	459.5	4.6522	488.3
0.3551	426.6	0.8923	458.4	5.7248	491.7
0.4556	442.7	1.4659	469.6	6.7689	491.9
0.5396	451	2.1157	476.9	7.9809	494.3

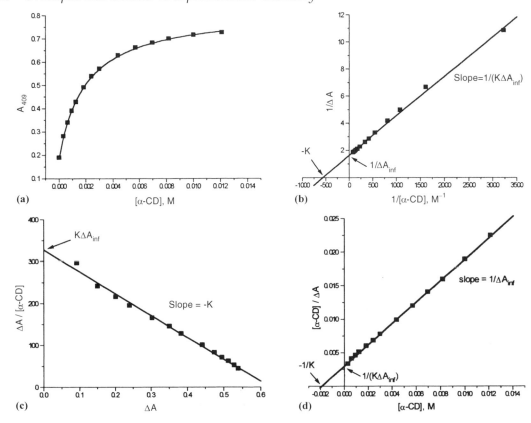

Figure D 27.   (a) Non-linear fitting; (b) Benesi–Hildebrand plot; (c) Scatchard plot; (d) Scott plot.

5. possible methods: (besides calorimetry which in principle always is possible if ΔH is large enough)

(a) ion-pair association between $^+Me_3N(CH_2)_2NMe_3^+$ and $C_2O_4^{2-}$:   conductivity

(b) ion-pair association between $^+H_3N(CH_2)_2NH_3^+$ and $^-O_2C(CH_2)_2CO_2^-$:   potentiometry

(c) complexation of Pb^{2+} with 1,7-dithia-12-crown-4:   potentiometry, polarography

(d) complexation of Ni^{2+} with 1,10-diaza-18-crown-6:   potentiometry

(e) inclusion of benzoic acid into β-cyclodextrin: potentiometry NMR-shift tiration UV/vis or fluorescence competition, HPLC

(f) inclusion of hydroquinone into β-cyclodextrin: as (e) except polarography instead potentiometry

(g) complexation of Cu^{2+} with cyclen (ethylenediamine):   UV/vis titration, potentiometry

(h) binding of Na^+ to 4,10-diaza-15.crown-5 in methanol: potentiometry, also with Na electrode

(i) inclusion of flavonols in an azoniacyclophane like **CP66**:   NMR-shift titration; if flavonoles deprotonate upon binding: also UV/vis and potentiometry

6. A distribution diagram for differently protonated forms of **D 7b** and benzoate anion is shown in Fig. D 28. Note, that only species with consecutive protonation constants separated by about four orders of magnitude

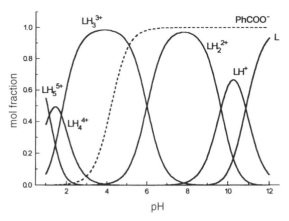

Figure D 28. A distribution diagram for differently protonated forms of **D-7b** and benzoate anion.

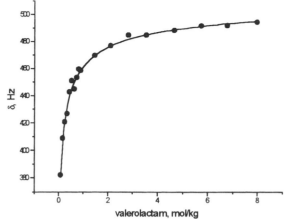

Figure D 29. A non-linear least-squares fitting of the experimental data to equation (D1-27).

are present at mol fractions close to unity in respective pH-intervals. Also the highly charged forms of the macrocycle (LH_4^{4+} and LH_5^{5+}) appear in pH interval where the anionic guest (benzoate anion) is already protonated; therefore, higher affinity of these forms is compensated by low population of the guest anionic form.

7. Carboxylate groups are here in α-positions to nitrogens; therefore, one would expect for them protonation constants typical for α-amino acids: $\log K_H \approx 2.3$ for carboxyl group and $\log K_H \approx 9.6$ for nitrogen. The distance between the amino acid fragments is rather large, and their pK values can be approximately considered as independent. With these assumptions the expected dominant forms are: LH_2 (neutral form), LH_4^{2+} at pH 2, LH_2 (zwitterionic) at pH 5 and 8, and L^{2-} at pH 11.

8. Equation (D1-27), which takes in this case the form

$$\delta_{obs} = \delta_S + \Delta\delta\{1 + 0.25(K_d/[S]_t) \times (1 - (1 + 8[S]_t/K_d)^{1/2})\}$$

where δ_S is the chemical shift of monomeric substrate, $\Delta\delta$ is the difference between shifts of dimer and monomer, and $K_d = 1/K$ is the dissociation constant of dimer should be used in this case. Figure D 29 shows the non-linear least-squares fitting of the experimental data to this equation, from which one obtains $\delta_{monomer} = 304 \pm 19$ Hz $= 5.07$ p.p.m, $\delta_{dimer} = 517 \pm 18$ Hz $= 8.62$ p.p.m, and $K = 5.37 \pm 1.55$ kg mol^{-1}.

REFERENCES

1 (a) Connors, K.A. (1987) *Binding Constants*, Wiley, New York; (b) Hartley, F.R.; Burgess, C.; Alcock, R.M. (1980) *Solution Equilibria*, Ellis Horwood, Chichester; (c) Polster, J.; Lachmann, H. *Spectrometric Titrations*, VCH, Weinheim (1989; (d) Beck, M.T.; Nagypál, I. (1990) *Chemistry of Complex Equilibria*, Ellis Horwood and Académiai Kiado, London, Budapest; (e) Martell, A.E.; Motekaitis, R.J. (1988) *The Determination and Use of Stability Constants*, VCH, New York; (f) Sommer, L.; Langova, M. (1988) *CRC Crit. Rev. Anal. Chem.*, **19**, 225.

2 (a) Johnstone, A.P.; Thorpe, R. (1996) *Immunochemistry in Practice*, Blackwell Science, Cambridge; (b) Deshpande, S. (1996) *Enzyme Immunoassays*, Chapman & Hall, New York.

3 (a) Benesi, H.A.; Hildebrand, J.H. (1949) *J. Am. Chem. Soc.*, **71**, 2703.; (b) Scatchard, G.; *Ann. N. Y. Acad. Sci.*, (1949), **51**, 660; (c) Scott, R.L. (1956) *Rev. Trav. Chim.*, **75**, 787.

4 (a) Fersht, A. (1985) *Enzyme Structure and Mechanism*, Freeman, New York; (b) Cantor, R.C.; Schimmel, P.R. (1980) *Biophysical Chem-*

istry, Vols. 1–3, W. H. Freeman Co., San Francisco, USA.

5 Rose, N.J.; Drago, R.S. (1962) *J. Am. Chem. Soc.*, **84**, 2037.

6 Deranleau, D.A. (1969) *J. Am. Chem. Soc.*, **91**, 4044.

7 Diederich, F. (1988) *Angew. Chem., Int. Ed. Engl.*, **27**, 362.

8 Perlmutter-Hayman, B. (1986) *Acc. Chem. Res.*, **19**, 90.

9 Hill, A.V. (1910) *J. Physiol. (London)*, **40**, 6.

10 Monod, J.; Wyman, J.; Changeux, J.-P. (1965) *J. Mol. Biol.*, **12**, 87.

11 Legget, D.J. (1985) (Ed.) *Computational Methods for the Determination of Formation Constants*, Plenum Press, New York.

12 Meloun, M.; Havel, J.; Högfeldt, E. (1988) *Computation of Solution Equilibria. A Guide to Methods in Potentiometry, Extraction and Spectrophotometry*, Ellis Horwood, Chichester.

13 Gans, P.; Sabatini, A.; Vacca, A. (1996) *Talanta*, **43**, 1739.

14 (a) Vetrogon, V.I.; Lukyanenko, N.G.; Schwing-Weill, M.-J.; Arnaud-Neu, F. (1994) *Talanta*, **41**, 2105; (b) Laouenen, A.; Suet, E. (1985) *Talanta*, **32**, 245.

15 Cox, B.G.; Schneider, H. (1992) *Coordination and Transport Properties of Macrocyclic Compounds in Solution*, Elsevier, Amsterdam.

16 Martell, A.E.; Motekaitis, R.J. (1990) *Coord. Chem. Rev.*, **100**, 323.

17 (a) Linder, P.W.; Torrington, R.G.; Williams, D.R. (1984) *Analysis Using Glass Electrodes*, Open University Press, Milton Keynes; (b) Bates, R.G. (1981) *CRC Crit. Rev. Anal. Chem.*, **10**, 247; (c) Albert, A.; Serjant, E.P. (1984) *The Determination of Ionization Constants*, 3rd ed., Chapman and Hall, London.

18 Wu, Y.C.; Koch, W.F.; Marinanko, G. (1984) *J. Res. Natl. Bur. Stand.*, **89**, 395.

19 Covington, A.K.; Bates, R.G.; Durst, R.A. (1985) *Pure Appl. Chem.*, **57**, 531.

20 Sigel, H.; Zuberbühler, A.D.; Yamauchi, O. (1991) *Anal. Chim. Acta.*, **255**, 63.

21 Irving, H.M.; Miles, M.G.; Pettit, L.D. (1967) *Anal. Chim. Acta.*, **38**, 475.

22 (a) Gran, G. (1950) *Acta Chem. Scand.*, **4**, 559; (b) Gran, G. (1952) *Analyst*, **77**, 661; (c) Rossotti, F.J.C.; Rossotti, H. (1965) *J. Chem. Ed.*, **42**, 375.

23 Papanastasiou, G.; Ziogas, I. (1995) *Talanta*, **42**, 827.

24 Kim, W.D.; HrcNir, D.C.; Kiefer, G.E.; Sherry, A.D. (1995) *Inorg. Chem.*, **34**, 2225.

25 Relly, S.D.; Khalsa, G.R.K.; Ford, D.K.; Brainard, J.R.; Hay, B.P.; Smith, P.H. (1995) *Inorg. Chem.*, **34**, 569.

26 Sabatini, A.; Vacca, A.; Gans, P. (1992) *Coord. Chem. Rev.*, **120**, 389.

27 Hofman, T.; Krzyzanowska, M. (1986) *Talanta*, **33**, 851.

28 Meloun, M.; Bartos, M. (1992) *Mikrochim. Acta*, **108**, 227.

29 Lomozik, L.; Jaskolski, M.; Gasowska, A. (1995) *J. Chem. Ed.*, **72**, 27.

30 (a) Sillen, L.G. (1962) *Acta. Chem. Scand.*, **16**, 159; (b) Ingri, N.; Sillen, L.G. (1962) *Acta. Chem. Scand.*, **16**, 173.

31 Sayce, I.G. (1968) *Talanta*, **15**, 1397.

32 Sabatini, A.; Vacca, A.; Gans, P. (1974) *Talanta*, **21**, 53.

33 Zuberbühler, A.D.; Kaden, Th.A. (1982) *Talanta*, **29**, 201.

34 Gans, P.; Sabatini, A.; Vacca, A. (1985) *J. Chem. Soc., Dalton Trans.*, 1195.

35 Motekaitis, R.J.; Martell, A.E. (1982) *Can. J. Chem.*, **60**, 2403.

36 Arena, G.; Rizzarelli, E.; Sammartano, S.; Rigano, C. (1979) *Talanta*, **26**, 1.

37 Wozniak, M.; Nowogrocky, G. (1978) *Talanta*, **25**, 633.

38 Motekaitis, R.J.; Martell, A.E. (1982) *Can. J. Chem.*, **60**, 168.

39 Potvin, P. (1994) *Anal. Chim. Acta.*, **299**, 43.

40 May, P.M.; Murray, K.; Williams, D.R. (1985) *Talanta*, **32**, 483.

41 (BSTAC and ES4EC) De Stefano, C.; Mineo, P.; Rigano, C.; Sammartano, (1993) *Ann. Chim. (Rome)*, **83**, 243; (ESAB) De Stefano, C.; Princi, P.; Rigano, C.; Sammartano, S. (1987) *Ann. Chim. (Rome)*, **77**, 643.

42 Freiser, (1980) ed. *Ion-Selective Electrodes in Analytical Chemistry*, Vols. 1–2, Plenum Press, New York.

43 Umesava, Y. (1990) ed. *Handbook of Ion-Selective Electrodes: Selectivity Coefficients*, CRC Press, Boca Raton, FL.

44 Koryta, J. (1991) *Ions, Electrodes and Membranes*, 2nd ed., Wiley, New York.

45 Lewenstam, A.; Hulanicki, A. (1990) *Selective Electrode Rev.*, **12**, 161.

46 Koryta, J. (1972) (a) *Anal. Chim. Acta.*, **61**, 329; (b) *Anal. Chim. Acta.*, (1977) **91**, 1; (c) *Anal. Chim. Acta.*, (1979), **111**, 1; (d) *Anal. Chim. Acta.*, (1982), **139**, 1; (e) *Anal. Chim. Acta.*, (1984), **159**, 1; (f) *Anal. Chim. Acta.*, (1986, **183**, 1; (g) *Anal. Chim. Acta.*, (1988),

206, 1; (h) *Anal. Chim. Acta.*, (1990), **233**, 1.

47 Kimura, K.; Shono, T. (1990) in *Cation Binding by Macrocycles*, Eds. Inoue, Y.; Gokel, G.W., Marcel Dekker, New York, p. 429.

48 Frensdorff, H.K. (1971) *J. Am. Chem. Soc.*, **93**, 600.

49 Lehn, J.M.; Sauvage, J.P. (1975) *J. Am. Chem. Soc.*, **97**, 6700.

50 (a) Arnold, K.A.; Gokel, G.W. (1986) *J. Org. Chem.*, **51**, 5015; (b) Arnold, K.A.; Echegoyen, L.; Gokel, G.W. (1987) *J. Am. Chem. Soc.*, **109**, 3713.

51 De Robertis, A.; Di Giacomo, P.; Foti, C. (1995) *Anal. Chim. Acta.*, **300**, 45.

52 Takisawa, N.; Hall, D.G.; Wyn-Jones, E.; Brown, P. (1988) *J. Chem. Soc., Faraday Trans. I*, **84**, 3059.

53 Gutknecht, J.; Schneider, H.; Stroka, J. (1978) *Inorg. Chem.*, **17**, 3326.

54 (a) Altman, J.; Lipka, J.J.; Kuntz, I.; Waskell, L. (1889) *Biochemistry*, **28**. 7516; (b) Moore, C.D.; Al-Misky, Q.N., Leconte, J.T.J. (1891) *Biochemistry*, **30**. 8357.

55 Hallé, J.-C.; Lelievre, J.; Terrier, F. (1996) *Can. J. Chem.*, **74**, 613.

56 Lüning, U. (1995) *Adv. Phys. Org. Chem.*, **30**, 63.

57 (a) Bianchi, A.; Micheloni, M.; Paoletti, P. (1991) *Coord. Chem. Rev.*, **110**, 17; (b) Andres, A.; Bazzicalupi, C.; Bencini, A.; Bianchi, A.; Fusi, V.; Garcia-Espana, E.; Giorgi, C.; Nardi, N.; Paoletti, P.; Ramirez, J.A.; Valtancoli, B. (1994) *J. Chem. Soc. Perkin Trans. 2*, 2367.

58 Shinkai, S. (1993) *Tetrahedron*, **49**, 8933.

59 Schneider, H.-J.; Güttes, D.; Schneider, U. (1988) *J. Am. Chem. Soc.* **110**, 6449.

60 van Staveren, C.J.; Aarts, V.M.L.J.; Grootenhuis, P.D.J.; Droppers, W.J.H.; van Eerden, J.; Harkema, S.; Reinhoudt, D. N. (1988) *J. Am. Chem. Soc.*, **110**, 8134 and references therein.

61 Alder, R.W. (1990) *Tetrahedron*, **46**, 683.

62 Smith, P.B.; Dye, J.L.; Cheney, J.; Lehn, J.-M. (1981) *J. Am. Chem. Soc.*, **103**, 6044 and references therein.

63 Cox, B.G.; Knop, D.; Schneider, H. (1981) *J. Am. Chem. Soc.*, **103**, 6044.

64 (a) Gelb, R.I.; Schwartz, L.M.; Johnson, R.F.; Laufer, D.A. (1979) *J. Am. Chem. Soc.*, **101**, 1869; (b) Gelb, R.I.; Schwartz, L.M.; Cardelino, B.; Fuhrman, H.S.; Johnson, R.F.; Laufer, D.A. (1981) *J. Am. Chem. Soc.*, **103**, 1750; (c) Cromwell, W.C.; Byström, K.; Eftink, M.R. (1985) *J. Phys. Chem.*, **89**, 326; (d)

Connors, K.A.; Lipari, J.M. (1976) *J. Pharm. Sci.*, **65**, 379.

65 Bard, A.J.; Faulkner, L.R. (1980) *Electrochemical Methods, Fundamentals and Applications*, Wiley, New York.

66 Crow, D.R. (1969) *Polarography of Metal Complexes*, Academic Press, New York.

67 Kapoor, R.C.; Aggarwal, B.S.; (1991) *Principles of Polarography*, Wiley, New York.

68 Nicholson, R.S.; Shain, I. (1964) *Anal. Chem.*, **36**, 706.

69 (a) Matsuda, H; Ayabe, Y. (1955) *Z. Electrochem.*, **59**, 494; (b) Matsuda, H. (1957) *Z. Elecrochem.*, **61**, 489.

70 Mabbott, G.A. (1983) *J. Chem. Ed.*, **60**, 697.

71 Kissinger, P.T.; Heineman, W.R. (1983) *J. Chem. Ed.*, **60**, 702.

72 DeFord, D.D.; Hume, D.N. (1951) *J. Am. Chem. Soc.*, **73**, 5321.

73 Gampp, H. (1987) *Anal. Chem.*, **59**, 2456.

74 Nyholm, L.; Wikmark, G. (1989) *Anal. Chim. Acta*, **223**, 429.

75 Montemayor, M.C.; Fatas, E. (1988) *J. Electroanal. Chem. Interfacial Electrochem.*, **246**, 271.

76 Kimura, E.; Sakonaka, S.; Yatsunami, T.; Kodama, M. (1981) *J. Am. Chem. Soc.*, **103**, 3041.

77 Peter, F.; Gross, M.; Hosseini, M.W.; Lehn, J.-M.; Sessions, R.B. (1981) *J. Chem. Soc. Chem. Commun.*, 1067.

78 Beer, P.D. (1989) *Chem. Soc. Rev.*, **18**, 409.

79 (a) Aragó, J.; Bencini, A.; Bianchi, A.; Domenech, A.; García-Espana, E. (1992) *J. Chem. Soc. Dalton Trans.*, 319; (b) Bianchi, A.; Domenech, A.; García-Espana, E.; Luis, S.V. (1993) *Anal. Chem.*, **65**, 3137.

80 Matsue, T.; Evans, D.H.; Osa, T.; Kobayashi, N. (1985) *J. Am. Chem. Soc.*, **107**, 3411.

81 Bersier, P.M.; Bersier, J.; Klingert, B. (1991) *Electroanalysis (N.Y.)*, **3**, 443.

82 Isnin, R.; Salam, C.; Kaifer, A.E. (1991) *J. Org. Chem.*, **56**, 35.

83 Robinson, R.A.; Stokes, R.H. (1970) *Electrolyte Solutions*, 2nd rev. ed., Butterworths, London.

84 Barthel, J.; Wachter, R.; Gores, H.-J. in Conway, B.E.; Bockris, J.O'M. (1979) (Eds.) *Modern Aspects of Electrochemistry*, Vol.13, Plenum Press, New York, 1.

85 Fuoss, R.M.; Accascina, F. (1959) *Electrolytic Conductance*, Interscience, New York.

86 Takeda, Y. in Inoue, Y.; Gokel, G.W. (1990) (Eds.) *Cation Binding by Macrocycles*, Marcel Dekker, New York, 133.

87 Fernández-Prini, R.; Justice, J.-C. (1984) *Pure Appl. Chem.*, **56**, 541.

88 Palepu, R.; Richardson, J.E.; Reinsborough, V.C. (1989) *Langmuir*, **5**, 218.

89 Salomon, M. (1990) *J. Solut. Chem.*, **19**, 1225.

90 (a) Harris, D.A.; Bashford, C.L. 1987 (Eds.) *Spectrophotometry and Spectrofluorimetry; a Practical Approach*, IRL Press, Oxford, (b) Lobinski, R.; Marczenko, Z.; (1992) *Crit. Rev. Analyt. Chem.*, **23**, 55; (c) Clark, B.J.; Frost, T.; Russel, M.A. (1993) (Eds.) *UV Spectroscopy: Techniques, Instrumentation, Data Handling*, Chapman & Hall, London.

91 (a) Taguchi, K. (1986) *J. Am. Chem. Soc.*, **108**, 2705; (b) Taguchi, K. (1992) *J. Chem. Soc. Perkin Trans. 2*, 17; (c) Buvári, A.; Barcza, L.; Kajtár, M. (1988) *J. Chem. Soc. Perkin Trans. 2*, 1687.

92 Cramer, F.; Saenger, W.; Spatz, H.-Ch. (1967) *J. Am. Chem. Soc.*, **89**, 14.

93 Maeder, M; Zuberbuehler, A. D. (1990) *Anal. Chem.*, **62**, 2220; Gampp, H.; Maeder, M.; Meyer, C.J.; Zuberbuehler, A. D. (1985) *Talanta*, **32**, 257.

94 Legget, D.J. in ref. 11, p. 159.

95 Politi, M.J.; Trans. C.D.; Gao, G.-H. (1995) *J. Phys. Chem.*, **99**, 14137.

96 Jaber, M.; Bertin, F.; Thomas-David, G. (1988) *Bull. Soc. Chim. Fr.*, 470.

97 Lakowicz, J.R. (1983) *Principles of Fluorescence Spectroscopy*, Plenum Press, New York.

98 Frankewich, R.P.; Thimmaiah, K.N.; Hinze, W.L. (1991) *Analyt. Chem.*, **63**, 2924.

99 Nigam, S.; Durocher, G. (1996) *J. Phys. Chem.*, **100**, 7135.

100 Eftink, M.R.; Ghiron, C.A. (1981) *Analyt. Biochem.*, **114**, 199.

101 Ueno, A.; Suzuki, I.; Osa, T. (1989) *J. Am. Chem. Soc.*, **111**, 6391.

102 Johnston, L.J.; Wagner, B.D. (1996) in: *Comprehensive Supramolecular Chemistry*, Vol. 8; Davies, J.E.D.; Ripmeester, J.A., Eds., Pergamon/Elsevier, Oxford, 537 ff.

103 Turro, N.J.; Okubo, T.; Chung, C. (1982) *J. Am. Chem. Soc.*, **104**, 3954.

104 Kuroda, Y.; Ito, M.; Sera, T.; Ogoshi, H. (1993) *J. Am. Chem. Soc.*, **115**, 7003.

105 Shionoya, M.; Furuta, H.; Lynch, V.; Harriman, A.; Sessler, L. (1992) *J. Am. Chem. Soc.*, **114**, 5417.

106 Forman, J.E.; Barrans, R.E.; Dougherty, D.A. (1995) *J. Am. Chem. Soc.*, **117**, 9213; this paper contains extensive citations of previous work.

107 Sakurai, T.; Saitou, E.; Hayashi, N.; Hirasawa, Y.; Inoue, H. (1994) *J. Chem. Soc., Perkin Trans. 2*, 1929.

108 Schuette, J.M.; Ndou, T.T., Warner, I.M. (1992) *J. Phys. Chem.*, **96**, 5309.

109 Schneider, H.-J. (1993) *Rec. Trav. Chim. Pays-Bas*, **112**, 412 and references cited therein.

110 Popov, A.I. (1991) in *Modern NMR Techniques and Their Application in Chemistry*, Popov, A.I. and Hallenga, K. Eds., Marcel Dekker, New York, 485.

111 Kumar, S.; Schneider, H.-J. (1989) *J. Chem. Soc., Perkin Trans. 2*, 245.

112 (a) Barrans, R.E.Jr.; Dougherty, D.A. (1994) *Supramol. Chem.*, **4**, 121; (b) Frassineti, C.; Gheli, S.; Gans, P.; Sabatini, A.; Moruzzi, M.; Vacca, A. (1995) *Analyt. Biochem.*, **231**, 374; (c) Wilcox, C. (1991) in *Frontiers in Supramolecular Chemistry and Photochemistry*, Schneider, H.-J.; Dürr, H. Eds; VCH Weinheim, 123 ff; (d) Schneider, H.-J.; Kramer, R.; Simova, S.; Schneider, U. (1988) *J. Am. Chem. Soc.*, **110**, 6442.

113 Zimmerman, S.C.; Weiming, W.; Zijian, Z.; (1991) *J. Am. Chem. Soc.*, **113**, 196.

114 Yamashoji, Y.; Fujiwara, M.; Matsushita, T.; Tanala, M. (1993) *Chem. Lett.* 1029; for earlier examples see e.g. Stöver, H.D.H.; Delville, A.; Detellier, C. (1985) *J. Am. Chem. Soc.*, **107**, 4167.

115 Ni, F., (1994) *Progr. NMR Spectr*, **26**, 517; Eis, M.; Allegrini, P.R. (1996) *Proc. Soc. Magn. Res. B.* **4** 1629; Eis, M.; Hoehn-Berlage, M. (1995) *J. Magn. Res. B.* **107**, 222; Meyer, B.; Weimar, T.; Peters, T. (1995) *Eur. J. Biochem.*, **246**, 705; and references cited therein.

116 Mayzel, O.; Cohen, Y. (1994) *J. Chem. Soc., Chem. Commun.*, 1901.

117 Lang, E.W.; Ludemann, H.-D. (1993) *Progr. NMR-Spectroscopy*, **25**, 507; Stilbs, P. *ibid* (1987), **19**, 1.

118 Vincenti, M. (1995) *J. Mass Spectrom.*, **30**, 925.

119 Wang, K.; Gokel, G.W. (1996) *Pure Appl. Chem.*, **68**, 1267.

120 Smith, R.D.; Light-Wahl, K.J. (1993) *Biol. Mass Spectrom.*, **22**, 493.

121 Smith, R.D.; Bruce, J.E.; Wu, Q.; Lei, Q.P. (1997) *Chem. Soc. Rev.*, **26**, 191.

122 (a) Hofstadler, S.A.; Bakhtiar, R.; Smith, R.D. (1996) *J. Chem. Ed.*, **73**, A82. (b) Bakhtiar, R.; Hofstadler, S.A.; Smith, R.D. (1996) *J. Chem. Ed.*, **73**, A118. (c) Hop, C.E.C.A.; Bakhtiar, R.; (1996) *J. Chem. Ed.*, **73**, A162.

123 Przybylski, M.; Gloker, M.O. (1996) *Angew. Chem., Int. Ed. Engl.*, **35**, 806.

124 (a) Neil, B.H.H.; Youngless, T.L.; Bursey, M.M. (1980) *Inorg. Nucl. Chem. Lett.*, **16**, 141. (b) Neil, B.H.H.; Fraley, D.H.; Bursey, M.M. (1981) *Inorg. Nucl. Chem. Lett.*, **17**, 121.

125 Teesch, L.M.; Adams, J. (1991) *J. Am. Chem. Soc.*, **113**, 812.

126 Johnstone, R.A.W.; Lewis, I.A.S.; Rose, M.E. (1983) *Tetrahedron*, **39**, 1597.

127 (a) Bonas, G.; Bosso, C.; Vignon, M.R. (1988) *Rapid. Commun. Mass Spectrom.*, **2**, 88. (b) Bonas, G.; Bosso, C.; Vignon, M.R. (1989) *J. Inclusion Phenom. Mol. Recognit. Chem.*, **7**, 637.

128 Burlingame, A.L.; Boyd, R.K.; Gaskell, S.J. (1994) *Analyt. Chem.*, **66**, 634R.

129 (a) Giraud, D.; Laprévote, O.; Das, B.C. (1994) *Org. Mass Spetcrom.*, **29**, 169. (b) Giraud, D.; Scherrens, I.; Lever, M.-L.; Laprévote, O.; Das, B.C. (1996) *J. Chem. Soc., Perkin Trans. 2*, 901.

130 Malhotra, N.; Roepstorff, P.; Hansen, T.K.; Becher, J. (1990) *J. Am. Chem. Soc.*, **112**, 3709.

131 (a) Bryant, A.J.; Blanda, M.T.; Vincenti, M.; Cram, D.J. (1991) *J. Am. Chem. Soc.*, **113**, 2167. (b) Cotter, R.J. (1980) *Anal. Chem.*, **52**, 1589A.

132 Man, V.F.; Lin, J.D.; Cook, K.D. (1985) *J. Am. Chem. Soc.*, **107**, 4635.

133 Ikonomou, M.G.; Blades, A.T.; Kebarle, P. (1991) *Anal. Chem.*, **63**, 1989.

134 (a) Mann, M.; Meng, C.K.; Fenn, J.B. (1989) *Anal. Chem.*, **61**, 1702; (b) Labowsky, M., Whitehouse, C.; Fenn, J.B. (1993) *Rapid. Commun. Mass Spectrom.*, **7**, 71; (c) Reinhold, B.B.; Reinhold, V.N. (1992) *J. Amer. Soc. Mass Spectrom.*, **3**, 207.

135 Li, Y.-T.; Hsieh, Y.-L.; Henion, J.D.; Ocain, T.D.; Schiehser, G.A.; Ganem, B. (1994) *J. Am. Chem. Soc.*, **116**, 7487.

136 (a) Camilleri, P.; Haskins, N.J.; New, A.P.; Saunders, M.R. (1993) *Rapid. Commun. Mass Spectrom.*, **7**, 949; (b) Selva, A.; Redenti, E.; Zanol, M.; Ventura, P.; Caseat, B. (1993) *Org. Mass Spectrom.*, **28**, 983; (c) Haskins, N.J.; Saunders, M.R.; Camilleri, P. (1994) *Rapid. Commun. Mass Spectrom.*, **8**, 423; (d) Cunniff, J.B.; Vouros, P. (1995) *J. Amer. Soc. Mass Spectrom.*, **6**, 437.

137 (a) Vincenti, M.; Pelizzetti, E.; Dalcante, E.; Soncini, P. (1993) *Pure Appl. Chem.*, **65**, 1507; (b) Vincenti, M.; Minero, C.; Pelizzetti, E.; Secchi, A.; Dalcante, E. (1995) *Pure Appl. Chem.*, **67**, 1075.

138 Brodbelt, J.S.; Dearden, D.V. in *Comprehensive Supramolecular Chemistry*, Vol. 8, Davies, J.E.D;

Ripmeester, J.A., eds., Pergamon/Elsevier, Oxford (1996), 567 ff.

139 Dearden, D.V. (1996) in *Physical Supramolecular Chemistry*, Echegoven, L.; Kaifer, A.E. eds., Kluwer Acad. Publ., 229.

140 (a) Comisarow, M.B. (1980) *Adv. Mass Spectrom.* **8**, 1698; (b) Marshall, A.G. (1985) *Acc. Chem. Res.* **18**, 316; (c) Buchanan, M.V., (1987) Ed. *Fourier Transform Mass Spectrometry: Evolution, Innovation, and Applications*, ACS Symposium Series Vol. 359, Washington D.C.

141 March, R.E.; Hughes, R.J.; Todd, J.F.J. (1989) *Quadrupole Ion Storage Mass Spectrometry*, Wiley, New York.

142 Busch, K.L.; Glish, G.L.; McLuckey, S.A. (1988) *Mass Spectrometry/Mass Spectrometry: Techniques and Applications of Tandem Mass Spectrometry*, VCH, New York.

143 Vincenti, M.; Dalcanale, E. (1995) *J. Chem. Soc., Perkin Trans. 2*, 1069.

144 Sekine, T.; Hasegawa, Y. (1977) *Solvent Extraction Chemistry*, Parts I and II, Marcel Dekker, New York.

145 Tachibana, M.; Furusawa, M.; Kiba, N. (1995) *J. Incl. Phenom. Mol. Recogn.*, **22**, 313.

146 Pedersen, C.J. (1970) *J. Am. Chem. Soc.*, **92**, 391.

147 Frensdorf, H.K. (1971) *J. Am. Chem. Soc.*, **93**, 4684.

148 (a) Kyba, E.P.; Helgeson, R.C.; Madan, K.; Gokel, G.W.; Tarnowski, T.L.; Moore, S.S.; Cram, D.J. (1977) *J. Am. Chem. Soc.*, **99**, 2564; (b) Koenig, K.; Lein, G.M.; Stuckler, P. Kaneda, T.; Cram, D.J. (1979) *J. Am. Chem. Soc.*, **101**, 3553.

149 Buncel, E.; Shin, H.S.; Bannard, R.A.B.; Purdon, J.G.; Cox, B.G. (1984) *Talanta*, **31**, 585.

150 (a) Kimura, K.; Maeda, T.; Shono, T. (1979) *Talanta*, **26**, 945; (b) Maeda, T.; Kimura, K.; Shono, T. (1979) *Fresenius Z. Anal. Chem.*, **298**, 363.

151 Takeda, Y.; Goto, H. (1979) *Bull. Chem. Soc. Jpn.*, **52**, 1920.

152 Armstrong, D.W. (1998) *Adv. Chromatogr.*, **39**, 239.

153 Fujimura, K.; Ueda, T.; Kitogawa, M.; Takayanagi, H.; Ando, T. (1986) *Anal. Chem.*, **58**, 2668, and references therein.

154 Horváth, C.; Melander, W.; Nahum, A. (1979) *J. Chromatogr.*, **186**, 371.

155 Armstrong, D.W.; Fendler, J.H. (1977) *Biochim. Biophys. Acta*, **478**, 75.

156 Armstrong, D.W.; Nome, F.; Spino, L.A.; Golden, T.D. (1986) *J. Am. Chem. Soc.*, **108**, 1418.

157 Hummel, J.P.; Dreyer, W.J. (1962) *Biochim. Biophys. Acta*, **63**, 530.

158 (a) Yoza, N. (1977) *J. Chem. Ed.*, **54**, 284; (b) Korpela, T.K.; Himanen, J.-P. (1984) *J. Chromatogr.*, **290**, 351.

159 Thuaud, N.; Gosselet, N.-M.; Sebille, B.; Veyron, N.; Tachon, P. (1996) *J. Incl. Phenom. Mol. Recognit. Chem.*, **25**, 267.

160 (a) Rundlett, K.L.; Armstrong, D.W. (1996) *J. Chromatogr. A.*, **721**, 173; (b) Chu, Y.-H.; Avila, L.Z.; Gao, J.; Whitesides, G.M. (1995) *Acc. Chem. Res.*, **28**, 461.

161 Hemminger, W.; Höhne, G. (1984) *Calorimetry - Fundamentals and Practice*, Verlag Chemie, Weinheim.

162 (a) Christensen, J.J.; Ruckman, J.; Eatough, D.J.; Izatt, R.M. (1972) *Thermochim. Acta*, **3**, 203; (b) Eatough, D.J.; Christensen, J.J.; Izatt, R.M. *ibid.*, 219; (c) Eatough, D.J.; Izatt, R.M.; Christensen, J.J. *ibid.*, 233.

163 Christensen, J.J. in Braibanti, A. (1980), ed. *Bioenergetics and Thermodynamics: Model Systems*, D. Reidel Publ. Dordrecht, Berton, 75.

164 Hansen, L.H.; Lewis, E.A.; Eatough, D.J. in Grime, J.K. (1985), ed. *Analytical Solution Chemistry*, Wiley, New York, Ch. 3.

165 (a) Stödeman, M.; Wadsö, I. (1995) *Pure Appl. Chem.*, **67**, 1059; (b) Wadsö, I. (1997) *Chem. Soc. Rev.*, **25**, 79.

166 Hallén, D.; Schön, A.; Shehatta, I.; Wadsö, I.; (1992) *J. Chem. Soc., Faraday Trans.*, **88**, 2859.

167 Arena, G.; Calí, R.; Maccarrone, G.; Purrello, R. (1989) *Thermochim. Acta*, **155**, 353.

168 Cram, D.J.; Lein, G.M. (1985) *J. Am. Chem. Soc.*, **107**, 3657.

169 Dijkstra, P.J.; Brunink, J.A.J.; Bugge, K.-L.; Reinhoudt, D.N.; Herkema, S.; Ungaro, R.; Ugozzoli, F.; Ghidini, E. (1989) *J. Am. Chem. Soc.*, **111**, 7567.

170 Tabushi, I.; Kimura, Y.; Yamamura, K. (1981) *J. Am. Chem. Soc.*, **103**, 6486.

171 Schneider, H.-J.; Kramer, R.; Rammo, J. (1993) *J. Am. Chem. Soc.*, **115**, 8980.

172 Tee, O.S.; Bennett, J.M. (1988) *J. Am. Chem. Soc.*, **110**, 269.

173 (a) Cornish-Bowden, A. (1979) *Fundamentals of Enzyme Kinetics*, Butterworths, London also the 2nd edn., (1995), Portland Press, London; (b) Segel, I.H. (1993) *Enzyme Kinetics; Behavior and Analysis of Rapid Equilibrium and Steady State Enzyme Systems*, Wiley, New York.

174 VanEtten, R.L.; Sebastian, J.F.; Clowes, G.A.; Bender, M.L. (1967) *J. Am. Chem. Soc.*, **89**, 3242.

175 Tee, O.S.; Bozzi, M.; Hoeven, J.J.; Gadosy, T.A. (1993) *J. Am. Chem. Soc.*, **115**, 8990.

176 Tee, O.S.; Gadosy, T.A.; Giorgi, J.B. (1996) *Can. J. Chem.*, **74**, 736.

177 Kretschmann, E.; Raether, H. (1968) *Z. Naturforschung*, **A23**, 2135; Liedberg, B.; Nylander, C.; Lundstrom, I., (1983) *Sensors and Actuators*, **4**, 299; Schuck, P. (1997) *Curr. Opin. Biotechnol.*, **8**, 498; Fisher, R. J.; Fivaslı, M. (1994) *ibid.* **5**, 389.

178 Karlsson, R.; Michaelsson, A.; Mattsson, L. (1991) *J. Immunol. Meth.*, **145**, 229.

179 Granzow, R.; Reed, R. (1992) *BioTechnology*, **10**, 390.

180 After BIACORE 3000 Product Information 1998.

181 Review see e.g. Chaiken, I., Ros, S. and Karlsson, R. (1992) *Anal. Biochem.*, **201**, 197; recent application see e.g. Alper, P.; Hendrix, M.; Serrs, P.; Wong, C.-H. (1998) *J. Am. Chem. Soc.*, **120**, 1965.

182 Karlsson, R.; Stahlberg, R. (1995) *Anal. Biochem.*, **228**, 274.

183 Chasseray, X.; Lo, S.M.; Lemordant, D. (1995) *J. Incl. Phenom. Mol. Recognit. Chem.*, **23**, 127.

184 Eastman, M.P.; Freiha, B.; Hsu, C.C.; Lum, K.C.; Chang, C.A. (1987) *J. Phys. Chem.*, **91**, 1953.

185 (a) Kotake, Y.; Janzen, E.G. (1988) *J. Am. Chem. Soc.*, **110**, 3699; (b) Kotake, Y.; Janzen, E.G. (1989) *ibid.* **111**, 7319.

186 Castrejon, S.E.; Yatsimirsky, A.K. (1997) *Talanta*, **44**, 951.

187 Tee, O.S.; Gadosy, T.A.; Giorgi, J.B. (1996) *Can. J. Chem.*, **74**, 736.

188 Purcell, J.M.; Susi, H.; Cavanaugh, J.R. (1969) *Can. J. Chem.*, **47**, 365; Tan, H.K.S. (1994) *J. Chem. Soc. Farad. Trans.*, **90**, 3521

E1. INDIRECT APPROACHES (COMPETITION EXPERIMENTS, STRUCTURE–ACTIVITY CORRELATIONS)

The chemical nature and shape of receptor sites can be explored by comparing, for instance, binding constants of different guest molecules, assuming optimal binding strength with optimal match both with respect to complementary functions and shape. In medicinal chemistry this is the traditional search method for a pharmacophor, relying here often on less well defined biological activity instead of on binding constants. In corresponding structure–activity correlations one has to use a large enough data set for arriving at safe conclusions, and one often refrains from constructing a model of the host–guest interaction at a molecular level. Before the recent development of protein crystallography coupled with computer aided simulations, active sites of enzymes were explored by comparing Michaelis–Menten constants or inhibition constants of a variety of ligands, leading for instance to the so-called lattice analysis.[1] The method is still valuable for receptors which are either not accessible to X-ray or NMR analysis, or are not available in pure form. Modern computational approaches based on affinity comparison of many ligands (as training, or learning sets for prediction of both energies and receptor site properties) are discussed in Section E7.

For synthetic host–guest complexes the variation of affinity constants, K, by guest modification was used early to obtain insight into binding modes with cyclodextrin associations. Thus, introduction of methyl groups into p-nitrophenole leads to K variations clearly pointing towards a penetration of the cavity with first the nitrogroup in nitrophenolate itself (Scheme E1).[2]

In simple competition experiments one can check whether, and possibly also to what degree, a known substrate is removed from a ligand binding site upon addition of other substrates. In a related experiment it was shown in one of the oldest supramolecular enzyme models that p-nitrophenyl acetate is not bound inside the α- and β-cyclodextrin cavities in the transition state, as addition of alcohols, alkanesulfonate or alkanoate ions, which are known to form inclusion complexes

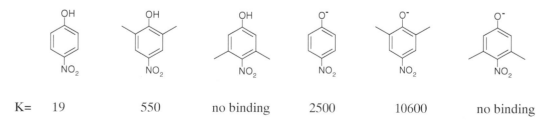

K=   19   550   no binding   2500   10600   no binding

Scheme E1. Binding constants $K(\mathrm{M}^{-1})$ of methyl substituted p-nitrophenol derivatives to α-cyclodextrin.[2] Reprinted with permission from *J. Am. Chem. Soc.*, Bergeron *et al.*, 1997, **99**, 5146. Copyright 1977 American Chemical Society.

with cyclodextrins did not slow down the reaction.[3] Competition experiments , including those with natural receptors employing radioimmuno assays (RIA) etc., are of course primarily aiming at evaluation of affinities; with a large enough data set one can, however, also learn about the structure of the receptor site.

E 2.   DIFFRACTION TECHNIQUES

Single crystal structural analysis has laid the foundations for most of what we know about intermolecular interactions, like van der Waals radii, directionality and interatomic distances. We have already discussed numerous non-covalent interactions, which could not have been recognized or even modeled without such parameters. The contributions of X-ray analyses to host–guest chemistry are most obvious in the development of ionophore complexes[4] and other supramolecular associations.[5,6] The particularly strong role of X-ray analysis and neutron diffraction is obvious in the characterization of hydrogen bonds (see corresponding examples in Section B 3). In Section B 1 we already pointed out the strategy for exploring geometric rules for intermolecular interactions on the basis of very many solid structures which are available (now also via the internet) for smaller molecules in the Cambridge database and for biomolecules in the Brookhaven protein database.

X-ray analysis for smaller molecules has been developed almost to a routine operation, but supramolecular complexes can pose technical problems.[7] These are primarily a consequence of the number of reflections: for normal organic molecules with a cell edge length of 10 Å there are typically around 3000 reflections. For supramolecular complexes even if the length only doubles to 20 Å the number of reflections increases to 20 000. At the same time the mean intensity of the reflections drops to 1/8, causing signal-to-noise problems. The other serious problem is the less dense packing in larger molecules, which leads to inclusion of more solvent molecules, to higher flexibility of the frameworks and to more disorder in the crystals. For details and modern developments which allow one to solve the phase problem for supermolecules with direct methods, as long as the resolution is better than 1.2 Å, we refer to the special literature.[7]

X-ray analyses were first applied to supramolecular structures such as cyclodextrins with the help of heavy atoms like complexed iodine and the use of Patterson methods for solving the phase problem.[8] In the meantime, hundreds of large supramolecular structures with only hydrogen or carbon atoms have been reported, usually with hydrogen refinement, at resolutions much better than 1 Å and Hamilton R factors well below 10%. Even the precise structure of very large structures, containing a supramolecular complex within another host[9] in the manner of a Russian matrioschka shown in Fig. E 1 were determined, and the method is applied now more and more to supramolecular networks which are the basis of new nanomaterials like thin films (Section I 2). It should be pointed out that such structures will usually fall apart in solution, which is why one has to resort to special solid state techniques, in particular the magic angle spinning method (MAS) if one wants to use NMR-spectroscopy for such materials.[10] On the other hand, it may well occur that a supramolecular complex is stable only in solution, but not in the solid state. This is so for the association with a benzidine-derived cyclophane, which in the crystal always contains exactly one mole of benzene; on the basis of this and of CPK (Corey–Pauling–Kottun) models the system was published as one of the first supramolecular complexes in the literature.[11] X-ray analysis revealed 27 years later that in the solid state the benzene molecule resides outside the cyclophane cavity.[12] Also years later again NMR studies showed that in aqueous solution the guest is in fact essentially immersed in

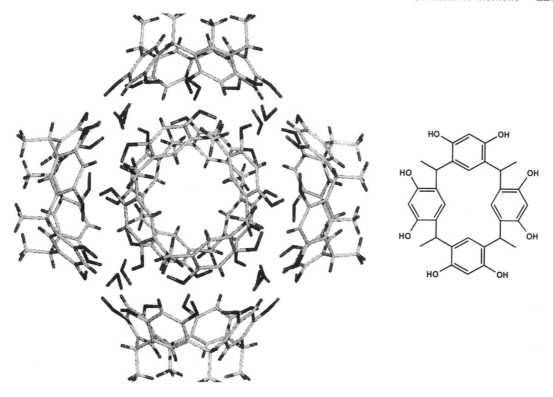

Figure E1. Solid state structure from X-ray analysis for a complex consisting of six resorcarenes and eight water molecules.[9] The hydrogen atoms could not be located by the X-ray analysis, but from the position of donor and acceptor atoms it followed that 60 hydrogen bonds hold the supermolecule together. NMR and MS experiments suggest that the complex does persist in benzene solution and in the gas phase. Reprinted with permission from *Nature*, Vol. **389**, MacGillvray *et al.*, pp 469. Copyright 1997 Macmillian Magazines Limited.

the cavity.[13] Also the binding mode of the guest can be different in the solid state and solution. Thus, solid state X-ray study of the complex formed between 4-fluorophenol and α-cyclodextrin revealed an unusual inclusion mode of the phenol with the hydrophilic OH-group inside the cavity,[14] but a more recent ROESY (rotating frame Overhauser) study[15] confirmed for aqueous solution a 'normal' inclusion of the guest with the OH-group outside the cavity. One must keep in mind that the non-covalent interactions leading to supramolecular complexes can differ considerably between solution and solid phase, not only where solvophobic interactions are the driving force like in

many cyclodextrin complexes, but also where hydrogen bonds can only materialize in the absence of competing solvents. Furthermore, short range, dispersive interactions can play a larger role in closely packed crystals than in solution complexes, whereas long range, Coulombic interactions can work over large distances in loosely bound aggregates.

Neutron diffraction allows one to locate also hydrogen atoms down to an accuracy of 0.001 Å, whereas in X-ray analyses one often needs at least low temperature measurements to determine exact H atom positions. Unfortunately, the method is much more laborious, requires a strong neutron source, much larger crystals of about 5 mg, and data collection times of up to several weeks in comparison to days or even hours with X-ray. In the presence of very many hydrogens it also may be necessary to use specific deuteration. Nevertheless, neutron diffraction provides insight particularly into details of hydrogen bonds which used to evade detection. The exact determination of the H atom location circumvents the ambiguity involved with the hydrogen bond evaluation based on acceptor–donor atom distances alone (Section B 3). Also features like the structural consequences of cooperative hydrogen bonding become visible in elongations of N–H and C=O bonds in amide-type of associations including Watson–Crick base pairs.[16] So-called flip-flop hydrogen bonds are seen by the occurrence of seemingly two half hydrogen atoms, like in cyclodextrin hydrates, with those 'half' protons in the static picture residing at a distance of about 1 Å from either the donor or the acceptor site, corresponding to the two-well potential of weaker hydrogen bonds shown in Scheme B8.

E 3. SUPERMICROSCOPY (STM, AFM, CFM, SNOM)

Supermicroscopic techniques[17] present some of the most exciting recent developments which can provide detailed pictures of surfaces at atomic or at least at molecular resolution. For this reason we discuss these techniques in the chapter on structural methods, although the possible direct measurements of intermolecular *forces* with these methods may stir even more interest amongst chemists. These methods can not be applied to study interactions of dissolved molecules and require special equipment and expertise. We refer here largely to special reviews[17] of this rapidly developing field, but want to make the reader aware of the fascinating possibilities for exploring not only supramolecular structure at the surface of crystals or films, but also for measuring directly non-covalent forces.

In scanning tunneling microscopy (STM) one uses the electron flow between the investigated surface and a tiny tip, which occurs by tunneling if the distance is within 5–10 Å. The tip is then moved across the surface, controlled by piezoelectronics (see the AFM set-up in Fig. E 2). With the tip at atom centers there will be a measurable electron flow or current; this decreases between the atoms. With organic material the limitation is that the tip must penetrate the material until it reaches a distance to a conducting layer of about   5–10 Å.

The atomic force microscope (AFM) has the tip mounted on a cantilever (Fig. E 2), the tilting of which can be measured with high accuracy. In contrast to STM, the AFM method measures not electric currents but forces between the tip and the surface, with local forces down to 10^{-15} Newton. The cantilever upon coming closer to the solid surface to be scanned is tilted by the forces, and the tilt angles are magnified by reflection of a focussed laser light beam. The direct measurement of intermolecular forces thus attained was done for the first time with the exceptionally strong biotin–streptavidin attraction (Section A 1). For softer materials one can use non-contact techniques, with some loss of resolution. Scanning near-field microscopy (SNOM) is also done in non-

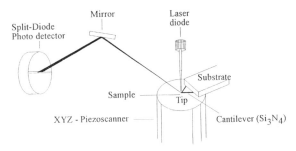

Figure E 2. Principle of an atomic force microscopy set-up with light deflection at the cantilever tip.[17a] Reprinted with permission of Elsevier Science from: Kaupp *et al.*, in *Comprehensive Supramolecular Chemistry*, Vol. 10, Figure 8, p. 698, copyright 1996 Pergamon/Elsevier Oxford etc.

contact mode with fiber tips vibrating close to their resonance frequency; in scanning the surface a frequency damping occurs which measures a lateral shear force between tip and surface. With SNOM one can in particular detect changes in chemical nature on the surface of organic crystals. Chemical force microscopy (CFM) uses specifically functionalized AFM tips to study adhesive forces with chemical specificity at nanoscale resolution.[18] The strong forces attracting antigens to antibodies can be measured this way.[19] One should keep in mind that the forces measured by interactions with the tip of supermicroscope systems will not necessarily correlate with interaction energies measured in solution with 'free' host and guest molecules: the very different distance dependence of non-covalent interactions can lead to a shift in dominating binding mechanisms, and the possible strain induced in the receptor molecule by interaction with the tip is not measured by the force. Tip–surface interactions generally can influence the result of supermicroscopic analyses.[17,20]

E 4. NMR METHODS

E 4.1. NMR SHIELDING (CIS VALUES) FOR THE EVALUATION OF COMPLEX STRUCTURE

The large NMR screening effects usually observed upon complex formation (Scheme E 2) are the most often applied indicators for complex geometry. With larger host–guest complexes the analysis of the spin systems requires two-dimensional techniques[21] with shift correlation methods such as COSY (correlated spectroscopy), TOCSY (total correlation spectroscopy), RELAY (related correlation spectroscopy), etc. or the somewhat more time-consuming HETCOR (heteronuclear correlation), HMQC (heteronuclear multiple quantum correlation), HSQC (heteronuclear single quantum correlation), ^{13}C-pro-

ton shift correlation methods, and finally also NOE applications (Section E 4.2): these are described in several textbooks.[21,22] Assignment of individual resonances even in very complex systems with strongly coupled signals of small shift differences is now possible, yielding a wealth of data relating to molecular structures. The multitude of shielding variations thus determined, and the heavily overlapping signals found, are frequently not used properly. Often conclusions are based on shift changes $\Delta\delta$ measured at arbitrary concentrations and not on on CIS values; the $\Delta\delta$ values will be a function of the chosen concentrations, and can even lack reproducibility. Furthermore, the interpretations often suffer from adhoc assumptions without sound reasoning .

With the exception of solid state investigations (see below) applications of NMR-spectroscopy to supramolecular complexes is mostly restricted to protons: these are on the periphery of molecules and therefore 'feel' non-covalent interactions such as anisotropy effects much more than the carbon atoms 'inside'. Measurements of ^{13}C NMR spectra are not only more time-consuming, but yield mostly information on *intra*molecular geometry variations; they therefore can shed light on conformational changes by induced fit.

The examples shown in Scheme E 2 illustrate ^{1}H-NMR CIS values which are typical of some supramolecular complexes.[23b] The most simple qualitative application of CIS values makes use of the shifts observed in structurally related supramolecular complexes, such as those given in the scheme. If such CIS values are consistently found in several structures, and if they 'make sense' in terms of known shielding mechanisms, one can at least draw conclusions as to whether there is intracavity inclusion or not. In particular, if aryl units are present either in the host or the guest entity one can rely upon the ring current induced anisotropy as the most convenient way to distinguish intracavity inclusion from other association modes. No other method is

(a) CIS of an aromatic guest on the azoniacyclophane host **CP66** (in D_2O)

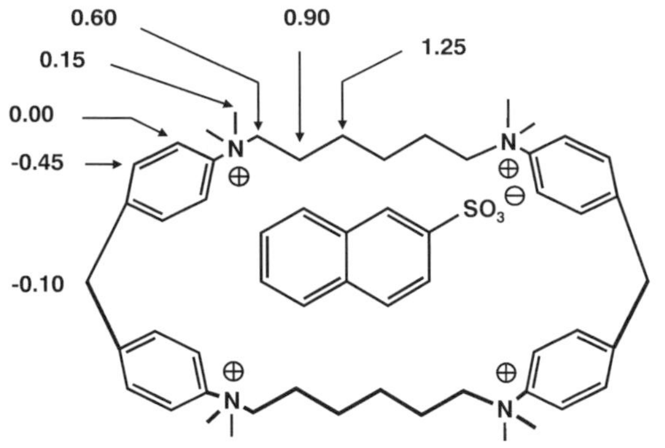

(b) CIS of **CP66** host on a steroid and on ADP as guest (in D_2O)

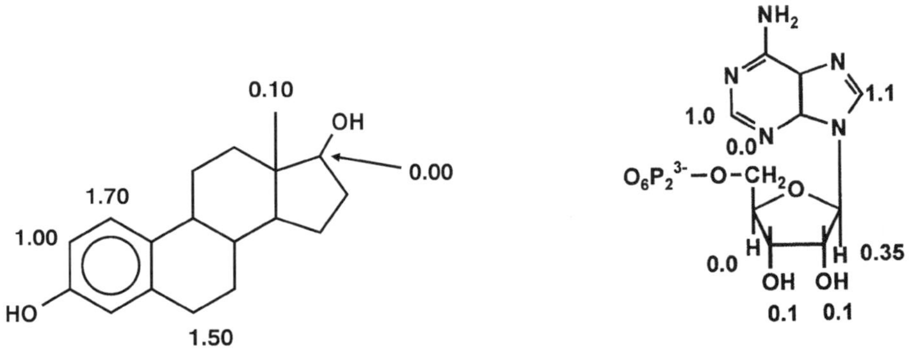

Complexes with cyclodextrins **CD** (in D_2O) :

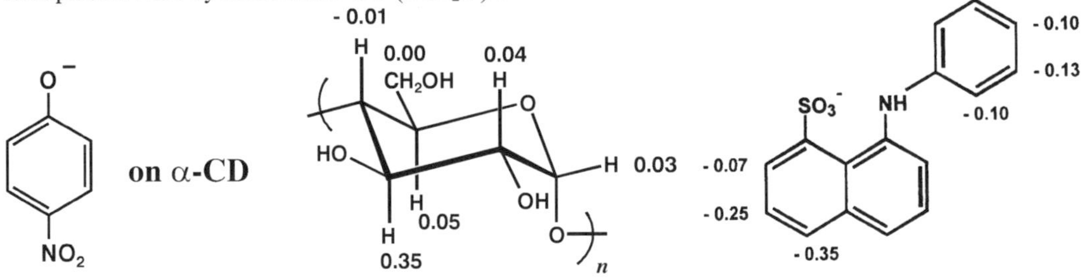

on α-CD

γ-CD (n = 8) **on** ANS as guest

Scheme E 2.   Some typical CIS values in supramolecular complexes.[23b] (all values in [ppm]; positive: upfield)

as simple and direct for providing experimental evidence for host–guest inclusion in solution, which often is taken granted without proof.

If no clear-cut decision based on ring-current induced shifts is possible, and in particular if better use shall be made of the available experimental CIS data, one should carry out calculations of the essential screening parameters for the possible host–guest complex geometries. Unlike X-ray analysis, but similar to other spectroscopic techniques such as electron diffraction, NMR-spectroscopy is a method in which quantitative structural analysis must always start with realistic model geometries. These are varied until one reaches agreement between calculated and observed shielding values. Computer aided molecular modeling (Section E 7) is now about the only acceptable way to generate the necessary number of conformations, and to judge at the same time whether these are not too far from the global energy minimum, or contain bad contacts between host and guest instead of the necessary fitting between complementary binding sites. Complexes like those with cyclodextrins, which are based to an essential degree on solvophobic forces, may require molecular dynamics simulations (Section E 7) in a water box; this has been instrumental in explaining conflicting results with respect to inclusion modes of aromatic substrates in cyclodextrins.[24]

Calculations of complexation induced shifts then is based on the application of shielding tensors,[23] which can contribute in such associations. Fortunately, high order shielding mechanisms due to, bond length or angle changes or to steric compression (*'van der Waals'* shifts) are typical for *intra*molecular effects only and can be neglected, as such energy-demanding distortions will usually not occur in host–guest complexes. The important contributions[23] then are anisotropy effects $\Delta\chi$, including aromatic ring currents, and linear electric field effects, LEF. The first ones, $\Delta\chi$, result from electric currents induced by the external magnetic field and therefore have the same absolute magnitude on all nuclei, depending only on their geometric situation with respect to the anisotropic function. For C=X functions (X=O, C etc.) these are calculated with equations after McConnell.[23]

$$\Delta\sigma = \frac{1}{12\pi} \sum_i \Delta\chi_i \frac{(1 - 3\cos^2\theta_i)}{\pi R^3}$$

where $\Delta\sigma$ is the shielding difference in [p.p.m.], $\Delta\chi$ the difference in bond susceptibilities perpendicular and parallel to the bond axis, θ the angle between inducing dipole and distance vector R, and point A the chosen center of the inducing dipole (see Fig. E 3 with the carbonyl unit as inducing anisotropic unit acting on a hydrogen atom H).

For aromatic units different methods can be used,[25] of which the Johnson–Bovey description using elliptic integrals for the π-electron double loops (Fig. E 4) has been found to be most suitable for the shift quantification of

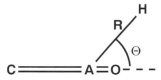

Figure E 3. Geometry description used for McConnell equation.

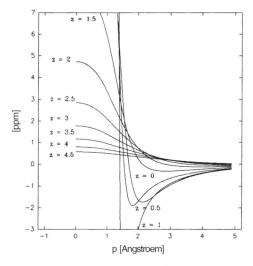

Figure E 4. The double loop description of the aromatic ring current induced by the external magnetic field, and nomograms for the evaluation of $\Delta\sigma$ (in [p.p.m.]) for nuclei at distances z and p (in Å) from the benzene ring center.

protons also above aromatic ring planes; this is a particularly frequent structural motif in host–guest complexes and also in biopolymers.[26] The corresponding equations[25] contain the radius a of the loop, the electron charge e and its mass m, the number n of π electrons, p and z as the polar coordinates of the observed nucleus with respect to the benzene ring center, θ as the spherical polar coordinate with the center of the loop, and the magnetic field constant μ_0

$$\Delta\sigma = \frac{\mu_0 n e^2}{4\pi 6\pi m a} \cdot \frac{1}{[(1+p)^2 + (z)^2]^{\frac{1}{2}}}$$

$$\cdot \left[K + \frac{1 - p^2 - z^2}{(1-p)^2 + z^2} E \right]$$

$$K = \int_0^{\frac{\pi}{2}} \frac{d\Theta}{\sqrt{1 - k^2 \sin^2\Theta}}$$

$$E = \int_0^{\frac{\pi}{2}} \sqrt{1 - k^2 \sin^2\Theta}\, d\Theta$$

$$k^2 = \frac{4p}{(1+p)^2 + z^2}$$

Tables or nomograms found in many NMR textbooks (Fig. E 4) can in principle be used for the evaluation of aromatic ring currents. However, such approximate methods can be misleading, in particular for nuclei which are in the vicinity of the shielding cone edges where shielding changes to deshielding. At these edges there is a steep and unsteady dependence between geometry and shielding (Fig. E 4); a geometric displacement of only 0.15 Å here can result in a shielding difference of approximately 0.7 p.p.m.

Linear electric fields exerted by full or partial charges through space polarize electron clouds; this leads to accumulation or depletion of shielding electrons around the observed nuclei. In Fig. E 5 an example is illustrated with the observed nuclei C and H of a C–H bond which is polarized by a charge q_i at an atom X. The electric field effects vary for different nuclei, being order of magni-

Figure E 5. Parameters for the calculation of electrical field effects.

tudes higher for C-13 or F-19 than for H-1. They are approximated by

$$\Delta\sigma_{\text{LEF}} = k \sum_i (P_{\text{CH}} l_{\text{CH}} r_i^{-2} q_i \cos\theta)$$

where q is the inducing charge, P_{CH} the polarizibility of the influenced C–H bond, l_{CH} its length, k a constant characteristic for the observed nucleus (e.g. 18 p.p.m. for protons, about 300 p.p.m. for carbon atoms); r and θ describe the geometry of the field gradient with respect to the polarized bond (see Fig. E 5), and point A the point of action of this vector on the polarized bond.

These semiempirical shielding calculations are also well suited for very large supramolecular complexes or even proteins,[27] and for the prediction of necessarily often very small intermolecular shift variations (Scheme E 2). In analogy to molecular mechanics calculations the errors in the predictions depend essentially on the parametrization of the factors in the used equations.[23,26,28] More rigorous quantum-chemical approaches aim at the *ab initio* prediction of usually larger shielding differences in smaller molecules. Based on Hartree–Fock approximations for solving the Schrödinger equation one calculates the energies of the molecular system and then the pertubation by an external magnetic field, using methods such as GIAO[29] (gauge including atomic orbitals) or IGLO[30] (individual gauge for localized orbitals). Calculational methods based on mechanical molecular models, or based also on the reading of ring anisotropy values from tables, can lead to substantial errors: these can be avoided by combining computerized application of the shielding equations with geometries generated by computer aided molecular modeling. Corresponding programs like SHIFT[26] (see also the website) also allow the evaluation of time-averaged NMR signals, and dynamic variations of the host–guest complex conformation.

Tables E 1 and E 2, and Scheme E 3, illustrate with some supramolecular complexes the possible agreement between experimental and calculated shifts, and at the same time the large and conformation-dependent contributions of linear electric field effects. Calculation of ring current effects alone can lead to fortuitously better agreement, but is obviously unrealistic at least in the presence of permanent charges in the molecules. Problems very similar to those which limit the value of molecular mechanics calculations

(Section E 7) are encountered in NMR shielding calculations: the reliability critically depends on the underlying parametrizations, the major problem being here also the choice of local charges and dielectrics. Furthermore, there is a multitude of possible optimal solutions on the hypersurface characterizing the shift–structure dependence. In the case of limited mobility between host and guest molecules, such as in complexes between phenylderivatives and a tightly fitting host like the cyclophane CP44 or α-cyclodextrin, one finds consistent CIS values for a variety of substrates.

The shifts observed in α-cyclodextrin complexes (Table E 2) invariably indicate the deep immersion of the aryl unit into the cavity, with the largest CIS on the H, 3 facing (by time average) more the aromatic shielding

Table E 1. Calculated (Eff_{calc}) and differences (ΔE) to experimental CIS values [ppm] for selected different positions of the guest TsOH in the CP44 cavity, see Scheme E 3. Contributions of ring current effects (Eff_{ar}), and linear field effects (Eff_{LEF}). The smallest ΔE of 0.3 ppm average is observed for $d = -0.69$ Å; with automatic fitting procedures typical ΔE values are around 0.2 ppm.[28b]

d^a	Guest proton	Eff_{ar}	Eff_{LEF}	Eff_{calc}	$\Delta E = Eff_{calc} - E_{exp}$
1.02	H_o	2.75	−0.11	2.64	1.07
	H_m	0.66	−0.43	0.23	−1.68
	Me	0.13	−0.22	−0.09	−0.74
$\Delta E_{average}$					1.16
0.89	H_o	2.13	−0.32	1.81	0.24
	H_m	1.68	−0.43	1.25	−0.66
	Me	0.34	−0.23	0.11	−0.54
$\Delta E_{average}$					0.49
0.53	H_o	2.69	−0.20	2.49	−0.92
	H_m	1.05	−0.44	0.61	−1.30
	Me	0.21	−0.22	−0.01	−0.66
$\Delta E_{average}$					−0.66
−0.69	H_o	1.41	−0.41	1.00	−0.57
	H_m	2.30	−0.41	1.89	−0.02
	Me	0.52	−0.22	0.30	−0.35
$\Delta E_{average}$					−0.35

^{a} Distance d in Å between center of CP44 cavity and guest phenyl moiety, see Scheme E 3. Positive values indicate that the SO_3 group is closer to the CP44 center than the CH_3 group. Experimental shifts in D_2O ($H_o = 1.57$; $H_m = 1.91$; Me = 0.65) and calculation details see ref.[28b].

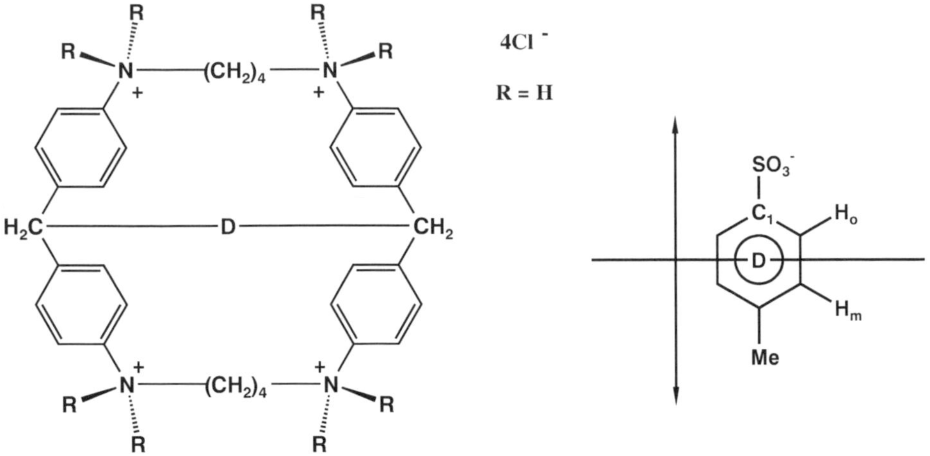

Scheme E 3. Complex of the azoniacyclophane **CP44** + *p*-toluenesulfonic acid (TsOH); for the evaluation of CIS values the center of the aromatic TsOH moiety is moved up and down, yielding differences *d* (see Table E 1) to the CP44 cavity center described by line D.[28b]

cone than the deshielding plane of the guest. As will be discussed in the next section, this geometry is furthermore quantified by NOE measurements.

A quite different view of shift changes is applied to hydrogen bond associations, where one can approximately judge strength by the deshielding of the proton involved in the bond[32] (Section B 3). The application of isotope effects on NMR shifts for the study of energy potentials and the identification of strong low-barrier hydrogen bonds (LBHB) is discussed in Section B 3. The location of the

proton between a COOH-donor and a nitrogen acceptor atom can also be inferred from ^{1}H and ^{15}N shifts as well as from changes in ^1JNH coupling constants and again isotope effects on shifts.[33]

E 4.2 NOE MEASUREMENTS FOR CONFORMATIONAL ANALYSIS OF SUPRAMOLECULAR COMPLEXES

Nuclear Overhauser effects (NOE)[34] provide significant information about interactions through space and association modes of

Table E 2. Cyclodextrin–aryl complexes.[a] Observed and calculated CIS values [ppm].[28b]

Proton	Eff_{ar}[b]	Eff_{LEF}[c] Gastei.	CT	CNDO	Eff_{calc}[d] Gastei.	CT	CNDO	Eff_{Lit}[e]	Eff_{exp}[f]
with p-nitrophenol									
H 3	− 0.36	− 0.06	− 0.41	− 0.16	− 0.42	− 0.77	− 0.52	− 0.26	− 0.35
H 5	− 0.03	− 0.04	0.48	− 0.11	− 0.07	0.45	− 0.14	0.08	− 0.05
with benzoic acid									
H 3	− 0.40	− 0.07	− 0.10	− 0.02	− 0.47	− 0.50	− 0.42	− 0.40	− 0.45
H 5	0.01	0.05	0.01	0.06	0.06	0.02	0.07	0.09	0.17

[a] Measurements in D_2O; [b] Eff_{ar} = ring current effect; [c] Eff_{LEF} = linear electric field effect; [d] Eff_{cal} = total calculated effect; [e] Eff_{Lit} = earlier calculations without LEF[31]; [f] exp. CIS; the different charge distributions used (Gasteiger, CT and CNDO) are discussed in Section E 7, see also Table E 5.

host–guest complexes. Besides the analysis of chemical shift changes NOEs therefore represent the most promising method for the structural analysis of supramolecular complexes and generally non-covalent interactions.[35] Unfortunately, several factors make NOE application not straightforward, and require a short discussion of the technical problems involved particularly with synthetic host–guest complexes. Thus, application of traditional NOE difference or of two-dimensional NOESY methods is limited by the unfavorable tumbling rates τ_c of compounds with molecular weight around 1000–2000 Dalton. The critical cross relaxation rate depends (besides on the viscosity of the solvent) on the product of τ_c and the spectrometer frequency, which for the necessary high signal dispersion should be around 500 MHz. In consequence, the conventional steady state NOE is positive only for molecular masses well below 1000, and sizeably negative for those well above 5000 Dalton, placing synthetic host compounds and their complexes near the zero transition between positive and negative NOEs. Spin-lock experiments[34] such as the rotating frame NOE (ROESY) or CAMELSPIN (cross-relaxation appropriate for minimolecules emulated by locked spins) overcome this serious limitation by application of an additional excitation pulse, relative to which one observes the then transverse instead of the conventional longitudinal magnetization enhancements. It should be mentioned that spin lock techniques require high stability, which is now standard with digital NMR instruments.

In the two-dimensional NOE experiment, which is necessary for the analyses of most multi-spin supramolecular complexes, the interesting parameter is the integrated intensity of the cross peaks between two protons. Such volume integration is possible with modern instruments and software and may be directly correlated with the internuclear distance, r, of the two observed protons via the known r^{-6} dependence. However, the underlying relaxation constants also depend on the tumbling and therefore also the shape of the molecule. Spin diffusion and contributions from third nuclei as well as processes other than transverse relaxation during spin-locking also complicate a rigorous quantitative interpretation. More quantitative studies may require variation of solvent viscosity, of the magnetic field dependence for securing a dipole–dipole relaxation mechanism etc. Measurements of build-up rates and dynamic coupling to molecular dynamic simulations, taking into account multiple conformations and different correlation times, are becoming standard with biopolymers,[36] but are quite limited with synthetic host–guest complexes due to their typically very small NOEs. In such complexes the NOE intensities are often weakened by rapid exchange between many binding modes with similar internucleus distances. With spin-lock techniques like ROESY, which are a must for medium-sized synthetic host–guest complexes, the cross peak intensities are more blurred by possible artifacts (see below) than the integrals observed in normal NOEDIF (NOE-Difference) experiments, or in two-dimensional NOESY spectra which are commonly applied to biopolymers. The integrals yield at least semiquantitative information about internuclei distances, or allow one to introduce upper limits for distances between protons separated by more than about 4 Å, which lead to the absence of NOE signals. By comparison between fixed intramolecular distances, r, preferably between vicinal aromatic or olefinic protons one can in principle deduce also intermolecular distances with some confidence. Table E3 illustrates the size of intermolecular cross peaks in a cyclodextrin complex; in view of the possible artifacts (see below) they represent only approximate integrals. In most cases one has to settle with designation like large, medium, small or no NOE; usually the NOE serves as cut-off constraints in computer aided molecular modeling (Section E7).

Table E 3.   Relative ROESY Integrals (in [%]) for the complex between α-CD and para-iodopheno-late, and distances [Å] from molecular mechanics calculations (CHARm) in italics.[28b] The NOE between the vicinal protons H_o and H_m ($r = 2.48$ Å) is set equal to 100%.

Proton	H_m		H_o	
CD-H-3	70 %	*2.32 Å*	35 %	*3.00 Å*
CD-H-5	59 %	*2.37*	<2 %	*>4.0 Å*

In Fig. E 6a,[37] which is an extract of the cross peaks from intramolecular NOEs with cyclo-dextrin protons, we illustrate some common artifacts in ROESY spectra which need to be taken into account before one relies on NOE data for structural analyses of these medium-sized molecules. The 2D matrix in Fig. E 6 a is essentially free from so-called T-noise which usually is recognized by longer traces instead of cross peaks; it shows, however, antiphase or 'negative' peaks not only at the diagonal (which cannot be used anyway) but also at the cross peaks due to NOE between the vicinal H 2 and H 3 protons (denoted 23 and 32). In the black-and-white representation this can be seen only by the presence of open contours instead of those which are filled by the positive and superimposed NOE areas as seen for the 46/64, 35/53 and 56/65 cross peaks. The reversed sign of such 'negative' peaks, which in the computer output of the 2D-NOE matrix are more easily recognized by their different color, arises from so-called J-cross talk due to magnetization transfer between several spin-coupled protons. Another artifact results from TOCSY transfer from H 3 to the vicinal H 2, which then transfers to H 1 which is also vicinal to H 2 (denoted in Fig. E 6 a T231) and from H 5 to H 4 to H 1 (T451). These artifacts can be reduced by T-ROESY and by reducing the irradiation power, which, however, reduces then the anyway often weak cross peak intensities. The intensity of ROESY peaks also depends on the field lock frequency used.

Obviously, experience and chemical reason-ing is needed for the assignment and inter-pretation of such ROESY cross peaks.

Nevertheless, the cross peaks observed for the interaction between benzoic acid and α-cyclodextrin in the corresponding area of the same ROESY experiment yield a clear picture of the inclusion mode. In Fig. E 6 b[37] we see strong cross peaks between the phenyl ortho proton H_o and both the H 3 and the H 5 protons of the host, indicating that both must flank the o-proton. In contrast, H_m shows only interaction with the cyclodextrin H 3, which even exhibits a small signal at the H_p position. All this is only compatible with an insertion of the benzoic acid with the carboxylic group 'head-on' inside and towards the narrower rim of the cavity.[38] As illustrated with the last figure of this chapter in the problems section the situation changes significantly upon ioni-zation of the carboxylic group.

Cyclodextrin complexes with disubstituted phenyl derivatives can occur in three different inclusion modes (Fig. E 7). Type I has a para-substituted phenol partially immersed into the cavity, leading to sizable contacts only between the guest H_o and the host H 3 proton. Type II with deeper immersion has the similar contacts between H_m and the CD protons H 3/H 5, whereas H_o can only 'feel' H 3. Type III has the opposite immersion mode with the phenolic group inside the cavity. Obviously, all three modes should lead to quite different NOEs and can be distinguished this way. Volume integration of the cross peaks shows intensities which, if one sets the intramolecu-lar NOE between the vicinal protons H_o and H_m equal to 100%, semiquantitatively corre-late with distances calculated for mode I by molecular mechanics (Table E 4). The data in Table E 4 indicate for all parasubstituted phenyl derivatives an inclusion mode between type I or II, clearly with the more hydrophilic group outside the cavity.

Hydrophobic instead of a dispersive bind-ing mechanisms are expected to lead to less tight intracavity binding. This is indeed borne

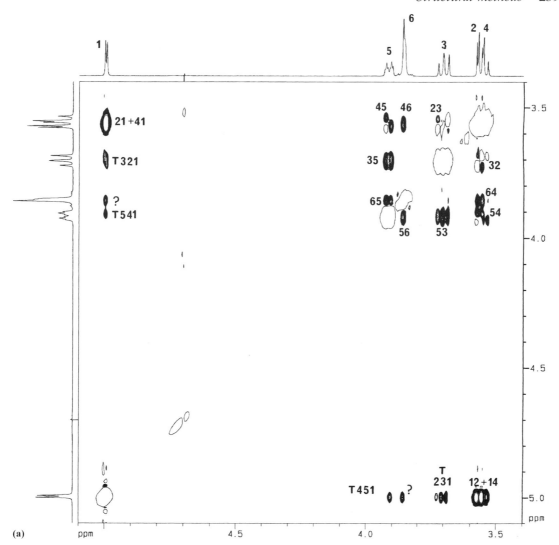

Figure E6. (a) T-ROESY matrix of the complex between α-cyclodextrin (4×10^{-3} M) and benzoic acid (5×10^{-3} M) in D_2O at 300 K; complexation degree 51% for the acid and 63% for CD ($K = 720\,M^{-1}$); aliphatic part (intramolecular interactions between the CD protons, other explanations see text). Measuring parameters: pd 1.6; Acquisition: spectral width 3000 Hz, 2 K/360 data points in F2/F1 direction, 90° pulse width 10.3 μs, pulse delay 1.8 s, spin-lock time 800 ms (3000 times $2 \times 180°$ pulses, duration of a 180° pulse 250 μs; power attenuation 26 dB below maximum power output, corresponding to a spin-locking field of 2 KHz), 16 scans, 16 + 4 dummy scans; a homospoil pulse of 5 ms was used for reduction of T1-noise, presaturation of the water signal; processing: 1K/1K with cosine window function shifted by $\pi/2$; positive T_1 noise subtraction and automatic polynomial base line correction; no symmetrization.[37]

out by ROESY measurements with complexes between cyclodextrins and less polarizable substrates.[39] The α-CD intracavity protons must show intermolecular NOEs with the

phenyl part of any substrate in view of the host–guest size matching. The wider cavity of β-CD would match the naphthyl part of the dye, shown in Fig. B10 which, however, in

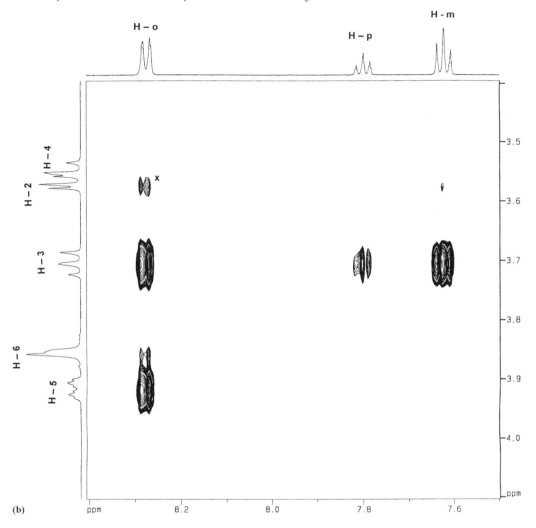

Figure E6.   (b) Intermolecular interactions in the T-ROESY matrix of the complex between α-cyclodextrin and benzoic acid, for conditions see Fig. E6(a).

contrast to the phenyl residue shows no intermolecular NOEs. Even the large γ-CD shows cross peaks not only with the naphthyl, but also with the phenyl part. Exactly the same conclusions emerge from an analysis of the complexation induced shifts, and demonstrate that strong complexation may require loose binding in the case of predominantly solvophobic interactions (see section B6).

In the transferred NOEs (trNOE or TNOE) technique[40] the NOE cross-relaxation between two protons of the complexed ligand is transferred to the uncomplexed ligand, yielding a negative NOE of a magnitude depending on the cross relaxation rate within the complex. If the exchange rate between free and complex is much faster than the relaxation rate of the free proton, and if the population of the free ligand is within certain limits determined by the cross-relaxation rates[40] one can obtain ratios of internuclei distances from trNOE experiments.

Table E4. NOE (ROESY) effects in α-cyclodextrin (CD) complexes of phenylderivatives with substituents X and Y in *p*-position[a] (compare CIS values in Table E2).[28b]

X =	O^-		OH		O^-		OH		CH_2CO_2H	
Y =	NO_2		NO_2		I		I		I	
H[a]	o	m	o	m	o	m	o	m	ar	CH_2
H3(CD)	+	+	+	+	+ +	+ +	+	+ +	+	
H5(CD)	−	+	−	+	−	+ +	−	+ +	−	−

[a] observed proton; + + = strong; + = medium; − = no NOE-cross peaks; conditions: [α-CD] = 5 mM, [Phenylderivative] = 10 mM.

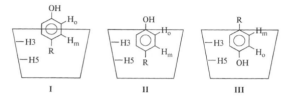

Figure E7. Binding modes of para-disubstituted phenyl compounds in cyclodextrins, with interactions sites of the protons observable by NMR-spectroscopy.

E4.3. SOLID STATE NMR-SPECTROSCOPY

NMR analyses in the solid state, usually relying on the observation of 13-C signals, are possible with cross polarization (CP, from protons to less sensitive nuclei like carbon)– magic angle spinning (MAS) techniques. Fast spinning of the solid sample around an axis which has an angle of 54.7° with respect to the external field reduces the direct dipolar coupling between nuclei so much that one can observe under proton broad band decoupling high resolution signals of nuclei like carbon atoms, which due to their spin quantum number of $I = \frac{1}{2}$ show spectra similar to broad-band proton-decoupled spectra in solution. Another nucleus frequently used in solid state investigations is deuterium, which as result of large quadrupole splitting also yields important information on dynamics in the solid state. For details on these more special methods we refer to NMR textbooks and special reviews.[41] An interesting solid state application of cesium-133 NMR-spectroscopy is the identification of a Cs^+ inside an

18-crown-6 ether either as electride (with a 'free' electron as gegenion), or a cesium anion (a so-called alkalide) as gegenion, with a shielding difference of around 200 ppm between the cesium ions.[42] Crystals of cyclodextrin hydrates showed shielding differences at C1 and C4 to solution spectra indicative of conformative changes in the 1.4-linkage.[43] Deuterium NMR quadrupole splitting shed light on the orientation of azodyes in solid cyclodextrin complexes.[44]

E5. CHIROPTICAL METHODS FOR STRUCTURE ELUCIDATION

Complexes in which either host or guest, or both are chiral lend themselves not only to evaluation of binding constants by measuring changes in optical activity (Section D3.3), but also to conformational analyses by observation of circular dichroism (CD)[45] changes. The effects are particularly strong if one of the partners, or both contain aromatic chromophores; the induced CD (ICD) upon complexation can be significant and shows regular pattern associated with different complex conformations. Figure E8 illustrates a Cotton effect decrease by complexation of a chiral host **E-1** with spherical, cavity filling guests like 1-trimethylammonium-adamantane at all wavelength, whereas flat naphthalene derivatives lead to a Cotton effect increase at low wavelength.[46] This is in line with a more rhomboid complex conformation of C_2 symmetry in the first case, and elongated thoroid conformations of D_2 symmetry in the second

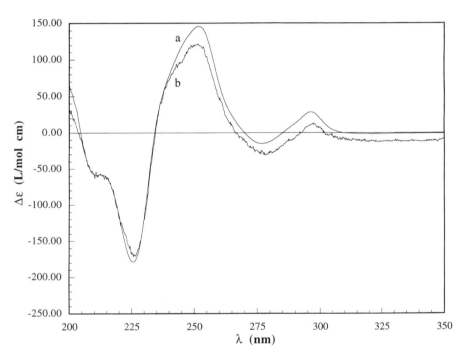

E-1 (X=H, Sp= 1,4-phenyl-)

Figure E 8. CD spectra of a macrocyclic host **E-1** and *N*-methylquinolinium chloride (upper part) and with 1-trimethylammonium-adamantane (lower part); trace a: host alone; trace b: host and guest, in water at pH 9.[46] Reprinted with permission from *J. Am. Chem. Soc.*, Forman *et al.*, 1995, **117**, 9213. Copyright 1995 American Chemical Society.

case, which were also deduced from X-ray and computer aided molecular modeling studies. Quantitative interpretations of ICD are based on the coupled oscillator mechanism or exciton optical activity[45a] and require analysis of all, in solution spectra often heavily overlapping transitions of the different single chromophores, as well as identification of the orientation of the transition dipole moments. The dipole–dipole interaction between transition moments of a guest and a host molecule can be quantified by the Kirkwood–Tinocco equation.[45,47] In large aromatic host systems like the one in Fig. E 8 a quantification can be tried with semiempirical molecular orbital calculations, needing here consideration of at least 19 $\pi–\pi^*$ transitions, neglecting all $n–\pi^*$ transitions.[46]

The CD method is most often applied to cyclodextrins, where the higher energy σ-bond oscillators of each glucose unit result in an induced oscillator which lies close to a line dissecting the C3 and the O5 atoms.[48] If for an aromatic guest molecule the orientation of a transition has an electric dipole moment parallel to the z-axis of the macrocyclic plane a positive CD is expected, if the transition is within the x, y plane of the macrocycle a negative CD will result.[48] In view of the almost prohibitive rigorous analyses of ICD with larger and anisotropic host–guest complexes, a semiquantitative scheme was proposed to deduce not only sign but also the approximate magnitude of the ICD effect in cyclodextrin complexes with aromatic guests. In the sector diagram (Fig. E 9 a) electric moments of the guest's transition along the macrocycle z-axis produce a positive ICD

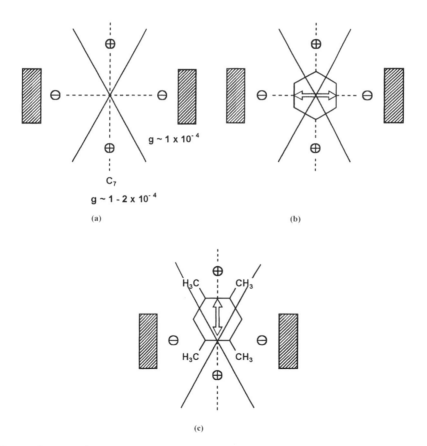

(a)

(b)

(c)

Figure E 9. Sector diagram for sign and magnitude of ICD (characterized with $g = \Delta\varepsilon/\varepsilon$) in cyclodextrin complexes.[49] The rectangles symbolize the β-cyclodextrin cavity wall (C7-symmetry); the + and − indicate the sign of the Cotton effect for electric moments of the observed guest transition falling into the corresponding sectors, the dissecting lines the borderline with $g = 0$, and the open arrows the orientation of the electric moments of the observed guest transition.[49] Reprinted with permission from *Acta Chim. Sci. Hung.*, Kajtar, M. *et al.*, 1982, **110**, 327. Copyright 1982 Akademiai Kiadó.

which is characterized by an anisotropy number $g = \Delta\varepsilon/\varepsilon = (1 \text{ to } 2) \times 10^{-4}$; orientation in the x, y plane lead to negative ICD of similar magnitude, whereas at a tilt angle of 30° between the guest moment orientation and the z axis yields an effect close to zero.[49] Monosubstituted benzene derivatives have transitions with electric moments both parallel and perpendicular to the α-cyclodextrin plane if they are included with the substituent outside or at the rim of the cavity; correspondingly one observes negative Cotton effects with g around 1×10^{-4} at longer wavelength, and positive Cotton effects with slightly higher g values at shorter wavelength (Fig. E 9 b).[49] In contrast, durene (1,2,4,5 tetramethylbenzene) cannot fully immerse even in the wider β-cyclodextrin cavity; the resulting conformation (Fig. E 9 c)[49] leads to a weak positive Cotton effect. Similar rules can be extended to chromophores outside the cavity and applied not only to cyclodextrins but also to other chiral macrocycles; generally the sign of the ICD reverses if a guest chromophore moves out of the cavity. Similar rules have been developed for the ICD prediction of supramolecular complexes with helical host molecules like nucleic acids.[50,51] Such schemes may be reminiscent of the well known octant rule used for conformational and configurational analyses within organic frameworks. Such rules, however, are more difficult to apply in view of the necessary assignments of electric dipole moment orientation for each observed transition frequency, and nevertheless lead usually only to semi-quantitative results.

E 6.   VIBRATIONAL SPECTROSCOPY

The use of vibrational spectra for the evaluation of equilibrium constants has already been discussed in Section D 3.1. The problems of the usually-overlapping bands of host and guest and their complexes persist in the application of IR and Raman techniques to structural analyses. Computerized deconvolution and spectra subtraction techniques are of great help but as for all spectra, manipulations are fraught by the generation of possible artifacts. Vibrational spectroscopy has the advantage of being applicable to complexes in gas, liquid and solid state. The extraction of structural information, however, is in most cases restricted to indirect and more qualitative data due to broad and overlapping bands as result of many vibrational levels, and the complexity of supramolecular structures.

Vibrational states of simple molecules, particular those of high symmetry, can in principle be rigorously quantified with normal coordinate analysis (NCA) using valence force fields.[52] In the field of supramolecular chemistry the method has until now been restricted to highly symmetric metal complexes with crown ethers.[53] The most interesting spectral changes are those reflecting directly the non-covalent metal–(crown ether–) oxygen interactions, also as they offer a hitherto largely unexploited experimental access to the underlying force constants. Due to the relative weakness of these interactions the absorptions are in the low near infrared (NIR) region, and require special NIR instruments. Bands at around 1100 cm^{-1} are due to vibrations of the ethyleneglycole units, those below 400 cm^{-1} to metal vibrations in the complex.[53] Assignments of the observed vibrations need, as a rule, measurements also with isotopically substituted complexes, like ^{40}Ca and ^{44}Ca salts etc. This way the D_{3d} symmetry of 18-crown-6 complexes, and a C_{2h} symmetry for 1,10-diaza 18-crown-6 complexes with calcium was established, as expected with a distinctly shorter wavelength for the metal–N in comparison to the metal–O interaction.[54]

For the investigation of solid host–guest complexes Raman spectroscopy has the advantage of not requiring special sample preparations; this is different for transmission techniques like conventional IR, UV/vis etc. Modern reflectance methods including FT IR overcome these limitations, which allows easy access to the study of solid complexes.[55] Qualitative conclusions with respect to binding mode and mechanisms in such complexes can be drawn from the observed spectral differences between 'free' (gas or solution) and bound states of guest molecules. As often in vibrational spectroscopy, one concentrates on key bands, such as those of carbonyl groups. Figure E 10 shows corresponding Raman spectra of solid complexes between various aldehydes and β- cyclodextrin,[56] with carbonyl frequencies $\nu_{C=O}$ around $1700\,\mathrm{cm}^{-1}$, $\nu_{C-C\,aryl}$ around $1600\,\mathrm{cm}^{-1}$, and $\nu_{C=C}$ around $1620\,\mathrm{cm}^{-1}$. Complexation induces, for example, a positive $\nu_{C=O}$ change by $6-12\,\mathrm{cm}^{-1}$, increasing in the order benzaldehyde < cinnamaldehyde < vanilline, indicative of the expected increasing participation of the carbonyl group as hydrogen bond acceptor in the complex in analogy to corresponding solvent effects.

The strongest spectral changes by non-covalent interactions are, of course, expected for stretching absorptions of donor-H bonds involved in hydrogen bonding. These frequency changes are therefore most often used, not only for the evaluation of binding constants (Section D 3.1) but also for getting insight into complexation modes. Hydrogen bonding results in weakening of the donor-H bond with concomitant shift $\Delta\nu$ to smaller frequency and broadening of the IR bands, for the D–H stretching and, to a lesser degree, also for the bending modes. As mentioned in Section B 3 there is no reliable correlation between the hydrogen bond length and its strength, correspondingly there is only a trend to larger $\Delta\nu$ values for stronger interactions.[57] Nevertheless, the observation of D–H regions in IR spectra is an important diagnostic tool which has been applied in particular to interactions involving amide functions. For oligo- or polyamides reliable informations can be better deduced from non-covalent interactions in so-called β-turns or hairpins than from truly intermolecular systems, which usually are too complex for a

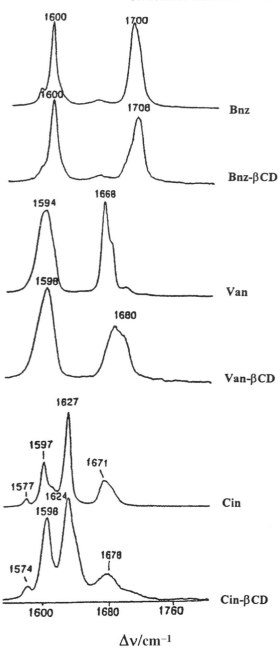

Figure E 10. Solid state Raman spectra of benzaldehyde (**E-2**, Bnz), cinnamaldehyde (**E-3**, Cin) and vanillin (**E-4**, Van); neat and complexed with β-cyclodextrin.[56] Reprinted with permission from *J. Carbohydrate Chem.* Moreira *et al.*, 1995, vol. **4&5**; p. 677, fig. 1; Copyright 1995 Marcel Dekker, Inc.

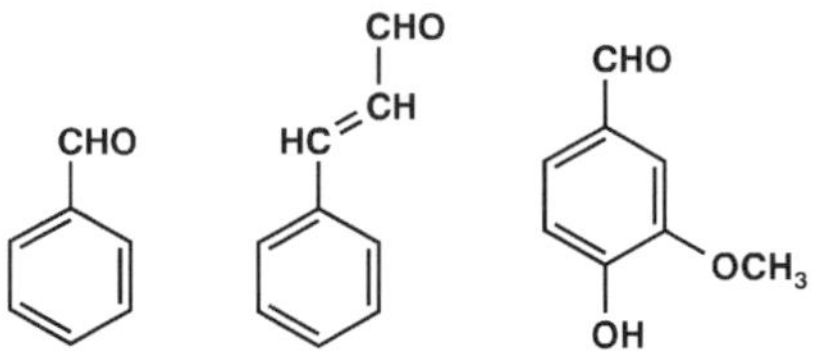

CHO

CHO

CHO

HC
CH

CHO

OCH₃

OH

(Bnz)   (Cin)   (Van)

E-2   E-3   E-4

$3300\,\text{cm}^{-1}$ of the urea derivative **E-6** (Fig. E 11 b) in chloroform is much more intense than for **E-5**, (Fig. E 11 a), indicating a clear preference for a 9-membered hydrogen bonded ring structure in comparison to a 10-membered ring, in line with NMR analyses.[58b] Convergent preorientation with more rigid spacer between amide chains as with structure **E-7** leads to rather stable 15-membered

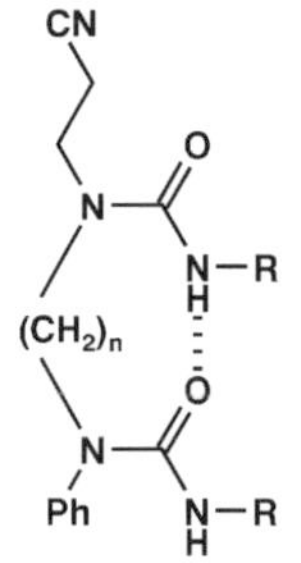

rigorous analysis. Structures like **E-5** to **E-8**, sometimes called molecular scaffolds, offer an opportunity to study also by IR techniques such interactions which are typical also for proteins.[58] Generally, N–H stretch bands around 3330–$3340\,\text{cm}^{-1}$ are observed in a solvent like dichloromethane or chloroform and are indicative of amide–amide hydrogen bonds, whereas free amides absorb around 3440–$3360\,\text{cm}^{-1}$.[58a] The N–H stretch bands at

CN

N

O

N–R

(CH₂)ₙ

N

Ph   O

N–R

E-5 (n=3); **E-6** (n=2)

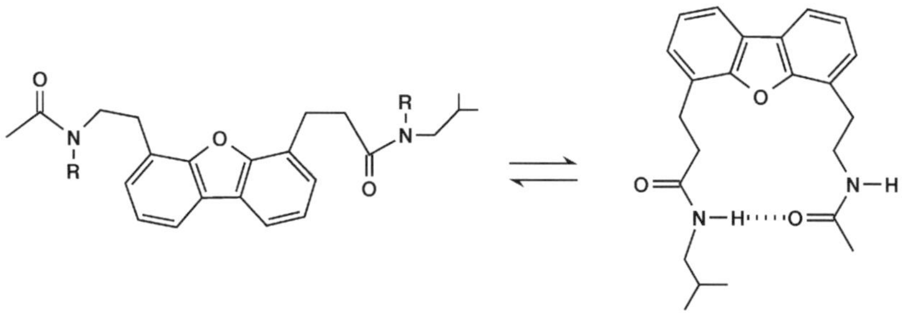

E-7   (R = H or R = C₂H₅)   (for R = H)

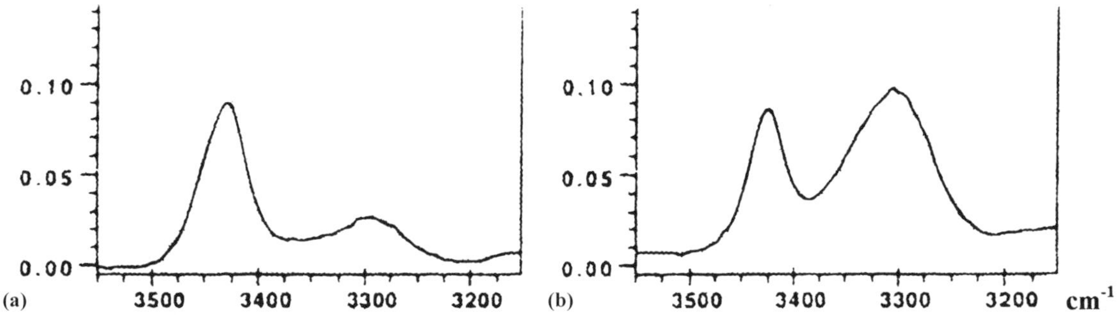

Figure E 11.   IR spectra in the N–H stretch band region of **E-5** (a) and **E-6** (b); in chloroform with solvent IR spectrum subtracted.[58b] Reprinted with permission from *J. Am. Chem. Soc.*, Nowick *et al.*, 1995, **117**, 89. Copyright 1995 American Chemical Society.

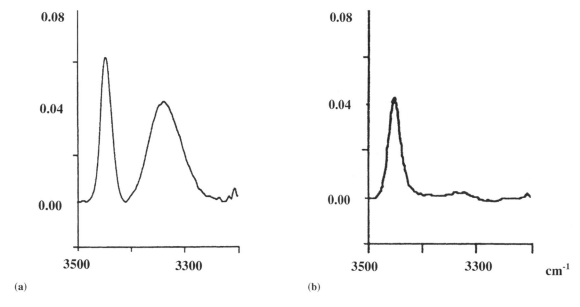

(a)　　　　　　　　　　　　　　　　　　(b)

Figure E 12. IR spectra in the N–H stretch band region of **E-7** with R = H (a) and **E-7** with R = C₇H₅ (b); in dichloromethane with solvent IR spectrum subtracted.[58c] Reprinted with permission from *J. Am. Chem. Soc.*, Tsang *et al.*, 1994, **116**, 3988. Copyright 1994 American Chemical Society.

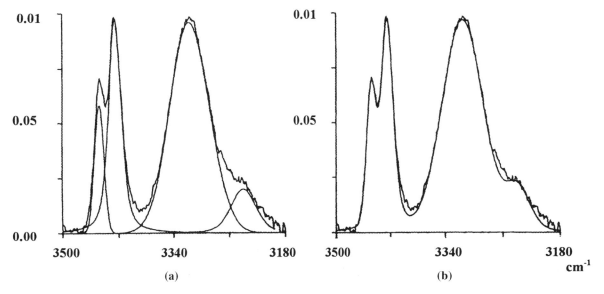

(a)　　　　　　　　　　　　　　　　　　(b)

Figure E 13. Illustration of deconvolution techniques with IR spectra in the N–H stretch band region of **E-8** (in dichloromethane, solvent IR subtracted); (left side, a): four calculated bands (maxima at 3448, 3425, 3316 and 3240 cm⁻¹) superimposed on the observed spectrum; (b): calculated and observed spectrum.[58d] Reprinted with permission from *J. Am. Chem. Soc.*, Dado *et al.*, 1993, **115**, 4228. Copyright 1993 American Chemical Society.

E-8

rings, as visible in the N–H stretching frequencies of the secondary amide with R = H in comparison to the one with R = C$_2$H$_5$ (Figure E 12 a *vs.* b).[58b] The presence of three amide functions in an open chain molecule like **E-8** leads already to many overlapping IR bands, which only tentatively can be assigned to single frequencies using deconvolution techniques illustrated with Fig. E 13.[58d] In all these cases *inter*molecular associations, which would complicate the IR pattern furthermore, were excluded by the low concentrations used and the weak complexation constants for such species ($K < 100 \, \text{M}^{-1}$, Section B 3).

E 7. COMPUTER AIDED MOLECULAR MODELING

Molecular simulations[59] are becoming a standard method in supramolecular chemistry for several reasons. Most obviously, the geometric fit between several molecules, which is the heart of molecular recognition, can be easily controlled by these now highly developed techniques. In contrast to molecular orbital methods, molecular mechanics calculations have the advantage of giving a direct picture of the interacting groups; they provide a better and practical lead for the chemist to understand and to design supramolecular complexes. In the early days of supramolecular chemistry mechanical molecular models were often used; these, however, present a particular problem for the evaluation of van der Waals contacts with space-filling representations. Furthermore, it is difficult to quantify with mechanical models for instance torsional angles, which by induced fit can change with small strain energies in order to allow optimal non-covalent interactions. The same holds for intermolecular distances,

which have a decisive influence on the strength of the supramolecular complexes, in particular with dispersive forces falling off steeply with distances.

Another incentive for using numerical models is the possibility of controlling the structures by experimental methods such as NMR-spectroscopy or X-ray diffraction. As pointed out in Section E 4, any quantitative use of NMR-spectral data must be based on calculated molecular models. Finally, computer visualizations of supramolecular structures have a strong esthetic appeal, besides fulfilling the obvious need to appreciate and understand their three-dimensional architecture. Even less expensive molecular mechanics programs usually provide for graphic representations, or can read in either atomic coordinates, including those from databases such as the Brookhaven Protein Data Bank (for biomolecules) or the Cambridge Crystallographic Database (mostly for smaller molecules, see website addresses in the internet). Many programs allow one to build structures on the computer screen and to convert the two-dimensional formulae into three-dimensional representations. Care, however, must be taken to check this conversion process, as it can be the source of mistakes particularly with chiral centers. Visualization software, often freely available, not only provides colorful static pictures, but also allows various animations which can greatly help comprehension of large supramolecular complexes.

Force field methods and in particular the potentials describing intramolecular interactions, are elaborated in specialized textbooks or reviews.[59] We will discuss here only features which are of special interest for supramolecular associations. Of course computer aided molecular modeling invariably is based on the minimization of potential *energies*. Nevertheless, we discuss these methods here as in practice their major application in supramolecular chemistry involves the evaluation of *structural* features including ligand or host design. The simulation of intermolecular forces is fraught by more difficulties than those involving covalent bonds. The potential energy, EQ, due to interactions between permanent charges q_1, q_2, etc., separated by distances r, is described with the Coulomb potential, equation (E 7-1). Interactions with dipoles with the moment μ are evaluated with equations (E 7-2 a) with angles as indicated in the figures, when the charge is at large distance from the dipole; alternatively one describes the dipole or polypole by point charges and then uses equation (E 7-2 b). The dipole–dipole interactions can be evaluated also by summation of point-charge interactions, or by equation (E 7-2 c).

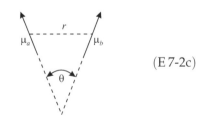

(E 7-2 a)

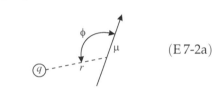

(E 7-2 b)

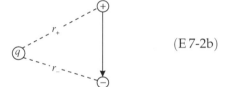

(E 7-2 c)

for pole–pole $EQ = \dfrac{q_1 q_2}{r\varepsilon}$ (E 7-1)

for pole–dipole $EQ = \dfrac{q\mu}{r^2 \varepsilon} \cos\phi$ (E 7-2 a)

or $EQ = \dfrac{q}{\varepsilon}\left(\dfrac{q_+}{r_+} + \dfrac{q_-}{r_-}\right)$ (E 7-2 b)

for dipole–dipole $EQ = \dfrac{\mu_a \mu_b}{r^3 \varepsilon} f(\theta)$ (E 7-2 c)

As discussed in Section B 2 hydrogen bonds $D(\delta-)\cdots H(\delta+)\cdots A(\delta-)$ in supramolecular systems are usually also described by Coulomb potentials with appropriate negative point charges at the donor and acceptor atom, and a positive charge at the hydrogen bridge atom. Some force fields are based also on the angles between the D, H and A atoms, or use modified 6–12 potentials (see equation E 7-4a); these can reflect orbital overlap contributions which, however, are usually small in weaker 'organic' hydrogen bonds.

Non-covalent forces beyond ion–ion interactions (equation E 7-1) are often referred to as *van der Waals* forces (Sections B 4, B 5). These include high order interactions based on still permanent charges such as those between two dipoles (equation (E 7-2c), or for instance those with quadrupoles, for which we indicate here only the increasingly steep fall-off with the distance, r (equations (E 7-3 a, b, c)).

A quadrupole (e.g. benzene) interacting with:

(a) a mono–pole $E = f(r^{-3})$ (E 7-3 a)

(b) a dipole $= f(r^{-4})$ (E 7-3 b)

(c) another quadrupole $= f(r^{-5})$ (E 7-3 c)

$$[\oplus\ \ominus\ \oplus\ \ a\ quadrupole]$$

Dispersive (London–Eisenschitz) interactions due to fluctuating dipoles like those between noble gases or between weakly polar C–H bonds (Section B 5) are usually quantified by

$$EV = ar^{-12} - br^{-6} \quad \textit{Lennard-Jones Potential}$$

(E 7-4 a)

or

$$EV = 10^4 e^{-4.6r} \quad - 50\,r^{-6} \quad \textit{Buckingham Potential}$$

$$\uparrow \qquad\qquad \uparrow$$

repulsive   attractive   term

(E 7-4 b)

In equations (E 7-4 a) and (E 7-4 b), the first term represents the repulsive potential and the second term describes the attractive potential.

These potentials are in their most simple form; in more elaborate force fields they are extended by additional terms in order to take care of their anharmonic nature.

The reliability of force fields depends essentially on the underlying parametrizations, and unwary users may easily be misled to extend potentials to molecules that have not been tested rigorously before. Best results can be expected from parametrizations that are based on a large variety of experimental data, including heats of formation, conformational equilibria and barriers, dipole moments, vibrational spectra, etc. For reliable simulations of intermolecular interactions advanced force fields like MM3,[60] OPLS,[61] AMBER,[62] or CHARMm[63] also partially use heats of sublimation and of solvation, liquid densities, and geometries in different states of matter for the parametrization of van der Waals forces. Another increasingly important approach is parametrization with the help of a large basis set molecular orbital calculations. With smaller molecules such calculations, including electron correlation, are affordable for many functionalities occurring in supramolecular structures, which themselves are too large for quantum chemical calculations.[64] Force fields often use different potentials, even though the calculational results at least with respect to geometries agree to a large extent. Thus, MM2 and MM3 uses an all-atom description as do most other fields, with the exception of OPLS which treats $C-H_n$ groups as one entity. As mentioned, electrostatic interactions are usually evaluated with atomic point charges, whereas MM3 is based on dipole–dipole calculations. Van der Waals forces are usually described with the Lennard–Jones 6–12 potential, but MM3, directed more towards small molecules, uses the more time-consuming exponential function of the Buckingham equation.

Before applying a particular force field, one is well advised to check with suitable model molecules whether it shows realistic results not only in terms of structures obtained by energy minimization, as this may to a large degree just reflect either the input geometry, or a normalized one with undistorted lengths and angles. Correct geometries may be obtained with quite unrealistic force constants as long as the bond and angle values are set to undistorted standard values. A better benchmark test is the comparison with experimental *energy* values, such as conformational equilibria of the compounds one wants to model. One should keep in mind that the basis of computer aided molecular modeling is *energy* minimization, and that applications in supramolecular chemistry rely also on energies expected from complex formation. Important information can be obtained with respect to strain energies eventually required to attain an optimal fit between the binding sites.

The computational analyses of most noncovalent interactions rely on the proper choice of local charges and dielectrics. Although only equation (E7-1) contains such a term ε explicitly, the interactions described by equations (E7-2) and (E7-3) also depend on the local dielectrics. The ambiguity involved in the description of the local dielectrics poses a particular problem for the simulation of supramolecular complexes, as local charges, polarizibilities and dielectric constants in and around polar molecules dominate their associations and cannot be observed directly. Instead, they can be derived from quantum chemical methods, calculating the electrostatic potential at many points around a molecule or, simpler and older, the atomic orbital overlap elements for Mulliken populations.[65] The frequently used Gasteiger atom charges rest on orbital electronegativities based on ionization potentials and electron affinities.[66] Semiempirical molecular orbital calculations can also be fitted to experimental data, leading to more efficient evaluation of charges in larger systems.[67] Depending on the calculational method, the atomic charges vary substantially; we illustrate this in Table E5 with values for *N*-methylacetamide which represents a building block common in many supramolecular and biological molecules. Note, that evaluations including polarization functions and solvation by water[68] (entries FQ and PSGVB in Table E 5) lead, as expected, to significantly higher charges. There is some obvious ambiguity in the choice of charge distributions and of the dielectric constant, ε, used. The latter can be used as a constant, for example a value of $\varepsilon = 2$ has been found to be practical for interactions in and with most organic molecules. Often one applies a distant-dependent function of ε in view of a ε change with increasing distance from the (local) charge. The variations of parameters can lead to unrealistic energies, as exemplified by the E/Z conformational equilibria around the C–N bond in amides where it has been shown that some force fields as well as the semiempirical molecular orbital method AM1 furnish quite wrong predictions.[69]

Stacking interactions (Section B 5) are usually simulated with potentials similar to equation (E 7-4), where parametrization of the 'softness' coefficients mostly stem from liquid simulations, but can be a source of considerable variations in the energetic results. Nucleobase and nucleoside interaction energies obtained with frequently used force fields

were compared with predictions from large basis set calculations (MP2 6-31G*),[73] and also with X-ray-derived geometries. The results also reflect the simulation reliability of the Watson–Crick hydrogen bonds, and correlate best with data from the AMBER4 force field. As expected smaller variations are seen in the van der Waals radii, R, which are generally close to the normal value for *van der Waals* contacts between the bases. As discussed in Section B 5 and B 6, the driving force for stacking can be to a large degree hydrophobic; in molecular modeling this is often taken care of by attributing a value of 0.2–0.4 kJ per square Å and mol, where the surface in $Å^2$ is a measure of the contact between lipophilic parts of the interacting molecules (Section B 6).

E 7.1. RECOGNITION BETWEEN MOLECULAR SURFACES : COMPUTATIONAL SEARCH METHODS FOR COMPLEMENTARITY IN SHAPE AND CHEMICAL NATURE

As seen above, the simulation of hydrophobic interactions already relies on the calculation of molecular surfaces. In the Conolly technique, which is standard in most computer aided molecular modeling packages, such

Table E 5. Calculated atomic charges in *trans* N-methylacetamide[a]

	FQ-DelPhi[a]		PSGVB-DelPhi[b]		Gasteiger[c]	CT[d]	CNDO[e]	AM1[f]	MM2[g]
	Gas	Aqueous	Gas	Aqueous	Gas	Gas	Gas	Gas	Gas
C–CH₃	−0.70	−0.71	−0.55	−0.57	0.01	−0.06	−0.09	−0.24	−0.04
CH₃	0.19	0.20	0.14	0.16	0.03	0.04	0.03	0.11	—
C=O	0.77	0.82	0.77	0.82	0.21	0.59	0.36	0.30	0.54
C=O	−0.56	−0.74	−0.55	−0.66	−0.28	−0.56	−0.35	−0.37	−0.45
N	−0.53	−0.54	−0.52	−0.54	−0.32	0.41	−0.21	−0.39	−0.52
N–H	0.30	0.35	0.31	0.36	0.15	0.24	0.11	0.22	0.27
C(N)	−0.24	−0.23	−0.26	−0.26	0.00	−0.01	0.10	−0.08	0.21
NCH₃	0.13	0.15	0.14	0.13	0.04	0.04	0.01	0.09	—

[a] FQ: fluctuating charge force field, charges calculated with polarization functions, ref. 68a. For methyl protons average values are given. [b] (PSGVB) Pseudo-spectral gaussian valence bond program (ab initio molecular orbital calculation with Del Phi = Poisson equation Solver program); b) ref. 68b,c) ref. 66) CT: Charge Templates, see ref. 63 (only used for biopolymers); CNDO:[70] [f] AM1:[71]; [g] calculated from bond dipole moments in MM2 version 1985/87.[72]

water-excluded surfaces are obtained by simulating a rolling ball over the surfaces; with the ball diameter to be chosen by the user. An obvious step further is the description of non-covalent interactions between molecules not by atom-to-atom analyses, but by surface parts which can be hydrophobic or lipophilic, or bear alternative negative or positive charges or fields, etc. Programs like GRID[74] allow to locate and to measure surfaces according to their chemical nature. With this kind of (rather time consuming) programs the surface is constructed by atom-to-atom potentials between the molecular model and a sampling molecule shifting over the surfaces.

An interesting approach to evaluate free energies in supramolecular complexes makes use of training sets with a series of known association constants with different guest compounds and a given receptor. This extension of classical QSAR strategies to three dimensions[75] does not require a known receptor structure to start with, but rather maps out the surface around the known guest or ligand molecules by applying suitable interaction potentials. This can be done also by distributing atomistic properties in the cavity surrounding the known ligands, particular if one uses directional hydrogen bonds to identify the position of receptor groups such as aminoacids in proteins.[76,77] Based on training sets which, however, may be twice as large as the test sets, calculated free energies for several protein–ligand complexes were obtained with an accuracy of typically $2.5\,\text{kJ}\,\text{mol}^{-1}$.[76]

Molecular surfaces with variable properties like different electrostatic potentials can also be constructed with the help of experimental affinities of a large enough training set of ligands, as discussed in Section E 1. Such 3 D QSAR programs like CoMFA[78] (Comparative Molecular Field Analysis) are of great significance for rational structure-based drug design. The large number of variables necessary to construct molecular surfaces with sufficient resolution for the optimal alignment of the molecules has lead to the application of neural network approaches[79] in this rapidly developing field.

E 7.2. APPLICATIONS TO SUPRAMOLECULAR COMPLEXES

Computer aided molecular modeling has become an indispensable tool in supramolecular chemistry. It can be applied at several different levels, starting with the rational design of synthetic ligands. For this, one will generate several starting structures which bear complementary binding functions connected by different spacers or frameworks. The fit between these can be controlled visually on the computer screen, preferably using transparent representations of van der Waals radii, such as so-called dotted surfaces. Many programs allow one to highlight contact positions (by 'bumps'), or hydrogen bonds, or to measure continuously distances. As a next step one can control which geometric deformations in host and guest might be necessary to achieve optimal contact between the interaction sites. Most of the distortions will reside in torsional angle changes, as the variation of bond length and angles costs too much energy; therefore these are usually kept constant. Semiquantitative information on the strain which may occur upon complex formation is then accessible by consideration of the torsions in the 'free' ligands and in the complex. The same evaluation is put on firmer basis by a corresponding simple computer experiment, in which one obtains the strain imposed on complexed ligands by energy minimization of the supramolecular complex; after separation of the complex into the components one compares the energy (for instance the 'final steric energy') of each ligand to its value in the energy minimum conformation of the single association partners before complexation. However, even this most simple application of force field minimization depends critically on the parametrizations used for the non-covalent interactions. In particular indiscriminate application of the 6-12 potential can lead to exaggeration of strain differences as consequence of the overly tight fit which may be only seemingly required in a given complex. As discussed in Chapter B the complex may be dominated by other forces with less steep potentials or by solvophobic interactions.

Larger complexes can exist in a multitude of metastable conformations; to find the most stable states presents a particularly severe problem in supramolecular associations in consequence of the much less directional non-covalent forces in comparison to covalent bonds. For simpler systems one can search for conformations significantly contributing to the association by 'driving' torsional angles, or by changing distances between selected atoms (use of 'constraints'). Many modeling packages offer special tools for the exploration of the conformational space, by systematic search, by Monte Carlo (MC) methods or by molecular dynamics (MD). Corresponding texts including manuals for many commercial program packages should be consulted for these applications, which also describe special algorithms for the identification of local minima.

Metal complexes in principle would require separate parametrizations for each element of the periodic table, which in practice is neither practical nor necessary.[80] Metal–donor atom distances and ligand conformations, including the, at first sight unexpected, affinity decrease for larger ions with the increase of the chelate ring size (Section A 5), are fairly well reproduced with force fields describing the metal–donor interactions with equations similar to those for covalent bonds.[80a] It has, however, been pointed out that even for alkali ion–crown ether complexes the oxygen lone pair polarization should be taken into account explicitly.[81] This applies even more to complexes with transition metal ions.[82] In view of the strong and directional forces operating in transition metal ions the metal ion–ligand

atom interaction is often represented by bond stretch and angle terms similar to covalent bonds. Experimental data such as heats of formation or force constants from vibrational spectra are less available for inorganic than for organic molecules and complexes, which makes the development of generally applicable force fields for coordination complexes more difficult.

As discussed in Chapter C the *medium* around interacting molecules has a profound effect on their associations, which should be described by appropriate expansions of force fields. Besides the simple possibility, mentioned above, of varying the ε term in equations (E7-1 and 2), more sophisticated approaches like the OPLS[61] functions parametrize simple Coulomb and Lennard–Jones potentials for interactions in water and other media, giving excellent agreement with experimental data, such as for heats of vaporization or of hydration, at least for electroneutral molecules. Suitable parametrization can be based on *ab initio* molecular orbital wavefunctions and on application of FEP methods.[83]

Classical force field calculations yield enthalpies ΔH, and not the more interesting free energies. This as well as the explicit consideration of solvent molecules can be achieved through so-called free energy perturbation (FEP) calculations (Section C 2 for an example).[84] For these one generates for the interacting molecules atomic coordinates by either MD or MC simulations in a set ('box') of several hundreds solvent molecules, using periodic boundary conditions. In the MD calculations, Newton's equations of motion are applied to each atom, with averaging over time. The problem is that one cannot use too large a size for each step, whereas the simulated reaction, even if it is only a conformational change, can take 10^{-3} s, which leads to very long sampling times. Although algorithms like SHAKE[85] accelerate the sampling considerably, by keeping bond lengths constant, the required cpu time still

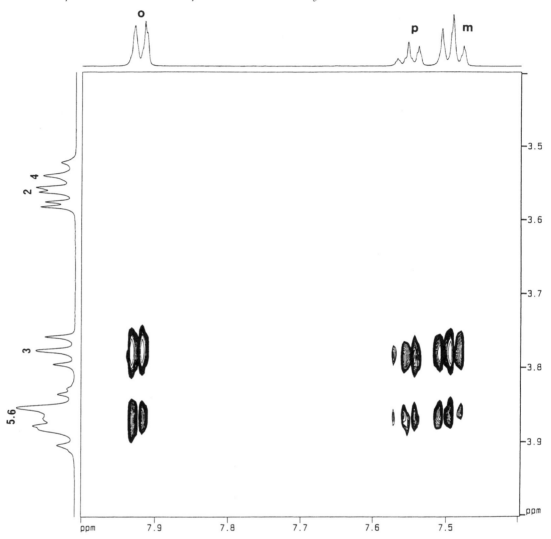

Figure E 14.   T-ROESY matrix of the complex between α-cyclodextrin $(5 \times 10^{-2}\,M)$ and benzoate anion $(2 \times 10^{-1}\,M)$ in D_2O at 300 K; complexation degree 17% for the anion and 68% for CD $(K = 13\,M^{-1})$; Intermolecular interactions as in Fig. 6(b); other explanations see text. Measuring parameters: pD 6.4 and as for Fig. E 6(a).[37]

poses severe problems for typical organic host–guest complexes. Also one avoids the calculation of too many interactions by defining cut-off distances for atoms which are beyond predefined distances. The field is in rapid development, and we refer to recent literature which compares performance of different free energy calculation methods.[86]

A new MC-based method using a continuum description of solvents instead of explicit solvent molecules is reported to allow much faster ΔG calculations.[87] Semianalytical treatments of solvation[88] (Section B 6) use the observation that for hydrocarbons the hydration free energy, composed of the energies of cavity formation and a solute–solvent van der

Waals term, G_{cav} and G_{vdW}, is linearly related to the total solvent accessible surface area SA_k, with an atomic solvation parameter σ_k which has been empirically derived[88] to be around $30\,J\,mol^{-1}\,Å^{-2}$

$$G_{solv(alkane)} = G_{cav} + G_{vdW} = \sum \sigma_k SA_k$$
(E 7-5)

Solvation free energies of moderately polar solutes can be calculated by adding a polar term G_{polar} on the basis of atomic charges and again empirically derived Born radii; with a dielectric constant of $\varepsilon = 78$ satisfactory description of hydration free energies were observed.[88] The description of solvophobic forces as enthalpic advantage, due to creation of one larger cavity instead of two cavities before complexation has been discussed already in Section B6 together with the equation (B 6-1), which relates to equation (E 7-5) above.

Some free energy perturbation calculations with larger host–guest complexes like that between a cyclophane and pyrene (complex **C 2** in Scheme C 1) have reproduced qualitatively the solvent effect but yielded large ΔG differences between calculations and experiment (e.g. 73 *vs.* $39\,kJ\,mol^{-1}$ in water, or 31 *vs.* $9.5\,kJ\,mol^{-1}$, in chloroform).[89] Other calculations with enantioselective complexes based largely on hydrogen bonds yielded surprisingly small deviations.[90] It should be noted, that as with the classical force field methods the reliability of free energy calculations depends on the choice of parametrization for the underlying potentials.

E 8.   EXERCISES AND ANSWERS

E 8.1.   EXERCISES

1. Which inclusion mode must be inferred from the ROESY spectrum in Fig. E 14?

2. Suggest suitable NMR experiments for the characterization of the cyclodextrin inclusion complexes shown on the website (list particular signals and expected effects).

E 8.2   ANSWERS

1. The spectrum shows not only the cross peaks seen in Fig. E 4(a) for the acid form, but additional interactions between the guest H-meta and the host H 5 protons, also stronger signals between H-para and H 3 or even H 5. This indicates the presence of an additional complex with the carboxylate oriented outside the cavity, in line with the higher hydrophilicity in comparison to the COOH group.

2. See website for answers.

REFERENCES

1 Prelog, V. (1964) *Pure Appl. Chem.*, **9,** 119; see also Graves, J.M.H.; Clark, A.; Ringold, H.J. *Biochemistry*, (1965), **4**, 2655.

2 Bergeron, R.J.; Channing, M.A.; Gibeily, G.J.; Pilor, D.M. (1977) *J. Am. Chem. Soc.*, **99**, 5146.

3 Tee, O.S.; Bozzi, M.; Hoeven, J.J.; Gadosy, (1993) *J. Am. Chem. Soc.*, **115**, 8990.

4 Goldberg, I. in *Inclusion Compounds*; Vol. 2, (1984) Atwood, J.L.; Davies, J.E.D.; MacNicol, D.D., eds; Oxford University Press, Oxford, 261.

5 Davies, J.E.D.; Finocchiaro, P.; Herbstein, F.H. *ibid.*, 407.

6 Goldberg, I. (1998) in *Crown Ethers and Analogs*; Patai, S.; Rappoport, Z. eds.; Wiley, Chichester, 359, 399.

7 see e.g. Dauter, Z.; Wilson, K.S. (1996) in *Comprehensive Supramolecular Chemistry, Vol. 8,* Davies, J.E.D.; Ripmeester, J.A., eds., Pergamon/Elsevier Oxford 1, and monographs or reviews cited therein.

8 Lindner, K.; Saenger, W. (1982) *Carbohydr. Res.*, **99**,103.

9 MacGillvray, L.R.; Atwood, J.L. (1997) *Nature*, **389**, 469.

10 Ripmeester, J.A.; Ratcliffe, C.A. in ref. 7, 323.

11 Stetter H.; Roos, E.-E. (1955) *Chem. Ber.*, **88**, 1390.

12 Hilgenfeld, R.; Saenger, W. (1982) *Angew. Chem.*, **94**, 788.

13 Wald, P.; Schneider, H.-J., unpublished results.

14 Shibakami, M., Sekiya, A. (1992) *J. Chem. Soc. Chem. Commun.*, 1742.

15 Alderfer, J.L.; Eliseev, A.V. (1997) *J. Org. Chem.*, **62**, 8225.

16 Jeffrey, G.A.; Saenger, W. (1994) *Hydrogen Bonding in Biological Structures.*, Springer Verlag, Berlin etc.

17 (a) Kaupp, G. in ref. 7, 381; (b) Frommer, J. (1992) *Angew. Chem. Int. Ed. Engl.*, **31/10**, 1298.

18 for leading references see: Vezenov, D.V.; Noy, A.; Rozsnyai, L.F.; Lieber, C.M. (1997) *J. Am. Chem. Soc.*, **119**, 2006.

19 Dammer, U.; Hegner, M.; Anselmetti, D.; Wagner, P.; Dreier, M.; Huber, W.; Guentherodt, H.-J. (1996) *Biophys. J.*, **70**, 2437.

20 Magonov, S.N.; Whangbo, M.-H. (1994) *Adv. Mater.*, **6**, 355.

21 Croasmun, W.R., Carlson, R.M.K. (1994) eds.; *Two-Dimensional NMR-Spectroscopy*; VCH, New York, Martin, G.; Zetzker, A.S. (1988) *Two-Dimensional NMR Methods for Establishing Molecular Connectivity.* VCH, New York, Weinheim.

22 Roberts, G.C.K. (1993) Ed., *NMR of Macromolecules – A Practical Approach*, IRL Press Oxford.

23 See (a) recent NMR textbooks; (b) examples in: Schneider, H.-J.: (1993) *Rec. Trav. Chim. Pays-Bas*, **112**, 412.

24 Amato, M.E.; Djedaini, F.; Pappalardo, G.C.; Perly, B.; Scarlata, G. (1992) *J. Pharm. Sci.*, **81**, 1157; Amato, M. E.; Lipkowitz, K. B.; Lombardo, G. M.; Pappalardo, G. C. (1996) *J. Chem. Soc. Perkin Trans.* **2**, 321.

25 For a review on aromatic ring current descriptions see : Haigh, C.W.; Mallion, R.B. (1980) *Prog. NMR. Spectrom.*, **33**, 303.

26 Schneider, H.-J.; Rüdiger, V.; Cuber, U.: (1995) *J. Org. Chem.*, **60**, 996.

27 For related NMR shielding analyses of proteins see: Asakura,T.; Williamson, M.P. (1993) *J. Magn. Res.*, **B 101**, 63, and references cited therein.

28 (a) Schneider, H.-J.; Buchheit, U.; Becker, N.; Schmidt, G.; Siehl, U. (1985) *J. Am. Chem. Soc.*, **107**, 7827; (b) Rüdiger, V. (1997) *Ph.D Dissertation, Universität des Saarlandes, Saarbrücken.*

29 Wolinski, K.; Haacke, R.; Hinton, J. F.; Pulay, P. (1997) *J. Comput. Chem.*, **18**, 816; Rauhut, G.; Puyear, S.; Wolinski, K.; Pulay, P. (1996) *J. Phys. Chem.*, **100**, 6310.

30 Kutzelnigg, W.; van Willen, C.; Fleischer, U.; Franke, R.; van Mourik, T. (1993) *NATO ASI Ser., Ser. C (Nuclear Magnetic Shieldings and Molecular Structure)*, **386**, 141.

31 Komiyama, M.; Hirai, H. (1981) *Polymer J.*, **13**, 171.

32 Perrin, C.L.; Nielson, J.B. (1997) *Annu. Rev. Phys. Chem.*, **48**, 511.

33 Smirnov, S.M.; Goluibev, N.S.; Denisov, G.S.; Benedict, H.; Schah. Mohammedi, P.; Limbach, H.-H. (1996) *J. Am. Chem. Soc.*, **118**, 4094.

34 Neuhaus, D.; Williamson, N.P., (1989) *The Nuclear Overhauser Effect in Structural and Conformational Analysis*, Verlag Chemie, New York.

35 Mo, H.; Pochapsky, T. C. (1997) *Prog. Nucl. Magn. Reson. Spectrosc.*, **30**, 1.

36 see e.g. Barsukov, I.L.; Lian, L.-Y. 9, 315 ff; Sutcliffe, M.J. *ibid.*, 359–390.

37 Simova, S.; Schneider, H.-J., *unpublished results.*

38 (a) Bergeron, R. J.; Clarke, R.J.; Coates, J.H.; Lincoln, S.F. (1988) *Adv. Carbohydr. Chem.*, **46**, 205; for further references see (b) Schneider, H.-J.; Hacket, F.; Rüdiger, V.; Ikeda, H.; (1998) *Chem. Rev.*, **98**, 1755.

39 Rüdiger, V.; Eliseev, A.; Svetlana, S.; Schneider, H.-J.; Blandamer, M.J.; Cullis, P.M.; Meyer, A. (1996) *J. Chem. Soc., Perkin Trans.* **2**, 2119.

40 Feeney, J.; Birdsall, B. in ref. 22, p. 183.

41 Ripmeester, J.A.; Ratcliffe, C.I. In *Comprehensive Supramolecular Chemistry*, Vol. 9, eds. Davies, J.E.D.; Ripmeester, J.A. (1996) Pergamon/Elsevier.

42 Dawes, S.B.; Ellaboudy, A.S.; Dye, J.L. (1987) *J. Am. Chem. Soc.*, **109**, 3508.

43 Ripmeester, J. A. (1993) *Supramol. Chem.*, **2**, 89.

44 Suzuki, M.; Tsutsui, M.; Ohmori, H. (1994) *Carbohydr. Res.*, **261**, 223.

45 (a) see Harada, N.; Nakanishi, K. *Circular Dichroism Spectroscopy – Exiton Coupling in Organic Stereochemistry*, University Science Books, Mill Valley, CA, (1983); b) Charney, E. (1985) *The Molecular Basis of Optical Activity – Optical Rotatory Dispersion and Circular Dichroism*; R. Krieger Publ. Co.; Malabar, FL; (c) Rosini, C.; Zandomeneghi, M.; Salvadori, P. (1993) *Tetrahedron Asymmetry*, **4**, 545.

46 Forman, J.E.; Barrans, R.E. Jr.; Dougherty, D.A. (1995) *J. Am. Chem. Soc.*, **117**, 9213.

47 Kodaka, M. (1993) *J. Am. Chem. Soc.*, **115**, 3702.

48 Harata, K.; Uedaira, H. (1975) *Bull. Chem. Soc., Japan*, **48**, 375.

49 Kajtar, M.; Horvath-Toro, C.; Kuthi, E.; Szejtli, J. (1982) *Acta. Chim. Sci. Hung.*, **110**, 327.

50 Kodaka, M. (1997) *J. Chem. Soc., Farad Trans.*, **93**, 2057.

51 Kubista, M.; Akerman, B.; Norden, B. (1997) *J. Phys. Chem.*, **92**, 2352.

52 Selected monographs on vibrational spectroscopy techniques: Schrader, B. Ed. (1995) *Infrared and Raman Spectroscopy: Methods and Applications*, Wiley, Weinheim (for NCA, see Bougeard, D.) 445 Colthup, N.B.; Daly, L.H.; Wiberley, S.E. (1980) *Introduction to Infrared and Raman Spectroscopy*, 3rd edn (1990), Aaron, H.S. *Top. Stereochem.*, **11**, 1.

53 Takeuchi, H.; Arai, T.; Harada, I. (1986) *J. Mol. Struct.*, **146**, 197; Raevsky O.A., Trepalin S.V., Zubareva V.E., Batyr D.G., (1986) Koord. Chimiya (Russ), **12**, 273 and references cited therein.

54 Trepalin, S.V.; Jarkov, A.V.; Solotnov, A.F. Raevsky, O.A. private communication.

55 Davies, J.E.D. In *Comprehensive Supramolecular Chemistry*, Vol. 8; Davies, J.E.D.; Ripmeester, J.A. Eds., Pergamon/Elsevier Oxford (1996), 33 ff.

56 Moreira de Silva, A.M.; Amado, A.M.; Ribeiro. Claro, P.J.A.; Empis, J.; Texeira-Dias, J.J.C. (1995) *J. Carbohydrate Chem.*, **14**, 677.

57 Jeffrey, G.A.; Saenger, W. (1994) *Hydrogen Bonding in Biological Structures*; Springer, Berlin, 50 ff.

58 (a) Gardner, R.R.; Gellman, S.H. (1997) *Tetrahedron*, **53**, 9881; (b) Nowick, J.S.; Abdi, M.; Bellamo, K.A.; Love, J.A.; Martinez, E.J.; Noronha, G.; Smith, E.M.; Ziller, J.W. (1995) *J. Am. Chem. Soc.*, **117**, 89; (c) Tsang, K.Y.; Diaz, H.; Graciani, N.; Kelly, J.W. (1994) *J. Am. Chem. Soc.*, **116**, 3988; (d) Dado, G.P.; Gelman, S.H. (1993) *J. Am. Chem. Soc.*, **115**, 4228 and references cited therein.

59 (a) Rappé; A.K.; Casewit, C.J.: (1997) *Molecular Mechanics Across Chemistry*, Freeman/Univ. Science Books; (b) Höltje, H.-D.; Folkers, G.: (1996) *Molecular Modelling*; VCH, Weinheim; (c) Boyd, D.B.; Lipkowitz, K.B. (1987) *J. Chem. Ed.*, **59**, 269, ibid. (1995), **72**, 1070; (d) *Reviews in Computational Chemistry* (Lipkowitz, K.B.; Boyd, D.B., Eds.), (1990) (Vol 1) (1997) to (Vol. 10). VCH, Weinheim, New York; (e) Burley, S.K.; Petsko, G.A. (1988) *Adv. Protein Chem.*, **39**, 125.

60 Allinger, N.L.; Yuh, Y.H.; Lii, J.-H. (1989) *J. Am. Chem. Soc.*, **11**, 8551, 8566, 8576 (MM3).

61 Jorgensen, W.L.; Tirado. Rives, J. (1988) *J. Am. Chem. Soc.*, **110**, 1657 (OPLS).

62 Cornell, W.D.; Cieplak, P.; Bayly, C.I.; Gold, I.R.; Merz, K.M.; Ferguson, D.M.; Spellmeyer, D.C.; Fox, T.; Caldwell, J.W.; Kollman, P.A. (1995) *J. Am. Chem. Soc.*, **117**, 5179 (AMBER 4).

63 Brooks, B.R.; Bruccoleri, R.E.; Olafson, B.D.; States, D.J.; Swaminathan, S.; Karplus, M. (1983) *J. Comp. Chem.* **4**, 187; Brooks, C.L.; Karplus, M. (1986) *Methods Enzymol.*, **127**, 369; Brünger, A.T.; Karplus, M. (1991) *Acc. Chem. Res.*, **24**, 54 (CHARMm).

64 For evaluation also non-covalent interactions with *ab initio* methods see in particular Maple, J.R.; Hwang, M.J.; Stockfisch, T.P.; Dinur, U.; Waldman, M.; Ewig, C.S.; Hagler, A.T. (1994) *J. Comp. Chem.*, **15**, 162 ("Class II field"), Halgren, T. (1992) *ibid* **15**, 162 ("MMFF field").

65 for comparison see (a) Sigfridsson, E.; Ryde, U. (1998) *J. Comp. Chem.*, **19**, 377; (b) Chipot, C.; Maigret, B.; Rivail, J.-L.; Scheraga, H. (1992) *J. Phys. Chem.*, **96**, 10276; (c) Bachrach, S.M. (1994) *Rev. Comp. Chem.*, **5**, 171.

66 Gasteiger, J.; Marsili, M. (1980) *Tetrahedron*, **36**, 3219.

67 Field, M.J.; Bash, P.A.; Karplus, M. (1990) *J. Comp. Chem.*, **11**, 700; Vasilyev, V.V.; Blyniuk, A.A.; Voityuk, A.A. (1992) *Int. J. Quantum. Chem.*, **44**, 897.

68 (a) Rick, S.W.; Berne, B.J. (1996) *J. Am. Chem. Soc.*, **118**, 677; (b) Tannor, D.J.; Marten, B.; Murphy, R.; Fiesner, R.A.; Sitkoff, D.; Niecholls, A.; Ringnalda, M.; Goddard III, W.A.; Honig, B. (1994) *J. Am. Chem. Soc.*, **116**, 11875.

69 Gundertofte, K.; Palm, J.; Petterson, I.; Stamvik, A. (1991) *J. Comp. Chem.*, **12**, 200.

70 Pople, J.A.; Beveridge , D.L. (1970) *Approximate Molecular Orbital Theory*, McGraw Hill, New York.

71 Dewar, M.J.S.; Zoebisch, E.G.; Healy E.F. Stewart, J.J.P. (1985) *J. Am. Chem. Soc.*, **107**, 3902.

72 Allinger, N.L.; Lii, J.H. (1987) *J. Comp. Chem.*, **8**, 1146; Allinger, N.L.; Kok, R.A.; Imam, M.R. (1988) *ibid.*, **9**, 591.

73 Hobza, P.; Kabelaz, M.; Sponer, J.; Mejzlik, P.; Vondrasek, J. (1997) *J. Comp. Chem.*, **18**, 1136.

74 Reynolds, C.A.; Wade, R. C.; Goodford, P.J.; (1989) *J. Mol. Graphics*, **7**, 103; Wade, R. C.; Goodford, P. J. (1993) *J. Med. Chem.*, **36**, 148.

75 Kubinyi, H. Ed.; (1993) *3D QSAR in Drug Design: Theory, Methods and Applications*, Escom, Leiden.

76 See Vedani, A.; Dobler, M.; Zbinden, P. (1998) *J. Am. Chem. Soc.*, **120**, 4471.

77 Boehm, Hans-Joachim (1996) *J. Comput.-Aided Mol. Des.*, **10**, 265.

78 Cramer, R.D., III; Patterson, D.E.; Bunce, J.D. (1988) *J. Am. Chem. Soc.*, **110**, 5959.

79 For leading references see Good, A.C.; Richards, W.G.; (1993) *J. Med. Chem.*, **36**, 433; Wagener, M.; Sadowski, J.; Gasteiger, J. (1995) *J. Am. Chem. Soc.*, **117**, 7769; M. Rarey, B. Kramer, T. Lengauer, and G. Klebe (1996) *J. Mol. Biology*, **261**, 470.

80 Reviews on force fields for metal complexes: (a) Hancock, R.D. (1989) *Progr. Inorg. Chem.*, **37**, 187 ff; *Acc. Chem. Res.*, (1990) **23**, 253; (b) Landis, C.R.; Root, D.M.; Cleveland, T. *Reviews in Computational Chemistry* (Lipkowikz, K.B. and Boyd, D.B. (1995) (eds)) VCH, Weinheim, Vol. VI, 73; (c) Comba, P. (1993) *Coord, Chem. Rev.*, **123**, 1; *Molecular Modeling of Inorganic Compounds*; VCH Weinheim, New York, (1995) d) Wipff, G., (1992) *J, Coord. Chem.*, **27**, 7; Wipff, G.; Troxler, L. in: *Computational Approaches in Supramolecular Chemistry*, Ed. Wipff, G. pp. 319–348; Kluwer, Dordrecht

(1994); (e) Hay B.P.; (1993) *Coord. Chem. Rev.*, **126**, 177.

81 Shirts, R.B.; Stolworthy, L.D. (1995) *J. Incl. Phenom. Mol. Recogn.*, **20**, 297.

82 Vedani, A.; Huhta, D.W. (1990) *J. Am. Chem. Soc.*, **112**, 4759 (force field YETI).

83 McDonald, N.A.; Carlsson, H.A.; Jorgensen, W.L. (1997) *J. Phys. Org. Chem.*, **10**, 563 and references cited therein.

84 Jorgensen, W.L. (1989) *Acc. Chem. Res.*, **22**, 124; Kollman, P. (1993) *Chem. Rev.*, **93**, 2395.

85 van Gunsteren, W.F.; Berendsen , H.J.C. (1977) *Mol. Phys.*, **34**, 1311 (SHAKE).

86 Radmer, R.J.; Kollman, P.A. (1997) *J. Comp. Chem.*, **18**, 902.

87 Kolossvary, I. (1997) *J. Am. Chem. Soc.*, **119**, 10233.

88 Still, W.C.; Tempzyk, A.; Hawley, R.C.; Hendrickson, T. (1990) *J. Am. Chem. Soc.*, **112** 6127 and references cited therein.

89 Mordasini Denti, T.Z.; van Gunsteren, W.F.; Diederich, F. (1996) *J. Am. Chem. Soc.*, **118**, 6044.

90 McDonald, D.Q.; Still, W.C. (1996) *J. Am. Chem. Soc.*, **118**, 2073.

DYNAMICS OF SUPRAMOLECULAR SYSTEMS

Dynamics of supramolecular systems have two fundamental aspects: first, the kinetics of formation and dissociation of supramolecular complexes and aggregates, like surfactant micelles; second, intramolecular dynamics, which may involve conformational changes and/or intracomplex mobility of bound guests. Due to the non-covalent character of supramolecular interactions, bond-making and bond-breaking processes usually have low activation barriers and intermolecular dynamics in these systems is generally fast, except for cases where association/dissociation is accompanied by considerable conformational changes or is hindered sterically. Also the kinetics becomes slow with strongly interacting systems, like spherands. For the same reason intramolecular mobility of supramolecular complexes is high compared to covalent molecules.

Complexation mechanisms for supramolecular complexes involving metal ions have been a long-standing subject in inorganic coordination chemistry, which allows us to refer here mostly to corresponding monographs or reviews.[1-4] The classical description of ligand substitution mechanisms is based on the types of reaction intermediates: formation of an intermediate with increased metal coordination number n is characteristic of associative mechanism (A): with decreased n, of dissociative mechanism (D). When bond-breaking and bond-making processes with leaving and entering ligands, respectively, occur *via* a single transition state one speaks of an interchange mechanism (I).[1] Metal sub-

stitution reactions have similar mechanistic alternatives: intermediate formation of a binuclear complex with bridging ligand (associative), intermediate formation of the free ligand and leaving metal cation (dissociative) or synchronous interchange of metal cations over a single transition state. Desolvation can be particularly slow with metal ions of high charge density; in the classical Eigen–Winkler mechanism, dissociation of aqua ligands is always rate-limiting, even for kinetically labile complexes. Rate constants (at room temperature, in (s^{-1}) units) for water exchange vary from 10^9 (with alkali cations) to 10^2–10^1 (e.g. with Fe^{3+} or Al^{3+}) to only 10^{-5} (for Ru^{3+} and Cr^{3+}). Rigid macrocyclic ligands like porphyrins slow down association/dissociation rates furthermore; in consequence one has often to use high temperature, non-aqueous solvents, high concentrations and long reaction times in the formation of such complexes.

With organic host–guest complexes as well as with alkali or earth alkali cations and even with many enzyme-substrate complexes the association rates are most often found to be diffusion-controlled (with $k_{diff} = 10^8$–10^9 $(M^{-1}s^{-1})$ in water). The complex dissociation rate constants k_{off} are then just a function of the association constant K, with $k_{off} = k_{diff}/K$. As equilibrium constants K are usually much easier to determine one often assumes completely or nearly diffusion controlled on-rates and refrains from more complicated rate measurements. In the comparatively few cases where kinetics have been determined, for instance with crown ether–alkali complexes, the constants usually have indeed indicated diffusion-controlled on-rates.[1–6] Exceptions with on-rate constants as low as $k = 10^{-5}$ ($M^{-1}s^{-1}$) even in aprotic solvents are observed if a major reorganization of the ligand conformation is necessary for the complexation.[7]

The derivatization of rate constants from concentration changes, preferably by non-linear least-square fitting to the underlying rate laws is described in many textbooks, as is the determination of the kinetic order. In many cases one will use excess of either host or guest, leading to simple pseudo-first order kinetics. The total order can and should then be determined by measuring the reaction rate with variable concentrations of the excess partner. In the following the dynamics of some supramolecular systems are discussed with emphasis on experimental methods.

F1. NMR METHODS

If either the *on* rates are slow at least on the NMR-time scale, or if the association strength is high enough one can measure the concentrations of an educt and the complex by integration of *separate* NMR signals. With carceplexes or cavitands (Section H1) complex formation is often so slow that one can measure concentrations during the reaction by *any* analytical, preferably spectroscopic technique. The applicability of the NMR signal integration method is limited by the known correlation between the exchange rate constant k_{exch} and the frequency difference Δv between the exchanging NMR signals (see NMR textbooks). One should stay well below the coalescence point, where

$$k_{exch} = \Delta v \, \pi / 2^{1/2}$$

in order to avoid systematic errors by under-estimation of smaller signals. Thus, at a magnetic field strength corresponding to 400 MHz (^{1}H-NMR) some of the exchanging signals should be separated for instance by

at least 1 p.p.m. = 400 Hz in case the equilibrium constant is $K = 10^6$ M^{-1} or larger, assuming an *on*-rate with $k_{diff} = 10^8 – 10^9$ $M^{-1}s^{-1}$.

Cases (or temperatures) where the exchange rate constants k differ not much from Δv are the realm of classical NMR shape analysis.[8] The actual line shape reflects not only the exchange rate k_A or k_B of one partner (k_A either for host or guest; k_B for the host–guest complex), but also their populations (or concentrations) p_A and p_B. If one measures not too much above the coalescence point one may use for either signal the approximation

$$k_A = (4\pi \, p_A^2 \, p_B^2 \, \Delta v^2)/\Delta_A$$

where Δ_A is the observed line broadening of signal A minus the half-height line width (all in Hz) without exchange broadening. Here Δ_A *may* be estimated also from another reference signal, and p_A and p_B from the equilibrium constant K measured by other means. Below coalescence one may use as approximation[8]

$$k = (\pi/2)^{1/2}(\Delta v_0^2 - \Delta v_{A,B}^2)^{1/2}$$

where $\Delta v_{A,B}$ is the separation of the two exchanging signals at a given temperature, and Δv_0 their separation without exchange (e.g. at very low temperature, or from model compounds).

However, accurate dynamic parameters are much better obtained from complete shape analysis of NMR signals, which with suitable computer programs does not involve any approximations (see NMR textbooks).[8] Figure F1 shows a typical dynamic ^{23}Na NMR spectrum of a complex between sodium cations and a bibrachial diazacrownether; the temperature dependence can be used to construct an Arrhenius or Eyring plot, which indicates activation parameters of $\Delta H^* = 41 \pm 0.8$ (kJ mol^{-1}, and $\Delta S^* = 66.4 \pm 2.5$ e.u. at 331 K and acetonitrile as solvent.[9] Although the temperature dependence of NMR-derived rate constants can be used to evaluate

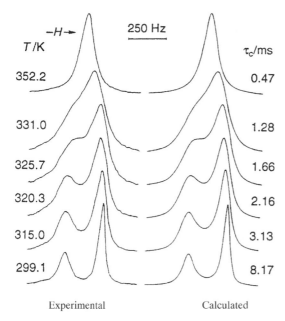

Figure F 1. A dynamic ^{23}Na NMR spectrum at 79.39 MHz of a complex between sodium cations and a bibracchial diazacrownether in acetonitrile; temperatures in K are indicated at the left side, calculated inverse rate constants on the right side; the complexed Na$^+$ signal appears upfield.[9] Reprinted with permission from *J. Chem. Soc., Dalton Trans.* Lucas *et al.*, **1994**, 423, by permission of the Royal Society of Chemistry.

enthalpy and entropy contributions on the basis of the Gibbs–Helmholtz equation $\Delta G = \Delta H - T\Delta S$, the relatively low accuracy of NMR signal integration, the limited temperature range accessible in NMR probes, problems with very accurate temperature control and the variation of intrinsic chemical shifts with temperature make it difficult to obtain very accurate parameters beyond ΔG this way. Before one starts such data collection one always should carry out a preliminary error analysis.

Solid state conformations of for instance crown ether alkali cation complexes can be studied by 13-C-CP/MAS (cross polarization/magic angle spinning), deuterium and proton NMR techniques.[10] These can not be discussed here, but have added important dynamic informations to static crystal structures obtained from X-ray studies, including large amplitude motions of macrocycles. Solid state relaxation techniques allow one to identify, for example guest flip-motions in cyclodextrins with activation energies as small as 4 kJ mol^{-1}. A combination of temperature-variable 13-C/MAS and 2-H NMR-spectroscopy has for the 18C6–KNCS complex in the solid state shown rotations of the –O–CH$_2$–CH$_2$– units around the macrocycle with barriers around 40 kJ mol^{-1}.

Two dimensional NOE (nuclear Overhauser effect) experiments (NOESY spectra, (Section E 4.2)) show cross peaks due not only to cross relaxation but also to spin exchange, and therefore can be used also to extract dynamics of intermolecular exchange. The two contributions can be distinguished in principle by rotating frame experiments, using NOESY or EXSY pulse sequences with variable mixing times. TOCSY transfer, multiple spin and differential off-resonance effects can complicate the extraction of exchange rate constants and require additional experiments. For the rather intricate details the reader is referred to the literature on 2D NMR-spectroscopy.[11] An interesting application is illustrated in Fig. F 2.[12] Here the problem was to measure dynamics of dissociation and association of a dimer complex being so strong that the monomer is invisible even at very low concentration. Introduction of different substituents Y at the narrow rim, with Y = Me at the 1,3-, and Y = C$_5$H$_{11}$ at the 2,4- phenyl rings of these calix[4]arenes was used to produce a C$_{2v}$ symmetry in the monomer, and a C$_2$ symmetry in the dimer as consequence of the opposite direction of the carbonyl groups in the urea units. Evaluation of several then nonidentical and exchanging protons by a sequence of NOESY experiments lead to exchange rate constants of $0.26 \pm 0.06\,\text{s}^{-1}$. Noticeably, the uptake of benzene into the dimer cavity occurs with a similar rate ($k = 0.47 \pm 0.1\,\text{s}^{-1}$), indicating that the dimer dissociation is the rate limiting step here. In

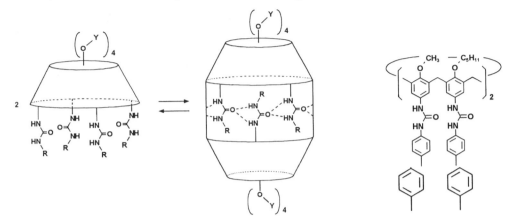

Figure F 2.   Dimerization of a tetra urea calix[4]arene.[12] Reprinted with permission from *J. Am. Chem. Soc.*, Mogck *et al.*, 1997, **119**, 5706. Copyright 1997 American Chemical Society.

solvents like CD_2Cl_2 which are not able to fill the dimer cavity properly, one only observes dimer formation upon addition of benzene traces; otherwise ill defined associations occur. With similar capsules from resorcinarenes , shown in Scheme I 8 , guest exchange times have been determined to vary from milliseconds to days.[13]

Relaxation measurements of hydrogen or carbon nuclei provide data not only for very fast complexation on- and off rates, but are mainly used for obtaining insight into motions of and within supramolecular complexes. Spin–spin, or transverse relaxation time, T_2, changes are the basis of the line shape analyses discussed above. Spin–lattice, or longitudinal relaxation time, T_1 changes depend on energy exchange of the observed nucleus transition with the surrounding and thus reflect nature and dynamics also of the host–guest interactions. Quantitative analyses require not only time consuming and delicate experiments but also evaluation of the underlying relaxation mechanisms, for instance by measuring T_1 as a function of temperature and of the applied magnetic field. For this reason these techniques are not often used in synthetic host–guest chemistry and will not be discussed in

detail here. The few studies carried out until today have, however, shown, for example, that the correlation time of phenyl compounds included in α-cyclodextrin complexes is four to six times *shorter* than the time of the CD receptor alone, indicating a weak coupling between host and guest with anisotropic and fast motion of the guest inside the cavity.[14]

With certain assumptions discussed in the literature one can describe the relaxation by the equation[15–17]

$$T_2^{-1} - T_1^{-1} = [H]_0(K_D + [G]_0)^{-1}k_1^{-1}\Delta v$$

where $[H]_0$ and $[G]_0$ are host and guest concentrations, K_D and k_{-1}^{-1} are dissociation equilibrium and rate constants, respectively, and Δv is the shift displacement in radian/s of the observed signal induced by complexation. For association of zinc porphyrin derivatives with the amino group of aminoacid methylesters the dissociation rate constants k_{-1} derived from relaxation times were reported[16] to be $1–25 \times 10^5 \, s^{-1}$, depending on the porphyrin substituent and the aminoacid. Comparison with the equilibrium constants K gave association rate constants between 1 and 16 $M^{-1}s^{-1}$.

F2.   FLUORESCENCE METHODS

Fluorescence lifetime measurements use single photon counting instruments and deconvolution of exponential decays with suitable fitting methods. They can extend the dynamic measuring range down to nanoseconds now. Thus, analysis of a double exponential fluorescence decay in cyclodextrins with covalently bound *N*-dansylleucine residues[18] indicates two species, one with short (e.g. 0.23 ns), the other with long lifetimes (up to e.g. $\tau_{1/2} = 0.77$ ns): Long lifetimes are characteristic for derivatives with stronger inclusion, as visible also by stronger fluorescence emission. Upon addition of a guest large enough to expel the dansyl moiety out of the cavity (Fig. F3, case **I**), one observes that the fraction with now longer lifetimes (up to 0.99 ns) decreases since the dye is more exposed to water. The opposite change is observed if the derivatives allow intracavity inclusion of both the fluorescence dye and the added guest compound (Fig. F3, case **II**).

Measurements fluorescence depolarization degree provide information about intramolecular mobility of fluorescent guests.[19] When rotational mobility of a fluorescent molecule is restricted, such as by a highly viscous medium, its fluorescence becomes partially polarized. This is expressed quantitatively as molecular anisotropy, r, or degree of polarization, p, defined by equations

$$r = (I_\parallel - I_\perp)/(I_\parallel + 2I_\perp)$$
$$p = (I_\parallel - I_\perp)/(I_\parallel + I_\perp)$$

where $I_\parallel$ and $I_\perp$ are the fluorescence intensities measured in parallel and perpendicular planes to the plane of polarization of the excitation light. For a spherical fluorescent molecule these parameters are related to the rate of rotation of the sphere R_S and the rate of fluorescent emission λ in the Perrin equation

$$r_0/r = (1/p - 1/3)/(1/p_0 - 1/3) = 1 + 6R_S/\lambda$$

where r_0 and p_0 refer to a fluorophore which conserves the fixed orientation (in very

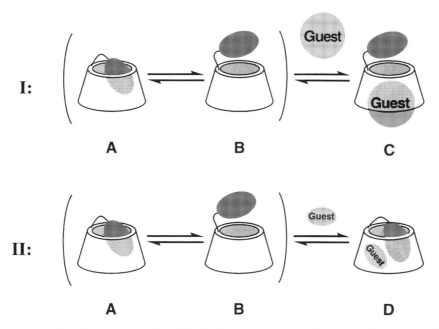

Figure F3. Conformational changes and equilibria of cyclodextrins with covalently attached fluorescence dyes.[18] Reprinted with permission from *J. Am. Chem. Soc.*, Ikeda *et al.*, 1996, **118**, 10980. Copyright 1996 American Chemical Society.

viscous medium). This equation can be transformed into the following expression[20]

$$r_0/r = 1 + k_B T \tau / \eta v$$

where η is the viscosity of the medium, v is the effective volume of the sphere and τ is the average lifetime of the excited state. The last expression allows one to determine microscopic viscosity in a host cavity using an appropriate fluorescent probe.

F3.   OTHER TECHNIQUES

Methods for rate determinations with supramolecular complexes such as stopped flow, ultrasonic absorption, or temperature-jump measurements (for constants as high as $k = 10^9$ (s^{-1})), will largely be in the hands of specialized laboratories, and only few applications can be discussed here.

A combination of high-pressure and stopped-flow measurements with α-cyclodextrin and azodyes such as mordant yellow ($R_3 = COO^-$, $R_4 = OH$; $R_5 = CH_3$ (Fig. F4) indicate a first and fast step (k around 10^4 M^{-1} s^{-1}), with a negative activation volume ΔV^* due to the necessary desolvation of the sulfonate head group and to the release of two high enthalpy water molecules out of the cyclodextrin cavity.[21] A second, slow step (k 0.2–1.8 s^{-1}) is assumed to consist of a

relatively slow intramolecular motion of the substrate in the intermediate complex, resulting in large negative ΔV^* values (-16 cm^3 mol^{-1}), which was ascribed to the formation of hydrogen bonds between the substrate COOH group and the cyclodextrin OH groups. With other substituents where such bonds are not formed, one finds only very small ΔV^* values.

Temperature jump and relaxation techniques were already used in 1967 with cyclodextrins and azodyes.[22] In line with later investigations[23] one observes generally a biphasic behavior as discussed above; the second rate determining step being essentially controlled by desolvation and conformational reorganization of the loose complex formed in the fast first step. Interestingly, the association rate constants k can be quite low and vary from 3 M^{-1}s^{-1} to 5×10^7 M^{-1}sec^{-1}, showing a larger substrate selectivity than the association constants K, which varied only between 270 and 1010 M^{-1}.[22]

Using the time dependent response of quartz balances rate constants around 200–2000 M^{-1}s^{-1} for association and around 6–140×10^{-6} s^{-1} for dissociation were observed for concanavalin binding to glycolipid monolayers.[24] Kinetics of solid state inclusions cannot be discussed here in detail. They can

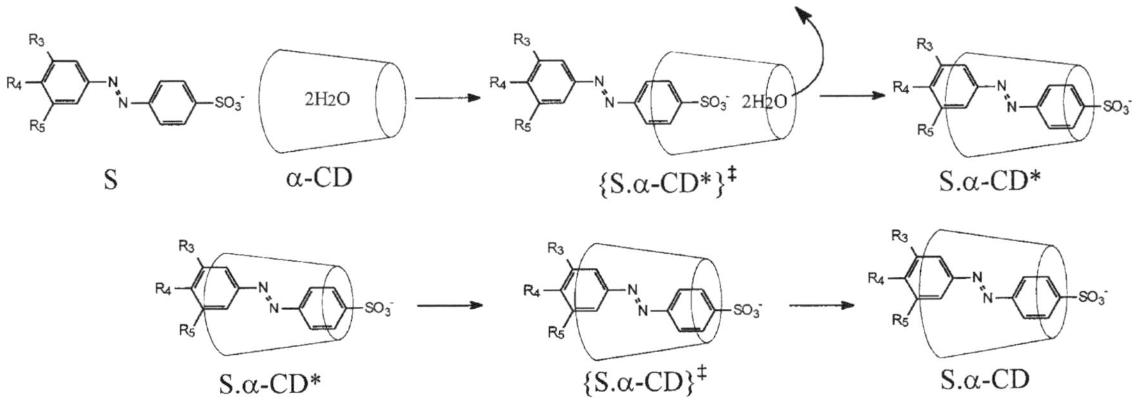

Figure F4.   Intermediates and transition states of α-cyclodextrin association with dyes S as observed by stopped flow and high pressure experiments.[21] Reprinted with permission from *Chimia*, Bugnon et al., 1996, **50**, 615. Copyright 1996, Chimia Zürich.

be measured, for example, by increase of mass with a suitable balance system, and show for instance for inclusion of acetone vapor in an organic diol layer anti-Arrhenius behavior which is generally attributed to phase changes.[25]

REFERENCES

1 Langford, C.H.; Gray, H.B. (1965) *Ligand Substitution Dynamics*, Benjamin, New York, Cox, B.G.; Schneider, H. (1992) *Coordination and Transport Properties of Macrocyclic Compounds in Solution*, Elsevier, Amsterdam.

2 Lincoln, S.F.; Merbach, A.E. (1995) *Adv. Inorg. Chem.* **42**, 1.

3 See also Petrucci, S.; Eyring, E.M.; Konya, G. (1996) in *Comprehensive Supramolecular Chemistry, Vol. 8*, Davies, J.E.D.; Ripmeester, J.A.; eds., Pergamon/Elsevier, Oxford, **483**.

4 Data collection also for rate constants: Izatt, R.M.; Pawlak, K.; Bradshaw, J.S.; Bruening, R.L. (1995), *Chem. Rev.*, **95**, 2529 and earlier reviews cited therein.

5 (a) Detellier, C.; Graves, H.P.; Briere, K.M. (1991) *Isotopes in Physical and Biomedical Science*, Elsevier Science, Amsterdam; (b) Popov, A.I.; Hallenga K., (1990) eds: *Modern NMR Techniques and their Application in Chemistry*, Marcel Dekker, New York; (c) For recent references see e.g. Li, Y.; Echegoyen, L. (1994) *J. Am. Chem. Soc.* **116**, 6832; (d) Chen, Z.; Mercier, L.; Tunney, J.J.; Detellier, C. (1996) SO NATO ASI Ser., Ser. C **485** (Physical Supramolecular Chemistry), 393.

6 Bouquant, J.; Delville, A.; Grandjean, J.; Laszlo, P. (1982) *J. Am. Chem. Soc.* **104**, 686.

7 Rebek, J, Jr.; Luis, S. V.; Marshjall, L.R. (1986) *J. Am. Chem. Soc.* **108**, 5011.

8 Sandström, J. (1982) *Dynamic NMR-Spectroscopy*, Academic Press, New York; Oki, M. (1985) *Applications of Dynamic NMR-Spectroscopy to Organic Chemistry*, VCH, Deerfield Beach FL.

9 Lucas, J.B.; Lincoln, S.F. (1994) *J. Chem. Soc., Dalton Trans.*, 423.

10 See e.g. Ripmeester, J.A.; Ratcliffe, C.I. (1996) in *Comprehensive Supramolecular Chemistry, Vol. 8*, Davies, J.E.D.; Ripmeester, J.A., eds., Pergamon/Elsevier, Oxford, 323; for recent examples see Buchanan, G.W.; Denike, J.K. (1995) *J. Am. Chem. Soc.* **117**, 2900.

11 Croasmun, W.R.; Carlson, R.M.K., (1994) eds.; *Two-Dimensional NMR-Spectroscopy*; VCH, New York.

12 Mogck, O.; Pons, M.; Böhmer, V.; Vogt, W. (1997) *J. Am. Chem. Soc.*, **119**, 5706.

13 Chapman, R.G.; Sherman, J.C. (1998) *J. Am. Chem. Soc.* **120**, 9818.

14 Behr, J.P.; Lehn, J.-M. (1976) *J. Am. Chem. Soc.*, **98**, 1743.

15 Meiboom, S.J. (1961) *Chem. Phys.* **34**, 375.

16 Sykes, B.D. (1969) *J. Am. Chem. Soc.*, **91**, 949.

17 Mizutani, T.; Murakami, T.; Ogoshi, H. (1996) *Tetrahedron Lett.* **37**, 5369.

18 Ikeda, H.; Nakamura, M.; Ise, N.; Oguma, N.; Nakamura, A.; Ikeda, T.; Toda, F.; Ueno, A. (1996) *J. Am. Chem. Soc.*, **118**, 10980.

19 Shinitzky, M.; Dianoux, A.-C.; Gitler, C.; Weber, G. (1971) *Biochemistry*, **10**, 2106.

20 Weber, G. (1953) *Adv. Protein Chem.*, **8**, 415.

21 Bugnon, P.; Lye, P.G.; Abou-Hamdan, A.; Merbach, A.E. (1996) *Chimia* **50**, 615.

22 Cramer, F.; Saenger, W.; Spatz, H.C. (1967) *J. Am. Chem. Soc.* **89**, 14.

23 Review: Connors, K. A. (1997) *Chem. Rev.* **97**, 1325.

24 Ebara, Y.; Okahata, Y. (1994) *J. Am. Chem. Soc.* **116**, 11209.

25 Barbour, L.J.; Caira, M. R.; Nassimbeni, L.R., (1993) *J. Chem. Soc., Perkin Trans. 2*, 2321.

G SURFACTANT-BASED SUPRAMOLECULAR SYSTEMS AND DENDRIMERS

G 1. MONOLAYERS

The most important physicochemical property of amphiphilic molecules, reflected also in the commonly used name *surfactants*, is their surface activity, that is the capacity to reduce surface tension by adsorption at the interface.[1] Molecules of typical surfactants have polar or charged head groups and large non-polar tails, usually long hydrocarbon chains. Structures of several often employed surfactants are shown in Scheme G 1. Surfactants form numerous supramolecular assemblies,[1–3] which can serve as hosts or can provide a special 'organized' environment for intermolecular interactions between other molecules.

Monomolecular liquid layers of surfactants at the water–air interface represent the simplest and most well-characterized[4] type of a surfactant assembly. Recognition of many guests such as nucleic acid bases and nucleotides, amino acids, and peptides, and sugars has been studied at the water–air interface employing monolayers of surfactants with functional polar groups.[5] Several examples which illustrate the involved host–guest interactions are shown in Fig. G 1 and G 2.

The composition and structure of host–guest complexes was studied by examination of multilayers transferred onto solid substrates by the Langmuir–Blodgett (LB) technique. In the LB technique a solid 'substrate', such as a glass plate, is raised up through the monolayer spread on water, and surfactant molecules are transferred to the substrate surface, which interacts with polar heads or non-polar tails of surfactant molecules depending on the surface polarity.[4] If the substrate surface is polar, the transfer of surfactant monolayer makes it hydrophobic and if now one repeats the procedure the new layer will be transferred by adsorption of nonpolar surfactant tails affording the bilayer, etc. Elemental analysis of the transferred layers, such as by X-ray photoelectron spectroscopy allows determination of the binding stoichiometry; FT-IR spectroscopy was used to determine the binding mode by observation of characteristic shifts of C=O, C=N and N–H peaks in host and guest molecules.[5] Binding constants for guests dissolved in water can be determined from the adsorption isotherms, which in the absence of guest–guest interactions in the surface layer were of the Langmuir type (mathematically coinciding with the binding isotherm for a 1:1 complexation process of the type D 1-10).

Surfactants with a diaminotriazine function are capable of binding nucleobases via complementary H-bonding, as illustrated in Fig. G 1(a) for thymine.[6] In spite of the fact that no base pairing in bulk water is usually observed, the binding constant in this case equals 300 M^{-1}, being of the order expected for such a complex in chloroform (Section B 3). Also effective hydrogen bonding to barbiturate monolayers was reported.[7]

Figure G 1(b) illustrates predominantly electrostatic recognition of adenosine monophosphate by a cationic surfactant bearing a guanidinium head group. A very large bind-

sodium dodecyl sulfate (SDS)

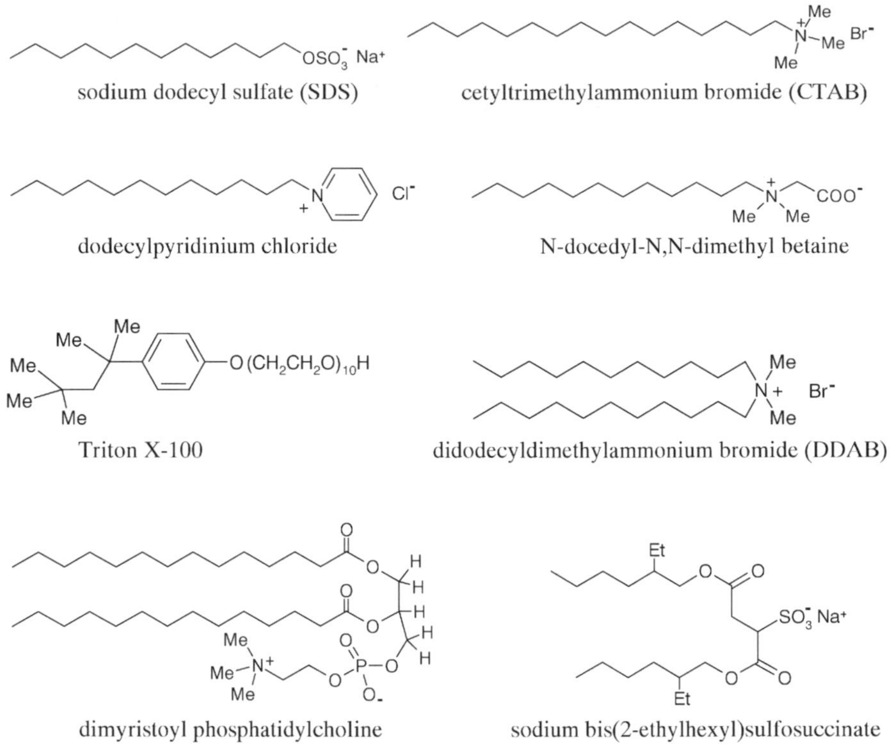

cetyltrimethylammonium bromide (CTAB)

dodecylpyridinium chloride

N-docedyl-N,N-dimethyl betaine

Triton X-100

didodecyldimethylammonium bromide (DDAB)

dimyristoyl phosphatidylcholine

sodium bis(2-ethylhexyl)sulfosuccinate

Scheme G 1.   Structures of some representative surfactants.

ing constant of $K = 3.2 \times 10^6$ M^{-1} was found in this case[5,8] far exceeding what one would expect for the ion pairing constant between, for example, a monocation and a monoanion in water ($K \approx 1$ M^{-1}, Section B 2). With the same monolayer one observed for ATP $K = 1.7 \times 10^7$ M^{-1}.[5,8] Multisite interactions in mixed surfactant monolayers are also possible, as illustrated in Fig. G 2 for the binding of flavin mononucleotide to a mixed monolayer. In this case also a very large binding constant $K > 10^7$ M^{-1} was found.[5]

The dramatically increased electrostatic attraction in the air-water interface is attributed to the close proximity of the hydrocarbon-like phase, with a low dielectric constant created by the aliphatic tails of adsorbed surfactants.[5] A similar situation should be expected for micellar or bilayered aggregates of the same surfactants, but in these systems the affinity is much smaller (Section G 2).

Evidently the interface exhibits some unique properties.

Monolayers at solid interfaces can be prepared by two methods. One is the LB technique described above. It has the advantage of possible preparation of multilayers in addition to monolayers and also of incorporation of various lipophilic additives in the film. However, the resulting films are very unstable and have gel-like structures. Another method involves chemisorption of the film components from the solution, yielding so-called self-assembled monolayers (SAMs).[9,10] Formation of SAMs occurs spontaneously by adsorption of various biphilic molecules on solid supports of appropriate chemical structure, like fatty acids on aluminum oxide. The most useful type of SAMs are monolayers of alkanethiols on gold, due to their stability, regular structure, ease of preparation and other features. The interaction of mercaptanes

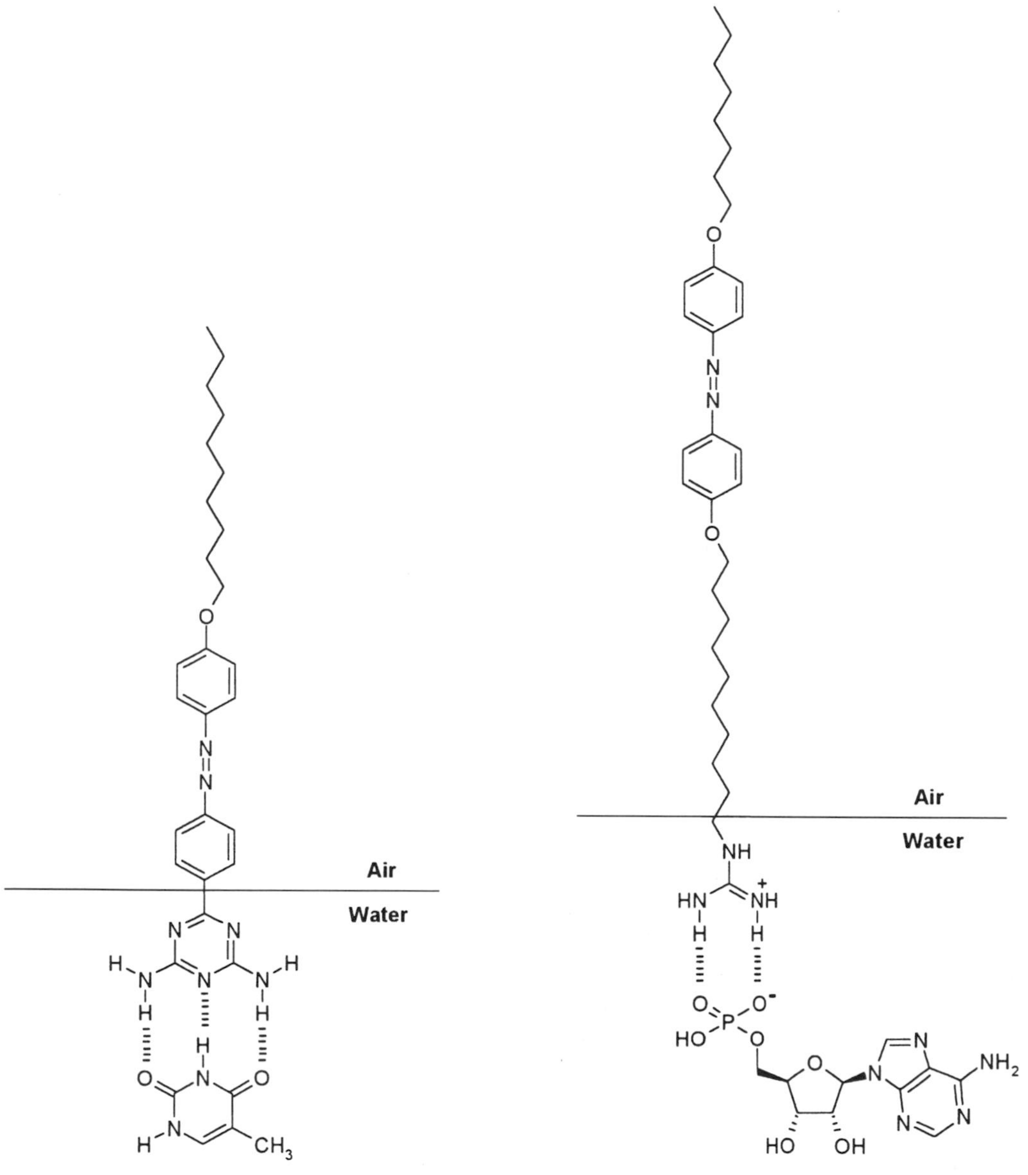

(a)   (b)

Figure G1. Recognition of thymine (a) and AMP (b) at the air–water interface.[5] Reprinted with permission from *Acc. Chem. Res.*, Ariga *et al.*, 1998, **31**, 371. Copyright 1998 American Chemical Society.

with gold can be considered as the oxidative addition of SH group to Au^0 with subsequent elimination of dihydrogen[10]

$$RSH + Au_n^0 \rightarrow RS\,Au^+Au_{n-1}^0 + 0.5\,H_2$$

The resulting monolayer is characterized by a regular structure with the hydrocarbon tails usually tilted by 62–64° against the surface. Monolayers containing thio-derivatives of host molecules are often used in sensors

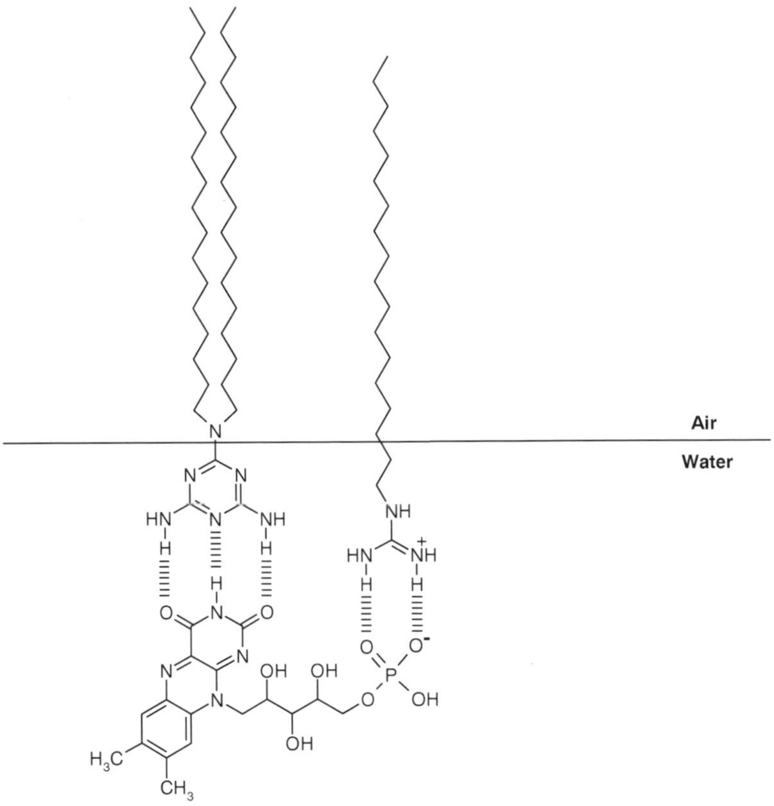

Figure G2. Recognition of a flavinmonoucleotide (FMN) at the water–air interface.[5]

(Chapter I). Thus, monolayers containing thiocyclodextrins serve as electrochemical sensors for ferrocene.[11] A series of fluorescent receptors to barbiturates, exemplified by a structure shown in Fig. G3, use multiple hydrogen bond host–guest interactions[12] which are well characterized in solution (Section B3). Formation constants of complexes like that shown in Fig. G3 were found to be of the order 10^4 M^{-1}, similar to those reported in chloroform solution. Instead of optical sensing one often uses mass-sensitive detectors such as the quartz balance in combination with receptors immobilized on gold surfaces.

In general, there is a practical problem, which restricts applications of monolayers as reaction media to analytical fields: the amount of material which can enter into the interface is very small. For separation or synthetic applications one should therefore look for large surfactant aggregates like micelles or vesicles, which can incorporate appreciable amounts of reactants, and which in their structural organization have much in common with monolayers.

G2. MICELLES AND BILAYERS

Micelles are aggregates of tens to hundreds amphiphilic molecules or ions (D) with a narrow distribution of the aggregation numbers (N) around a mean value[1-3]

$$N\,D \rightleftharpoons D_N$$

With such high aggregation numbers the micelle concentration (D_N) becomes an extremely sharp function of the monomer concentration; only with small aggregation numbers and/or when polydispersity is high do the

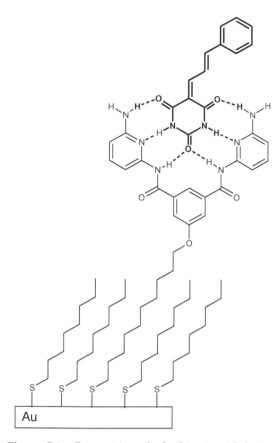

already formed micelles in $1/N$ degree, $[D] = (1/K)^{1/N}[D_N]^{1/N}$, and practically does not change with increasing micelle concentration remaining equal to the CMC. Thus, below the CMC all dissolved amphiphile is monomeric, but above the CMC the concentration of monomers remains approximately constant and equal to the CMC, whereas all excess of added amphiphile converts into micelles. This behavior usually can be adequately described as a phase transition (like precipitation) instead of association described by the mass action law. In terms of this pseudo-phase model the CMC is considered as the solubility of surfactant in the monomeric form.

At relatively low surfactant concentrations above the CMC, micelles have spherical form with polar head groups at the surface and a hydrophobic interior. Several models of micellar structure were proposed which differ in the packing mode of hydrocarbon chains inside the micellar core.[13] A schematic view of the cross-section of a spherical micelle, which reflects some common features of different models, is shown in Fig. G4. The radius of the hydrophobic core is approximately equal to

Figure G3. Recognition of a barbituric acid derivative by a fluorescent hydrogen-bonding receptor on the surface of self-assembled thiolate monolayer on gold.[12] Reprinted with permission from *J. Am. Chem. Soc.*, Motesharei *et al.*, 1998, **120**, 7328. Copyright 1998 American Chemical Society.

curves described by the following equation become more flat.

$$[D_N] = K[D]^N \ (N \geq \approx 10)$$

As a result the concentration of micelles changes abruptly from practically zero to a detectable level in a narrow range of amphiphile concentrations known as *critical micelle concentration* (CMC).[1,2] In practice the sharpness of transition to micelles depends on numerous factors, including the method employed. In turn, the monomer concentration is proportional to the concentration of

Figure G4. Schematic view of the cross-section of a spherical micelle.

the length of the fully extended hydrocarbon chain tail, but there are many bent chains and a considerable portion of the micellar surface is also covered by methylene and even some terminal methyl groups. With increasing surfactant concentration the aggregation number usually increases, and spherical micelles are transformed into rod-like micelles and finally into a liquid-crystalline phases. Surfactant monomers are in rapid exchange between micelles and the bulk phase.

Surfactants bearing two chains, like DDAB or dimyristoyl phosphatidylcholine (Scheme G1), have very low solubilities in the monomeric form. Instead of micelles, such surfactants produce upon dispersion in aqueous solutions bilayer aggregates in form of vesicles (or liposomes),[14] Fig. G5. As a rule, the initially formed liposomes have a multibilayer structure and a wide size distribution; subsequent sonication and size separation can produce uni-bilayer vesicles with a narrow size distribution. Vesicular solutions are generally unstable, in contrast to micellar solutions, and after preparation should be used within several days. Bilayers are more rigid and more ordered than micelles. Another type of surfactant aggregates are reversed micelles which are formed in organic solvents including hydrocarbons with some surfactants, like sodium bis(2-ethylhexyl)sulfosuccinate (Scheme G1). In spite of important differences in their structures, all types of surfactant

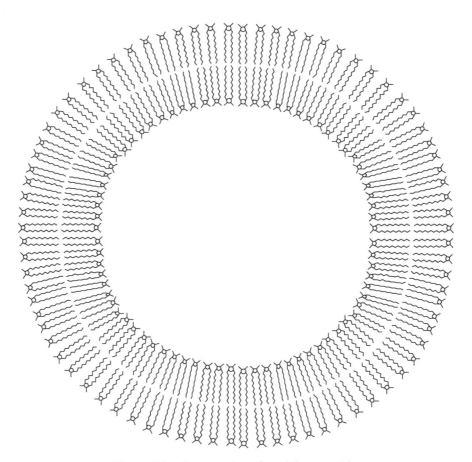

Figure G5.   Cross-section of uni-bilayer vesicle.

aggregates have, as a common feature, pseudo-interfacial regions, which separate the bulk aqueous phase, and hydrocarbon-like apolar region formed by hydrophobic tails. In this sense surfactant aggregates resemble adsorption monolayers.

Host properties of surfactant aggregates were studied in more detail with micelles. The well known phenomenon of solubilization of non–polar solutes in micellar surfactant solutions[15] is a practically important manifestation of binding of such solutes to micelles. Binding of non-polar solutes S, considered in terms of the pseudo-phase model of micellar solutions is as a partition of S between two 'phases', micelles and water; it is rather unspecific and largely is determined by the guest hydrophobicity. Thus, in contrast to macrocyclic hosts like cyclodextrins or cyclophanes, micelles do not discriminate structural isomers of similar hydrophobicity, as illustrated in Table G 1. Partition constants within series of chemically similar compounds correlate with partition constants between octanol/water.[16] Analysis of large data sets in terms of the multiparameter correlation of the type used for analysis of solvent effects (Section C 2, equation (C 2-6)) shows the applicability of equation (G1-1)[17a,b]

$$\log K_x = c + rR_2 + s\pi_2^H + a\sum \alpha_2^H + b\sum \beta_2^0$$
$$+ vV_x \qquad \text{(G 1-1)}$$

Table G 1. Logarithms of partition constants K_x of isomeric butanols between sodium dodecyl sulfate micelles and water and their binding constants to cyclodextrins.

	SDS	α-cyclodextrin	β-cyclodextrin
OH	2.44	1.95	1.22
OH	2.53	1.44	1.62
OH	2.41	1.42	1.19
OH	2.23	0.64	1.68

where K_x is the partition constant calculated with molar fraction standard states in both phases, R_2 is the excess molar refraction, π_2^H is the solute dipolarity/polarizability, $\Sigma\alpha_2^H$ and $\Sigma\beta_2^0$ are the overall solute hydrogen-bond acidity and basicity, respectively, V_x is the characteristic volume, r,s,a,b and v represent the corresponding sensitivity coefficients, and c is a constant. Both for anionic and cationic surfactants the dominant terms in equation (G 1-1) are the last two, coefficient b being negative and v positive. The same equation is applicable to partition constants in various water–organic solvent systems; comparison of the coefficients show indeed a large degree of similarity in the behavior of micellar and alcoholic media, for instance SDS micelles resemble isobutyl alcohol, and cetylpyridium micelles behave like pentanol.[17b]

Binding of charged guests to micelles of ionic surfactants involves considerable contributions from electrostatic interaction with the surface charge of the micelles. This effect is manifested, in particular, in micelle-induced shifts of acid–base dissociation constants of guests with ionogenic groups.[18] Typically micelles of cationic surfactants lead to a decrease of pK_a values of acids by 1–2 units due to stabilization of dissociated form (for equilibria of the type $AH \rightleftharpoons A^- + H^+$), or to destabilization of protonated forms (for equilibria of the type $R_3NH^+ \rightleftharpoons R_3N + H^+$); micelles of anionic surfactants cause an opposite effect. The electrostatic contribution can be estimated from the surface potential, which can be determined experimentally (from shifts in pK_a of highly hydrophobic completely bound to micelles dyes or from electrophoretic measurements), or derived theoretically by solving the Poisson–Boltzmann equation.[19] Alternatively the binding of ions can be treated as ion exchange between guest ions and a fraction of surfactant counterions tightly bound in the so-called Stern layer of the micelle.[20]

Recognition of organic guests was studied in micellar systems with functionalized surfactants or highly hydrophobic hosts, which are completely bound to micelles of common surfactants. Binding of AMP and ATP anions to micelles of $CH_3(CH_2)_{13}NHC(=NH_2)NH_2^+$ and CTAB as well as to bilayers of surfactants like $[CH_3(CH_2)_{15}]_2NCO(CH_2)_4NHC(=NH_2)NH_2^+$ was studied for comparison with the binding of these guests to charged mono-layers (Section G 1).[21] Binding constants of the order of 10^2–10^4 M^{-1} were observed, far below those in bilayers and close to those expected for purely electrostatic binding of these charge type ions (binding of ions to micelles usually is not treated as complexa-tion, see above, but respective binding con-stants can be estimated from known micellar properties; assuming the surface potential of 100 mV, which is a typical value for common ionic micelles,[19] one obtains for a dianion (dominant form of AMP in neutral solution) the electrostatic binding free energy of about -20 kJ mol^{-1} which corresponds to a binding constant 2.3×10^3 M^{-1}).

Hydrogen bonding between nucleobases incorporated into surfactant micelles is weaker than in monolayers or in organic solvents, but is much stronger than in water. Binding between the hydrophobic thymine derivatives **G-1** or **G-2** and adenine deriva-tives **G-3** in aqueous SDS solutions occurs with equilibrium constants 20–50 M^{-1}, being higher for more hydrophobic deriva-tives.[22] No complexation between these species occurs in the absence of SDS in water. Binding of nitrogen heterocycles and amino acid methyl esters to the capped porphyrin receptor **G-4** was studied in aqueous micellar media and some organic solvents.[23] Interac-tions involved are coordination of one of the guest nitrogen donor atoms to Zn(II) and hydrogen bonding with OH-groups of the steroid roof. The observed binding constants in SDS micelles are close to those in dichlor-omethane, but subtraction of the ligand–micelle partitioning term shows that binding free energies inside the micelle are similar to those in methanol.

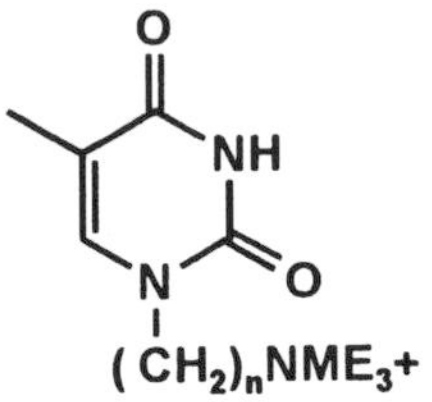

G-1 (n=4; 6; 8; 10) 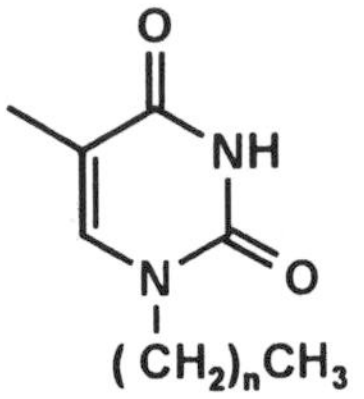   **G-2** (n=1; 3; 5; 7)

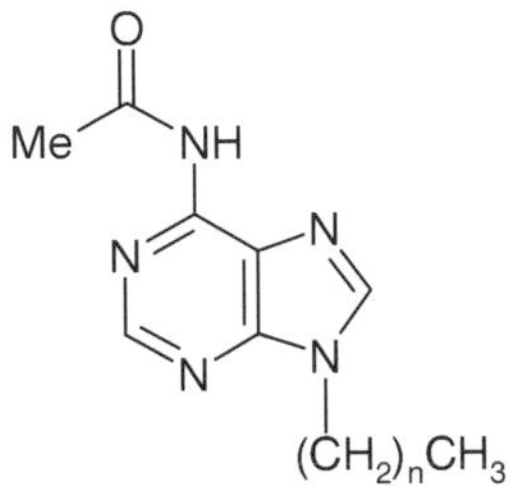

G-3 (n = 1; 2; 3; 4)

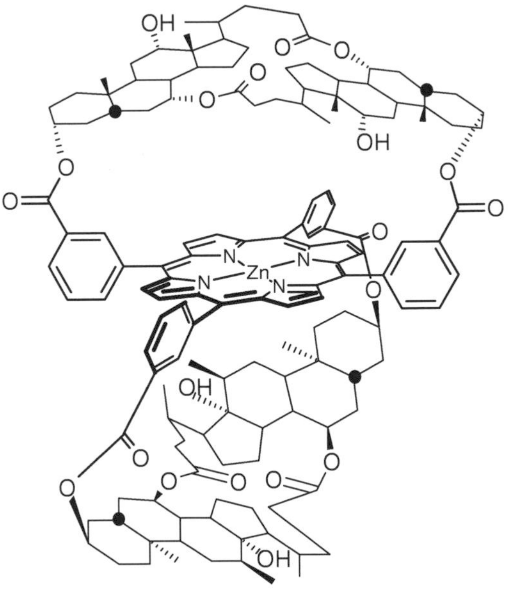

G-4

These examples demonstrate that micelles in spite of their rather simple organization carry out some important functions like a globular protein: they enable hydrogen bond-ing between host and guest molecules in aqueous solution and serve themselves as hosts for binding of hydrophobic and charged guests. On the other hand, bilayers are good models for biological membranes; they find important applications in studies of membrane transport phenomena (Section I6) and for sensor construction (Section I4).

G3. DENDRIMERS

The new class of hyperbranched polymers called dendrimers have recently attracted increasing attention,[24,25] in particular with respect to molecular recognition and self-assembly.[26] The general structure of dendrimers involves a core unit from which the branching starts, with several concentric 'generations' of branching layers and peripheral terminal groups. Recent advances in synthesis of dendrimers allows preparation of large molecules with molecular weights higher than 10^3 kDa and generation numbers

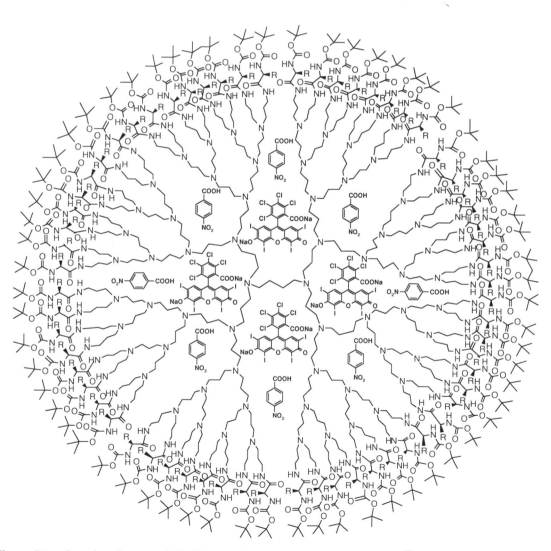

Figure G 6. Complex of a capped dendrimer with guest molecules (R = CH$_2$Ph).[27] Reprinted with permission from *J. Am. Chem. Soc.*, Jansen *et al.*, 1995, **117**, 4417. Copyright 1995 American Chemical Society.

up to 10. From the viewpoint of recognition chemistry two aspects are of interest: the potential of dendrimers to encapsulate guest molecules in cavities inside a dendrimer, and complexation of guests to surface functional groups, which are present in high local concentrations.

An impressive example of encapsulation of small molecules was reported for the system illustrated in Fig. G6.[27] A poly(propylene imine) dendrimer in methylene chloride/triethylamine is treated with Bengal Rose and 4-nitrobenzoic acid, and than with an activated ester of *t*-BOC-protected phenyl alanine, which caps the amino end-groups. The resulting inclusion complex looses none

of the entrapped small guest molecules upon prolonged dialysis in 5% water–acetone, indicating formation of a densely packed surface layer. Deprotection of phenyl alanine by hydrolysis with formic acid selectively liberates 4-nitrobenzoic acid; further hydrolysis of the surface amino acid by 12 M HCl liberates also the Bengal Rose dye.

Fast reversible complexation of hydrophobic guests (naphthalene derivatives and steroids) in water has been achieved with 'dendrophanes', which contain a cyclophane-type core and hydrophilic dendritic periphery.[28] Formation constants of inclusion complexes are of the same order or even smaller, than for parent core cyclophanes, but

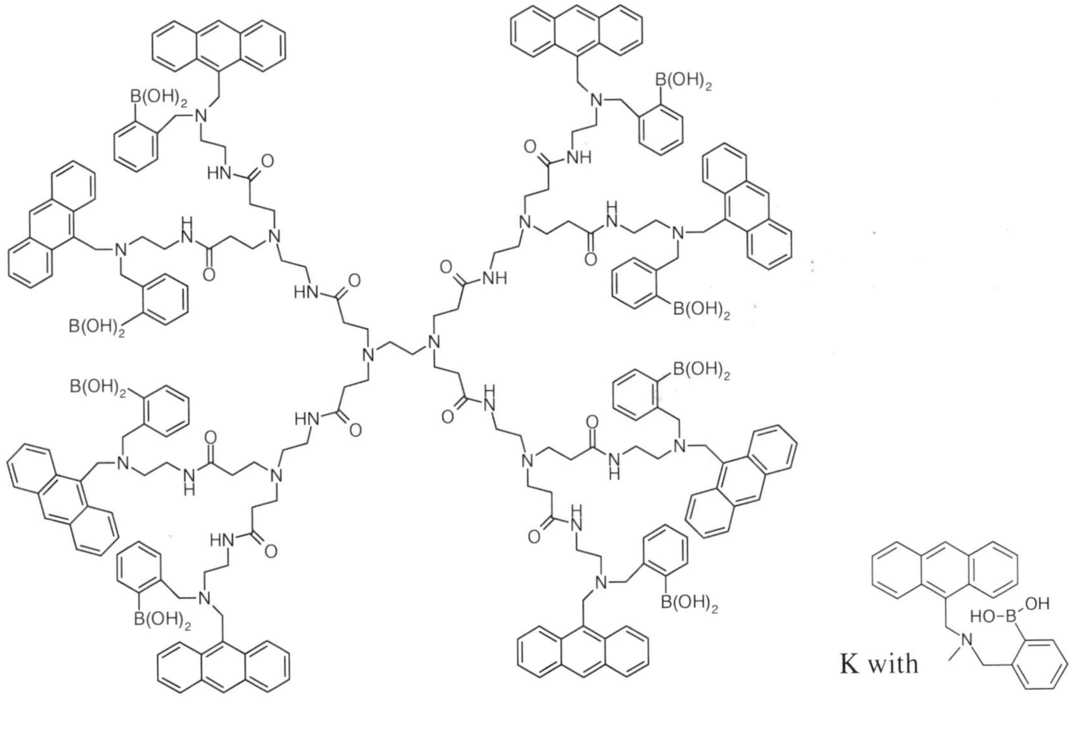

K with

	log *K* with dendrimer	log *K* with monomer
D-galactose	4.43	2.32
D-fructose	4.23	2.86
D-glucose	2.87	2.31

Scheme G2. Complexation of monosaccharides in methanol by a dendritic octaboronic acid and the monomeric fragment.[30]

NMR results indicate differences in guest orientation, which can be important for complexation selectivity. These structures can be considered as unimolecular micelles. A dendrimeric analog of reversed micelle was prepared by amino end-group acylation of a poly(propylene imine) dendrimer with palmitoyl chloride.[29] The resulting compound binds dyes like Bengal Rose in hexane.

As an example of surface recognition by a dendrimer, we cite the complexation of monosaccharides in methanol by a dendritic

octaboronic acid containing antracene units for fluorescence detection (Scheme G2).[30] In some cases formation constants are, as expected, considerably higher with the dendrimer, than with the respective monomeric unit, due to formation of an intramolecular 2:1 complex instead of an 1:1 complex formed with the monomer. Also complexation with the dendrimer seems to be more selective. Many other examples of recognition by dendrimers can be found already in reviews.[24-26]

REFERENCES

1 (a) Rosen, M.J. (1989) *Surfactants and Interfacial Phenomena*, 2nd edn., Wiley, New York; (b) Myers, D. (1992) *Surfactant Science and Technology*, 2nd edn, VCH Publishers, New York.

2 Tanford, C. (1980) *The Hydrophobic Effect: Formation of Micelles and Biological membranes*, 2nd edn., Wiley, New York.

3 Fuhrhop, J.-H.; Köning, J. (1994) *Membranes and Molecular Assemblies: The Synkinetic Approach*, The Royal Society of Chemistry, Cambridge, UK.

4 Adamson, A.W. (1990) *Chemistry of Surfaces*, 5th edn., Wiley, New York.

5 Ariga, K.; Kunitake, T. (1998) *Acc. Chem. Res.*, **31**, 371.

6 Kurihara, T.; Ohto, K.; Honda, Y.; Kunitake, T. (1991) *J. Am. Chem. Soc.*, **113**, 7342.

7 Bohanon, T.M.; Denzinger, S.; Fink, R.; Paulus, W.; Ringsdorf, H.; Weck, M. (1995) *Angew. Chem. Int. Ed. Engl.*, **34**, 58.

8 Sasaki, D.Y.; Kurihara, K.; Kunitake, T. (1991) *J. Am. Chem. Soc.*, **113**, 9685.

9 Ulman, A. (1991) *An Introduction to Ultrathin Organic Films: From Langmuir–Blodgett to Self-Assembly*, Academic Press, San Diego, CA.

10 Ulman, A. (1996) *Chem. Rev.*, **96**, 1533.

11 Rojas, M.; Königer, R.; Stoddart, J.F.; Kaifer, A.E. (1995) *J. Am. Chem. Soc.*, **117**, 336.

12 Motesharei, K.; Myles, D.C. (1998) (*J. Am. Chem. Soc.*, **120**, 7328.

13 Dill, K.A.; Koppel, D.E.; Cantor, R.S.; Dill, J.D.; Bendedouch, D.; Chen, S.-H. (1984) *Nature*, **309**, 42.

14 New, R.R.C. (1990) (ed.) *Liposomes: a Practical Approach*, IRL Press, Oxford.

15 Elworty, P.H.; Florence, A.T.; MacFarlane, C.B. (1968) *Solubilization by Surface Active Agents*, Chapman & Hall, London.

16 Treiner, C.; Mannebach, M.-H. (1987) *J. Colloid Interface Sci.*, **118**, 243.

17 (a) Abraham, M.H.; Chadha, H.S.; Dixon, J.P.; Rafos, C.; Treiner, C. (1995) *J. Chem. Soc. Perkin Trans.*, **2**, 887; (b) Abraham, M.H.; Chadha, H.S.; Dixon, J.P.; Rafos, C.; Treiner, C. (1997) *J. Chem. Soc. Perkin Trans. 2*, 19.

18 Bunton, C.A.; Romsted, L.S.; Sepulveda, L. (1980) *J. Phys. Chem.*, **84**, 2611.

19 (a) Frahm, J.; Diekmann, S. (1984) in *Surfactants in Solution*; Vol. 2, Lindman, B.; Mittal, K.L. eds., Plenum Press, New York, 897; (b) Gunnarsson, G.; Johnsson, B.; Wennerstrom, H. (1980) *J. Phys. Chem.*, **84**, 3114; (c) Bunton, C.A.; Moffatt, J.R. (1988) *J. Phys. Chem.*, **92**, 2896.

20 (a) Romsted, L.S. (1997) in *Micellization, Solubilization and Microemulsions*, Vol. 2, Mittal, K.L., ed., Plenum Press, New York, 509; (b) Bunton, C.A.; Nome, F.; Quina, F.H.; Romsted, L.S. (1991) *Acc. Chem. Res.*, **24**, 357.

21 Onda, M.; Yoshihara, K.; Kayano, H.; Ariga, K. Kunitake, T. (1996) *J. Am. Chem. Soc.*, **118**, 8524.

22 Nowick, J.S.; Cao, T.; Noronha, G. (1994) *J. Am. Chem. Soc.*, **116**, 3285.

23 Bonar-Law, R.P. (1995) *J. Am. Chem. Soc.*, **117**, 12397.

24 (a) Newkome, G.R.; Moorefield, C.N.; Vögtle, F. (1996) *Dendritic Molecules; Concepts, Synthesis, Perspectives*, VCH, Weinheim; (b) Zeng, T.W. Zimmerman, S.C. (1997) *Chem.*

Rev. **97**, 1681; (c) Escamilla, G.H. ; Newkome, G.R. (1994) *Angew. Chem., Int. Ed. Engl.* **33**,.1937; (d) Issberner, J.; Moors, R.; Vögtle, F. (1994) *ibid.*, **33**, 2413.

25 Tomalia, D.A.; Durst, H.D. (1993) *Top. Curr. Chem.*, **165**, 193.

26 Zimmerman, S.C. (1997) *Curr. Opinion Coll. Interface Sci.*, **2**, 89.

27 Jansen, J.F.G.A.; Meier, E.W.; de Brabander-van der Berg, E.M.M. (1995) *J. Am. Chem. Soc.*, **117**, 4417.

28 (a) Mattei, S.; Seiler, Y.; Diederich, F.; Gramlich, V. (1995) *Helv. Chim. Acta*, **78**, 1904; (b) Wallimann, P.; Diederich, F. (1996) *Helv. Chim. Acta*, **79**, 779.

29 Stevelmans, S.; van Herst, J.C.M.; Jansen, J.F.G.A.; van Boxtel, D.A.F.J.; de Brabander-van der Berg, E.M.M.; Meier, E.W. (1996) *J. Am. Chem. Soc.*, **118**, 7398.

30 James, T.D.; Shinmori, H.; Takeuchi, M.; Shinkai, S. (1996) *J. Chem. Soc. Chem. Commun.*, 705.

SHAPE RECOGNITION AND SOLID STATE INCLUSION COMPLEXES

In this chapter we will very briefly discuss the underlying principles of some typical supramolecular complexes in host structures, which are highly organized either by crystallization in the solid state, or by enforced cavities in dissolved receptor molecules. The self-organization of such supermolecules is governed by non-covalent forces between complementary shaped surfaces, with rules and examples as discussed and illustrated in Section I2. The common feature of the superstructures mentioned in this Chapter H is that their complexation properties with, most often small, guest molecules is often, but not always or entirely, governed by recognition of the guest shape, and not or to a lesser degree by specific non-covalent forces. In this sense

one can speak of true container structures,[1] which, however, can also be formed in reversible and controllable processes (Section I2). For solid state inclusion complexes, which are of growing industrial importance, we refer here essentially to monographs[2,3] and reviews[4] dealing with these special topics. Zeolites represent the extreme end of rigid preorganization; they are naturally occurring, but today mostly synthetically accessed host compounds,[5,3b] which combine well defined cavities (see Fig. H1 and structure on the website) with catalytic activity. Some examples for the construction of synthetic nanoporous materials by self-organization will be discussed below and in Section I2.

H1. CARCERANDS AND CAVITANDS

The cavity of carcerands like **H1** protects entrapped guest molecules from intermolecular reactions which are responsible for kinetic destabilization of highly reactive intermediates such as cyclobutadiene[7] or benzyne[8] (Fig. H2). The openings are large enough to allow, at elevated temperature, entrance and departure of small molecules like α-pyrone, which upon UV-irradiation generates cyclobutadiene inside the cavity. Similarly, benzocyclobutenedione served as precursor for o-benzyne. The latter could be spectroscopically characterized, although it slowly reacts also with the host structure.

Cycloveratrylene **H2** forms similarly closed cavities, which selectively encapsulate, slowly on the NMR time scale, small molecules like

Figure H1. Example for a zeolite with an eight-membered ring channel opening of a 0.41 nm diameter (zeolite A-LTA; general composition $(Al_2O_3)_m(SiO_2)_n$ $q H_2O$ with tetraeders around Si and Al atoms).[6] Reproduced from *Atlas of Zeolite Structure* with permission of the Laboratorium für Kristallographie, ETH Zürich, Meier, W.H. *et al.* 1996 (http:\\www. iza-sc.ethz.ch/IZA-sc/Atlas/data/models/).

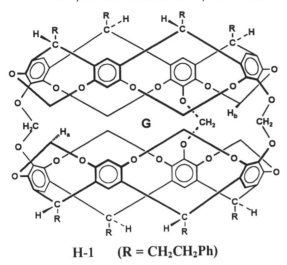

H-1   **(R = CH₂CH₂Ph)**

Figure H 2. Structure of a hemicarcerand **H-1** (R = –CH₂CH₂Ph) capable of stabilizing cyclobutadiene even above room temperature inside the cavity.[7]

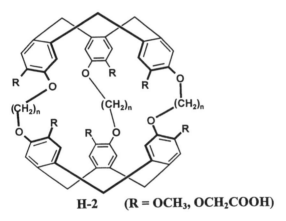

H-2   **(R = OCH₃, OCH₂COOH)**

Figure H 3. Cycloveratrylene **H-2** ($n = 3$, 5) for inclusion of small substrates like methane or xenon (with $n = 3$) in lipophilic solvents (with R = O–CH₃), or of ammonium compounds in water (with R = O–CH₂COOH).[9–11]

methane[9] or even xenon (Fig. H 3). In the presence of solvent molecules like 1,1,2,2-tetrachloroethane, which are too large to enter, methane or xenone is complexed with K around 3×10^3 M⁻¹, the ¹²⁹Xe NMR resonance is remarkably deshielded by 160 p.p.m.[10] Derivatives of **H 2** with carboxylic side chains allow to study inclusion of

ammonium compounds from tetramethylammonium chloride to acetylcholine in water; the observed[11] dependence on the guest size indicate that loose association is preferred over tight binding in such cavities, similar to observations with cyclodextrins (Section B 6). The cavity size of **H-2** ($n = 3$ or $n = 5$) differs by only 9%, the binding strength for, for instance, tetramethylammonium chloride, which is already well accomodated in the smaller host ($n = 3$) with $\Delta G = -15\,\mathrm{kJ\,mol^{-1}}$ (in D₂O), increases to $\Delta G = -20\,\mathrm{kJ\,mol^{-1}}$ with the wider cavity ($n = 5$).

H 2.   CLATHRATES AND SOLID STATE INCLUSION COMPLEXES

Many molecules, particularly those with a limited flexibility and irregular shape, form in the solid state lattices with voids, which for the always preferred dense packing and therefore higher stability[12] are filled with guests, often solvent molecules, with stochiometrically well defined ratios.[13] The structure of such crystal inclusion compounds or clathrates is ruled primarily by non-covalent interactions between the host molecules, although clathrate formation may also be induced by suitable guest molecules. As an indication, one often observes better crystallization in presence of guest molecules. The prediction of crystalline structures from molecular topology is still at its infancy,[3,13a,14] and relates to the recent development of controlled crystal growth.[15] Many clathrates like those from urea, cholic acids, hydroquinone, dianine, triphenylmethane etc., were already discovered by chance in the last century. Hydroquinone can form solid giant super-octahedral cubes with fullerenes.[16] Gas hydrates, of which chlorine hydrate is the oldest example,[17] are often stable at low temperatures and high pressures. Generally, bulky molecules have a tendency to crystallize with empty cavities.[13b] In the so-called Werner clathrates large anions around a transition metal center generate voids, which

Host: Ph₃CNH—CH—C—O—C—CH—NHCPh₃

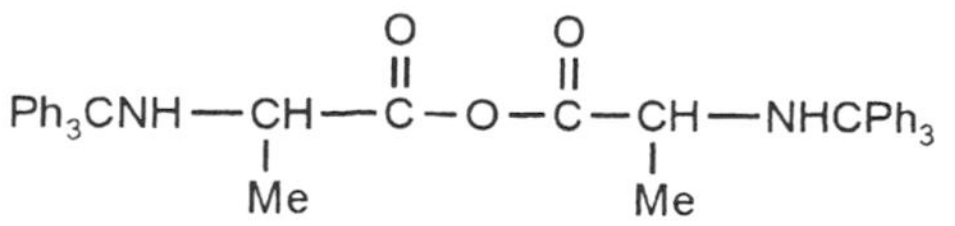

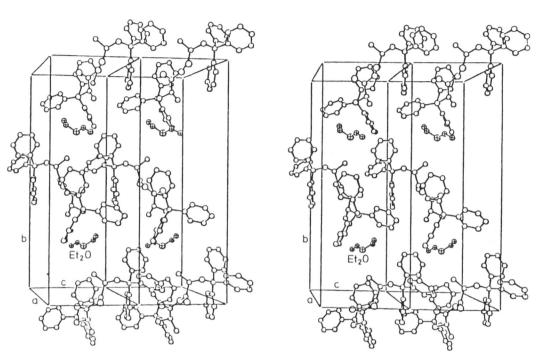

Figure H 4. X-ray structure of inclusion of diethylether with *N*-trityl-alanine anhydride (stereoview).[13a] Reprinted with permission of Springer Verlag N.Y. (*Top. Curr. Chem.*, I. Goldberg *et al.*, 1988, Fig. 17, p. 149)

often are filled in a non-stochiometric way with other molecules.[18] The propeller-shaped triphenylmethyl (trityl) group itself prevents dense packing in the crystal, providing a typical case for a void generation required to complex small neutral guest compounds; Fig. H 4 shows such an inclusion complex structure of diethylether with *N*-trityl-alanine.[13a]

Clathrate formation may be controlled by 'passive' control of crystal packing like with the trityl derivatives, but also by properly arranged functional groups. In old examples like clathrates with urea or hydroquinone the functions stabilize the lattice by host–host interactions. In so-called coordinatoclathrates bulky groups ensure cavity formation, whereas appended functions interact also, or

primarily with the entrapped guest. This way one reaches recognition not only by shape but also by complementarity with respect to functions.[13b] The 1,1'-binaphthyl-2,2'-dicarboxylic acid (**H-3**) provides for cavity formation due to its scissor shape between the orthogonal naphthylunits, and for hydrogen bonding both between the host molecules and, for example 2-butanol as guest molecule (Fig. H 5). The chirality of such host structures suggests the use of such systems for enantiomer separations. In addition, the regio- and stereoselectivity of guest molecule reactions within such cavities can be controlled, with often high optical yields particularly in the case of photo-induced processes.[19] Several reviews and recent publications can be con-

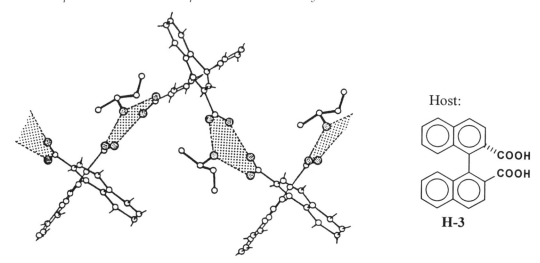

Host:

H-3

Figure H 5. X-ray structure of a complex between 1,1′-binaphthyl-2,2′-dicarboxylic acid **H-3** and 2-butanol; shaded region: hydrogen bond loops; H-atoms omitted if not involved in hydrogen bonding.[13b] Reprinted with permission of Springer Verlag N.Y. (*Top. Curr. Chem.*, Weber *et al.*, 1988, **149**, Fig. 17b, p. 45)

sulted on nanoporouos new materials, in which non-covalent interactions between properly designed building blocks and networks (Section I 2) generate any desired kind of voids. These materials can be of future use not only for the control of chemical reactions and for separations, but also for shape-selective catalysis and optoelectronic applications.[20] The three dimensional structure of such crystalline material can be qualitatively understood, also on the basis of Kitaigorodskii's Aufbau principle[21] from one-dimensional aggregates via two-dimensional monolayers to three dimensional solids, but still encounters severe problems in rigourous predictions[22] (Sections E 2 and E 7).

Layered solids can form 'lamellae' in which an enormous number of small molecules can intercalate.[23] The specific surface S (in $(m^2 g)^{-1}$) of such a material can be calculated from equation (H 2-1), in which ab is the surface of one unit cell (usually known from X-ray), n is the number of sides exposed to the interlayer, N is the Avogadro number, and M the molecular weight

$$S = N \cdot ab \times 10^{-18} \; n/M \qquad (H 2-1)$$

One can also calculate S from the density ρ of the layer material and the interlayer distance d (in nm)[23]

$$S = 2 \times 10^3/(\rho \times d) \qquad (H 2-2)$$

Typical numbers for such clay-like material are $S = 970 \; m^2/g$ for the zirconium phosphate $\alpha\text{-}Zr(HPO_4)_2H_2O$, indicating the great potential of such materials for reversible adsorption techniques. Again, there are usually non-covalent interactions mostly between polar groups of the layers and the intercalated molecules together with well defined space restrictions responsible for the selective adsorption processes.

H 3. MOLECULAR IMPRINTING

An approach to combine shape and functional selectivity by formation a cavity complementary to a desired template molecule in a solid support (usually a polymer) is known as molecular imprinting.[24] The general idea relates to host preorganization; it consists in preparation of a rigid binding site by polymerization of a monomer possessing suitable functional groups complementary to func-

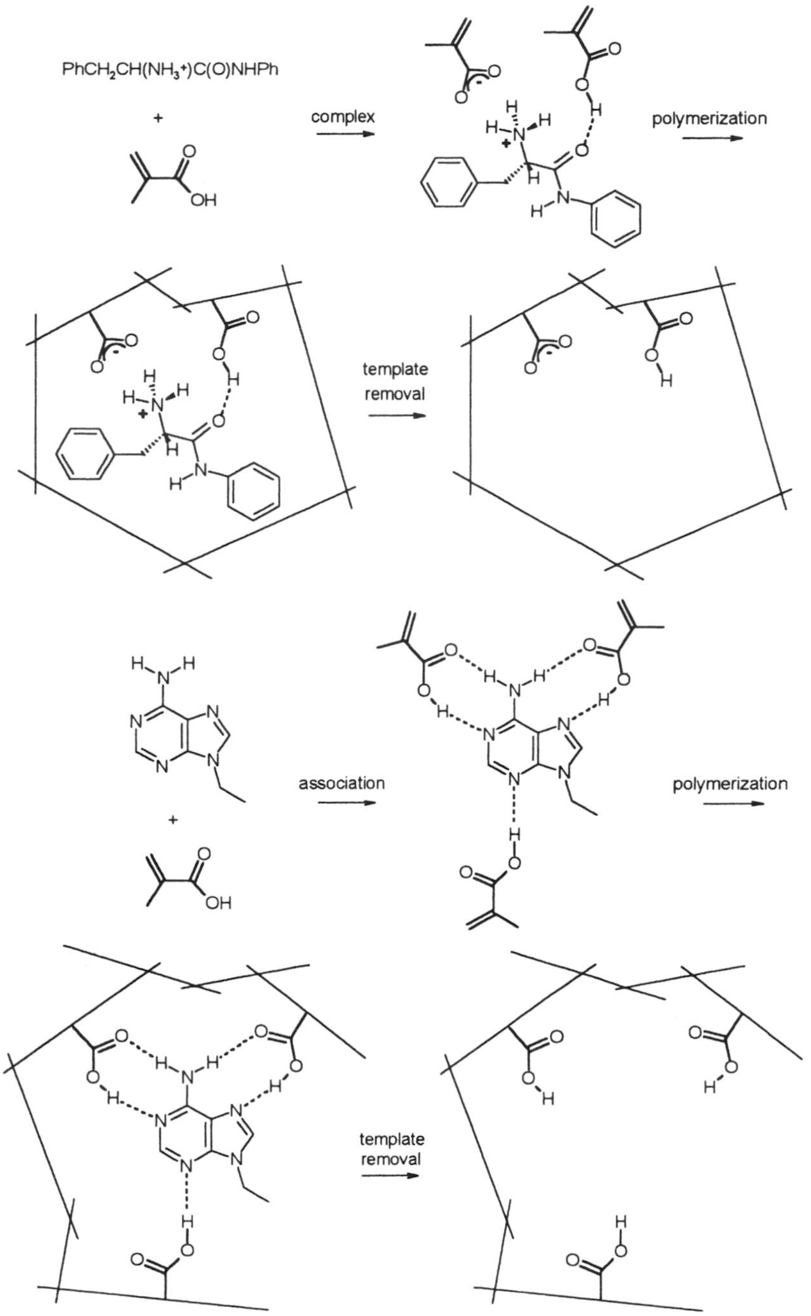

Scheme H1. Creation of selective binding sites by imprinting in polymethacrylic acid with (a) electro-statically[25] and (b) hydrogen bound[26] templates. (a) Reprinted with permission from *J. Am. Chem. Soc.*, Sellergren *et al.*, 1988, **110**, 5853. Copyright 1988 American Chemical Society; (b) Reprinted with permission from *J. Am. Chem. Soc.*, Spivak *et al.*, 1997, **119**, 4388. Copyright 1997 American Chemical Society.

tions in the target guest molecule (template). One expects that the polymerization will fix the optimum three dimensional arrangement of functional groups created in the complex, and that in addition the polymer chains will be packed around the guest in a complementary shape thus enhancing the binding selectivity. When polymerization is completed, the template is removed e.g. by extraction and the polymer matrix can be used as a selective sorbent for separation or chromatography as well as for other applications including catalysis.[24]

Scheme H1 illustrates the preparation of selective binding sites in the matrix of poly-methacrylic acid by imprinting templates which use electrostatic (Scheme H1a)[25] or hydrogen bonding (Scheme H1b)[26] interactions. In the first case the chiral cationic template L-phenylalanine anilide is employed. The resulting imprinted polymer possesses fairly high enantioselectivity in chromatographic separation of L- and D-phenylalanine anilide ($\alpha = k'_L / k'_D = 3.2$; k'_L and k'_D are the

capacity factors, see Section D4.3, for the respective enantiomers) and increased affinity to the template guest as compared to structurally similar molecules (other derivatives of phenylalanine). In the second case the template is the nucleobase derivative 9-ethyladenine, the binding constant of which to the resulting polymer $K = 7.6 \times 10^4 \ M^{-1}$ in chloroform was evaluated from a Langmuir adsorption model. This K value can be compared with $K = 160 \ M^{-1}$ for the complex of 9-ethyladenine with butyric acid in the same solvent. The binding selectivity is high as is evident from comparison of chromatographic capacity factors for the template molecule and analogues.

Besides organic polymers, imprinting can be accomplished with various other matrixes such as silica gels, surface layers of inorganic materials and biopolymers (proteins and carbohydrates) and imprinted matrixes find the growing applications, principally, in separation and analysis.[24]

REFERENCES

1 Cram, D.J.; Cram, J.M. (1994) *Container Molecules and Their Guests*, Royal Society of Chemistry, Cambridge, UK.

2 Atwood, J.L.; Davies, J.E.D.; MacNicol, D.D., (1984–1992) eds; *Inclusion Compounds*, Oxford University Press, Oxford, *Vol. 1–5*.

3 a) MacNicol, D.D.; Toda, F.; Bishop, R.; (1996) eds.; *Comprehensive Supramolecular Chemistry*, Vol. 6, Pergamon/Elsevier, Oxford; b) Alberti, G.; Bein, T. (1996) eds. *Comprehensive Supramolecular Chemistry*, Vol. 7, Pergamon Elsevier, Oxford.

4 Weber, E., (1988) Ed.; *Top. Curr. Chem.* **149**.

5 See e.g. a) Dyer, A. (1988) *An Introduction to Molecular Sieves*, Wiley, New York; b) Chen, N.Y.; Degnan, Th. F.; Smith, C.M. (1994) *Molecular Transport and Reaction in Zeolites*: *Design and Application of Shape Selective Catalysis*; VCH, New York; c) Suib, S.-L. (1993) *Chem. Rev.*, **93**, 803; d) Chapman, R.G.; Sherman, J.C. (1997) *Tetrahedron*, **53**, 159.

6 Meier, W.M.; Olson, D.H.; Baerlocher, C. (1996) *Atlas of Zeolite Structure Types* Lab. für Kristallographie, ETH CH 8092, Zürich.

7 Cram, D.J.; Tanner, M.E.; Thomas, R. (1991) *Angew. Chem., Int. Ed. Engl.*, **30**, 1024.

8 Beno, B.R.; Sheu, C.; Houk, K.N.; Warmuth, R.; Cram, D. (1998) *Chem. Commun.*, **301**, and references cited therein

9 Garel, L.; Dutasta, J.P.; Collet, A. (1993) *Angew. Chem., Int. Ed. Engl.*, **32**, 1169.

10 Bartik, K.; Luhmer, M.; Dutasta, J.-P.; Collet, A.; Reisse, J. (1998) *J. Am. Chem. Soc.*, **120**, 784.

11 Garel, L.; Lozach, B.; Dutasta, J.-P.; Collet, A. (1993) *J. Am. Chem. Soc.*, **115**, 11652.

12 Kitaigorodsky, A.I. (1984) *Mixed Crystals*, Springer, Heidelberg; Kitaigorodsky, A.I. (1984) *Molecular Crystals and Molecules*, Academic Press, New York.

13 a) Goldberg; I., in ref. 4, p. 1; b) Weber, E.; Czugler, M. in ref. 4, p. 45.

14 Desiraju, G.R. in ref. 3a, p. 1; Weber, E. *ibid*, p. 535 ff.

15 Adachi, L.; Berkowitch-Yellin, Z., Weissbuch, I.; van Mil, J.; Shimon, L.J.W.; Lahav, M.; Leiserowitz, L. (1987) *Angew. Chem.*, **97**, 476; Weissbuch, I.; Popovitz-Biro, R.; Weinbach, S.; Lahav, M.; Leiserowitz, L. in ref. 3a, p. 885 ff.

16 Ermer, O.; Röbke, C. (1993) *J. Am. Chem. Soc.*, **115**, 10077.

17 Davy, H. (1911) *Phil. Trans. Roy. Soc. London*, **101**, 155; for a recent review see e.g. Dyadin, Y.A.; Belosludov V.R. in ref. 3a, p. 789; Sloan, E.D. Jr.; (1998) *Clathrate Hydrates of Natural Gases*, 2nd edn., Marcel Dekker, New York.

18 Lipkowski, J in ref. 3a, p. 691 ff.

19 Toda, F. 211 ff. ; Toda, F.; Tanaka, K, (1988) *Tetrahedron Lett.* **29**, 551.

20 Wuest, J.D. (1995) *Struct. Energ. React. Chem. Ser.*, **1**, 107; Vaugeois, J.; Simard, M.; Wuest, J.D. (1995) *Coord. Chem. Rev.*, **145**, 55; Swift, J.A.; Russel, V.A.; Ward, M.D. (1997) *Adv. Mater.*, **9**, 1183; Russel, V.A.; Evans, C..; Li, W. Ward, M.D. (1997) *Science*, **276**, 575.

21 Kitaigorodskii, A.I. (1961) *Organic Chemical Crystallography*, Consultants Bureau, New York.

22 For references and a recent approach with Monte-Carlo calculations restricted mostly to electrostatics see Perlstein, J. (1994) *J. Am. Chem. Soc.*, **116**, 455.

23 Alberti, G.; Costantino, U. in ref. 3b, p. 1 ff; Whittingham, M.S.; Jacobson, A.J., (1982) eds. *Intercalation Chemistry*, Academic Press, New York.

24 Wulff, G. (1995) *Angew. Chem. Int. Ed. Engl.*, **34**, 1812.

25 Sellergren, B.; Lepistö, M.; Mosbach, K. (1988) *J. Am. Chem. Soc.*, **110**, 5853.

26 Spivak, D.; Gilmore, M.A.; Shea, K.J. (1997) *J. Am. Chem. Soc.*, **119**, 4388.

I 1. CHIRAL DISCRIMINATION

The separation of enantiomers[1] may well represent the economically most important application of supramolecular chemistry. Chiral recognition relies on non-covalent interactions between dissymmetric molecules; the understanding of intermolecular forces is therefore the basis of a rational design of chiral phases for chromatographic separations, for example. The interaction of a chiral substrate, or selector A with a racemic mixture L_S, L_R of a compound to be separated is characterized by an equilibrium leading to diastereomeric complexes L_SA and L_RA, which may differ in rate of formation, in stability constants K_S and K_R; in spectroscopic and in many other properties

$$A + L_S \rightleftharpoons AL_S \qquad K_S = e^{-\Delta G(S)/RT}$$
$$A + L_R \rightleftharpoons AL_R \qquad K_R = e^{-\Delta G(R)/RT}$$

The efficiency of a chiral selector for enantiomeric separation is conveniently characterized by the free energy difference $\Delta\Delta G = \Delta G(S)-\Delta G(R)$. Enantiomeric differentiation requires recognition at an asymmetric center, which in principle is described by a three-point interaction, which may be either attractive or repulsive. In the latter case the intermolecular association must be secured by other parts of the molecules A and L. One can therefore design selector host compounds in which binding and chiral selection sites are quite separated (see Scheme I 1), thus achieving a fine-tuning of affinity and selectivity, as we see it in many natural receptors. A classical example for this is the complex of the chiral binaphthyl crown ether **I-1**, which

has a C_2 symmetry, with aminoacid derivatives (Scheme I-2).[2] Here the crown ether unit provides the driving force for the association of the aminoacid $^+NH_3$-terminus, whereas the interactions of the dissymmetric naphthyl units with, for instance the substituents R at the aminoacid α-C atom lead to stereoselection. The size of the substituents in the phenylalanine methyl ester decrease in the order Ph > COOMe > H which according to Cram's rules are accommodated with the least repulsive effects as shown in Scheme I-2, leading to preferential binding of the S-aminoacid derivative with the SS host compound. In addition an interaction between the ester group and the near naphthyl unit may stabilize this conformation.

Spectroscopic differentiation of enantiomers[3] does not necessarily require a difference between K_S and K_R. It can be sufficient that the intrinsic properties like NMR shifts of the diastereomeric complexes AL_S and AL_R differ; the affinity constant K itself, however, should be large enough to distinguish antipodes with small spectral differences. This explains why for the determination of optical yields chiral lanthanide shift reagents[4] are particularly successful, which with a large range of electron pair donor atoms such as oxygen or nitrogen form more stable complexes than organic compounds usually do. Significant improvements in signal separation between the diastereomers can be achieved by selecting solvents, which by the rules discussed in Section C 2 can enhance the binding constants.

Separation of enantiomers, for example by chromatographic techniques in contrast to

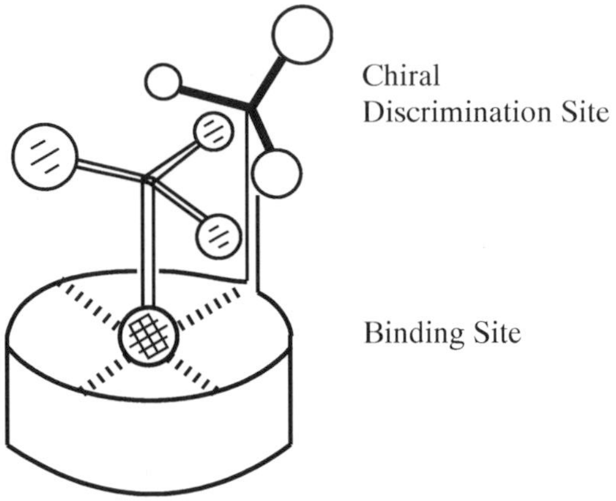

Chiral
Discrimination Site

Binding Site

Scheme I1.   Separation between binding and selection sites in supramolecular complexes.

spectroscopy always requires differences between K_S and K_R, which, however, can be quite small due to the repetitive principle of chromatography. Also in contrast to spectroscopic differentiations, the overall affinity should not be too large in view of the possibly too slow complexation and decomplexation rates in the mobile phase. The chiral selector may be either in the eluent, where it forms the diastereomeric complexes of different affinity to non-chiral stationary phase, or it is immobilized. Stationary chiral phases have obvious practical advantages, both with respect to the amount of material used and (for preparative applications) to the necessary separation between selector and substrate. The immobilization of course requires additional steps; it also may lead to variation of the binding

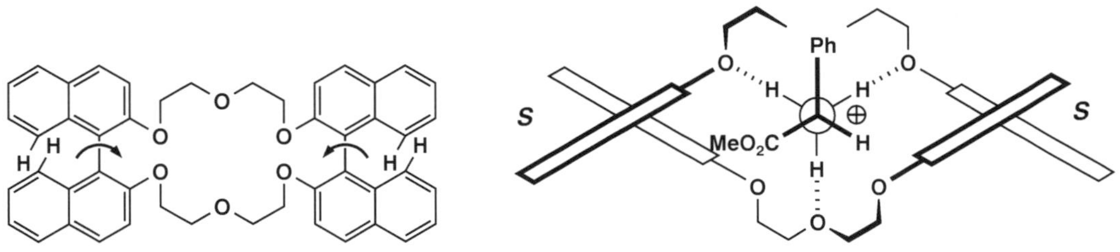

I-1

Scheme I2.   Chiral recognition of the phenylalanine methyl ester.[2] Reprinted with permission from *Acc. Chem. Res.*, Cram *et al.*, 1978, **11**, 9. Copyright 1978 American Chemical Society.

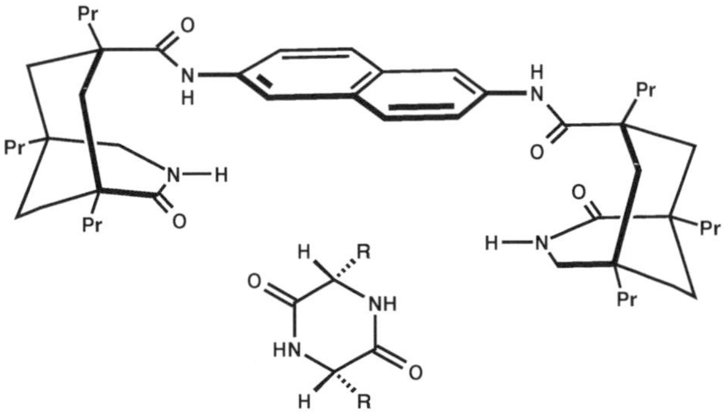

Pr = CH_2CH_2Me; R=CH_2CHMe_2 **I-2**

Scheme I3.   (*continued*)

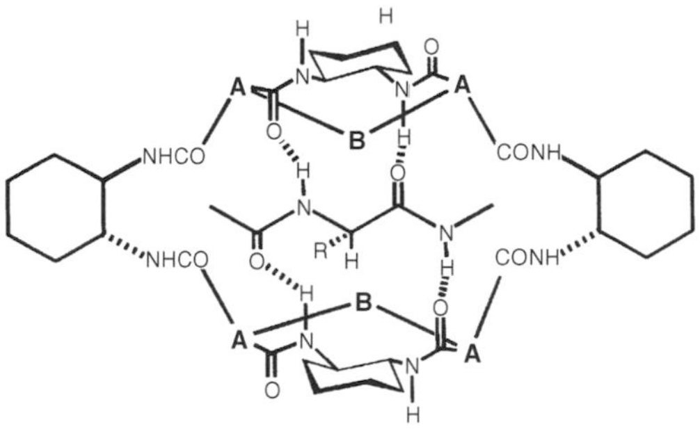

I-3

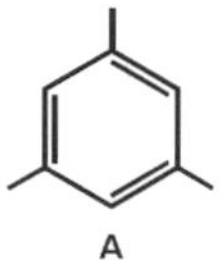

I-4

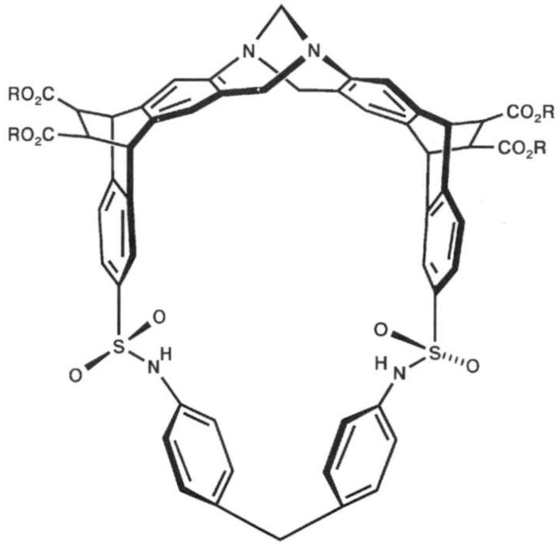

A

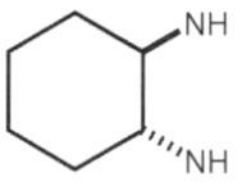

B

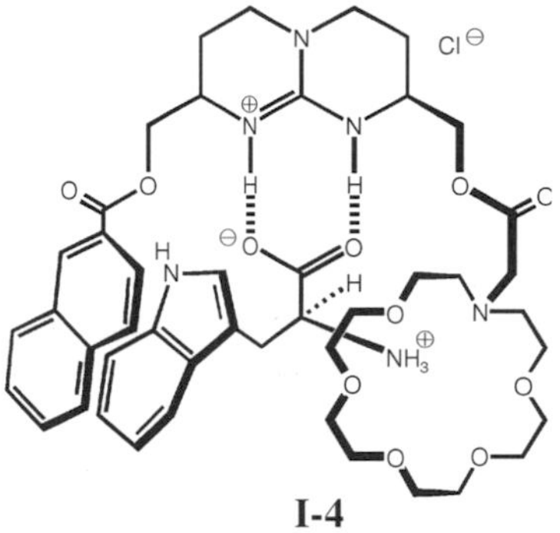

I-5

O$_2$N — ... — C—NH—CH—CO—X

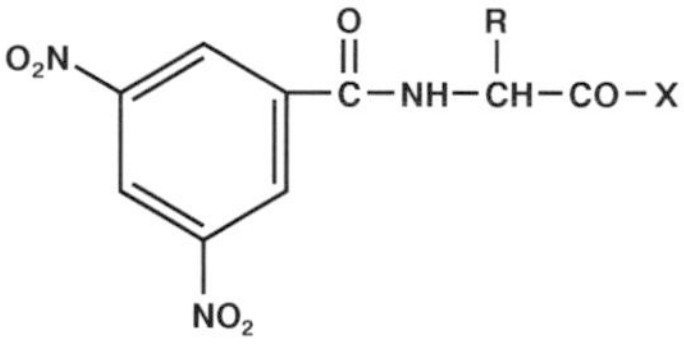

I-6

—NH—CH—C—X

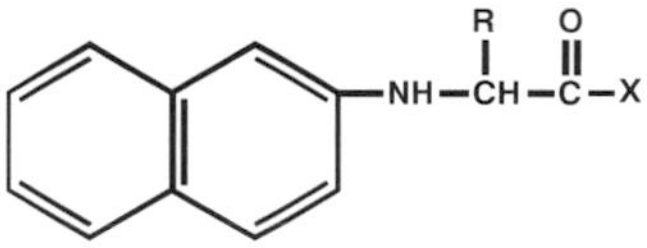

I-7

R = Ph, MeCHEt etc.

X = O-(CH$_2$)$_n$-O-Si≡

Scheme I 3. Examples for chiral selector molecules.

mechanisms, if the conformation of the complexes differ from that in the free state in solution.

Many systems for efficient chiral discrimination rely on hydrogen bonds and therefore work well only in solvents like chloroform.

The concave Kemp acid derivative **I-2** in Scheme I3 binds a leucine-derived diketopiperazine with the high preference of $\Delta\Delta G = 11.3$ kJ mol^{-1} of one enantiomer.[5] The particularly attractive host **I-3** is derived in a remarkably short synthetic route from the chiral *cis*-1,2-diaminocylcohexane. In chloroform **I-3** shows with *N*-acyl, NH-methyl aminoacids and some corresponding dipeptide derivatives relatively high selectivities.[6] One observes, for example for the alanine derivative $\Delta\Delta G$ of 5.5 kJ mol^{-1}, and for the valine derivative 11 kJ mol^{-1}. Since at the same time the total association free energy increases for both enantiomers the increase from Ala to Val suggests that not only the four hydrogen bonds shown in Scheme I3 but also the side chains contribute to binding. With receptor **I-4** elegant use is made simultaneously of ion pairing (or hydrogen bonding) at the guanidinium function, of the –NH$_3^+$ group complexation at the crown ether unit, and of stacking interactions at the naphthalene part; this way one can extract with dichloromethane from aqueous racemic Phe or Trp mixtures one pure enantiomer.[7]

Enantioselective recognition in protic solvents, particularly in water, usually has to rely on other interactions than on hydrogen bonds. Receptors like **I-5** have a largely hydrophobic dissymmetric cavity, and complex lipophilic substrates such as menthol with moderate enantioselectivity, but with binding constants of up to 4500 M^{-1}.[8] Very frequently used host compounds for chromatographic separations of lipophilic enantiomers are cyclodextrins, which often work more efficiently after chemical modification.[9] Cyclodextrins with covalently attached fluorophores (Section I4) show, for instance, with borneol enantiomers binding constant differences of up to 100%.[10] Such systems might be used for the design of chiroselective optical sensors. Chiral host compounds can also be used for enantiomer analyses by electrochemical methods in analogy to ion selective electrodes.[11] Lipophilic interactions are essential in the very efficient chromatographic separations achieved with aminoacid derivatives such as **I-6** and **I-7**.[1,12] The π-electron-poor dinitro compound **I-6** favors electron-rich analytes; conversely **I-7** is used for electron-poor compounds. For separation of non-aromatic analytes these can be functionalized prior to separation with corresponding aromatic acid halides, etc. Molecular modeling can be used to predict the outcome of the separations and to design new selectors, in particular by computer aided search for sites of maximum chiral discrimination in both analyte and selector molecule.[13] For optimal separation these interaction sites should as much as possible be congruent in the supramolecular complexes.

We close this Section with an example of discrimination with inherently achiral concave molecules similar to those discussed in Section I2 below (Scheme I4). These can self-associate to dissymmetric capsules, which dissociate slowly on the NMR time

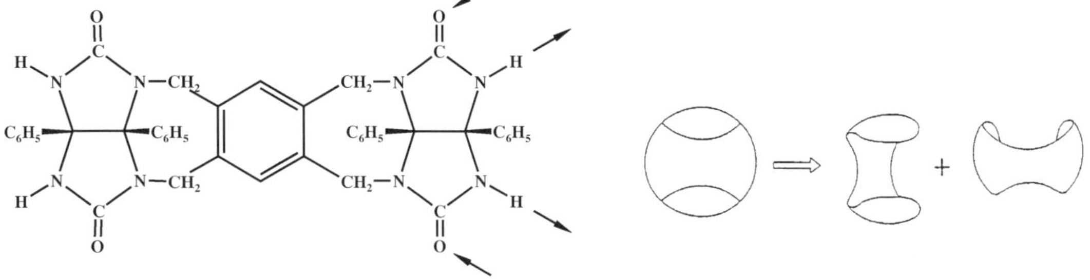

Scheme I4. Self-assembling achiral concave molecules, forming dissymmetric capsules from symmetric precursors.[14] Reprinted with permission from *Science*, 1998, **279**, p. 1021. Rivera *et al.* (Chiral spaces: dissymmetric capsules through self-assembly). Copyright 1998 American Association for the Advancement of Science.

scale, and allow to observe different signals for entrapped small chiral substrates, although the $\Delta\Delta G$ values are only up to 1.7 kJ mol^{-1}.[14]

12.   SELF ORGANIZATION PROCESSES AND APPLICATIONS

12.1. TEMPLATE ASSOCIATION AND SUPRAMOLECULAR SYNTHESIS

The combination of specific complementary functions within a tailor-made host requires many specific methods of preparative chemistry. We only give some leading references to principles, which are characteristic of supramolecular systems. Preparation of the desired macrocyclic hosts[15] frequently present a particular synthetic challenge, which traditionally is solved by high dilution techniques, in order to slow down unwanted intermolecular reactions which lead to polymers.

Of particular promise are synthetic approaches using templates.[16] These strategies are based themselves on properly designed non-covalent interactions, and therefore will be discussed here with some typical examples. Earlier the most often used templates were metal cations, which lead to pre-association of host parts such as ethyleneglycol or ethylenediamine units. Many crown ethers and their aza analogs can be obtained only, or with substantially improved yields, by the use of cesium[17a] (for crown ethers), other alkali cations (for calixarenes[17b]), or of transition metals[17c] (for azamacrocycles, phtalocyanines etc). This way, even catenanes are accessible in good yields by combinations of copper-binding bipyridyl units and crown ether fragments (Scheme I 5).[18] Metal coordination together with suitable, if possible rigid, spacers may provide entries into large new macrocycles. The free axial ligand site in zinc porphyrins can by complexation with s-tripyridyl-triazine lead to substantial improvement in yield and selectivity of large porphyrinophanes.[19]

Another approach rests on the complexation of secondary ammonium ions $R_2NH_2^+$ in large crown ethers such as 24C 8 (Scheme I 6). Reactions at the ends of R will then give in appreciable yields the intramolecular catenanes.[20] The structural identity of such catenanes is most easily proven by the parent peaks in the mass spectra.

More recently, also organic templates are used for efficient preorganization of building blocks around the template (Scheme I 7).[20] Such condensation lead in good yields to new rotaxanes, as the electron-deficient bis-bipyridyl unit and the electron-rich ethyleneglycolhydroquinone derivative pre-associate to rather stable complexes.

The intriguing carceplexes (Section H 1) are obtained from two bowl-shaped resorcarene fragments by condensation with chlorobromomethane; good yields of up to 87% are observed only, if a suitable template such as pyrazine as guest is added.[21] It has been shown that pyrazine indeed pre-associates with two resorcarene monomers, and that with other guest molecules the pre-transition state equilibrium constants correlate with the carceplex formation rates (Scheme I 8).[21b]

Non-covalent interactions with peptides and polynucleotides are not only the molecular basis of life, but are also used to construct synthetic analogs. Cyclic oligonucleotides with interesting binding capacities towards nucleic acids can be obtained by macrocyclisation combining oligothymidines with crown ether fragments, around a oligoadenine template, making use of Watson–Crick A–T base pairing (Scheme I 9).[22]

12.2.   NEW MATERIALS AND BIOMOLECULES

Amphiphilic oligopeptides, which are covalently bound on a topologically tailored polymer basis, lead to associations similar to secondary structures in proteins, called Template assembled synthetic proteins (TASP)[23] (Scheme I 10); their preparation relies on the use of protected peptides with the subsequent

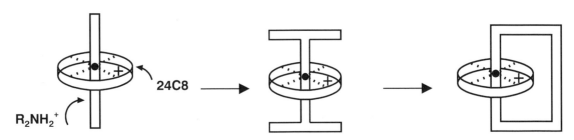

Scheme I 5. Template synthesis of a catenane.[18] Reprinted with permission from *Acc. Chem. Res.*, Sauvage *et al.*, 1990, **23**, 319. Copyright 1990 American Chemical Society.

Scheme I 6. Catenane synthesis via complexation between crown ether and secondary ammoniumions.[20]

assembly of deprotected oligomers. The template can be, for instance, a cyclic peptide; the ensuing supramolecular structures composed of α-helices or β-sheets can form ion channels etc.

Particularly close to the biological model are condensations of mono- or oligonucleotides on an oligonucleotide template.[24] The strong Watson–Crick base pairing G–C sequences yield with activated nucleotides

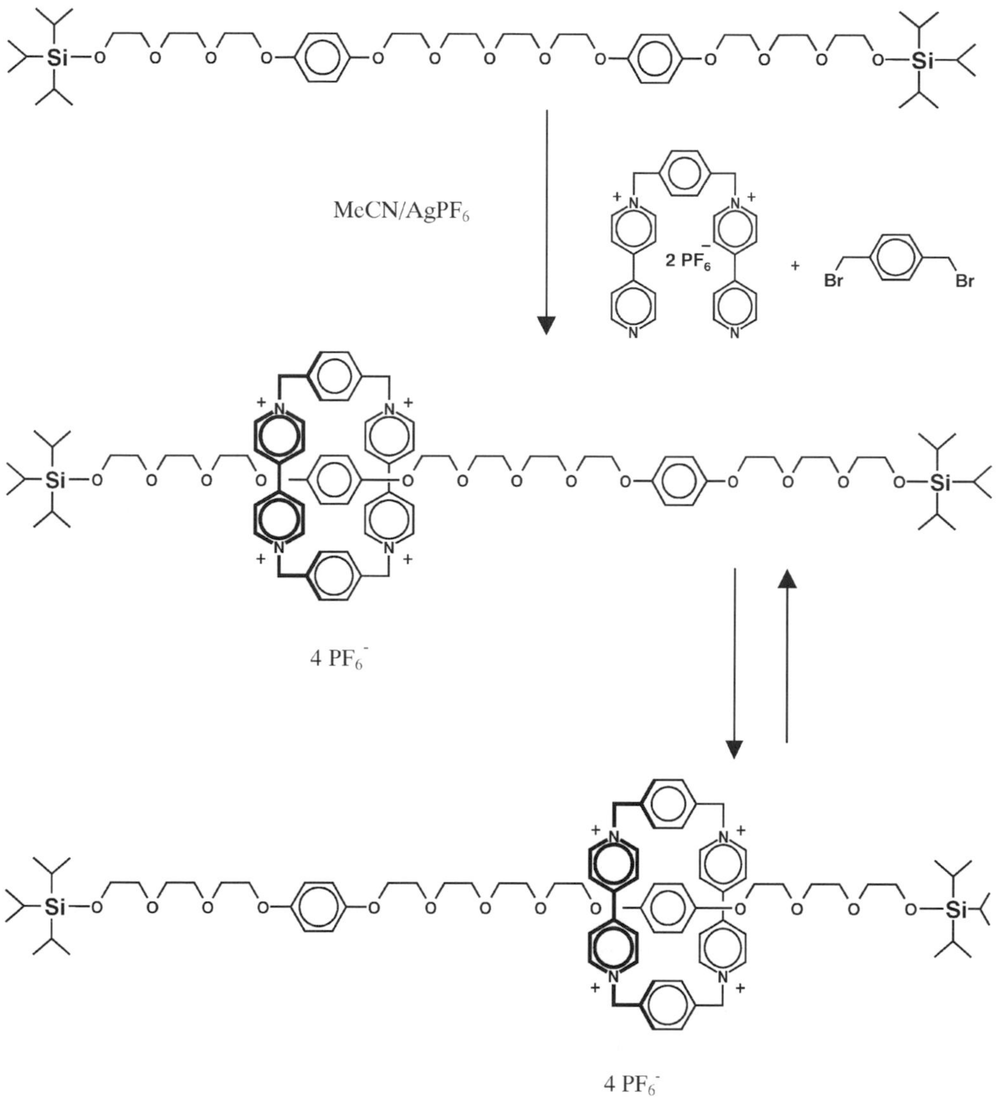

Scheme I7. Rotaxane synthesis via pre-association of components.[20]

in the form of 5′-phosphoro-2-methylimida-zolides palindromic sequences of up to 14 base pairs, with the Watson–Crick complementary bases oriented in the correct antiparallel 3′,5′-direction (Scheme I11). The new single strand produced, however, is usually not able to act as new template, for which reason the reactions are not self-replicating.

Coordination around transition metals provides another strategy towards construction of large supramolecular networks. Self-association of oligomeric bipyridyl units around copper(I) gives rise to metalorganic analogues of double-stranded nucleic acids, as result of the tetrahedral coordination around Cu(I) (Scheme I12, see also website).[25] The cooperativity of helix formation after incorporation of the metal is evident from the absence of any non-helical ligands after mixing. Monomers with exocyclic bipyridyl units in a macrocycle

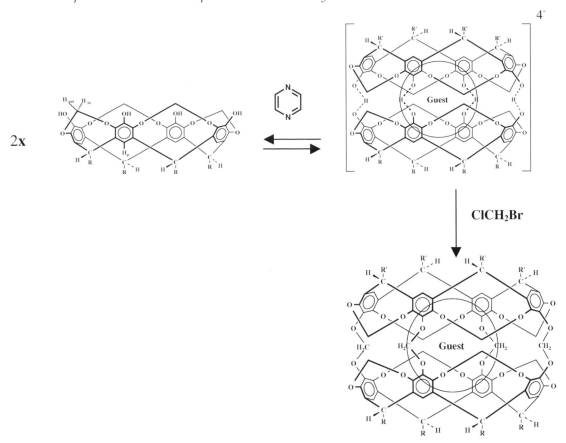

Scheme I8. Pre-transition state association of a resorcarene derivative ($R = CH_3$) with pyrazine as guest with subsequent formation of a carceplex.[21b] Reprinted with permission from *J. Am. Chem. Soc.*, Chapman *et al.*, 1995, **117**, 9081. Copyright 1995 American Chemical Society.

form with silver ions interesting helical coordination polymers.[26] The reaction of *para*-phenylen spacers bearing chelating substituents with Fe(III) ions yield adamanantane-shaped large endoreceptors encapsulating, for instance ammonium ions. (see website).[27]

Association between two or more concave host parts can provide new superhosts for guest molecules, from which a change of the microenvironment, like that of the pH, can release the guest molecule. The lactame rings in the glycoluril derivative in Scheme I4 can built up 8 hydrogen bonds to form a tennis-ball-type structure, which can encapsulate methane.[28]

Self-association by hydrogen bond networks has become one of the most fruitful concepts for the construction of large supramolecular aggregates.[29–31] The rather strong directional character of the hydrogen bond (Section B3) makes corresponding monomers particularly suitable candidates for the controlled build-up of such aggregates. Amides are ideal building blocks as they can form with the NH_2 group two donor, and with the CO group also two acceptor hydrogen bonds. Depending on their substitution they can form tape- or layer-like structures (Scheme I13). Thus, melamine and cyanuric acid form very stable lattices due to the build-up of nine hydrogen bonds around each monomer

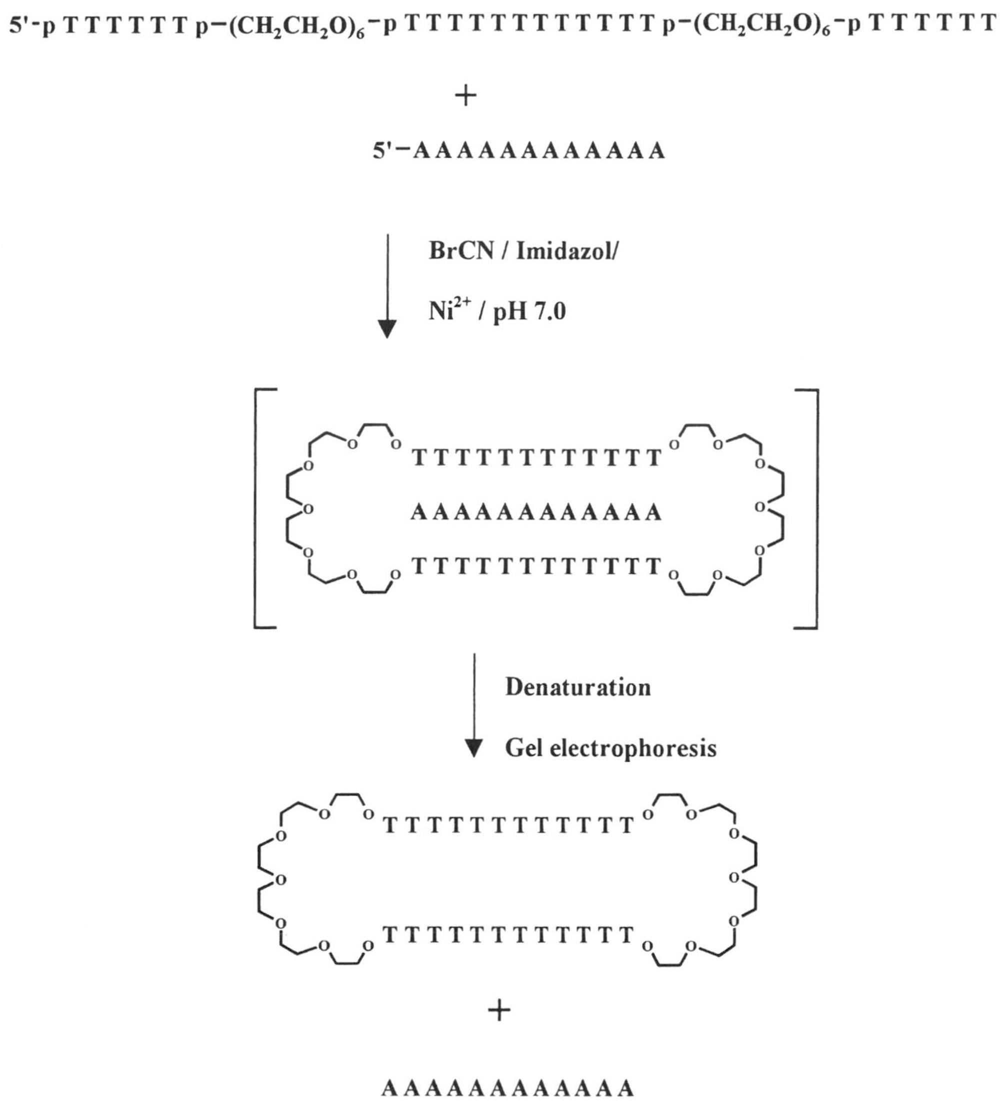

5'-p T T T T T T p−(CH₂CH₂O)₆−p T T T T T T T T T T T T p−(CH₂CH₂O)₆−p T T T T T T

$+$

5'−A A A A A A A A A A A

BrCN / Imidazol/

Ni²⁺ / pH 7.0

[T T T T T T T T T T T T
  A A A A A A A A A A A
  T T T T T T T T T T T T]

Denaturation

Gel electrophoresis

T T T T T T T T T T T T

T T T T T T T T T T T T

$+$

A A A A A A A A A A A

Scheme I9. Template assembled synthesis of cyclic oligonucleotides.

(Scheme I13).[29] If both hydrogen bond donor and acceptor sites are at the periphery of a molecule like in the form of lactames, self-association can lead to an intriguing cubic diamond lattice in the solid state.[32]

Such aggregates have an intriguing analogy to spontaneous formation of membranes and other biological superstructures. They provide an increasingly popular entry to nano-materials,[29,30] including monolayers, bilayers, thin films and tubes. Even linear aggregates can, dependent on their monomer structure, form a large variety of cylinders, helices, tubes, rods, tapes, or spheres. With anthracene-resorcinol monomers a combination of stacking and hydrogen bonding leads to columns which can take up in their cavities linear alkyl chains.[33]

The problem of low kinetic stability of the non-covalently bound aggregates can be over-

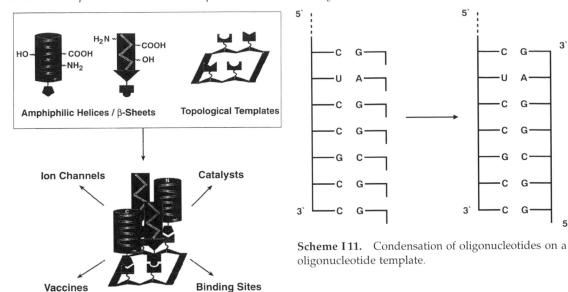

Scheme I 11. Condensation of oligonucleotides on a oligonucleotide template.

Scheme I 10. Template assembled synthetic proteins (TASP).[23] Reproduced by courtesy of the authors.

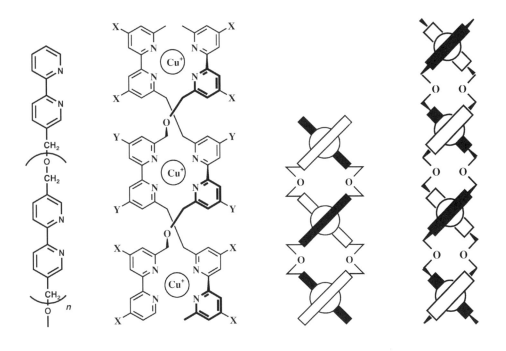

Y = X = H

Scheme I 12. Self-association of oligomeric bipyridyl units around copper(I) affording metalorganic analogues of double-stranded nucleic acids.[25] Reprinted with permission from *Pure Appl. Chem.*, Lehn, J.-M. 1994, **66**, 1961, copyright 1994, IUPAC, Blackwell Science.

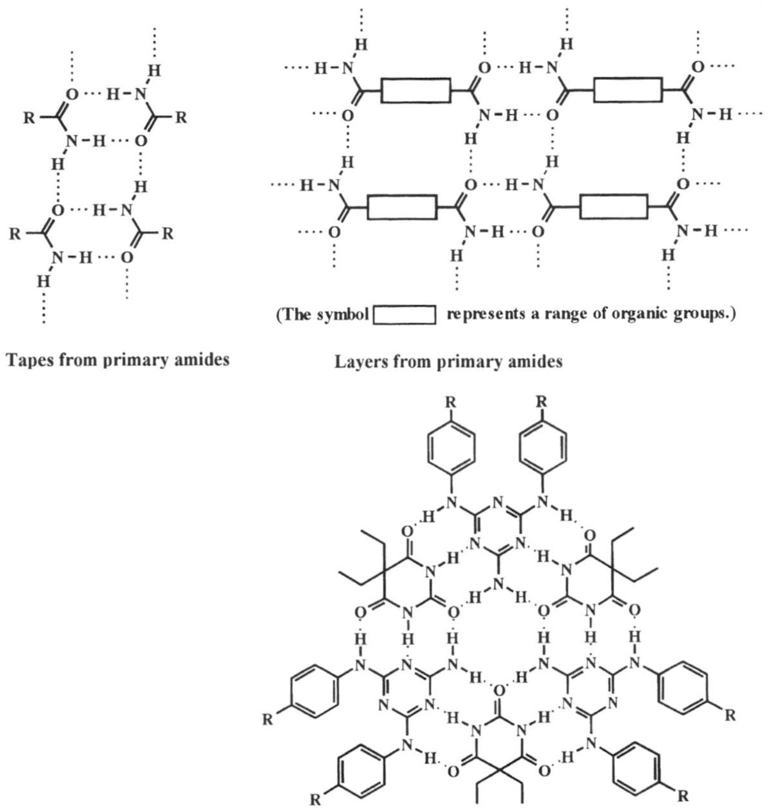

Tapes from primary amides

Layers from primary amides

(The symbol ☐ represents a range of organic groups.)

Scheme I 13. Superstructures formed by hydrogen bonding.

come by formation of covalent bonds subsequent to preassociation, as illustrated in Scheme I 14.[34] Here, one makes use of the self-assembly in chloroform of a cyclopeptide, bearing two olefinic side chains. In presence of a ruthenium-carbene catalyst one can achieve olefin metathesis to a stable tricyclic structure. Such strategies in principle allow the design of kinetically stable entities including protein modifications, or channels which then can be incorporated into membranes etc.

The abundant availability of non-covalent forces around properly chosen, often easily accessible, monomers has made the field of crystal engineering, one of the most promising new chemical technologies.[31] The selection of optimal monomers requires a detailed knowledge of geometries and energies of non-covalent interactions, which will be exploited in formation of network assemblies. The foundations for the rational design of nano-materials including uses of computer aided molecular modeling are covered in Section E 7 of this book. Molecular recognition on the surface of crystals also offers new ways for the resolution of racemates and, by stereochemical correlations, for configurational assignments. In contrast to the traditional way of resolution via formation of diastereomeric crystals one can amplify optical activity by enantioselecetive occlusion of dissolved racemic aminoacids through the enantiotopic faces of achiral (centrosymmetric) glycine crystals.[35]

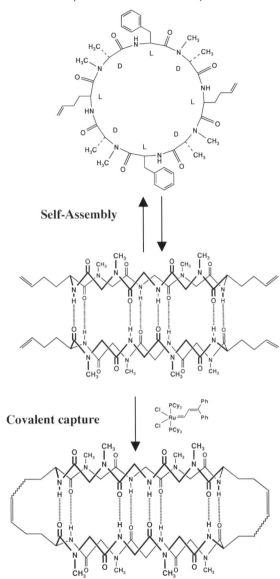

Self-Assembly

Covalent capture

Scheme I 14.   Covalently fixed self-assembled cyclo-peptide.[34] Reprinted with permission from *J. Am. Chem. Soc.*, Clark *et al.*, 1995, **117**, 12364. Copyright 1995 American Chemical Society.

12.3.   SELF-REPLICATION AND AUTOCATALYSIS

Autocatalysis, in which a reaction product enhances the reaction rate, has been known for a long time, for instance in the hydrolysis

of anhydrides. Self-replication is an intriguing supramolecular process, in which a reaction product not only enhances rates, but can react with the educts in a highly selective manner, so that the same catalyzing molecule is produced with high selectivity in a cyclic process.[24,36] A simple but elegant system is shown in Scheme I15: the anil template A forms a ternary complex with the two mono-meric educts M1 and M2, based on two salt bridges and/or hydrogen bonds, which brings the amino and aldehyde functions in proper vicinity for condensation to imine of the same structure as the template A.

Since the product will associate strongly to the educt template with increasing reaction progress, reminiscent of product inhibition with enzymes, one usually observes a depen-dence of reaction rate on the square root of the template concentration. Unless the starting (ternary) complex is more stable than the resulting binary aggregate the problem of increasing self-association can be overcome only if after each replication the usually formed associate is separated by change of reaction conditions, such as the pH, or by some transport mechanism. Ideally, the pro-duct would bind and therefore stabilize the transition state of the reaction, and complex not or to a lesser degree the educt template.

Synthetic U-shaped systems in which hydrogen bonding in organic solvents should lead to formation of another U-shaped associ-ate and thus to self-replication have been described;[37] the efficiency of this system has been discussed critically.[38] The idea is to produce from two educts A and B a template T, which by non-covalent interactions with single A and B should give a pre-transition state ternary complex T · A · B. The necessary convergent U-shape was reached with a Kemp's triacid imide, with a protected ami-noadenosine as a partner (Scheme I16).

The many possibilities for selective and catalytic syntheses by molecular recognition between nucleotides are illustrated in Scheme I17.[39] In the cross-catalytic pathway (a) the

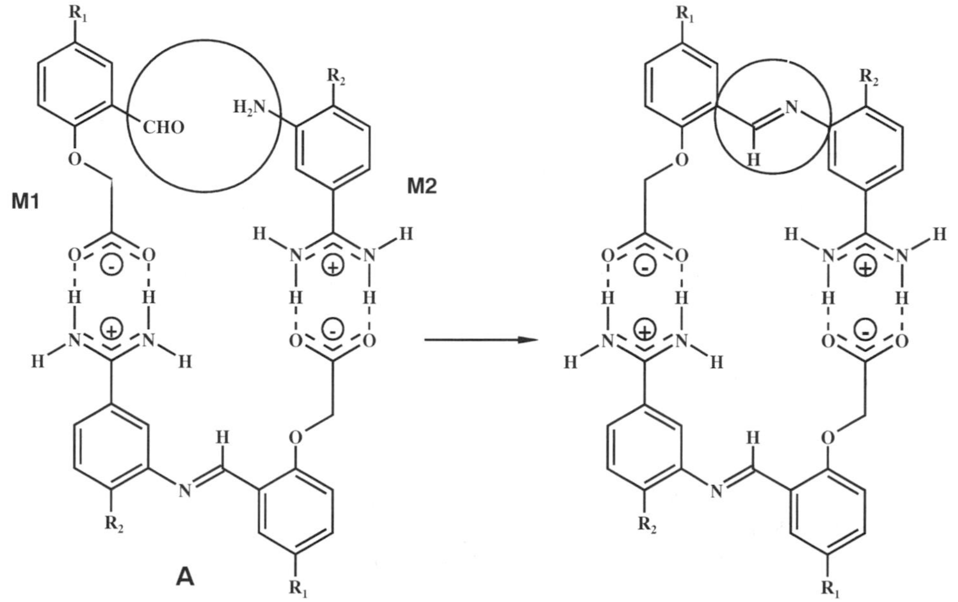

Scheme I 15.   Self-replication of a Schiff base template.[36a]

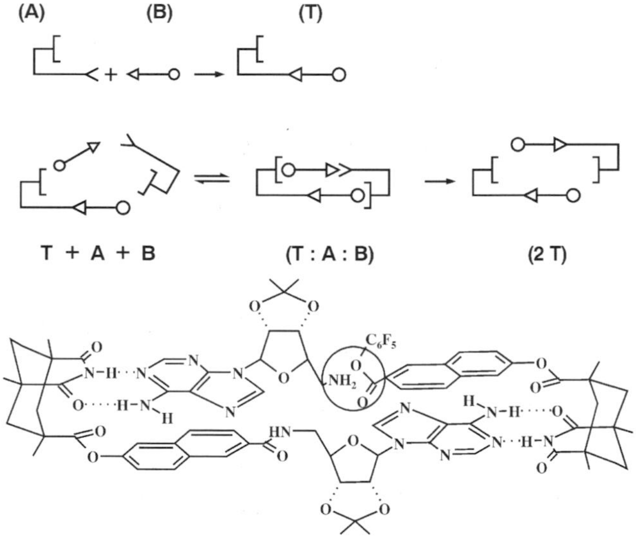

Scheme I 16.   Self-replication of an amide molecule via hydrogen bonding assembly.[37]

template AA or BB can react with single A or B to the two ternary complexes A · A · BB or B · B · AA which then by the proximity of suitable functions in A and B can react to give new AA and BB molecules. If the resulting duplex AA·BB dissociates then in single AA and BB strands each circle will double the number of available templates. If AB can be formed in the same way like AA or BB (like in some oligonucleotide coupling reactions) an autocatalytic pathway (b) leads also to AB. Only if all coupling reactions proceed with similar rates one observes the products from the cross-catalytic pathway.

Self-replication of surfactant aggregates occurs via autocatalytic alkaline hydrolysis of fatty acid esters[40] and anhydrides.[41] In these reactions a water insoluble substrate, like ethyl caprylate, undergoes initially a very slow hydrolysis by aqueous sodium hydroxide until the concentration of amphiphilic hydrolysis product sodium caprylate reaches its critical micelle concentration. After that the reaction rate increases very rapidly and, of course, all further carboxylate hydrolysis product appears in form of micelles. In spite of apparent simplicity of the system, the mechanism of 'micellar autocatalysis' is still under dispute, mainly because the micelles of an anionic surfactant should inhibit rather than catalyze the alkaline hydrolysis, which proceeds between a neutral hydrophobic molecule and an anionic OH$^-$ nucleophile of the same charge as the micelle (Section I3.2). It seems that the reason for autocatalysis is the strong solubilization effect of the

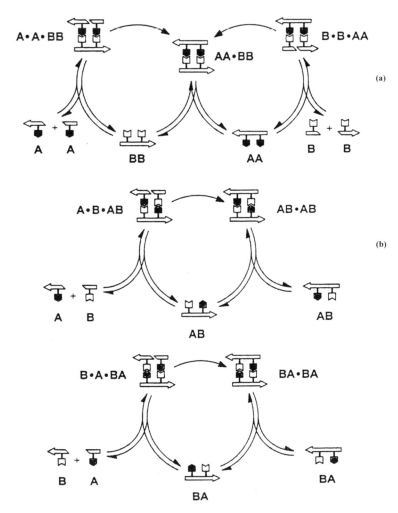

Scheme I17. Self-replication of nucleotides.[39] Reproduced by permission of the authors.[39]

amphiphilic product on the highly hydrophobic substrate rather than classical micellar catalysis.[42]

13. SUPRAMOLECULAR CATALYSIS

13.1. HOST–GUEST SYSTEMS

The difference between supramolecular and 'normal' catalysts, which in particular in the form of metal complexes play already a major synthetic role, is not always clear-cut. Supramolecular catalysts as we understand it here have one or several of the following features in common with enzymes: they (a) involve initial non-covalent substrate-catalyst binding, manifested in typical saturation kinetics (Section D6); (b) provide for transition state stabilization and/or activation of reacting groups; (c) can exert stereochemical control as a result of such non-covalent binding events; (d) can provide a rate-enhancing micromedium; (e) can contain functions as covalently or non-covalently bound cofactors, which accelerate reactions, either with or without covalent formation of intermediates. Most of the hitherto known examples represent only first steps towards true catalytic systems with high efficiency and turnover. As with the study of non-covalent binding mechanisms, where systems with intramolecular interactions often provide more straightforward information concerning the forces involved (Chapter B), intramolecular reactions may provide better insight into the mechanism of important catalytic steps. Many supramolecular systems need until now especially designed substrates such as activated esters, which mainly help to understand the underlying reaction mechanisms. However, we also show selected examples which already hold promise for practical applications, in particular with respect to steering selectivity of organic reactions. Another promising feature of supramolecular catalysis is that reactions can in most cases be carried out in water instead of organic solvents, and that

it can be designed in principle for *any* kind of reaction. Catalytic antibodies, microorganism or enzyme mutants also offer new ways; they need, however, special engineering and screening procedures and usually suffer from limitations with respect to applicability, availability and stability. Catalysis through host–guest complexation has been extensively reviewed, usually with an emphasis on the biomimetic aspect.[43–47]

We have already discussed complexation-induced changes in the guest reactivity in connection with kinetic determinations of formation constants (Section D6). Rate acceleration by host compounds was reported for numerous reactions, such as cleavage of esters, carbonates, amides and phosphates, nucleophilic and electrophilic aromatic substitution, decarboxylation, redox processes, and others. Like in nature there are two types of supramolecular catalytic systems: hosts with and without functional groups, which can participate by intermediate covalent bond making with the substrate (covalent catalysis as in the case of serine proteases). In both cases, the host cavity serves as a binding site, which accommodates substrates and affects their reactivity due to one or several factors such as bringing together and orienting reacting species, and producing changes in solvation, polarization effects, acid–base properties, etc. A decrease in the activation entropy term, $T\Delta S^{\neq}$, by an enzyme-like catalyst can also contribute significantly to rate accelerations.[48–50] For bimolecular reactions the loss of translational entropy alone can amount to $\Delta S^{\neq} \approx -150$ J mol^{-1}K^{-1}, in addition there can be considerable losses of rotational degrees of freedom (see Section A1.1 for related numbers on equilibria). These disadvantages can partially be compensated owing to conversion of a bimolecular reaction between reactants into a monomolecular process inside the initial non-covalent substrate–catalyst complex, which also can bring the reaction partners already in the correct, most productive orientation. Of course, non-covalent interactions occur also in uncatalyzed reactions when the reaction partners approach (or leave) each other. The difference to enzyme-like catalysts is, that here the binding forces stabilizing a pre-transition state complex are not restricted to the reaction sites of the molecules: a suitable host can make use of *all* available interaction possibilities of the *whole* substrate molecule, including parts remote from the reaction center. It is convenient to treat catalytic effects in terms of a unified approach based on the idea of different complexation strength between substrate and catalyst in the initial and in the transition states of the reaction.[44,46]

Let us consider first the simplest case of a single substrate reaction, which occurs in accordance to Scheme I 18 (similar to a scheme given by equations (D 6-4)–(D 6-6) and follows the Michaelis–Menten-type rate equation (similar to (D 6-7))

$$k_{obs} = (k_0 + k_C K_S[H])/(1 + K_S[H]) \quad (I3\text{-}1)$$

Note, that with enzymes one usually works with excess substrate concentration, in contrast to most supramolecular catalysts. Thus with enzymes and the neglect of typically much slower uncatalyzed reaction one can use the traditional form of the Michaelis–Menten equation (equation (D 6-11) in Ch. D 6), usually applied for analysis of initial reaction rates since the reaction is not first-order in the substrate in this case.

In terms of transition state theory, chemical transformation of free (S) and complexed (HS) substrate molecules involves a pseudo-equilibrium formation of the respective transition states $T^{\neq}$ and $HT^{\neq}$ with the respective equilibrium constants $K_0^{\neq}$ and $K_C^{\neq}$, Scheme I 19. Evidently, one can introduce (assuming all species in Scheme I 19 to be in equilibrium) the equilibrium constant for formation of $HT^{\neq}$ from the host molecule and substrate transition state $T^{\neq}$

$$K_T = K_S K_C^{\neq}/K_0^{\neq} \quad (I3\text{-}2)$$

Using transition-state theory expression (I 3-3) for the rate constant k

$$k = (k_B T/h)K^{\neq} = (k_B T/h) \exp\left(-\Delta G^{\neq}/RT\right) \quad (I3\text{-}3)$$

where k_B is Boltzmann's constant, h is Planck's constant, and $\Delta G^{\neq}$ is the activation free energy, one can rewrite equation (I3-2) as follows

$$K_T = K_S k_C/k_0 \quad (I3\text{-}4)$$

Another useful form of the thermodynamic cycle of Scheme I19 is shown in Scheme I20 as a free energy diagram. Note, also, that Schemes I18 and I19 do not require the reaction to be catalytic and can be applied equally to stochiometric reactions which often occur in examples discussed below in this chapter.

It follows from equation (I3-4) that the change in the reactivity upon complexation

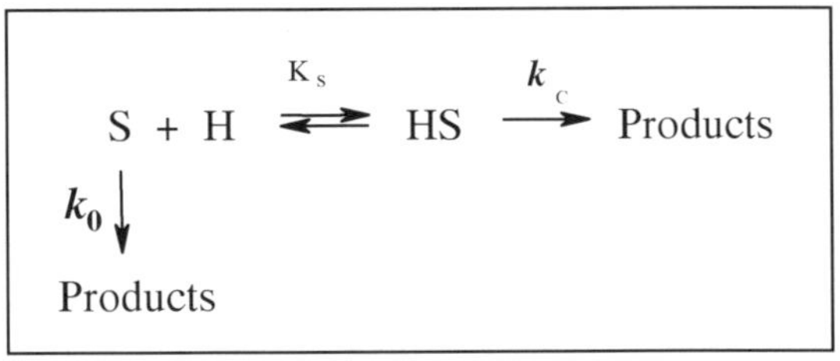

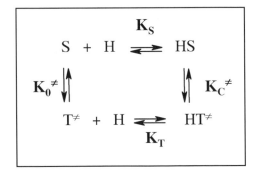

Scheme I18. Kinetic scheme for a one-substrate reaction in the presence of host catalyst.

Scheme I19. Thermodynamic cycle involving ground-state and transition-state complexation of the substrate to host catalyst in a one-substrate reaction.

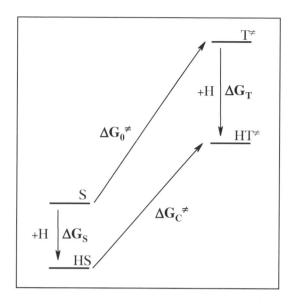

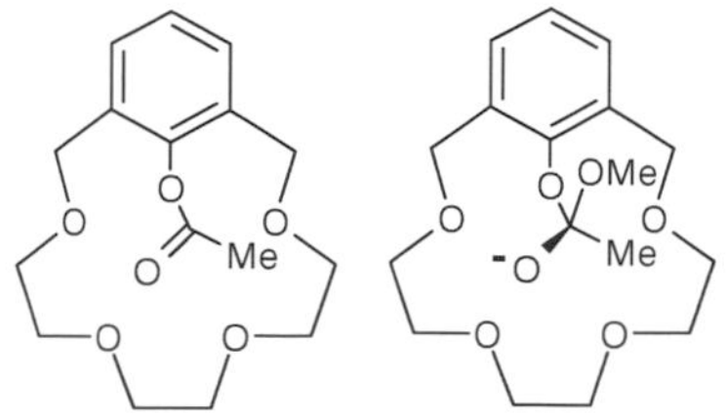

Na⁺:	$k_C/k_0 > 12$	$K_S < 3\ \mathrm{M^{-1}}$	$K_T = 39\ \mathrm{M^{-1}}$
K⁺:	$k_C/k_0 = 29$	$K_S = 11\ \mathrm{M^{-1}}$	$K_T = 320\ \mathrm{M^{-1}}$
Sr²⁺:	$k_C/k_0 = 40$	$K_S = 20\ \mathrm{M^{-1}}$	$K_T = 1200\ \mathrm{M^{-1}}$

Scheme I21. Substrate and transition state (approximated by the structure of tetrahedral intermediate) complexation in metal-promoted methanolysis of aryl crown acetates (from published data[51]).

Scheme I20. Free-energy diagram for the ground-state and transition-state complexation of the substrate to host catalyst.

(k_C/k_0) can be expressed as the ratio of formation constants K_S and K_T

$$k_C/k_0 = K_T/K_S \qquad (I3\text{-}5)$$

Thus, independently of the reaction mechanism, the effect of complexation on the reaction rate can be treated in terms of relative stability of host–guest complexes in the initial and transition states of the reaction. This approach is useful as one can apply the principles of molecular recognition developed for interactions between normal ground-state molecules also to transition states, the structures of which for many reactions are known at least in general terms.

As an example, let us consider the results for kinetics of methanolysis of crown ether aryl acetates by methoxide in the presence of alkali and alkaline earth metal cations in methanol, reaction (I3-6).[51]

$$\mathrm{AcO\text{–}arylcrown} + \mathrm{MeO^-} \rightarrow \mathrm{AcOMe}$$
$$+ \mathrm{arylcrown\text{–}O^-} \qquad (I3\text{-}6)$$

Structures of a typical substrate and the respective tetrahedral intermediate, resem-

bling the transition state are shown in Scheme I21 together with acceleration effects by three cations and with ground and transition state complexation constants calculated with equations (I3-1) and (I3-4). In this case the transition state has structure of a lariat ether with an anionic pendent group, which strongly enhances cation binding in comparison with neutral ground-state crown ether (Section A7). Note that the ground-state complexation of the substrate is unnecessary for catalysis: equation (I3-5) shows that high values of K_S decrease the catalytic effect, and that the best catalyst will have a maximal value of K_T and a minimal value of K_S. In other words, a supramolecular catalyst must be a good receptor for the transition state rather than for the ground state of the substrate, in general analogy to enzymes. As a rule, however, K_T and K_S values relate to each other, as exemplified in Scheme I21, because some of the host–guest pairwise interactions are always present both in the initial and transition states. The structural and energetic correlation between transition states, neighboring intermediates and educts/products is discussed in many organic chemistry textbooks under the Hammond postulate, or the reactivity/selectivity principle.

The approach outlined above has been successfully applied for the interpretation of biocatalysis,[44,46] where the most impressive manifestations for the requirement of stronger binding of the transition rather than initial state of the substrate are highly efficient transition-state analogue inhibitors[52–54] and catalytic antibodies.[55]

Schemes I 18–I 20 are applicable also to two-substrate reactions, when complexation of the second substrate is insignificant. When catalysis occurs through formation of a ternary complex between the catalyst and both substrates, Scheme I 22 must be considered. Assuming for simplicity the independent binding of the substrates S_1 and S_2 one obtains from Scheme I 22 the expression for K_T similar to equation (I 3-4)

$$K_T = K_{S1} K_{S2} k_C/k_0 \qquad (I\,3\text{-}7)$$

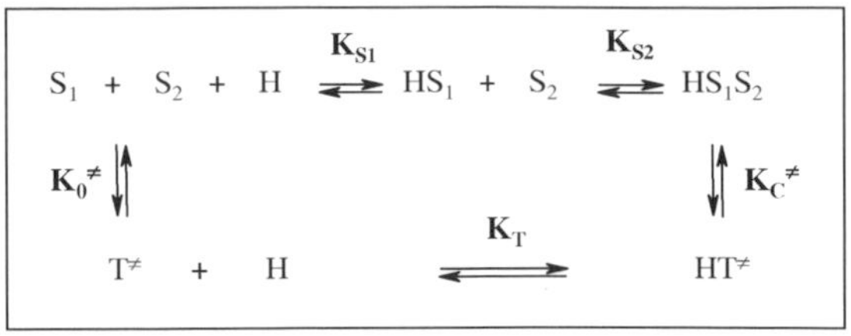

However, no direct comparison of the binding strength towards transition state given by K_T (M^{-1}) with the binding strength towards the substrates (given by the product $K_{S1}K_{S2}$ (M^{-2})) is possible due to the problem of different dimensions, exactly the same which we met previously in the analysis of thermodynamics of multi-site interactions (Section A1). The ratio $K_T/K_{S1}K_{S2}$, which has the molarity dimension and is equal to the ratio of rate constants of catalytic and spontaneous reactions, k_C/k_0, relates to the so-called 'effective molarity' (EM) defined as the ratio of rate constants of intramolecular and chemically related intermolecular reactions (e.g. lactonization and esterification). Reported experimental EM values range from 10^{-3}–10^{-2} to 10^{10}–10^{12} M[48] and their theoretical interpretation,[44,48–50,56] is a difficult problem, the discussion of which is beyond the limits of this book. It should be noted that although the ratio $k_C/k_0 = K_T/K_{S1}K_{S2}$ does not show the acceleration affect directly, it is certainly proportional to the observed acceleration, which again is related to stronger binding of the transition state as compared to the binding of the substrate.

Scheme I 22. Thermodynamic cycle involving ground-state and transition-state complexation of the substrates to host catalyst in a two-substrate reaction.

Actually, two-substrate reactions are not very common in catalysis by non-functional hosts, but they represent the most important group of reactions catalyzed by surfactant micelles, which is discussed in the second part of this chapter. The features of two-substrate reactions in the host–guest systems are illustrated here by results shown in Scheme I 23.

Acylation of hydroxy pyridine **I-8** by N-acetylimidazole in toluene is accelerated some 16-fold in the presence of 0.45 mM cyclic porphyrin host **I-10**.[57] The catalytic effect can be attributed to bringing together the reactants by intracavity coordination of each of them on different porphyrin metal sites, or can be alternatively (and more quantitatively) treated in terms of Scheme I 22. The reaction proceeds through a tetrahedral intermediate **I-9** structurally similar to the transition state. Scheme I 23 shows that **I-9** can be conveniently bound inside the host cavity as a bidentate ligand. The value of $K_T = 1.3 \times 10^6$ M^{-1} can be calculated with binding and rate parameters given in Scheme I 23 in accordance with equation (I 3-7). This can be compared to the binding constant 2.3×10^7 M^{-1} of ligand **I-11** chosen as a kind of transition-state analogue, which acts as an effective inhibitor of catalysis by the host **I-10**. Thus, the catalytic effect can be viewed as a sort of chelate effect: the bidentate transition state binds to the host stronger than two monodentate substrates. The same host (see also website) can by the same principle

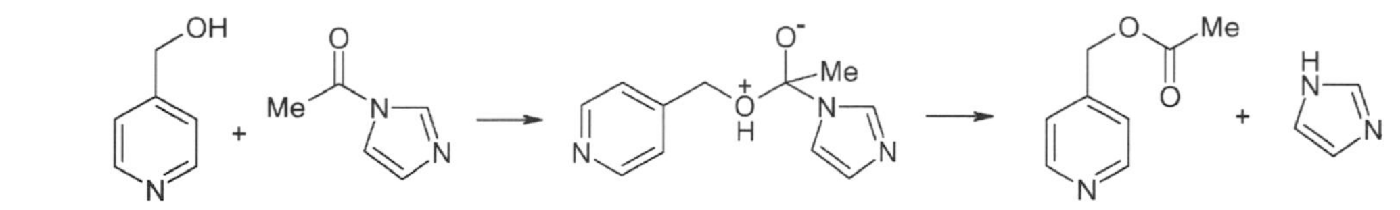

I-8

$K_{S1} = 1800 \text{ M}^{-1}$ $K_{S2} = 350 \text{ M}^{-1}$
$k_C \approx 6 \times 10^{-5} \text{ s}^{-1}$ $k_0 = 3.4 \times 10^{-5} \text{ M}^{-1}\text{s}^{-1}$

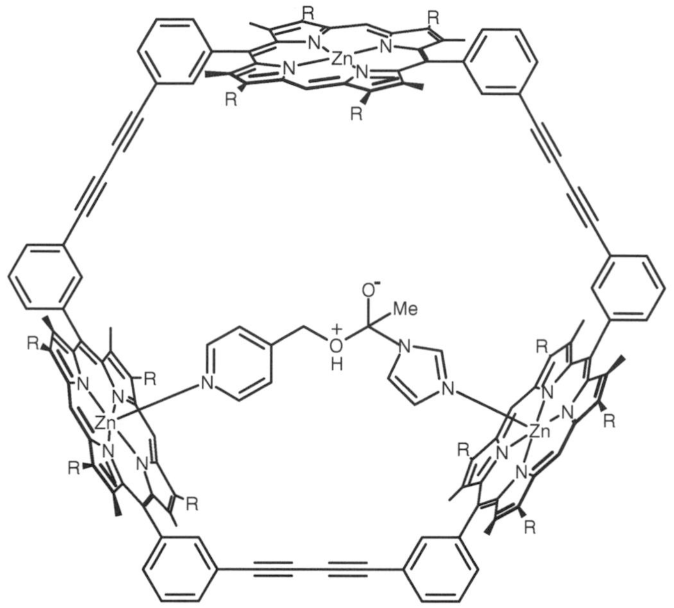

I-9

I-10

$K_T = 1.3 \times 10^6 \text{ M}^{-1}$

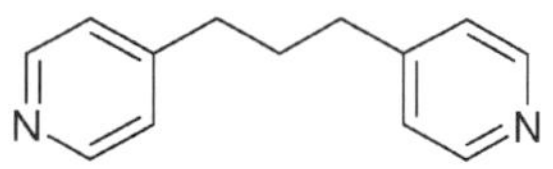

I-11

$K = 2.3 \times 10^7 \text{ M}^{-1}$

Scheme I 23. Catalysis of an acyl transfer reaction by a cyclic porphyrin trimer.[57] Reprinted with permission from *J. Am. Chem. Soc.*, MacKay *et al.*, 1994, **116**, 3141. Copyright 1994 American Chemical Society.

catalyze a Diels–Alder cycloaddition;[58] the significant change of stereochemistry towards a usually disfavored *exo*-addition illustrates the promising possibility to steer reaction selectivity by supramolecular catalysis.

Host molecules bearing reactive functional groups represent the most common group of supramolecular catalysts. The formal kinetic scheme in this case usually is the Scheme I 18, where k_0 refers to a pseudo-first-order reaction of the substrate S with some components, which are however often chemically different from the catalytic functionality. For example,

in the cyclodextrin-promoted hydrolytic reactions discussed below (see Scheme I 26), S is an ester and the reactive group is a hydroxyl group of a cyclodextrin, whereas the spontaneous reaction involves a combination of water and alkaline hydrolysis processes. In such cases the ratio k_C/k_0 shows the observed acceleration above the background reaction, which cannot be related to transition state complexation since catalytic and non-catalytic reactions have different transition states. Formal application of equation (I 3-4) will give in such cases an apparent parameter, which can be nevertheless considered to be proportional to K_T and is useful for correlation analysis.[46] A more direct and often employed approach to evaluate the role of host–guest complexation in this type of catalytic reactions is to compare the activity of a functional host (H-F) with a simple molecule (R-F) unable to host–guest complexation, but possessing the same catalytic functionality (F). The respective kinetic scheme and the thermodynamic cycle will in this case take the form shown in Scheme I24b.[53] Expression for K_T, which follows from Scheme I24b, is similar to equation (I3-4)

$$K_T = K_S k_C / k_{2,0} \qquad \text{(I3-8)}$$

but the physical meaning of K_T is different: now this is a dimensionless constant (note that $k_{2,0}$ is the second-order rate constant) for an exchange reaction between two transition states, one for the reaction of S with a simple molecule R-F and another one for the reaction of S with the host H-F. If both transition states are chemically identical (in other words, if group F has the same intrinsic reactivity in both molecules, R-F and H-F), the second transition state will have a lower energy when the host moiety will provide some additional non-covalent interactions with the $T^{\neq}$ moiety (in other words, when the binding of S to H-F in the ground state will be retained at least to some extent in the transition state) and will have the same energy when no such interaction occurs. The first situation corresponds to 'productive' binding and the second to 'non-productive' binding of the substrate (Section D6). Another meaning of K_T stems from the fact that the product $K_S k_C$ represents itself the second-order rate constant of the catalytic reaction under conditions far from the 'saturation' ($K_S[H] \ll 1$ in equation (I3-1)). Therefore, as evident from equation (I3-4), K_T is equal to the experimentally observed acceleration effect under conditions of second-order kinetics for both the supramolecular catalyst and the reference simple reagent molecule.

$$\mathrm{S + H\text{-}F} \underset{}{\overset{\mathbf{K_S}}{\rightleftharpoons}} \mathrm{H\text{-}F(S)} \xrightarrow{k_C} \text{Products}$$

$$\mathrm{S + R\text{-}F} \xrightarrow{k_{2,0}} \text{Products}$$

(a)

$$\begin{array}{ccc} \mathrm{H\text{-}F + S + R\text{-}F} & \overset{\mathbf{K_S}}{\rightleftharpoons} & \mathrm{H\text{-}F(S) + R\text{-}F} \\ \mathbf{K_0^{\neq}} \updownarrow & & \updownarrow \mathbf{K_C^{\neq}} \\ \mathrm{H\text{-}F + R\text{-}T^{\neq}} & \underset{\mathbf{K_T}}{\rightleftharpoons} & \mathrm{H\text{-}T^{\neq} + R\text{-}F} \end{array}$$

(b)

Scheme I24. Kinetic scheme (a) and the thermodynamic cycle (b) for a reaction with host catalyst bearing a catalytic functional group.

As an example of this type of systems let us consider phosphorylation reactions of a cationic acyclic host **I-12** bearing a hydroxyl and a simple alcohol **I-14** with a cyclic phosphate **I-13**, which was studied in DMF solution containing diisopropylethylamine as a base (Scheme I-25).[59] Association between substrate **I-13** and host **I-12** was confirmed by ^{31}P NMR titration in the absence of the base; $K_S = 2900\ \mathrm{M}^{-1}$. The ratio of the apparent second-order rate constants for reactions of **I-13** with **I-12** ($K_S k_C$) and with **I-14** ($k_{2,0}$) equals 3.8×10^5 and in accordance with equation (I3-8) this is the value of K_T, which corresponds to intramolecular stabilization of the transition state by 32 kJ mol^{-1} due to its interaction with the host guanidinium groups in **I-12**. This interaction energy seems reasonable taking into account that complexation of **I-13** with the non-functional host **I-15** enhances the rate of phosphorylation of **I-14** by factor of 5000, that is, **I-15** interacts with transition state anion stronger by 21 kJ mol^{-1} than with the ground state anion. Indeed, the negative charge on the phosphate group must increase

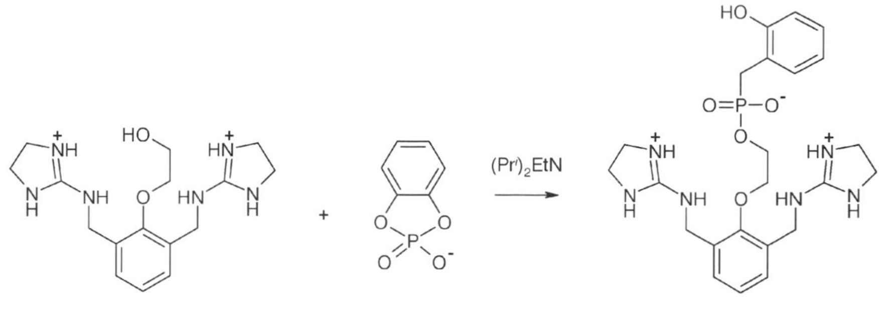

Scheme I 25. Phosphorylation of the cationic host **I-12** bearing a hydroxy function, and of the uncharged alcohol **I-14**.[59]

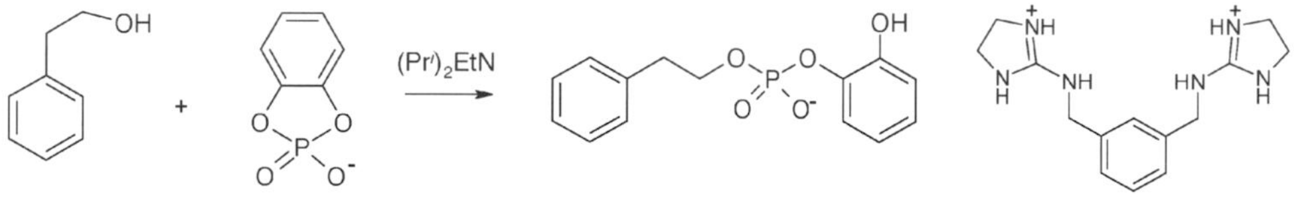

from -1 to about -2 upon the nucleophilic attack of the alcohol in the transition state and, therefore, the stabilization effect of **I-15** should be about half of the effect provided by **I-12**.

One of the oldest and most comprehensively studied reactions in the field is ester hydrolysis in the presence of cyclodextrins.[47] The general reaction mechanism is shown in Scheme I 26: it involves inclusion of the ester into cyclodextrin cavity (depending on the ester structure either acyl or alcohol moiety can be included) and nucleophilic attack on the bound substrate by a deprotonated secondary hydroxy group of cyclodextrin affording the acylated cyclodextrin through the tetrahedral intermediate (**T I**). Deacylation of acylcyclodextrin is slow; thus, in fact, the reaction is not catalytic.

Secondary hydroxy groups of cyclodextrins are unusually acidic with pK_a of about 12.5; their deprotonated forms can act as nucleophiles only towards esters with sufficiently

good leaving groups;[47,60] most often aryl esters are used as the substrates. Early studies with substituted phenyl esters as substrates already disclosed a selectivity for *meta*-substituted isomers.[47] This was explained in terms of an orientation effect: inclusion of *meta*-isomers positioned the ester carboxyl group closer to secondary hydroxyls. In terms of the transition-state stabilization approach, effects of substituents on the reactivity are dictated by their effects on substrate complexation in ground and transition states. Figure I 1 shows the effects of *para*- and *meta*-substituents in phenyl acetate on the efficiency of binding (K_S) and cleavage (k_C / k_0) of substituted esters by β-cyclodextrin. Only reactions with alkyl substituents are considered as these cause a minimum effect on the intrinsic substrate reactivity, but considerably affect the complexation due to their hydrophobicity and size. Evidently two different trends are observed: *meta*-substituents increase both binding and reactivity, while

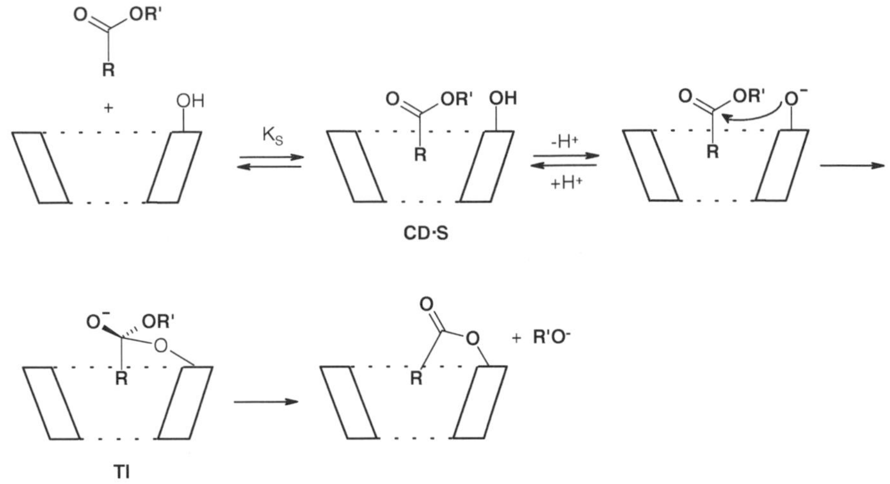

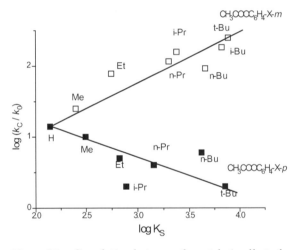

Scheme I 26.   General mechanism of cyclodextrin-promoted ester cleavage.

Figure I 1.   Correlation between the catalytic effect of β-cyclodextrin on the hydrolysis of *meta*- and *para*-substituted phenyl acetates and complexation strength (from published data[46]).

para-substituents also increase substrate binding, but decrease the reactivity. The latter trend is what one would expect in the case of non-productive binding. The non-productive binding of such substrates, amongst which *para*-nitrophenyl acetate was investigated in most detail, was proven recently by the absence of competitive inhibition with, for instance alcohols known to occupy the cyclo-

dextrin cavity and therefore serve, as expected, as inhibitors for the cleavage of *meta*-nitrophenyl acetate.[61]

Positive slope of correlation for *meta*-isomers indicates that in this case more hydrophobic substrates are bound even better in the transition state than in the ground state. Evidently, location of the substrate should be different between the ground-state complex (CD · S in Scheme I 26) and the transition state, which can be approximated by the structure of the tetrahedral intermediate (T I in Scheme I 26). Indeed, molecular modeling predicts that formation of T I leads to some displacement of the substrate outside the cyclodextrin cavity from its position in the CD · S complex.[62a] Optimization of substrate structure for best binding in the transition state allowed to reach very large acceleration effects with substrates possessing large acyl moieties.[62] Scheme I 27 illustrates the evolution of substrate specificity in β-cyclodextrin ester cleavage. Note that the binding specificity measured by K_S varies only in the limits of one order of magnitude, while the kinetic specificity measured by k_C/k_0 changes by seven orders of magnitude which reflects the strong variation in the transition-state binding. The systems discussed above usually

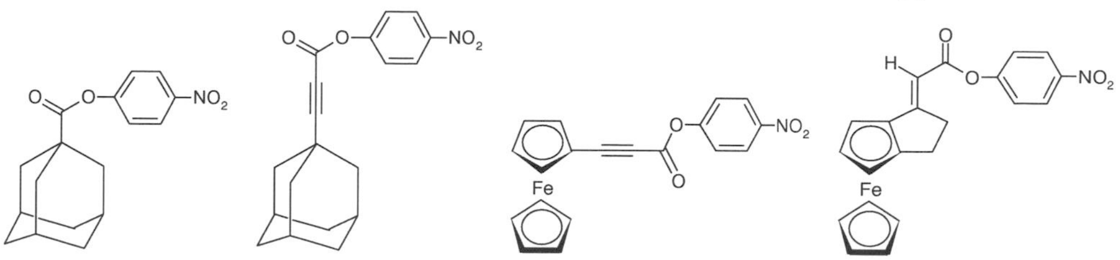

k_C/k_0:	0.3	2.15×10^3	1.4×10^5	5.9×10^6
$K_S(M^{-1})$:	210	3300	200	175

Scheme I 27. Kinetic parameters for the cleavage of a series of *para*-nitrophenyl esters with variable acyl moiety by β-cyclodextrin in mixed aqueous–organic solvents.[62,64]

materialize only the first step of the catalytic reaction, for example the acyl transfer from the phenylacetates to the CD unit. The covalent implementation of cofactors, for hydrolases in particular of imidazole units in cyclodextrin[63] or in micellar systems (see below), can lead not only to substantial acceleration of the acyl transfer[44], in analogy to chymotrypsin; it also allows liberation of the transferred acyl group, and thus turnover of the catalyst has been achieved.

Another example of support by a cofactor is the flavo-thiazolo-cyclophane **I-16**, designed as a model for pyruvate oxidase.[65] The host catalyzes oxidation of 2-naphthaldehyde to 2-naphthoate in the form of its methyl ester, reaction (I3-9) in methanol with $K_S = 43$ M^{-1} and with $k_C = 0.24$ s^{-1}. Such rate parameters are comparable to catalytic constants of a number of less active enzymes.

(I 3-9)

The similar, but not macrocyclic compound **I-17** reacts much slower and does not show any 'saturation' upon variation of substrate concentration. The second-order rate constant for **I-17** equals 0.058 M^{-1}s^{-1} and is only four times higher than the rate constant observed for the mixture of thiazolium ion **I-18** and 7-

arylflavin **I-19** (the reaction is zero order in flavin): 0.015 M^{-1}s^{-1}. The results clearly show the importance of the macrocyclic binding site. Other than in many enzymes the cofactors here are covalently bound to the active (substrate binding) site; the recycling of the flavin is possible by electrochemical reoxidation of the cofactor, and up to 100 cycles can be performed on a preparative scale.[66]

Hydrolytic metalloenzyme[67] models have been studied extensively; other than most other artificial systems they do of course not need recycling of any cofactors, usually show turnover and can catalyze reactions not of especially designed model substrates, but of 'normal' substrates such as non-activated amides or phosphates. These present a particular challenge also in view of their high hydrolytic stability and hold promise for future applications, for instance in biotechnology. Although the major contribution towards the catalytic activity stems from the metal ions itself, their coordination with suitable organic ligands already improve their possible use, by providing kinetic stabilization against dissociation, thus allowing immobilization or attachment of recognition groups (see below). Furthermore, one often needs special ligands enforcing a certain coordination sphere in order to achieve high activity; a classical case are cobalt(III) cyclen complexes[68] in which two coordination sites are occupied by water

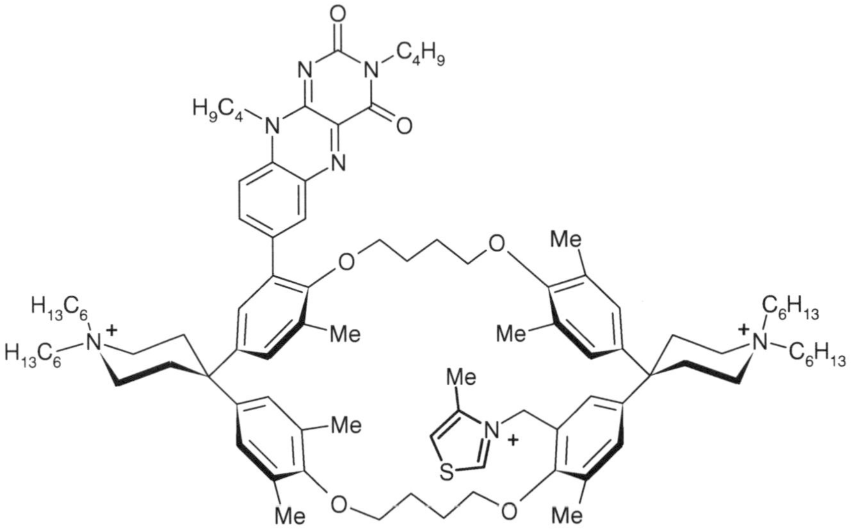

I-16

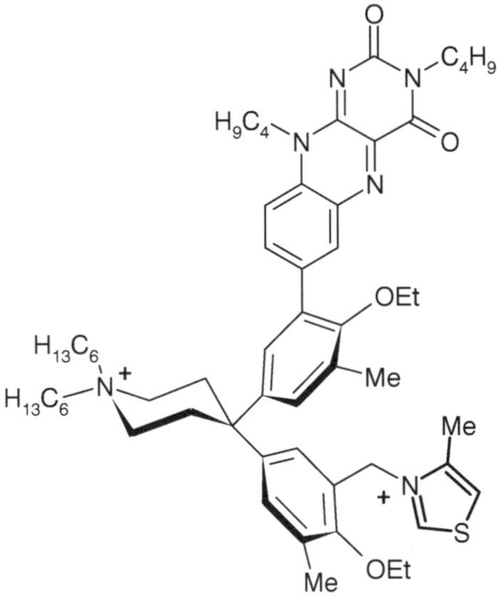

I-17

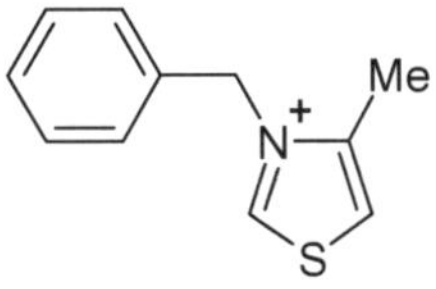

I-18

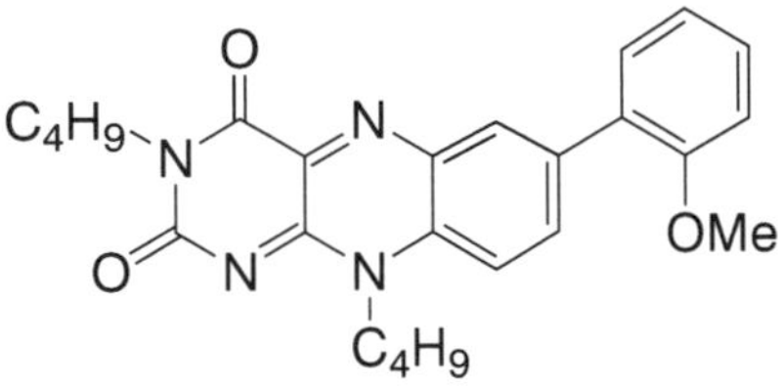

I-19

molecules with pK_a values between 5 and 8. The acidification of water by metal ions provides highly nucleophilic hydroxyl anions at physiological pH, leading to rate enhancements in the hydrolysis of amides, of esters, of nitriles and of phosphates,[68] for instance by up to measurable (not extrapolated) factors of

10^7 against double-stranded DNA after suitable modification of the complexes.[69]

The example in Scheme I28 illustrates another feature of supramolecular catalysts by covalent attachment of a functional nucleophilic hydroxyalkyl group.[70] Interaction with the zinc ion lowers the pK_a of the ROH group

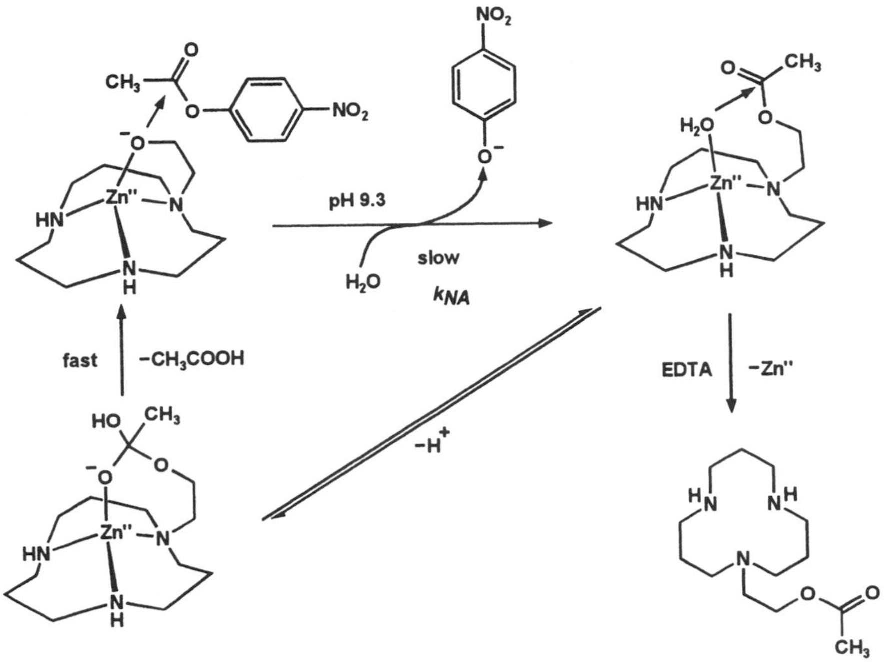

I-20

Scheme I28. A supramolecular catalyst with a covalently bound hydroxyalkyl group.[70] Reprinted with permission from *J. Am. Chem. Soc.*, Kimura *et al.*, 1994, **116**, 4764. Copyright 1994 American Chemical Society.

to 7.4. The system **I-20** (Scheme I28) is an obvious analog to serine proteases like chymotrypsin, in which, however, the serine hydroxyalkyl operates in neutral form with general-base assistance of the histidine imidazole: both in enzyme and in model system **I-20** the acyl group is first transferred to the catalyst nucleophile; the resulting ester – which can be trapped by adding EDTA taking away the Zn ion – is rapidly hydrolyzed.

The particularly high charge density in lanthanide(III)[71] cerium(IV)[71a] or zirconium(IV)[72] ions lead to stronger Lewis catalyst activation of carbonyl or phosphoryl groups in ester or amides, and stabilization of the negatively charged leaving group. These effects can equally be viewed as electrostatic stabilization of the transition state, which bears an additional negative charge after attack of the nucleophilic hydroxide anion. Supramolecular complexes allow to organize

several metal ions, which as in corresponding enzymes such as alkaline phosphatase or some nucleases can exert simultaneously functions such as nucleophile activation and Lewis catalysis/transition state stabilization.[73] Thus, a rate enhancement factor of 10^{13} is seen in the cleavage of the covalent enzyme model **I-21 a**, whereas the mononuclear complex **I-21 b** shows only the same reactivity as with outside La ions.[74] In the binuclear copper complex **I-22 a** oxygen is reversibly bound, which in a structure like **I-22 b** as a particularly potent nucleophile can attack bound dimethlyformamide, liberating formic acid within seconds instead of years without a catalyst.[75] In contrast to enzymes, cofactors in most synthetic counterparts are until now covalently bound to the catalyst. Scheme I29 shows for the europium (III)-catalyzed hydrolysis of bis(*p*-nitrophenyl)phosphate as substrate that the combina-

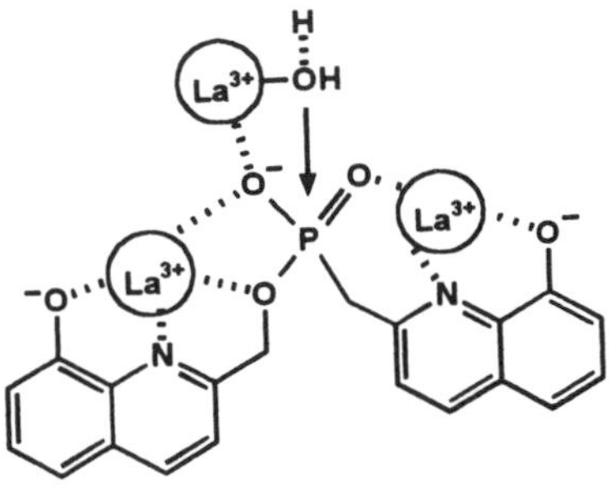

I-21 a

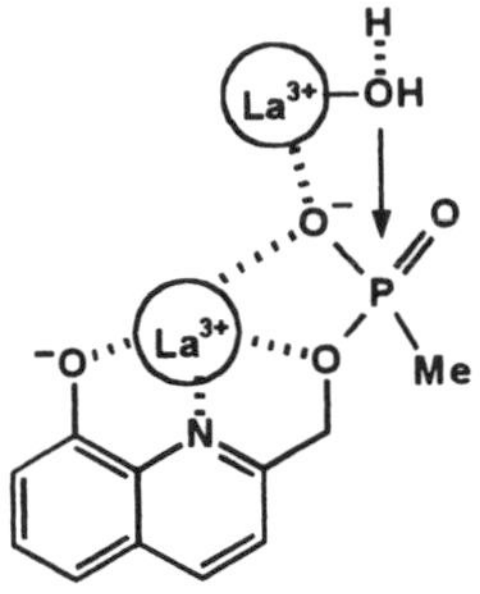

I-21 b

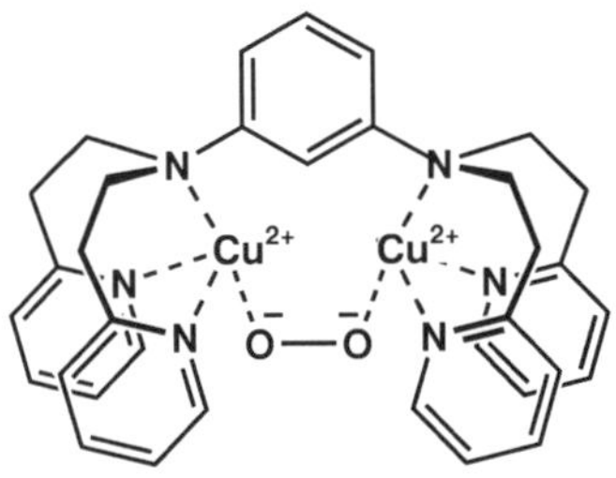

I-22a

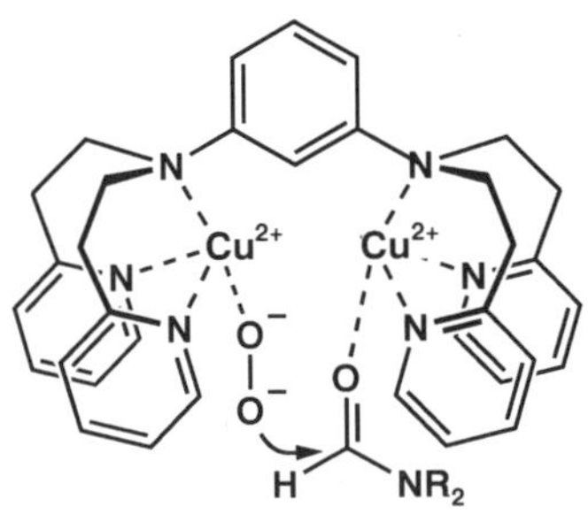

I-22b

tion of carboxylic and a imidazole group in one cofactor **I-23d** leads to similar, although still moderate rate enhancements as observed with the covalently bound cofactors in **I-23b** and **I-23c**.[76]

Selectivity of supramolecular catalysts can be achieved, for instance by attaching oligonucleotide side chains at the ligands which by Hogsteen base-pairing recognize special nucleotide sequences[71a], thus opening the way to synthetic restriction enzymes etc. Similarly, oligopeptides which can also recognize DNA fragments can be equipped with metal binding units, leading to selective scission of the biopolymers.[77] Other related examples make use of abasic sites in DNA, lacking a nucleobase, for selective recognition and subsequent repair of the strand by adding a ligand containing diaminopurine for recognition, and a polyamino linker which helps to cleave the damaged DNA.[78] As discussed in Section I1 fragments securing selectivity must not necessarily participate in the binding itself, but can discriminate reactants, in the case of catalysis and/or transition states, also by repulsive interactions. The particularly interesting control of stereochemistry[79] is illustrated in Scheme I30. Reduction of phenylglyoxylic esters with the macrocyclic NADH analog **I-24 a** occurs with up to 90% enantioselectivity, if magnesium salts are added. Structure **I-24 b** shows a model for the pre-transition state complex which also predicts the sign of the optical induction correctly. The phenyl ring of the substrate is located above the dihydropyridin; non-covalent interactions with the metal ion play a decisive role in holding substrate and catalytic unit together. Many other systems hold promise for other applications, even in fields like selective oxidations with air.[80]

I3.2. MICELLAR CATALYSIS

Catalysis by micelles and other surfactant aggregates is, perhaps, the oldest area in

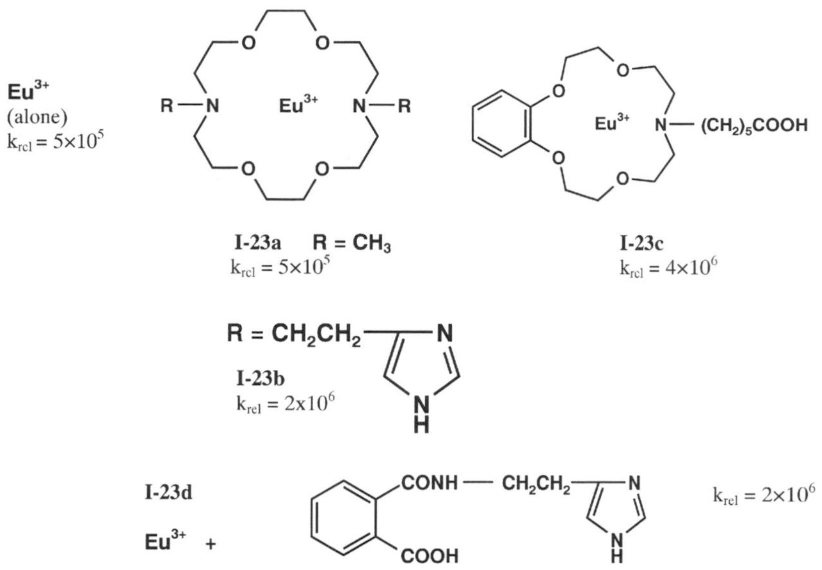

Scheme I 29. Relative rate constants k_{rel} of the europium (III)-catalyzed hydrolysis of bis(*p*-nitrophenyl) phosphate with different ligands and cofactors (with $[Eu^{3+}] = [Ligand] = 5$ mM, 50°C, pH 7.0).[76]

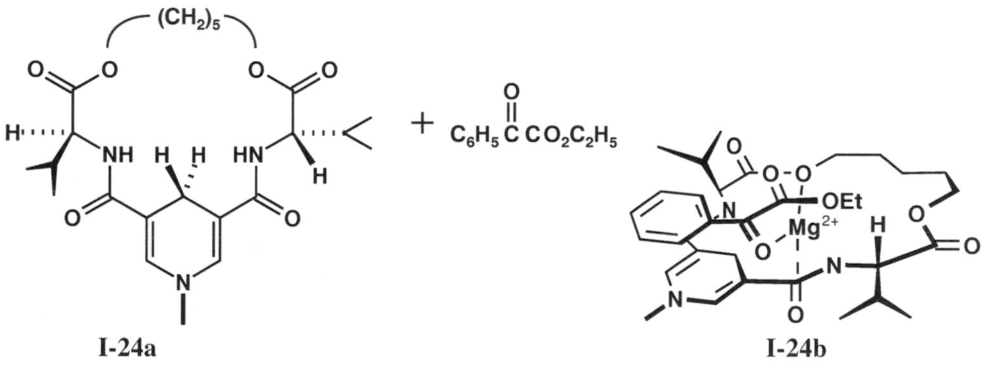

Scheme I 30. A enantioselective reduction catalyst.[79] Reprinted with permission from *Angew. Chem., Int. Ed. Engl.*, Kellog, R.M., 1984, **23**, 782, copyright 1984, Wiley.

supramolecular catalysis. Earlier work is summarized in a book;[81] more recent reviews on general[82,83] as well as on physicochemical,[84] applied analytical,[85] and biomimetic[45] aspects are available. Micellar effects, typically with accelerations of the order 10^2–10^3, sometimes up to 10^6, were reported for reactions of very different types. As a rule the acceleration is due to bringing together

reacting species in a micellar pseudo-phase; it is observed, accordingly, for reactions between non-polar molecules, which are included in micelles hydrophobically, between ions of similar charges, which are attracted by oppositely charged micellar surface, or between nonpolar molecules and ions charged oppositely to the micelle. Also, micellar inhibitory effects are observed for reactions between neutral hydrophobic substrates and ions of similar charge with micelle, for example.

Reaction kinetics in micellar solutions can be conveniently analyzed in terms of an approach, which considers micelles as a separate 'phase' inside which the chemical reaction proceeds simultaneously with the reaction in bulk solvent phase.[86] For a reaction of general type

$$A + B \rightarrow \text{Products} \qquad (I3\text{-}10)$$

assuming fast equilibrium distribution of A and B between phases in accordance with respective partition constants (subscripts 'm' and 'w' refer to micellar and aqueous bulk 'phases')

$$P_A = [A]_m/[A]_w \qquad P_B = [B]_m/[B]_w$$

one obtains the following equation for the observed second-order rate constant

$$k_{2,\text{obs}} = (k_m V_m^{-1} K_A K_B[D] + k_w)/\{(1 + K_A[D]) \\ (1 + K_B[D])\} \qquad (I3\text{-}11)$$

where V_m is the surfactant molar volume, $[D]$ is the concentration of surfactant in micellar form

$$[D] = [D]_t - \text{CMC}$$

where CMC is the critical micelle concentration (Section G 2), and K_A, K_B are the apparent binding constants defined as

$$K_A = P_A V_m \qquad K_B = P_B V_m$$

Other assumptions are that the volume fraction of the micellar 'phase' is small ($[D]V_m \ll 1$) and that reaction (I3-10) follows the second-order kinetics both in micellar and bulk phases with respective second-order rate constants k_m and k_w. The equation (I3-11) predicts an optimum value of $k_{2,\text{obs}}$ as a function of surfactant concentration since the denominator in (I3-11) depends on concentration of surfactant linearly, but the denominator quadratically. Such an optimum, indeed, is a characteristic feature of micelle-catalyzed reactions.[84] More important, equa-

tion (I3-11) predicts also that even when $k_m = k_w$, that is when the micellar medium by itself does not affect the reaction rate, a considerable rate enhancement can be observed provided the binding constants of reactants are big enough. Indeed, standard mathematical analysis of equation (I3-11) gives for the optimum surfactant concentration

$$[D]_{\text{opt}} = 1/(K_A K_B)^{0.5} \qquad (I3\text{-}12)$$

and for the observed rate constant at the optimum concentration (neglecting contribution from the bulk reaction)

$$(k_{2,\text{obs}})_{\text{opt}}/k_w = (k_m/k_w)K_A K_B/ \\ \{V_m(K_A^{0.5} + K_B^{0.5})^2\} \qquad (I3\text{-}13)$$

It follows from (I3-13) that, when $k_m = k_w$, and with $K_A \approx K_B = K$ the optimum micellar effect is

$$(k_{2,\text{obs}})_{\text{opt}}/k_w = K/(4V_m)$$

For common surfactants like sodium dodecylsulfate (SDS) or cetyl(= hexadecyl)trimethylammonium bromide V_m is of the order of $0.3\,\text{M}^{-1}$ and, therefore, the micellar effect is of the order of K, which for substrates like substituted benzenes is about $100\,\text{M}^{-1}$. Thus, considerable acceleration is observed even in this trivial case just due to increasing reactant concentrations in the micellar 'phase' (concentration effect). Analysis of large amount of kinetic data in micellar systems shows that this effect is the main cause of catalysis even when very large accelerations up to 10^4–10^6 are observed.[82,84,86] In the important case of ion–molecular reactions the binding of the neutral reactant to micelles is considered as partition between pseudo-phases, but binding of the ionic reactant as an ion exchange with surfactant counter-ions in micellar surface.[87]

For monomolecular reactions the pseudo-phase approach leads to a Michaelis–Menten type rate equation,[86] which was proposed also in earlier kinetic studies of micellar effects based on a kinetic scheme similar to Scheme

I22 with H=micelle.[88] In this case the concentration effect does not contribute to catalysis; an acceleration is possible in this case only if $k_m > k_w$ which occurs rarely. If a reaction of the type (I3-10) is reversible and the concentration effect is the dominating one, micelles should accelerate the forward reaction and not affect considerably the backward reaction, thus producing a shift in the reaction equilibrium.

The general features of micellar effects on reactivity discussed above are illustrated with results shown in Fig. I2 for a reversible Schiff base formation reaction between 6-aminopenicillanic acid and p-dimethylaminobenzaldehyde (reaction equation (I3-14)) studied in the presence of sodium dodecylsulfate (SDS) micelles in acidic solutions.

I2(b)) increase 200 and 100 times, respectively. The rate constant of the backward reaction undergoes a two-fold decrease. Fitting of the experimental forward rate constant *vs.* SDS concentration profile to equation (I3-11) shows[89] that the binding constants of reactants are about 80 and 200 M^{-1} for the aldehyde and amino acid respectively and that $k_m = k_w$. According to equation (I3-12) the optimum surfactant concentration is expected to be 0.008 M, which is close to the experimentally observed value, Fig. I2(b). Also the acceleration effect of about 90 ($V_m = 0.35$ M^{-1} for SDS) expected according equation (I3-13) is close to the experimentally observed value.

The observation that usually k_m equals k_w indicates generally low selectivity of micellar

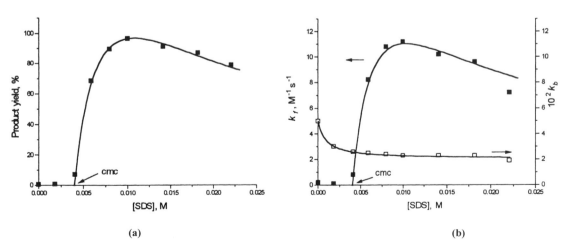

(I3-14)

When the surfactant concentration reaches CMC, both the product yield (Fig. I2(a)) and the rate constant of the forward reaction (Fig.

catalysis, besides, of course binding specificity, which is dictated mainly by substrate hydrophobicity and charge (Section G2). On

Figure I2. Effects of sodium dodecylsulfate (SDS) micelles on rate and equilibrium of reaction (I3-14) in aqueous acid solution.[89] Reprinted with permission from *Analyt. Biochem.,* Yatsimirsky *et al.,* 1995, **229**, 249. Copyright 1995 Academic Press.

the other hand, acceleration effects can be fairly high with strongly bound substrates. Especially effective are systems where one of the reactants is a surfactant bearing a reactive functional group. Functionalyzed surfactants are used also as components of co-micelles with common surfactants. Such systems are similar by their design to the functional synthetic hosts discussed in Section I3.1, and can also be considered as enzyme mimics. Many surfactants bearing co-enzymes and their analogs were prepared.[45]

Several examples of functional surfactants are given in Table I1. The use of chiral nucleophiles in micellar and vesicular aggregates deserves special comments. First attempts to induce chiral discrimination by using chiral non-functionalized surfactants, like with chiral tetraalkylammonium heads, with achiral nucleophiles demonstrated only marginal effects.[94] However, the alternative approach when a chiral nucleophile, which already possesses some enantioselectivity, is used in micelles or vesicles formed by common achiral surfactants leads to considerable increase in chiral discrimination. Much attention was focused on chiral histidine derivatives bearing imidazole as nucleophile capable to cleave activated *p*-nitrophenyl esters. Non-micellar *L*-histidine derivatives with *N*-acylated D- or L-phenylalanine *p*-nitrophenyl esters as substrates show small enantioselectivity, $k_L/k_D \leq 1.5$, which increases up to $k_L/k_D \leq 50$ with hydrophobic

Table I1. Functional surfactants.[90–93]

No.	Surfactant	Function	Ref.
1		NAD$^+$ analog	90
2		Nucleophile for the cleavage of carboxyl and phosphoric acid esters	91
3		Chiral nucleophile for the cleavage of activated esters	92
4		Chiral nucleophile for the cleavage of activated esters	92
5		Metal-coordinated chiral nucleophile for cleavage of amino acid esters	93

derivatives like **I-25** and also with hydrophobic substrates with long-chain *N*-acyl moieties strongly bound to surfactant aggregates.[95] Surfactants 3 and 4 in Table I1 differ in structure of hydrocarbon moieties in such a way that the former produces micellar, and the latter vesicular aggregates (Section G2). With *N*-acylated D- or L-phenylalanine *p*-nitrophenyl esters as substrates large accelerations of the order of 10^4 times are observed with enantioselectivities $k_L/k_D = 2.7$ and 4.4 for micellar and vesicular aggregates respectively.[92] Higher enantioselectivity in vesicles is explained by higher rigidity and higher degree of 'organization' of bilayers as compared to micelles. Chiral surface active copper(II) complexes, like no. 5, Table I1, catalyze the hydrolysis of unprotected amino acid esters, like *p*-nitrophenyl phenylglycine (PhGlyPNP), which are coordinated to Cu(II) through the amino group and undergo intramolecular nucleophilic attack by a coordinated alcoholate. Micelles formed by such complexes in water show high enantioselectivity in hydrolysis of PhGlyPNP with k_R/k_S ratios about 10, which contrasts to $k_R/k_S = 1.2$ for a short chain non-micellizing complex. Like in the preceding system, inclusion of the catalyst into bilayers enhances considerably the enantioselectivity, with the dioctadecyldimethylammonium bromide vesicles (Table I1) the ratio k_R/k_S reaches 26.

The peculiarity of bilayers as compared to micelles is, besides their higher structural ordering, the close analogy to biological membranes. This makes them very attractive for modeling of membrane-bound enzymes. As an example of such systems we consider a vesicular model of cytochrome P-450, a membrane-bound enzyme, which catalyzes hydroxylation of alkanes and epoxidation of alkenes, Scheme I31.[96] The vesicle formed from anionic surfactant dihexadecylphosphate incorporates a Mn(III)porphyrin **I-26**, which serves as the active site of epoxidation, and a surface active Rh(III) complex **I-27**, which serves as a redox cofactor (ferredoxin

I-25

in the natural enzyme) mediating electron transfer from formate to the porphyrin in conjugation with epoxidation. The system shows high turnover and, surprisingly, becomes inactive when a cationic vesicle instead of anionic is used. The latter was explained by the necessity of protonation of the porphyrin active site during the catalysis, which is hindered by a positive surface charge.

14. ANALYTICAL APPLICATIONS

Earlier Sections in Chapters A and B have already demonstrated the potential of supramolecular host compounds for selective, reversible binding of analytes. From the countless examples of analytical applications in the literature, of which separation techniques were already discussed in connection with determination of association constants (Section D4), we can only show here some principles, based on spectroscopic and on electrochemical methods.[97] All these techniques, which are of particular promise for sensor technology,[98] rely on an optimally chosen host compound which usually is immobilized in a device. Upon binding of an analyte the device changes an intrinsic property, be it a mass-dependent frequency such as with the quartz balance, or it emits an optical or an electrochemical signal. Piezoelectric sensors[99] are now widely applied, also to the enantioselective monitoring of peptide derivatives by hydrogen bonding to peptide-complementary polymers coated on a transducer; the same device can also be used for

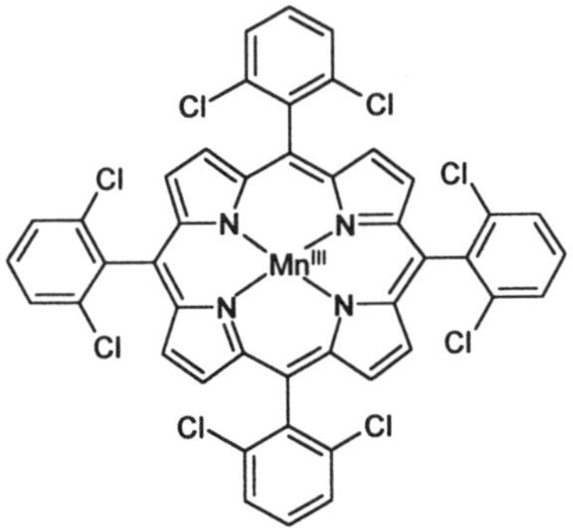

I-26

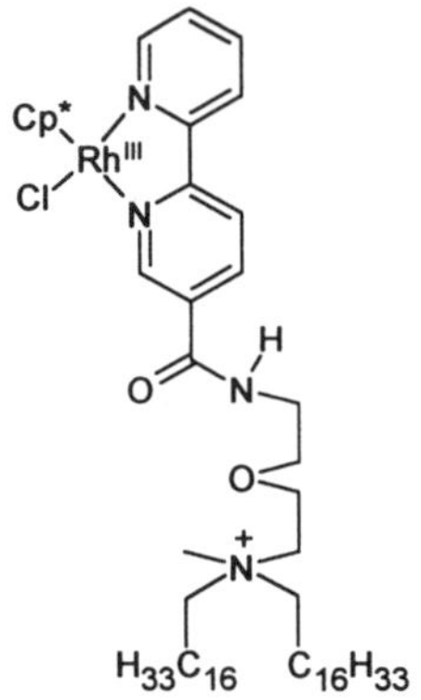

I-27

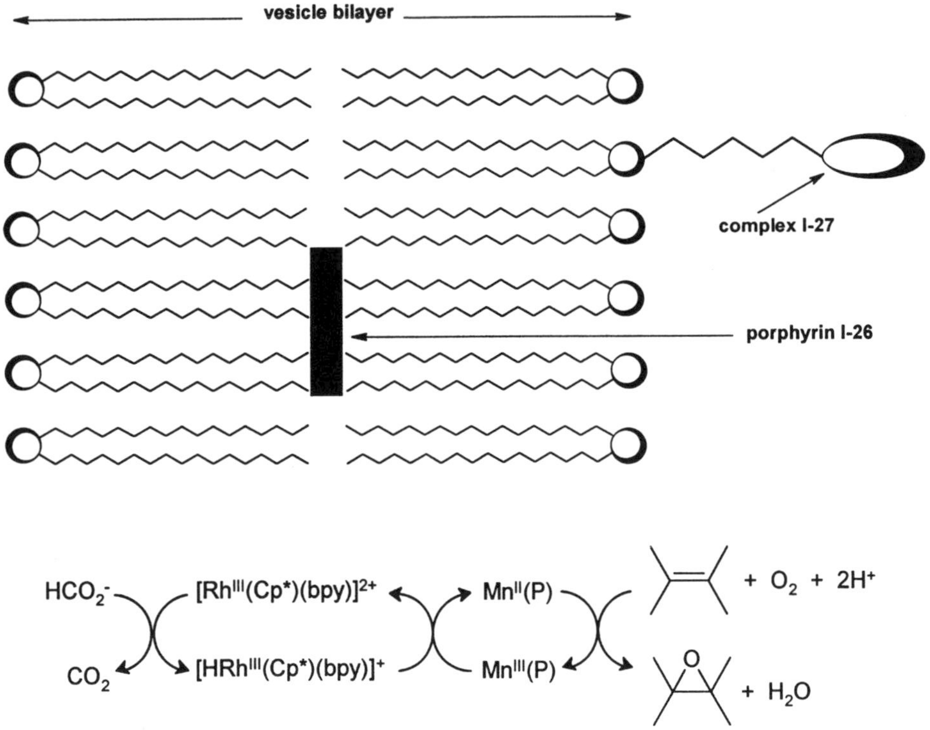

Scheme I 31. A vesicular cytochrome P450 model.[96]

monitoring by reflectometric interference spectroscopy on such thin films.[100] Mass-sensitive sensors like the quartz-balance are based on non-covalent interactions of the analyte usually with a suitable sensor molecule immobilized via a sulfur–gold bond on the surface of a quartz crystal; the so-called Raleigh frequency of its crystal surface acoustic waves is extremely sensitive against mass changes occurring upon binding of the analyte.

14.1. OPTICAL DEVICES

Detection methods based on changes in UV/vis absorbances or on fluorescence emissions have a broad applicability,[101] also in view of now available techniques for remote control by fiber optics. With few exceptions like porphyrins[102] the usual host molecules do not provide convenient optical signal changes upon binding of an analyte. Although one can use for detection dyes such as methylorange or anilinosulfonates which, as described in Chapter D, compete with the analyte binding, it is more practical to make such units part of the host itself. The chromophore in **I-28** shows color changes due to deprotonation of the phenolic unit upon highly selective binding of lithium ions. The enantioselectivity of biaryl crown ethers discussed in Section I 1 above is used in structure **I-29** for optical discrimination of chiral amines. Fluorescence techniques have the advantage of particularly high sensitivity, and find applications with fluoroscent ionophores etc.[103,104] Crown ethers or cryptants with fluorescent labels such as **I-30** show fluorescence quenching by the oxygen lone pairs, which disappears upon metal binding.[103–105] The related remove of quenching by nitrogen lone pairs is the basis of chelation-induced fluorescence upon protonation by phosphoric acid or by transition metal binding to polyamine units connected to fluorophores (**I-31**).[102,103] Fluorophores with a different binding site for a proton and for sodium cation such as **I-32** offer a way to logical AND gates, where strong fluorescence emission occurs only if *both* a proton AND a metal ion is present.[106] The cryptand **I-33** binds Ni(II), Cu(II), or Zn(II) ions in the lower region A, thereby stopping photoinduced electron transfer from the N lone pairs; in contrast to the usual fluorescence quenching by transition metal ions one therefore

observes a strong emission.[107] The system presents a simple logical OR gate; occupation of the *N* lone pairs by adding protons *or* metal ions lead to fluorescence emission.

I-28

With metallohosts one can take advantage of both the additional binding forces between the metal and the analyte and of the detection by optical changes due the metal center.[108] One also can make use of the photoelectronic properties of metal cations such as lanthanides.[103–105] Thus, the dicopper complex **I-34** changes its color to deep blue after binding imidazole derivatives.[109] Fluorescence detection of ribosenucleoside triphosphates can be achieved with apparent binding constants as high as $5 \times 10^5 \ M^{-1}$ by reaction with a boric acid derivative covalently fixed on poly(allylamine).[110]

An elegant principle for detection of organic analytes rests on fluorophores which are covalently bound to cyclodextrins (CD), such as in host **I-35**.[111] Addition of the analyte can expel the dye, measurable as fluorescence emmission decreases by often 90% as a consequence of quenching by the aqueous outside (case I in Scheme F3). Alternatively, one can observe *enhanced* fluorescence *I* if *both* the dye and the analyte find room in the cavity, which is the case e.g. with γ-cyclodextrins (case II in Scheme F3). Azodyes like MO (**I-35a**) are less accessible to protons with the azo group inside a cyclodextrin cavity; in consequence one observes a strong change from yellow to red upon addition of analytes which expel the dye residue from the cavity. In a series of bile acids one finds binding constants between 265 M^{-1} and 71900 M^{-1}, depending only on position and orientation of

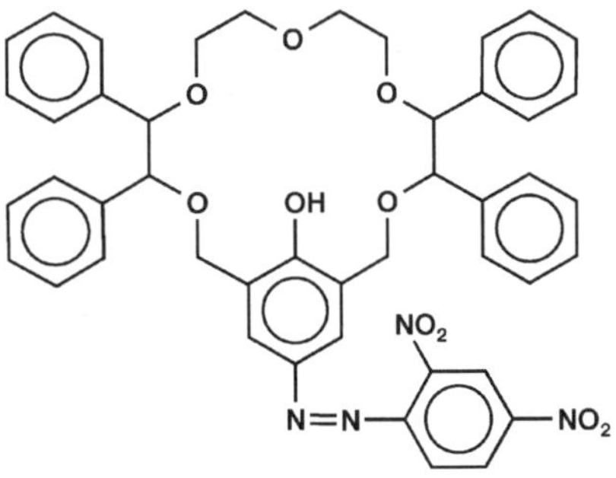

I-29

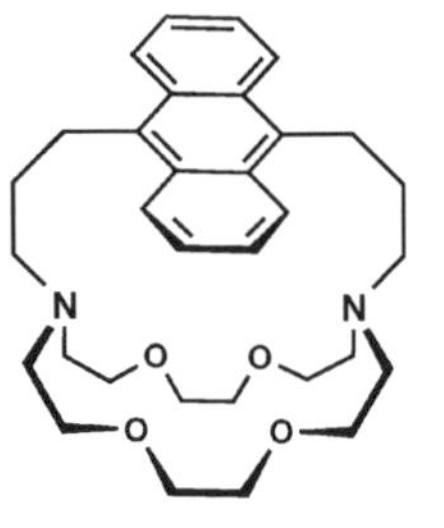

I-30

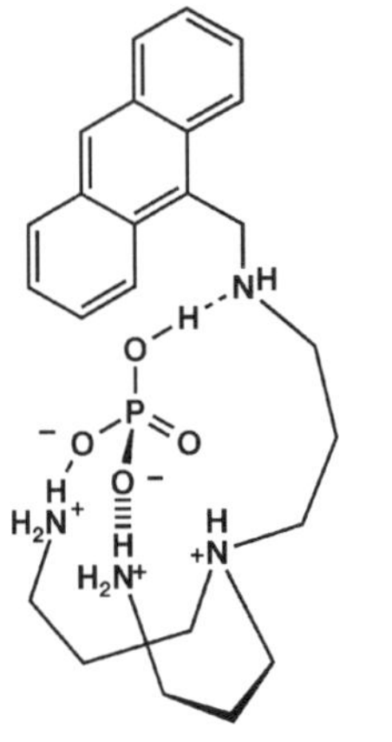

I-31

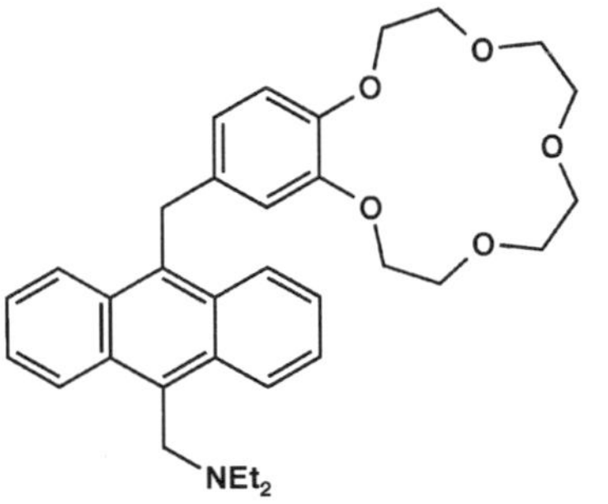

I-32

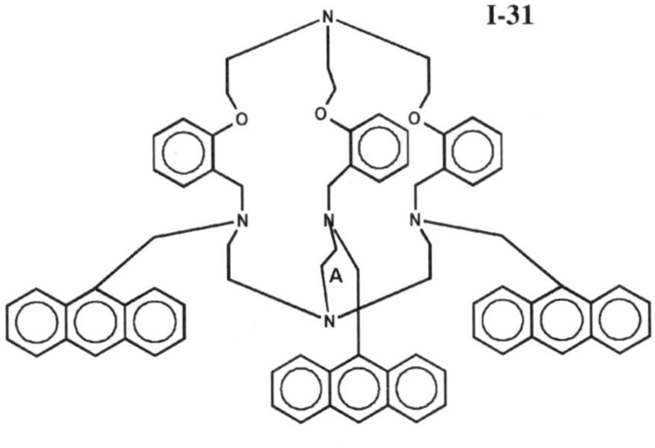

I-33

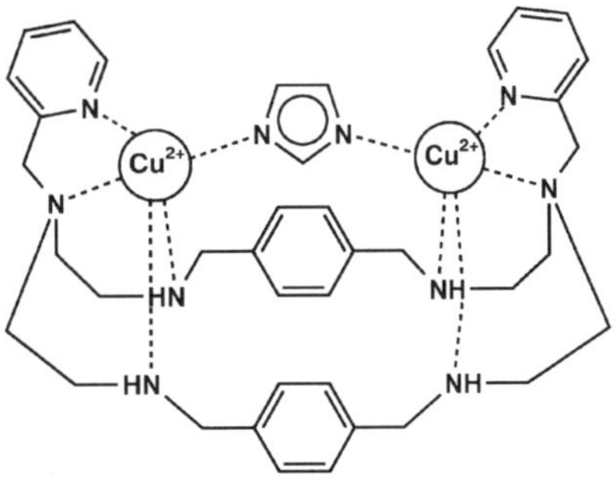

I-34

which are defined as $\Delta I = I - I_0$, where I and I_0 are fluorescence intensities measured with host alone and with host $+ 1$ mM guest. The binding constants are substantially higher than those observed with un-substituted cyclodextrins, suggesting that the fluorophore will act as a cap on the CD cavity.

Combinatorial chemistry may provide an extremely fast and economical access to host or guest compounds, as long as these can be prepared by repetitive synthetic steps. This is typical for peptides, and allows to obtain in small quantities thousands of compounds by automated techniques. A major problem then is automated screening of these compound libraries; besides the usual biological tests the screening can be extended to their molecular basis by binding studies with a defined receptor instead of a whole cell[112] (see also

the hydroxy groups (Scheme I 32). The strong complexes make it likely that there is also substantial contact between parts of the analyte sticking out of the cavity and the outside dye. There is also a distinct difference in complexation induced fluorescence changes,

the SPR method (surface plasmon resonance) discussed in Section D 7). One can for instance attach a different dye to the substrates which are to be used to screen a library of host compounds. Thus, [5]Met-enkephalin was colored red by conjugation with the dye Disperse-Red-1, and [5]Leu-enkephalin was colored blue with another dye. A library of hexapeptide-containing hosts was synthesized by the statistical combinatorial method using cheno(12-deoxy)-cholic acid or an open chain 1.5-disubstituted analog as supporting scaffold (Scheme I 33). Immobilization of the steroid on polymer beads provides then 10^4 different peptide sequences as hosts. The beads turn either red or blue only if one of the colored substrates is bound; these few beads are then selected, the corresponding peptide sequence is identified by HPLC, and can be resynthesized by conventional techniques. Although this approach does not quantify the underlying host–guest interaction energy and relies on a yes-or-no binding test, it already provides qualitative insight into peptide–peptide interactions, which can be varied also by the spacing between the host peptide chains. In another case a tripeptide library was prepared with 29 different aminoacids, yielding a maximum of $29^3 = 24\,389$ different peptides.[113] Each peptide is grown on a different polymer bead; one then treats the beads with a solution of the colored nickel saleno cyclodextrin **I-36** (Scheme I 33). Of the thousands of beads only few took up the color; after hydrolysis one then found by gas chromatographic analysis that these beads invariably contained the dipeptide sequence L-Phe-D-Pro or D-Phe-L-Pro. NMR titrations with the underlying β-cyclodextrin in water show that these peptides indeed bind relatively well ($K = 120$–180 M^{-1}).

I4.2. ELECTROCHEMICAL DEVICES

Supramolecular electrochemical devices have been aptly reviewed in the literature.[114,115] In ion sensitive electrodes (ISE's), which have

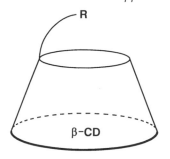

β-Cyclodextrin
with monosubstitution
R in 6-position

I-35

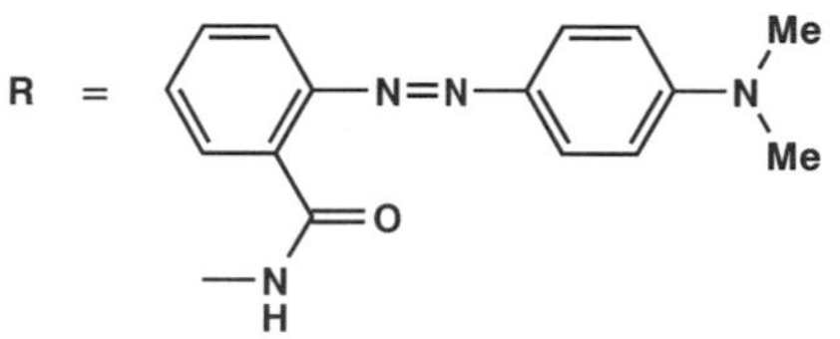

R =
I-35a   Substituent R
Methylorange (MO)

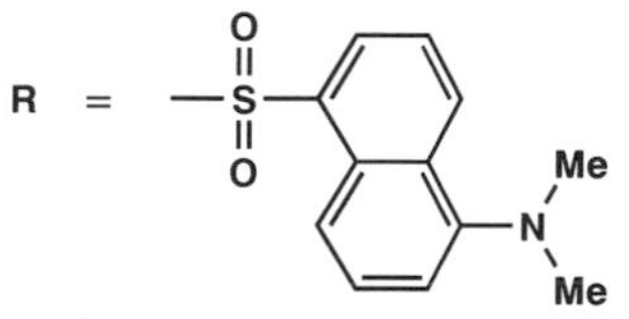

UV/vis Dye

R =
I-35b   Substituent R
Dansylamide (DNS)

Fluorescence Dye

been already mentioned in Section D 2.1, one uses ionophores implemented in membranes, which similar to the H_3O^+-sensitive glass electrode leads to a potential depending on concentration gradients due to selectively diffusing ions. This technique as well as the related, more robust ion sensitive field effects transistor (ISFET) is documented in special monographs and reviews.[97,98]

In Section A 6 we mentioned several transition metal based hosts for anion recognition, in particular a cobaltocenium (**A-39**) and ruthenium(II) bipyridyl (**A-40**) acyclic host,

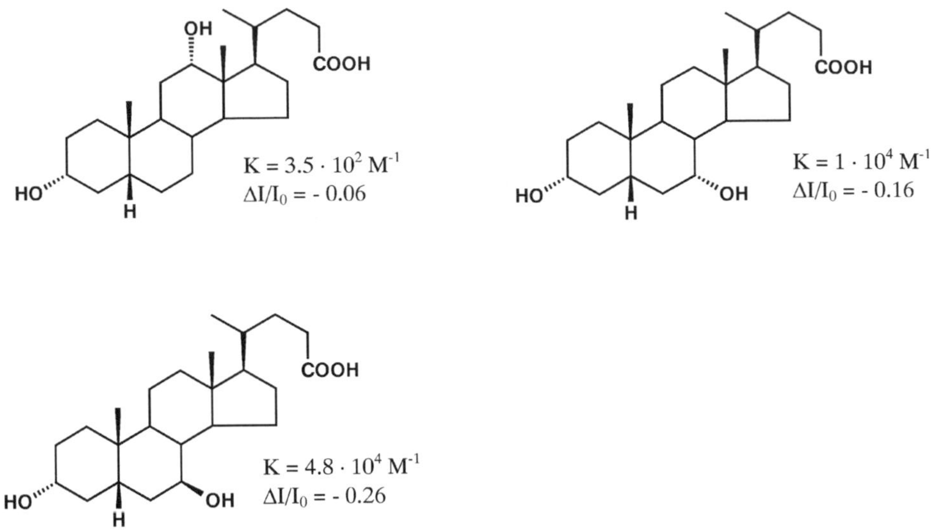

$K = 3.5 \cdot 10^2 \text{ M}^{-1}$
$\Delta I/I_0 = -0.06$

$K = 1 \cdot 10^4 \text{ M}^{-1}$
$\Delta I/I_0 = -0.16$

$K = 4.8 \cdot 10^4 \text{ M}^{-1}$
$\Delta I/I_0 = -0.26$

Scheme I 32.   Binding constants, K, and relative changes of fluorescence emission $\Delta I/I_0$ (see text) for some steroids with the cyclodextrin derivative.[111]

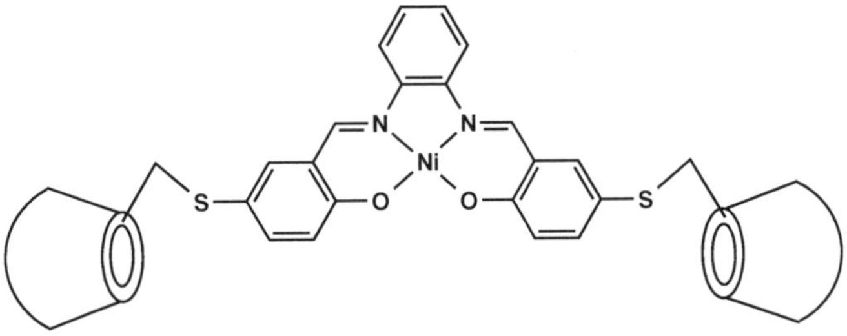

I-36

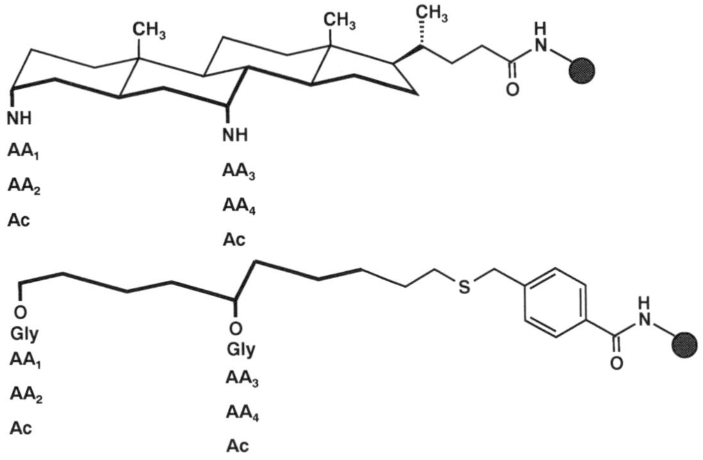

Scheme I 33.   Host structures used for screening affinities of combinatorial libraries.[113]

which can serve also as the electrochemical sensors for anions. Structures **I-37** and **I-38** (Scheme I 34) are more sophisticated macrocyclic derivatives of ruthenium(II) bipyridyl, which allow sensing of inorganic anions by electrochemical as well as by optical methods.[116] Anion binding by both hosts involves electrostatic and hydrogen bonding contributions. Host **I-37** shows as expected pronounced specificity to the more basic phosphate anion, evident both in binding and in the complexation-induced signal (Scheme I 34). The host is able to sense the phosphate anion in the presence of ten-fold excess of sulfate and chloride. Surprisingly, host **I-38**, which is a cyclic analog of **A-40**, is

specific for chloride: in DMSO it forms with Cl^- a very stable complex ($K = 4.0 \times 10^4$ M^{-1}), but practically does not bind $H_2PO_4^-$. At the same time, the acyclic host **A-40** binds phosphate stronger than chloride; the selectivity inversion is attributed to the rigid structure of the macrocycle.[116b]

The calix[4]arene derivative **I-39** allows the design of a redox active system with a set of binding sites for soft metal ions, and another set for hard ions.[117] If Fe(III) ions are added the hydroxamate groups converge for interaction with this hard ion, pushing away the bispyridyl units. Upon electrochemical reduction to Fe(II) these soft binding sites converge, with a concomitant color change.

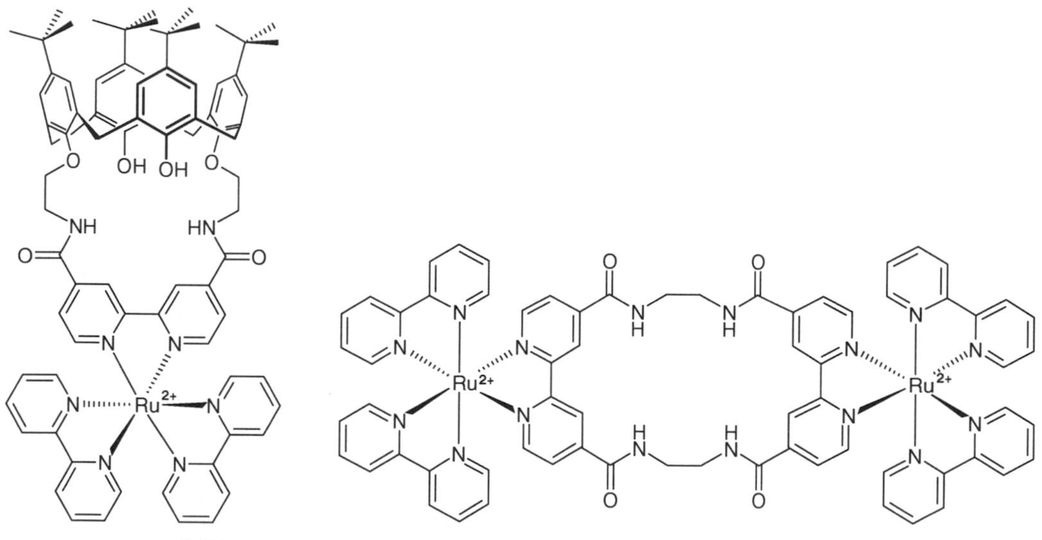

I-37

I-38

Anion	ΔE,[a] mV	K,[b] M^{-1}
$H_2PO_4^-$	175	2.8×10^4
HSO_4^-	15	
Cl^-	70	1.6×10^3
Br^-	60	

[a] Complexation-induced cathodic shifts of reduction potential of **I-37** in MeCN;
[b] equilibrium constant K in DMSO.

Scheme I 34. Redox-active sensor systems for anions.[116]

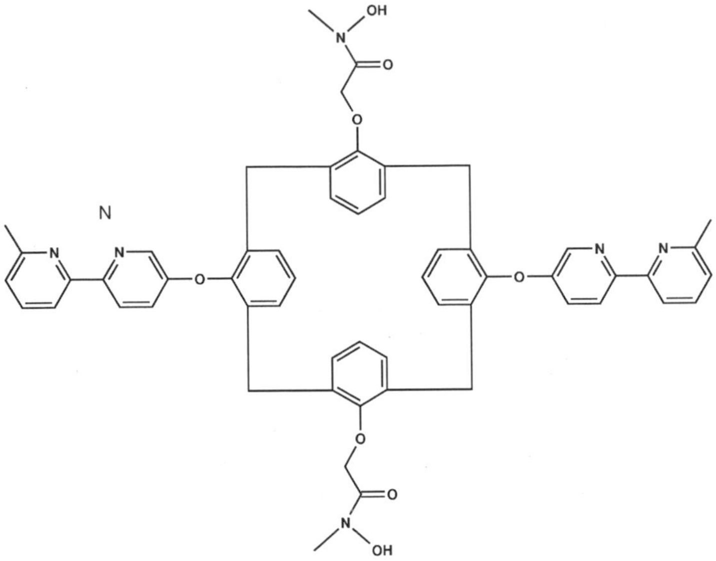

I-39

15. MOLECULAR SWITCHES AND SUPRAMOLECULAR PHOTOCHEMISTRY

Many of the supramolecular systems shown, particularly in the sections on allosteric effects (Section A 9), self-assembly (Section I 2), and analytical applications (Section I 4) represent molecular devices or machines. In this Section we can only discuss some of the underlying *chemical* principles, referring to other chapters in this book; for the engineering aspects of this exciting and rapidly developing field we refer to reviews and monographs.[118–124] The fascinating aspect of molecular electronics is not only the possible miniaturization of all kind of devices, but also the speed of switching events, which in case of tunneling effects (which may, however, also generate problems) can go down to femtoseconds, 10^{-15} s.

The formation of rotaxanes and catenanes by template-directed syntheses is based on non-covalent interactions between the components, which then are mechanically interlocked in the products (Section I 2; Scheme I 7 already shows an example of such a shuttle

device). Non-covalent forces can then lead to preferred locations of the movable parts on certain 'stations', either by attractive interactions between certain locations of L 1 and L 2, or by corresponding repulsions. As an example[125] we cite a rotaxane based on a thread containing two symmetrical diglycine peptide units ('A' in Scheme I 35) separated by a lipophilic alkane spacer B. The bead of the rotaxane, or 'shuttle' stays at the peptide part A in chloroform as solvent due to hydrogen bonds between the amide groups of shuttle and thread. Addition of a hydrogen bond-disrupting solvent such as DMSO moves the shuttle towards the central lipophilic part B (the $(CH_2)_n$ parts), and also enhances the velocity of the moving process. The different locations can be clearly distinguished by the NMR upfield shifts generated by the phenyl rings of the moving bead.

An example of a pH gradient moving cucurbituril as shuttle along a polyamine chain is illustrated in Scheme I 36.[126] In this so-called pseudorotaxane (which lacks blocking stoppers such as the trityl unit in Scheme I 35) the aniline nitrogen atom is protonated

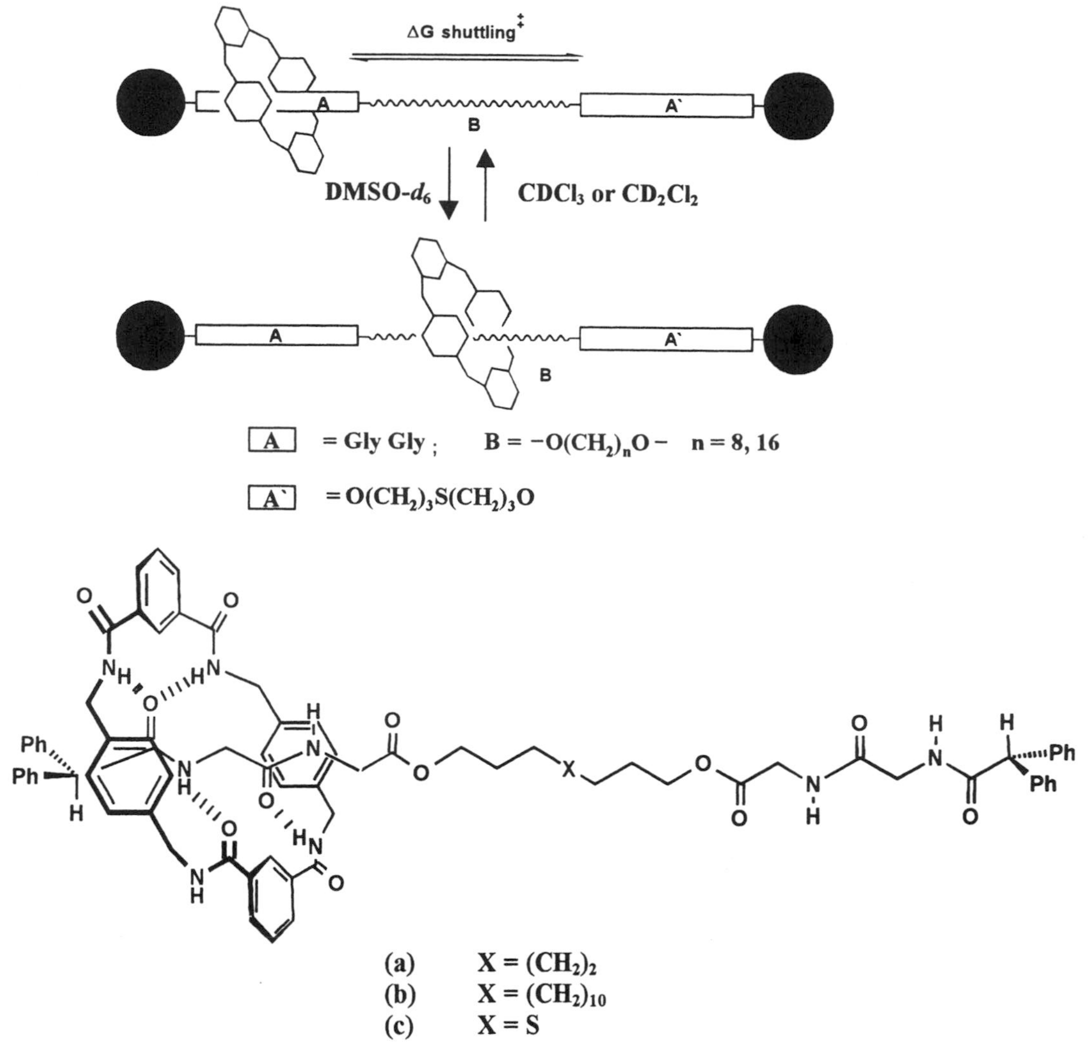

Scheme I 35. A shuttle moved in a rotaxane by solvent effects.[125] Reprinted with permission from *J. Am. Chem. Soc.*, Lane *et al.*, 1997, **119**, 11092. Copyright 1997 American Chemical Society.

only at pH below 6.7; in more basic solution this leads to preferential binding of the cucurbituril amide oxygens at the terminal 1,4-diammonium centers.

Other switches with built-in shuttles are based on the macrocyclic methylviologen (MV) rotaxane and catenane systems[127] (Section I2). The electron-rich thread of the rotaxane in Scheme I37 (case (a)) can be oxidized electrochemically;[128] upon this the positively charged MV shuttle can move away from the oxidized thread. Experimentally one observes a substantial anodic shift of the half-wave potentials in the rotaxane as compared to the free 1,4-phenylendiamine derivatives.

Photo-oxidation of electron rich threads also leads to electrostatic repulsion in a MV-

A

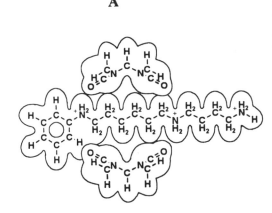

pH > 6.7↓↑ pH < 6.7

A

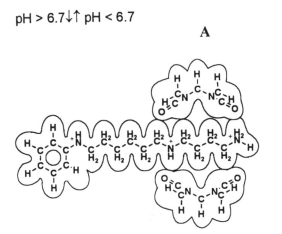

Scheme 136. The macrocycle cucurbituril A (see structure B-15) as shuttle moved by pH gradients in a pseudorotaxane (the drawing shows the macrocycle A as cutout only).[126]

based radical ion pair rotaxane (Scheme 137, case (b)); charge recombination from back-electron transfer, however, is so fast that a bead/shuttle movement cannot occur; instead, charge transfer processes are observed to the ferrocene stoppers which are used in these rotaxanes.[129,130] Permanent photoreduction of an electron acceptor thread based on 1,5-dioxynaphthalene substituted with ethyleneglycol units can be reached by adding an external electron-transfer photo-

sensitizer and a sacrificial electron donor to the rotaxane formed with the MV-cyclophane shown also in Scheme 137. The subsequent unthreading process is characterized by luminescence of the thread, which is quenched as long as the shuttling cyclophane is close to the dioxynaphthalene thread part.[131] The appearance of optical signals like luminescence then can form the basis for new *information processing and storage* systems.

Supramolecular   photochemistry   has become a mature field for which the reader is referred to specialized reviews and monographs.[118–123,130] Light has been used early to change conformations of host compounds, and thus to switch off or on the binding of suitable guests. Classical examples are azo-crown ethers.[132] The azo-crown shown in Scheme 138 allows in addition *active transport* of potassium ions through chloroform as membrane mimic (Section 16 discusses liquid bulk membrane transport).[133] The acidic receiving phase extracts metal cations which are taken up from the basic source phase, as long as the azo-crown ether is in the *cis* form. Light converts the thermodynamically more stable *trans* to the photostable *cis* isomer, which binds the cation better with the help of the phenolate group. Thus, the K^+ transport is pH driven, but can be activated or deactivated by light.

Crown ether styryl dyes also allow photo-controlled metal ion binding as shown in Scheme 139. The anionic sulfate 'cap' in the closed state leads to substantial increase of complexation strength. At the same time the process is accompanied by distinct changes of an optical signal.[134]

Supramolecular light harvesting devices are artificial analogs of the natural photosynthetic system. For a detailed discussion we refer to special introductions[118,135] and mention here only some underlying principles. As shown in Scheme 140 the light is collected by so-called antennae which channel the excitation energy to an acceptor P, where it can be used for charge separation. This photoin-

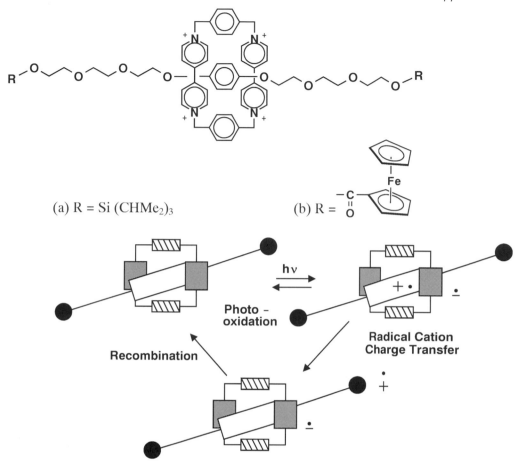

(a) R = Si (CHMe₂)₃

(b) R = $-\underset{O}{\overset{}{C}}$—Fe

Scheme I 37.   Rotaxanes switches by redox (a) or photochemical (b) signals.[128–130] Reprinted with permission from *J. Am. Chem. Soc.*, Benniston *et al.*, 1995, **117**, 5275. Copyright 1995 American Chemical Society.

duced charge separation is the basis for *energy conversion* from sun light to electrical energy. In order to avoid detrimental charge recombination a sensitizer is used prior to electron injection. The whole system can be constructed by non-covalent bonds between the different units, which for instance are composed of bipyridyl ruthenium complexes as antenna-sensitizer.

The crystal structure of the catenane in Scheme I 41 shows the electronrich hydroquinone ring exactly in the center of methylviologen-macrocycle, indicating multiple π–π- as well as edge-to-face interactions. Electrochemical and spectroscopic studies provide evidence for an efficient photoinduced electron transfer from the excited ruthenium center to the electron-deficient methylviologen ring.[136]

The system described in Scheme I 42 is based on a charge transfer complex between an immobilized eosin and azobenzene derivatives. Irradiation leads to the E–Z azobenzene isomers in solution, of which only the E isomer binds strongly to the immobilized eosin, thus converting optical to either microgravimetric or piezoelectric signals.[137]

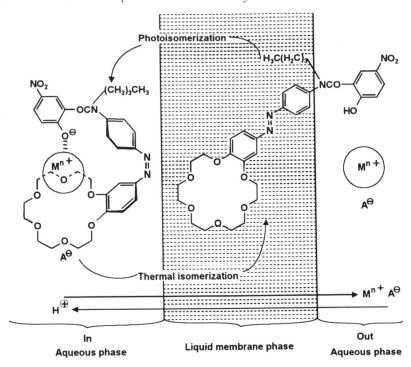

Scheme I 38. Active metal ion transport by a photochemically switched crown ether. (see ref. 133)

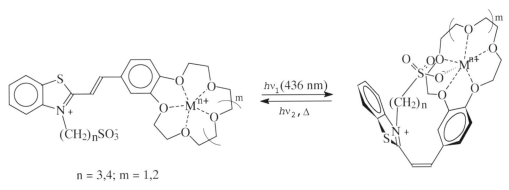

n = 3,4; m = 1,2

Scheme I 39. A photoswitched crown ether styryl dye.[134]

We finally describe briefly a new chemically switched multifunctional system (Scheme I 43). The redox-active aminoferrocene unit becomes oxidizing upon reaction with Zn(II) ions as cofactor. After addition of Zn(II) ions to a 1:1 mixture of the ligands *Fccrypt* and *Fcdpa* the latter oxidizes the *Fccrypt-Na* complex, which releases the then less well bound Na^+ ions. The process can be reversed by removing Zn^{2+} ions, for example by cyclam. The tuning of redox potentials with suitable ligands can thus allow to regulate e.g Na^+ ion levels by redox switches and Zn(II) concentration changes.[138]

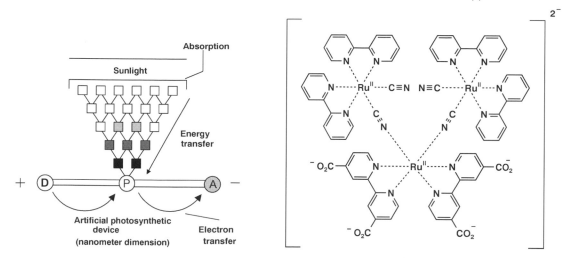

Scheme I40. Energy conversion with a supramolecular antenna system.[135] Reprinted with permission of Elsevier Science from: Balzani *et al.*, in *Comprehensive Supramolecular Chemistry*, Vol. 8, Figure 1, p. 383. Copyright 1996 Pergamon/Elsevier Oxford etc.

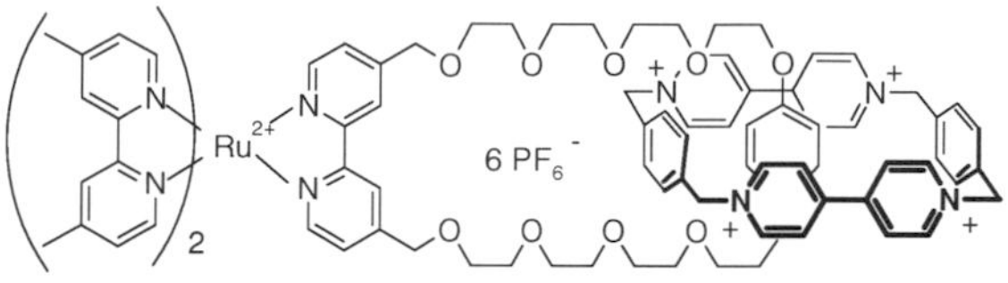

Scheme I41. Photoinduced electron transfer in a metalated catenane.[136]

I6. MEMBRANE TRANSPORT

Transport of ions and molecules across membranes is an important biological process. It has also considerable applications in separation and analysis. In this Section we shall discuss some features of ion transport through vesicle bilayers (Section G2); this imitates biological lipid bilayer membranes, and through more simple so-called bulk liquid membranes, which are used for studying carrier-mediated transport.

Lipid bilayer membranes are impermeable to inorganic ions because charged species cannot cross the hydrophobic membrane interior. There are, however, several ways to assist membrane ion transport, as illustrated in Fig. I3.[139] The carrier mechanism (a) involves complexation of cation with an ionophore bearing a lipophilic periphery at one side of the membrane, transport of the lipophilic complex across the membrane and its dissociation at the opposite side. The ion channel mechanism (b) involves a large transporter that connects both sides of the membrane; it performs the same complexation–dissociation processes, but does not move together with the cation. Through such a mechanism operates the biological channel peptide gramicidin. The channel can be formed by an aggregate of transporter molecules shown in (c), as was proposed for the antibiotic amphotericin. Finally, some amphiphilic compounds can induce a local disruption (defect formation) of the membrane, illustrated in (d), to such an extent that diffusion of an ion becomes possible.

The carrier-mediated transport of metal cations can be conveniently studied in a simple cell similar to that shown in Scheme I38. Here a liquid water-immiscible phase, 'bulk liquid membrane' (usually chloroform or dichloromethane), separates two aqueous phases: one (the source phase) contains a salt solution and another one (receiving phase) is

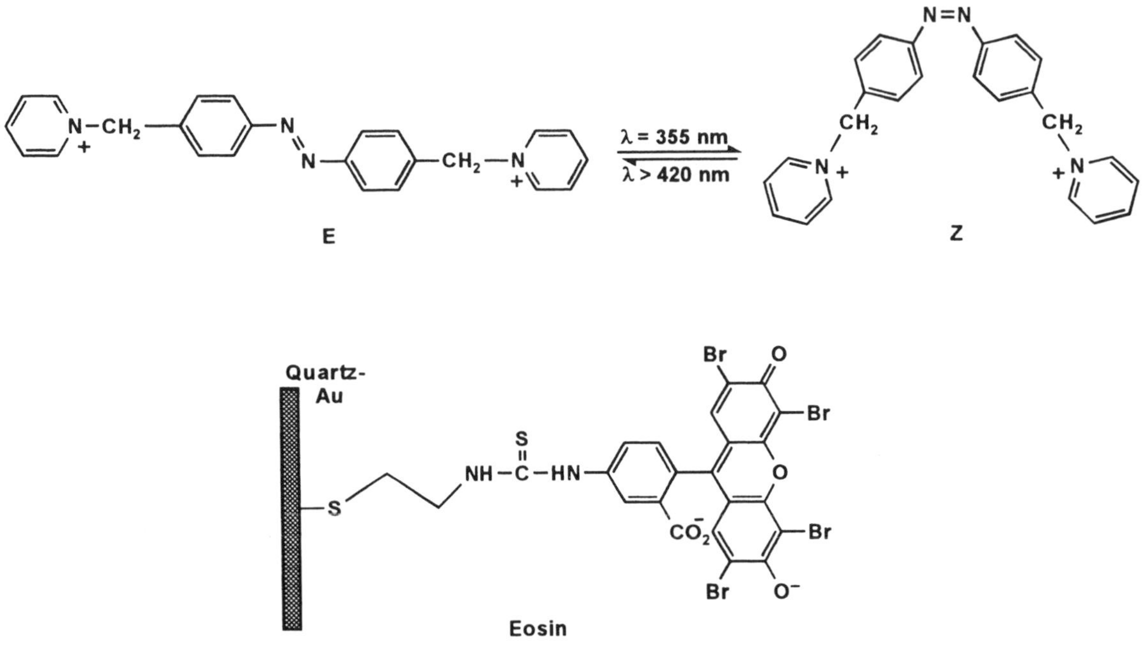

Scheme I 42.   Light conversion to gravimetric or piezoelectric signals.[137]

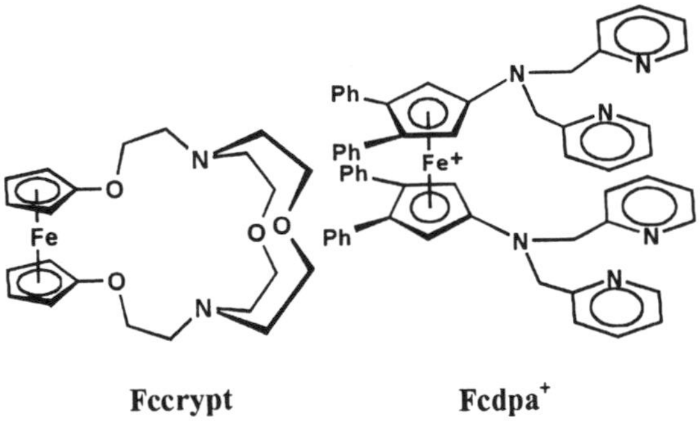

Scheme I 43.   A system for intermolecular chemical regulation.[138]

pure water. A lipophilic macrocyclic ionophore, like a crown ether, serves as carrier. At the surface layer between the membrane and source phase the ionophore binds metal cations and moves by diffusion as an ion pair with the salt counterion across the membrane mimic to the receiving phase, where the ionophore complex dissociates and releases the transferred salt.

The rate of cation transport through the bulk liquid membrane is proportional to the carrier concentration; it shows saturation behavior with respect to salt concentration in the source phase, and an optimum with respect to stability constants for different ionophores.[140,141] All these features are described quantitatively by application of Fick's first law for the dependence of the flux of a species on the diffusion coefficient, on membrane thickness and on the concentration gradient together with the mass-action law for complexation with the carrier ionophore.[141] Since the rate of liquid membrane transport is often used as a simple test for the efficiency of ionophore complexation, it is worth noting that the highest rate is observed with not too strongly binding carrier ligands. The reason for the optimum with respect to the complexation strength is, that strongly complexing ligands become saturated already at very low metal concentrations, and that therefore the gradient of metal ions over the membrane

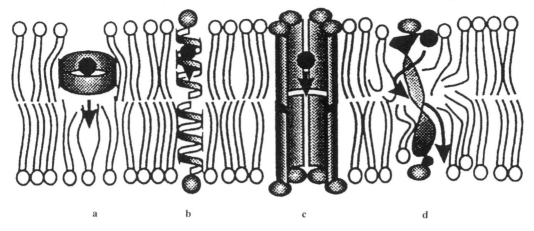

Figure I 3. Schematic mechanisms for ion transport across bilayer membranes: a, a carrier mechanism; b, an ion channel; c, an aggregate pore; d, a membrane-disrupting event.[139] Kluwer Academic Publishers, *Physical Supramolecular Chemistry*, Eds. Echegoyen, L., Kaifer A.E. (Nato ASI Ser. C) 1996, p. 39, Fyles, T.M. Fig. 1, copyright 1996, with kind permission from Kluwer Academic Publishers.

rapidly decays. In addition, the decomplexation rate with strong ligands may be too low. The transport rate strongly depends on the salt anion. Thus, the rate of transport of potassium salts taken at the same 0.002 M concentration in the source phase through chloroform liquid membrane with dibenzo-18-crown-6 as the carrier increases by a factor of $\sim 10^8$ on going from KF or KOAc to potassium picrate.[140] This effect is explicable by the higher capacity of more lipophilic anions to form a neutral ion pair, which is the actually moving species in the lipophilic membrane phase.

An important variation of bulk liquid membranes are supported liquid membranes, which consist of an organic solvent of low vapor pressure (usually *o*-dichlorobenzene, phenylcyclohexane or *o*-nitrophenyl octyl ether) immobilized on a porous polymer film, separating the source and receiving phases.[141] Such membranes, in principle, can be more useful for practical applications; they suffer, however, from low stability due to the loss of solvent and other factors such as mechanical instability.

For the study of artificial transmembrane channels two types of lipid bilayers (like

phosphatidylcholine, see Table G1) are employed: vesicles (Section G2) and planar bilayers (black lipid membranes).[142] Planar bilayers allow a more direct study of ion transport by conductometry, potentiometry, and some other techniques; on the other hand vesicles tolerate wide variations of solution composition and allow measurement of transport kinetics by monitoring transporting species initially entrapped inside vesicles by various, such as spectroscopic techniques, including NMR.[139] A useful method consists in measurements of coupled transport of protons and cations in the presence of a proton carrier, which can be monitored by using of a suitable pH indicator or by pH-state techniques.[139,143,144]

Ion transporters are large molecules, which must have a length sufficient to span the membrane which is approximately 30–40 Å thick. They should have a polar core that serves as a channel for ion diffusion, surrounded by a lipophilic exterior necessary for accommodation of the transporter inside the membrane. Surprisingly, rather simple amphiphilic molecules of type **I-40**[145] possess considerable transport capability, probably by assembling inside the membrane into a

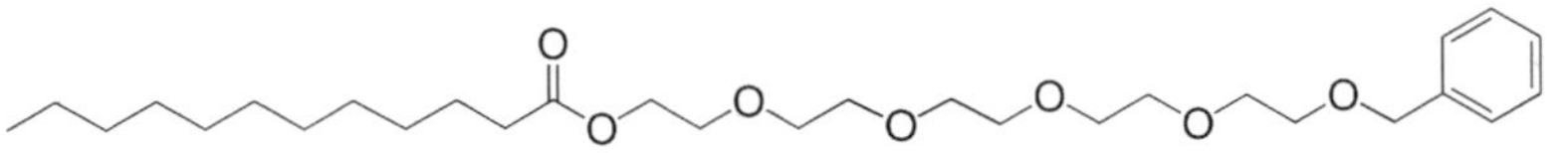

I-40

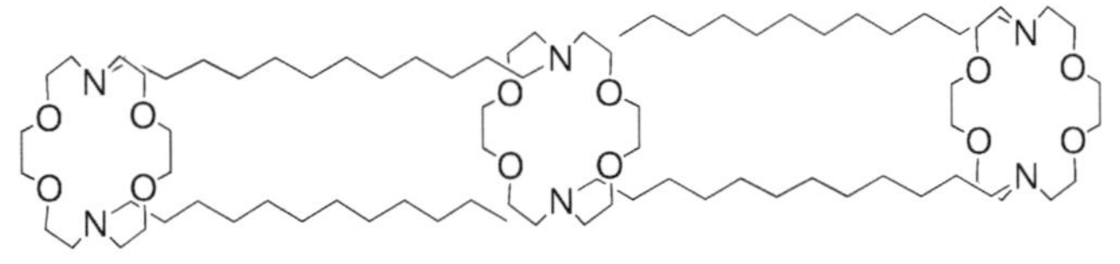

I-41

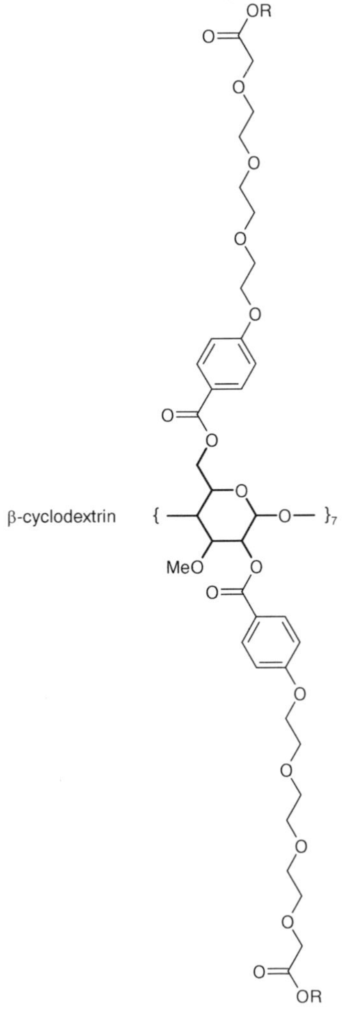
β-cyclodextrin

I-42

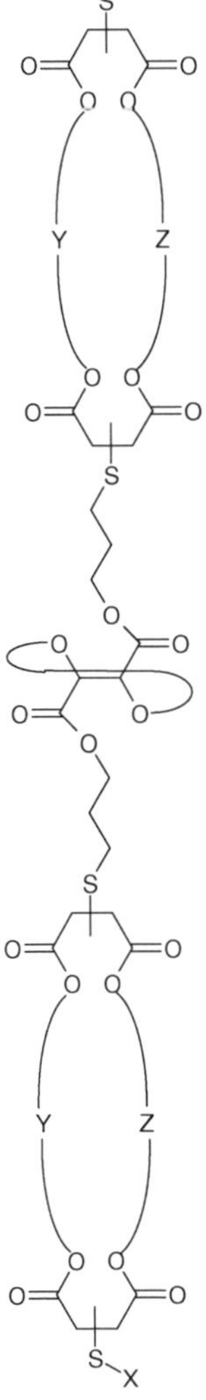

I-43

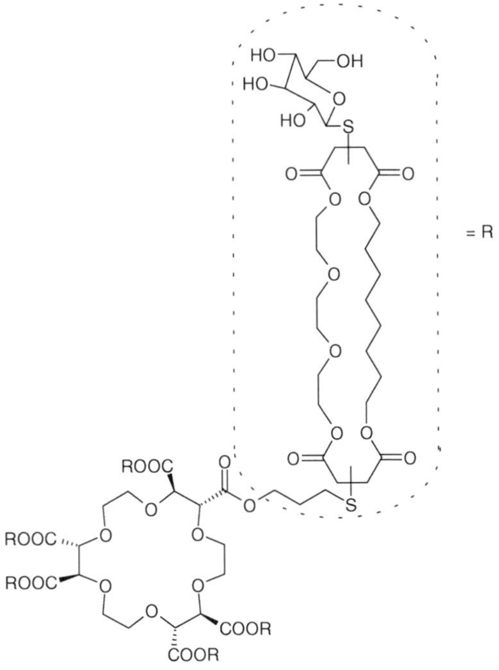
= R

I-44

transmembrane pore. Transporters working through a channel mechanism were prepared on the basis of crown ether or cyclodextrin units. One of the most active systems is the tris-crown ether **I-41**.[146] This compound of about 30 Å length exhibits high Na^+/K^+ selectivity and has a sodium cation transport rate only four times less than gramicidin. Structure **I-42** illustrates so-called bouquet molecules, based either on cyclodextrin or a crown ether.[147] A family of channel compounds of the general structure **I-43**, exemplified by compound **I-44**, was also prepared.[144] They have hydrophilic head groups, a combination of hydrophobic and hydrophilic groups in the wall units, and a crown ether central unit; the overall length is compatible with the bilayer thickness. According to different criteria some of these compounds indeed work as channels, but others function as carriers. In spite of the obvious difference between channel and carrier mechanisms their discrimination is a difficult problem; several experimental criteria are discussed in refs.[139,144] The area is in fast development; for details on chemical models for transmembrane channels see review.[146]

The cases discussed above involve mostly *passive* transport, driven by a concentration gradient between the phases which are separated by a membrane. *Active* transport, which moves species against the gradient, needs a source of energy supplied directly (e.g. by ATP hydrolysis), or indirectly by coupling to the transport of another species along their gradient (in biological systems this is usually Na^+ or H^+). In the most simple case of so-called proton-driven cation-transport one uses as a carrier a crown ether with pendent ionogenic group (an amino[148a] or phenolic[133] group or simply an aza-crown ligand[148b]) which transports cations against their gradient between basic source and acidic receiving phases. At the basic interface the ionogenic site is deprotonated and the ligand forms an ion-pair complex with the cation; then the complex is transported across the membrane to the acidic interface, where protonation of the ionogenic site releases the cation and the carrier returns to the membrane as an ammonium ion pair (in case of amino functionalized ligand) or as a neutral species. An example of an active transport by this mechanism in combination with a light-switched system is shown in Section I5, Scheme I38.

I7.   EXERCISES AND ANSWERS

I7.1.   EXERCISES

(solutions to the problems are described in references, if cited)

1. Design elements for rigid rod-like structures.[149]

2. Select components like those in Scheme I5 for the formation of catenanes.[18]

3. What would be needed to obtain tube-like structures which in principle provide at the same time channels for electrons and for ions? (Hints: most tubular structures are based on combinations with good stacking units, whereas ion transport needs ionophores).[150]

4. Design different shuttles for a rotaxane with functions which could 'read' in a peptide 'thread' selectively the following aminoacids: Lys, Asp, Tyr, Phe, Trp.

5. Design a cryptand containing a photoswitchable unit which is expected to show different extraction constants for cations of different size.

6. Which immobilized host ligands would be suitable for a mass-sensitive or optical detector (mostly in aqueous solution) of: (a) mononucleotides; (b) polycyclic aromatic hydrocarbons; (c) lanthanide ions; (d) carbon dioxide (in air); (e) ammonia (in air).

7. Which kind of ligand should be introduced into acidic aminoacids of peptide chains in order to allow photoresponsive conformational changes (also of enzyme activity etc.)?

8. Design a host for sequence-selective recognition of peptides in water; if possible also provide for enantioselectivity and for an optical measurement signal.

9. Propose explanations of catalytic effects observed in the following systems:

(a) formation of *N*-methylquinolinium cation (**B-20**) from quinoline and methyl iodide in water in the presence of cyclophane **B-18a** (Section B 4).[151]

(b) hydrolysis of and phosphate transfer reactions from ATP in the presence of macrocycle **I-45**.[152]

I-45

(c) enantioselective cleavage of 4-nitrophenyl esters of amino acids and peptides by thiol macrocycle **I-46** (an analog to the enzyme papain.)[153]

where

I-46

10. Enolization of indan-2-one **I-47** in aqueous solutions is catalyzed by cyclodextrins (CDs) via proton abstraction by deprotonated secondary hydroxyl groups.[154] The reaction kinetics follows Scheme I 18 and the respective rate law (I 3-1), with the following parameters (at 25°C and pH 11.6).

I-47

Host	$k_C(s^{-1})$	$K_S(M^{-1})$
α-CD	262	16.8
β-CD	57.9	154
HP-β-CD	70.4	191

(HP-β-CD is a β-cyclodextrin derivative alkylated by 2-hydroxypropyl groups at primary hydroxyls).

Under the same conditions the pseudo-first-order rate constant of spontaneous enolization of **I-47** equals $12\,s^{-1}$, and the second-order rate constant of the enolization catalyzed by trifluoroethanol (an alcohol with $pK_a = 12.4$, close to pK_a values 12.1–12.3 of the employed cyclodextrins) is $k_{2,0} = 206\,M^{-1}s^{-1}$.

Analyze the efficiency of cyclodextrin catalysts in comparison to a small alcohol molecule unable to host–guest complexation with the substrate in terms of Scheme I 24. Simulate (calculate) k_{obs} *vs.* catalyst concentration profiles (e.g. for 0–20 mM) for the cyclodextrins and trifluoroethanol with the parameters given above, and discuss the relative efficiency of catalysis at low and high catalysts concentrations, i.e. ≤ 5 mM and ≥ 20 mM.

17.2.   ANSWERS

4. The shuttle should in a suitable macrocycle contain following functions or elements: for Lys: acidic groups; for Asp: basic groups; for Tyr: stronger basic as well as aromatic groups; for Phe: aromatic moieties; for Trp: larger aromatic moieties.

5. see ref. 155.

6. (a) for mononucleotides: cyclophanes or cyclodextrins with positively charged elements such as ammonium ions;

(b) for polycyclic aromatic hydrocarbons: large cyclophanes or planar aromatic receptors including e.g. porphyrins; particularly electrondeficient systems containing e.g. pyridinium moieties.

(c) for lanthanide ions: e.g. azacrown ethers or cryptands, also with carboxylate as pendent side chains.

(d) for carbon dioxide (in air): basic material containing e.g. $Ca(OH)_2$, for mass sensitive detection immobilized on quartz surface.

(e) for ammonia (in air): for optical detection e.g. a immobilized azodye with acidic, e.g. phenolic groups.[156]

7. Couple amino azobenzenes.[157]

8. Possible: a host equipped, for instance, with a crown ether for binding the peptide $-NH_3^+$ terminus, and an ammonium- or guanidinium group for the COO- terminus. The spacer between these functions must have a length according to desired peptide length, and bear substituents with functions complementary to aminoacid side groups. Fluorescent substituents on the spacer can provide for optical detection.[158] Enantioselectivity may be reached with asymmetric units within the spacer (e.g. by using a peptide as spacer).

9. (a) The transition state of the alkylation reaction schematically shown in structure **I-48** has a partial positive charge on the quinoline molecule and therefore binds to cyclophane **B-18a** more strongly than the neutral substrate due to contribution of the cation-π interaction (Fig. B 7).

I-48

(b) The positively charged macrocycle attracts ATP anion electrostatically and/or by hydrogen bonding; one of remaining non-protonated nitrogens acts as a nucleophile, affording an intermediate phosphoramidate (and ADP), from which phosphate can be transferred to water or to other external nucleophiles.

(c) The initial non-covalent binding occurs via complexation of the substrate ammonium group to the crown ether moiety of the catalyst; the deprotonated mercapto group serves as an intramolecular strong nucleophile, as shown in structure **I-49**. Due to the presence of a chiral center in the catalyst molecule, the initial complex and transition state are diastereomeric, which explains the reaction enantioselectivity.

I-49

10. Assuming that anions of trifluoroethanol and cyclodextrins have the same intrinsic reactivity as enolization catalysts one can apply in this case Scheme I 24 considering $CF_3CH_2O^-$ as R-F and cyclodextrin anions as H-F species, respectively. It follows from equation (I 3-8) that the dimensionless transition-state binding constants $K_T = K_S k_C/k_{2,0}$ equal 21.4, 43.3 and 65.3 for

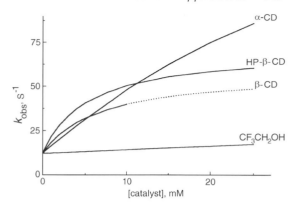

Figure I 4. Profiles of the k_{obs} *vs.* Catalyst concentration for the enolization of indan-2-one **I-47**.

α-, β- and HP-β-CD respectively (statistical correction for the presence of six and seven equivalent glucose units in α- and β-CD respectively would decrease K_T values to 3.6; 6.2 and 9.3, however, this does not change the conclusions below). Evidently, all K_T values are higher than unity, which indicates stabilization by non-covalent binding in the transition state. Increasing values of K_T for hosts with increased cavity sizes most probably reflect inclusion of a progressively larger portion of the rather large hydrophobic guest substrate. The same trend is observed also for the ground state binding, but K_S values increase considerably more rapidly than the K_T values. As a result k_C, which depends on the ratio K_T/K_S, becomes smaller in spite of stronger transition-state stabilization for larger cyclodextrins.

Figure I 4 shows profiles of the k_{obs} vs. catalyst concentration (the solubility of β-CD in water is only 10 mM; the dotted line shows expected profile for higher concentrations). One can see that at low concentrations the most active catalyst is that possessing the highest K_T (or $k_C K_S$), but that finally at high concentrations it is that possessing the highest K_T/K_S (or k_C). Thus, with a given transition-state stabilization free energy, the most efficient catalysts are those providing a rather small ground-state stabilization; this is why the biocatalysts are evolutionary optimized to have high k_C rather than small K_m values.[159]

REFERENCES

1 See e.g. Pirkle, W.H.; Pochapsky, T.C. (1989) *Chem. Rev.*, **89**, 347.

2 Cram, D.; Cram, J.M. (1978) *Acc. Chem. Res.*, **11**, 9; for a related approaches see Prelog, V. (1989)

Angew. Chem., Int. Ed. Engl., **28**, 114; Georgiadis, T.M.; Georgiadis, M.M.; Diederich, F. (1991) *J. Org. Chem.*, **56**, 3362.

3 Parker, D. (1991) *Chem. Rev.*, **91**, 1441.

4 Peters, J.A.; Huskens, J.; Raber, D.J. (1996) *Prog. Nucl. Magn. Reson. Spectrosc.*, **28**, 283; Geraldes, Carlos F.G.C. (1993) *Methods Enzymol. (Metallobiochemistry, Pt. D)*, **227**, 43.

5 Jeong, K.-S., Tjivikua, T., Muehldorf, A., Deslongchamps, G., Famulok, M., Rebek, J. (1991) *J. Am. Chem. Soc.*, **113**, 201.

6 Yoon, S.S.; Still, W.C. (1993) *J. Am. Chem. Soc.*, **115**, 823.

7 Galan, A.; Andreu, D.; Echavarren, A.M.; Prados, P.; de Mendoza, J. (1992) *J. Am. Chem. Soc.*, **114**, 1511.

8 Webb, T.; Wilcox, C.S. (1993) *Chem. Soc. Rev.*, 383 and references cited therein.

9 König, W. (1993), *Trends Anal. Chem.*, **12**, 130; Schurig, V. (1994) *J. Chrom. A.*, 666, 111.

10 Hamasaki, K.; Usui, S.; Ikeda, H.; Ikeda, T.; Ueno, A. (1997) *Supramol. Chem.*, **8**, 125.

11 Salmon, M.; Aguilar, M.M. (1994) *Curr. Top. Electrochem.*, **3**, 53; Tsukube, H.; Sohmiya, H. (1993) *Supramol. Chem.*, **1**, 297.

12 Pirkle, W.H.; Welch, C.J. (1994) *Tetrahedron: Asymmetry* **5**, 777; Pirkle, W.H.; Pochapsky, T.C. (1990) *Chromatogr. Sci.*, **47**, 783.

13 Lipkowitz, K.B.; Coner, R.; Peterson, M.A. (1997) *J. Am. Chem. Soc.*, **119**, 11269 and references cited therein.

14 Rivera, J.M.; Martin, T.; Rebek, J. (1998) *Science*, **279**, 1021.

15 Parker, D. (1996) ed; (a) *Macrocyclic Syntheses*, Oxford University Press, Oxford; (b) Cooper, S.R. (1992) ed.; *Crown Compounds*, VCH, Weinheim; (c) Gokel, G.W.; Korzeniowski, S.J., (1982) eds.; *Macrocyclic Polyether Synthesis*, Springer, Berlin; (d) Vögtle, F.; Weber, E., (1985) eds. *Host–Guest Chemistry – Macrocycles–Synthesis, Structures, Applications*, Springer, Berlin; (e) Inoue, Y.; Gokel, G.W. (1990) eds., *Cation Binding by Macrocycles*, Marcel Dekker, New York.

16 Sauvage, J.-P.; Hosseini, M.W. (1996) ed., *Comprehensive Supramolecular Chemistry, Vol. 9, Templating, Self-Assembly and Self-Organization*, Pergamon/Elsevier, Oxford; Hoss, R.; Vögtle, F. (1994) *Angew. Chem., Int. Ed. Engl.*, **33**, 375.

17 (a) Vögtle, F.; Weber, E. in ref. 15d.; (b) Gutsche, C.D. in *Calixarenes (Monographs in Supramolecular Chemistry)* J.F. Stoddart, ed. (1989) Royal Society of Chemistry Cambridge UK; Vicens, J.; Böhmer, V., (1990) eds., *Calixarenes*, Kluwer, Dordrecht; Böhmer, V. (1995) *Angew. Chem., Int. Ed. Engl.*, **34**, 713; (c) Martell, A.M. in ref. 15b, 99; Chen, D.; Motekaitis, R.J.; Murase, I., Martell, A.M. (1995) *Tetrahedron*, **51**, 77.

18 J.-P. Sauvage, (1990) *Acc. Chem. Res.*, **23**, 319; Chambron, J.-C.; Dietrich-Buchecker, C.; Sauvage, J.-P. (1993) *Top. Curr. Chem.*, **165**, 131ff.

19 Anderson, H.L.; Sanders, J.K.M. (1990) *Angew. Chem., Int. Ed. Engl.*, **29**, 1400.

20 D. Philp, J.F. Stoddart, (1996) *Angew. Chem., Int. Ed. Engl.*, **35**, 1155.

21 (a) Sherman, J.C.; Knobler, C.B.; Cram, D.J. (1991) *J. Am. Chem. Soc.*, **113**, 2194; (b) Chapman, R.G.; Sherman, J.C. (1995) *J. Am. Chem. Soc.*, **117**, 9081; see also Chapman, R.G.; Sherman, J.C. (1998) *J. Am. Chem. Soc.*, **120**, 9818.

22 Rumney, S.; Kool, E.T. (1992) *Angew. Chem., Int. Ed. Engl.*, **31**, 1617.

23 Mutter, M.; Vuilleumier, S. (1989) *Angew. Chem., Int. Ed. Engl.*, **28**, 535; Tuchscherer, G.; Mutter, M.J. (1995) *Biotechnology*, **41**, 197.

24 Orgel, L.E. (1995) *Acc. Chem. Res.*, **28**, 109.

25 Lehn, J.-M. (1994) *Pure Appl. Chem.*, **66**, 1961.

26 Kaes, C.; Hosseini, M-W.; Rickard, E.F.C.; Skleton B.W.; White, A.H. (1998) *Angew. Chem., Int. Ed. Engl.*, **37**, 920.

27 Saalfrank, R.W.; Burak, R.; Breit, A.; Stalke, D.; Herbst-Irmer, R.; Daub, J.; Porsch, M.; Bill, E.; Müther, M.; Trautwein, A.X. (1994) *Angew. Chem., Int. Ed. Engl.*, **33**, 1621.

28 Branda, N.; Wyler, R.; Rebek, (1994) *J. Science*, **263**, 1267.

29 (a) Whitesides, G.M.; Simanek, E.E.; Mathias, J.P.; Seto, C.T. (1995) *Acc. Chem. Res.*, **28**, 37; (b) Macdonald, J.C.; Whitesides, G.M. (1994) *Chem. Rev.*, **94**, 2383; (c) Lawrence, D.S.; Jiang, T.; Levett, M. (1995) *Chem. Rev.*, **95**, 2229.

30 (a) Fuhrhop, J.H.; Helfrich, W. (1993) *Chem. Rev.*, **93**, 1565; (b) Ringsdorf, H. (1988) *Angew. Chem., Int. Ed. Engl.*, **27**, 113; (c) Kunitake, T. (1992) *ibid* **31**, 709; (d) Bein, T. (1992) ed., *Supramolecular Architecture*, American Chemical Society, Washington DC; (e) Tour, J.M. (1996) *Chem. Rev.*, **96**, 537.

31 (a) Desiraju, G.R. (1995) *Angew. Chem., Int. Ed. Engl.*, **34**, 2311; (b) MacNicol, D.D.; Toda, F.; Bishop, R. (1996) eds., *Comprehensive Supramolecular Chemistry, Vol. 6*, Pergamon/Elsevier Oxford.

32 Simard, M.; Su, D.; Wuest, J.D. (1991) *J. Am. Chem. Soc.*, **113**, 4696.

33 Endo, K.; Ezuhara, T.; Koayanagi, M.; Masuda, H.; Aoyama, Y. (1997) *J. Am. Chem. Soc.*, **119**, 499.

34 Clark, T.D.; Ghadiri, M.R. (1995) *J. Am. Chem. Soc.*, **117**, 12364.

35 (a) Lahav, M.; Leiserowitz, L. in ref. 31b, 885; (b) Weissbuch, I.; Popovitz-Biro, R.; Lahav, M. Leiserowitz, L.; Kjaer, K.; Als-Nielsen, J. (1997) *Adv. Chem. Phys.*, **102**, 39; (c) Weissbuch, I.; Addadi, I.; Leiserowitz, L.; Lahav, M. (1988) *J. Am. Chem. Soc.*, **110**, 561.

36 (a) Sievers, D., Achilles, T., Burmeister, J., Jordan, S., Terfort, A., Kiedrowski, G. von, in *Self-Production of Supramolecular Structures – From Synthetic Structures to Models of Minimal Living Systems*, (G. Fleischaker, S. Colonna, P.L. Luisi eds.), Kluwer Dordrecht, (1994), 45; (b) Wintner, E.A., Conn, M.M., Rebek, J. (1994) *Acc. Chem. Res.*, **27**, 198; (c) Hoffmann, S., (1992) *Angew. Chem., Int. Ed. Engl.*, **31**, 1013. (d) Eschenmoser, A. (1994) *Origins Life Evol. Biosphere*, **24**, 389; (e) Kanavarioti, A., J. (1992) *Theor. Biol.*, **158**, 207.

37 Conn, M.M.; Wintner, E.A.; Rebek, J. (1994) *J. Am. Chem. Soc.*, **116**, 8823.

38 (a) Menger, F.A.; Eliseev, A.V.; Khanjin, N.A.; Sherrod, M.J. (1995) *J. Org. Chem.*, **60**, 2870; (b) Reinhoudt, D.N.; Rudkevich, D.M.; de Jong, F. (1996) *ibid.*, **118**, 6880.

39 D. Sievers, G. v. Kiedrowski in print; *Chem. Eur. J.*

40 Bachmann, P.A.; Luisi, P.L.; Lang, J. (1992) *Nature*, **357**, 57.

41 Walde, P.; Wick, R.; Fresta, M.; Mangone, A.; Luisi, P. (1994) *J. Am. Chem. Soc.*, **116**, 11649.

42 Buhse, T.; Nagarajan, R.; Lavabre, D.; Micheau, J.C. (1997) *J. Phys. Chem. A*, **101**, 3910.

43 (a) Breslow, R.; Dong, S.D. (1998) *Chem. Rev.*, **98**, 1997; (b) Breslow, R. (1995) *Acc. Chem. Res.*, **28**, 146.

44 Kirby, A.J. (1996) *Angew. Chem., Int. Ed. Engl.*, **35**, 707.

45 Murakami, Y.; Kikuchi, J.; Hisaeda, Y.; Hayashida, O. (1996) *Chem. Rev.*, **96**, 721.

46 Tee, O.S. (1994) *Adv. Phys. Org. Chem.*, **29**, 1.

47 Bender, M.L.; Komiyama, M. (1978) *Cyclodextrin Chemistry*, Springer, New York.

48 Kirby, A.J. (1980) *Adv. Phys. Org. Chem.*, **17**, 183.

49 Jencks, W.P. (1975) *Adv. Enzymol.*, **43**, 219.

50 (a) Page, M.I. (1973) *Chem. Soc. Rev.*, **2**, 295; (b) Page, M.I. (1977) *Angew. Chem*, **89**, 456.

51 Cacciapaglia, R.; van Doorn, A.R.; Mandolini, L.; Reinhoudt, D.N.; Verboom, W. (1992) *J. Am. Chem. Soc.*, **114**, 2611.

52 Wolfenden, R. (1972) *Acc. Chem. Res.*, **5**, 10.

53 Lienhard, G.E. (1973) *Science*, **180**, 149.

54 Mader, M.M.; Bartlett, P.A. (1997) *Chem. Rev.*, **97**, 1281.

55 (a) Lerner, R.A.; Benkovic, S.J.; Schultz, P.G. (1991) *Science*, **252**, 659; (b) Stewart, J.D.; Benkovic, S.J. (1993) *Chem. Soc. Rev.*, **21**, 213.

56 Menger, F.M. (1985) *Acc. Chem. Res.*, **12**, 128.

57 Mackay, L.G.; Wylie, R.S.; Sanders, J.K.M. (1994) *J. Am. Chem. Soc.*, **116**, 3141.

58 Walter, C.J.; Anderson, H.L.; Sanders, J.K.M. (1993) *J. Chem. Soc., Chem. Commun.*, 458.

59 Muche, M.-S.; Göbel, M.W. (1996) *Angew. Chem., Int. Ed. Engl.*, **35**, 2126.

60 Menger, F.M.; Ladika, M. (1987) *J. Am. Chem. Soc.*, **109**, 3145.

61 Tee, O.S.; Bozzi, M.; Hoeven, J.J.; Gadosy, T.A. (1993) *J. Am. Chem. Soc.*, **115**, 8990.

62 (a) Breslow, R., Grazniecki, M.F.; Emert, J.; Hamaguchi, H. (1980) *J. Am. Chem. Soc.*, **102**, 762; (b) Trainor, G.L.; Breslow, R. (1981) *J. Am. Chem. Soc.*, **103**, 154; (c) Breslow, R.; Trainor, G.L.; Ueno, A. (1983) *J. Am. Chem. Soc.*, **105**, 2739.

63 Ikeda, H.; Kojin, R.; Yoon, C.J.; Ikeda, T.; Toda, F. (1989) *J. Incl. Phenom. Mol. Recogn.*, **7**, 117.

64 Komiyama, M.; Inoue, S. (1980) *Bull. Chem. Soc. Jpn.*, **53**, 2330.

65 Mattei, P.; Diederich, F. (1996) *Angew. Chem., Int. Ed. Engl.*, **35**, 1341.

66 Mattei, P.; Diederich, F. (1997) *Helv. Chim. Acta*, **80**, 1555.

67 Selected monographs and reviews : Bertini, I.; Gray, H.B.; Lippard, S.J.; Valentine, J.S., (1994) eds, *Bioinorganic Chemistry*, University Science Books, Mill Valley, CA, Hendry, P.; Sargeson, A. M. in *Progress in Inorganic Chemistry: Bioinorganic Chemistry*, *Vol. 38*, Lippard, S.J. ed., (1990), **38**, 201; Sigman, D. S.; Mazumder, A.; Perrin, D. M., (1993) *Chem. Rev.*, **93**, 2295; Christianson, D.W.; Lipscomb, W.N. (1989) *Acc. Chem. Res.*, **22**, 62; Kimura, E. (1994) *Progr. Inorg. Chem.*, **41**, 443.

68 Chin, J. (1991) *Acc. Chem. Res.*, **24**, 145.

69 Hettich, R.; Schneider, H.-J. (1997) *J. Am. Chem. Soc.*, **119**, 5638.

70 Kimura, E.; Nakamura, I.; Koike, T.; Shinoya, M.; Kodama, Y.; Ikeda, T.; Shiro, M. (1994) *J. Am. Chem. Soc.*, **116**, 4764; for a related system see: Morrow, J.R.; Aures, K.; Epstein, D. (1995) *J. Chem. Soc., Chem. Commun.*, 2431.

71 (a) Komiyama, M. (1995) *J. Biochem.*, **118**, 665; (b) Komiyama, M., Takeda, N., Takahashi, Y., Uchida, H., Shiiba, T., Kodama, T., Yashiro, M. (1995) *J. Chem. Soc., Perkin Trans.*, 269; (c) Rammo, J.; Schneider, H.-J. (1996) *Liebigs Ann. Chem.*, 175 and references cited therein.

72 (a) Ott, R.; Krämer, R. (1998) *Angew. Chem., Int. Ed. Engl.*, **37**, 1957; (b) Moss, R.A.; Zhang, J.; Ragunathan, K.G. (1998) *Tetrahedron Lett.*, **39**, 1529.

73 Sträter, N.; Lipscomb, W.N.; Klabunde, T.; Krebs, B. (1996) *Angew. Chem., Int. Ed. Engl.*, **35**, 2024; Göbel, M.W. (1994) *Angew. Chem., Int. Ed. Engl.*, **33**, 1141.

74 Tsubouchi, A.; Bruice, T.C. (1995,) *J. Am. Chem. Soc.*, **117**, 7399.

75 Murthy, N.N.; Mahroof-Tahir, M.; Karlin, K.D. (1993) *J. Am. Chem. Soc.*, **115**, 10404.

76 Roigk, A.; Schneider, H.-J., unpublished results.

77 Ranganathan, D.; Patel, B.P.; Mishra, R.K. (1994) *J. Chem. Soc., Chem. Commun.*, 107; Shullenberger, E.F.; Eason, P.D.; Long, E.C. (1993) *J. Am. Chem. Soc.*, **115**, 11038.

78 Belmont, P.; Boudali, A.; Constant, J.-F.; Demeunynck, M.; Fkyearat, A.; Michon, P.; Serratrice, G.; Lhomme, J. (1997) *New J. Chem.*, **21**, 47.

79 Kellog, R.M. (1984) *Angew. Chem., Int. Ed. Engl.*, **23**, 782.

80 See e.g. Berkessel, A.; Bolte, M.; Schwenkreis, T. (1995) *J. Chem. Soc., Chem. Commun.*, **5**, 535; Coichev, N.; van Eldik, R. (1994) *New J. Chem.*, **18**, 123.

81 Fendler, J.H.; Fendler, E.J. (1975) *Catalysis in Micellar and Macromolecular Systems*, Academic Press, New York.

82 (a) Bunton, C.A. (1991) *Kinetics and Catalysis in Microheterogeneous Systems: Surfactants in Science Series, Vol. 38*, Gratzel, M.; Kalyanasundaram, K., eds., Marcel Dekker, New York, 13; (b) Bunton, C.A.; Savelli, G. (1986) *Adv. Phys. Org. Chem.*, **22**, 213.

83 Taciolu, S. (1996) *Tetrahedron*, **52**, 11113.

84 Bunton, C.A.; Nome, F.; Quina, F.H.; Romsted, L.S. (1991) *Acc. Chem. Res.*, **24**, 357.

85 Pérez-Bendito, D.; Rubio, S. (1993) *Trends Anal. Chem.*, **12**, 9.

86 (a) Berezin, I.V.; Martinek, K.; Yatsimirsky, A.K. (1973) *Russ. Chem. Rev. (Engl. Transl.)*, **42**, 787; (b) Martinek, K.; Yatsimirsky, A.K.; Levashov, A.V.; Berezin, I.V. (1977) *Micellization, Solubilization and Microemulsions, Vol. 2*, Mittal, K.L., ed., Plenum Press, New York, 489.

87 Romsted, L. S. in *Surfactants in Solution*, Lindman, B.; Mittal, K.L. (1984) eds., Plenum Press, New York, **1015.**

88 Menger, F.M.; Portnoy, C.E. (1967) *J. Am. Chem. Soc.*, **89**, 4698.

89 Yatsimirskaya, N.T.; Sosnovskaya, I.N.; Yatsimirsky, A.K. (1995) *Analyt. Biochem.*, **229**, 249.

90 Murakami, Y.; Aoyama, Y.; Kikuchi, J.; Nishida, K.; Nakano, A. (1982) *J. Am. Chem. Soc.*, **104**, 2937.

91 Moss, R.A.; Kim, K.Y.; Swarup, S. (1986) *J. Am. Chem. Soc.*, **108**, 788.

92 Murakami, Y.; Nakano, A.; Yoshimatsu, A.; Fukuya, K. (1981) *J. Am. Chem. Soc.*, **103**, 728.

93 Cleij, M.C.; Scrimin, P.; Tecilla, P.; Tonellato, U. (1996) *Langmuir*, **12**, 2956.

94 Bunton, C.A.; Robinson, L.; Stam, M.R. (1971) *Tetrahedron Lett.*, 121.

95 Ueoka, R.; Matsumoto, Y.; Moss, R.A.; Swarup, S.; Sugii, A.; Harada, K.; Kikuchi, Y.; Murakami, Y. (1988) *J. Am. Chem. Soc.*, **110**, 1588.

96 Schenning, A.P.H.J.; Hubert, D.H.W.; Esch, J.H.; Feiters, M.C.; Nolte, R.J.M. (1994) *Angew. Chem., Int. Ed. Engl.*, **33**, 2468.

97 See e.g. *Comprehensive Supramolecular Chemistry, Vol. 10*; Reinhoudt, D.N. (1996) ed., Pergamon/Elsevier, Oxford; Nabeshima, T. (1996) *Coord. Chem. Rev.*, **148**, 151.

98 See e.g. Spichiger-Keller, U.E. (1998) *Chemical Sensors and Biosensors for Medical and Biomedical Applications*, Wiley-VCH, Weinheim.

99 First description of these mass-sensitive devices: Sauerbrey, G. (1959) *Z. Phys.*, **155**, 206.

100 see e.g. Bodenhöfer, K.; Hierlemann, A.; Seemann, J.; Gauglitz, G.; Koppenhoefer, B.; Göpel, W. (1997) *Nature*, **387** 577.

101 Misumi, S. (1993) *Top. Curr. Chem.*, **165**, 163; Löhr, H.-G.; Vögtle, F. (1985) *Acc. Chem. Res.*, **18**, 65; Inouye, M. (1996) *Coord. Chem. Rev.*, **148**, 265.

102 For the use of porphyrin analogs for anion detection see: Sessler, J.L.; Cyr, M.; Furuta, H.; Kral, V.; Mody, T.; Morishima, T.; Shionoya, M.; Weghorn, S. (1993) *Pure. Appl. Chem.*, **65**, 393.

103 (a) Czarnik, A.W. (1994) *Acc. Chem. Res.*, **27**, 302; (b) Czarnik, A.W. (1992) *Fluorescence Chemosensors for Ion and Molecule Recognition*, ACS Symposium Series **538**, Washington DC.

104 De Silva, A.P. (1997) *Chem. Rev.*, (1997), **97**, 1515; de Silva, A.P. *J. Chem. Ed.*, **74**, 53.

105 Fages, F.; Desvergne, J.P.; Bouas-Laurent, H.; Marsau, P.; Lehn, J.-M.; Kotzyba-Hibert, F.; Albrecht-Gary, A.M.; Joubbeh, M. (1989) *J. Am. Chem. Soc.*, **111**, 8672.

106 de Silva, A.P.; Gunaratme, H.Q.N.; McCoy, C.P. (1997) *J. Am. Chem. Soc.*, **119**, 7891.

107 Ghosh, P.; Bharadwaj, P.K.; Mandal, S.; Ghosh, S. (1996) *J. Am. Chem. Soc.*, **118**, 1553.

108 (a) Canary, J.W.; Gibb, B.C. (1997) *Progr. Inorg. Chem.*, **45**, 1; (b) Busch, D.H. (1993) *Chem. Rev.*, **93**, 847; (c) Stang, P.J.; Olenyuk, B. (1997) *Acc. Chem. Res.*, **30**, 502; (d) Stang, P.J. (1998) *Chem. Eur. J.*, **4**, 19.

109 Fabrizzi, L.; Pallavivini, P.; Parodi, L.; Perotti, A.; Taglietti, A. *J. Chem. Soc., Chem. Commun.*, (1995), 2439; review: Fabrizzi, L.; Poggi, A. (1995) *Chem. Soc. Rev.*, **24**, 197.

110 Patterson, S.; Smith, B.; Taylor, R.E. (1997) *Tetrahedron Lett.*, **38**, 6323.

111 Ueno, A.; Osa, T. (1991) *Photochemistry in Organized and Constrained Media*, Ramamurthy, V.; ed.; VCH, New York, 739; for specific examples see Hamasaki, K.; Usui, S.; Ikeda, H.; Ikeda, T.; Ueno, A. (1997) *Supramol. Chem.*, **8**, 125.

112 For reviews see (a) Still, W.C. (1996) *Acc. Chem. Res.*, **29**, 155; (b) Gennari, C.; Nestler, H.P.; Piarulli, U.; Salom, B.; (1997) *Liebigs Ann. Recueil.*, 637.

113 Maletic, M.; Wennemers, H.; McDonald, D.Q.; Breslow, R.; Still, W.C. (1996) *Angew. Chem., Int. Ed. Engl.*, **35**, 1490.

114 Review on electrochemistry of supramolecular systems: Boulas, P.L.; Gomez-Kaifer, M.; Echegoyen, L. (1998) *Angew. Chem., Int. Ed. Engl.*, **37**, 216.

115 Review on carrier-based ion-selective electrodes and bulk optodes: Bühlmann, P.; Pretsch, E.; Bakker, E. (1998) *Chem. Rev.*, **98**, 1593.

116 (a) Beer, P.D. (1996) *J. Chem. Soc., Chem. Commun.*, 68; (b) Beer, P.D. (1998) *Acc. Chem. Res.*, **31**, 71.

117 Canevet, C.; Libman, J.; Shanzer, A. (1996) *Angew. Chem., Int. Ed. Engl.*, **35**, 2657.

118 (a) Balzani; V, (1987) *Supramolecular Photochemistry* ed., Reidel, Dordrecht; (b) Balzani,V.; Scandola, F. (1991) *Supramolecular Photochemistry*, Horwood, Chichester.

119 (a) Lehn, J.-M. (1988) *Angew. Chem., Int. Ed. Engl.*, **27**, 89; (b) *ibid.*, (1990), **29**, 1304.

120 Carter, F.L.; Siatkowsky, R.E.; Woltjien, H., eds. (1988) *Molecular Electronic Devices*. North Holland, Amsterdam.

121 Kuhn, H. (1991) *Molecular Electronics*, Lazaraev, P.I., ed.; Kluwer, Dordrecht, 175.

122 Drexler, K.E.; (1992) *Nanosystems: Molecular Machinery, Manufacturing, and Computation*, Wiley, New York.

123 Reinhoudt, D.N. (1996) ed., *Comprehensive Supramolecular Chemistry, Vol. 10: Molecular Devices and Applications of Supramolecular Technology*, Pergamon/Elsevier, Oxford.

124 Dürr, H.; Bouas-Laurent, E. eds. (1990) *Photochromism. Molecules and Systems.* Elsevier, Amsterdam

125 Lane, A.S.; Leigh, D.A.; Murphy, A. (1997) *J. Am. Chem. Soc.*, **119**, 11092.

126 Mock, W.L.; Pierpont, J. (1990) *J. Chem. Soc., Chem. Commun.*, 1509.

127 Ashton, P.R.; Ballardini, R.; Balzani, V.; Boyd, S.E.; Credi, A.; Gandolfi, M.T.; Gómez-López, M.; Iqbal, S.; Philp, D.; Preece, J.A.; Prodi, L.; Ricketts, H.G.; Stoddart, J.F.; Tolley, M.S.; Venturi, M.; White, A.J.P.; Williams, D.J. (1997) *Chem. Eur. J.*, **3**, 152.

128 Cordova, E.; Bissell, R.; Kaifer, A.E. (1995) *J. Org. Chem.*, **60**, 1033.

129 Benniston, A.C.; Harriman, A.; Lynch, V.M. (1995) *J. Am. Chem. Soc.*, **117**, 5275.

130 Benniston, A.C. (1996) *Chem. Soc. Rev.*, 427.

131 Ballardini, R.; Balzani, V.; Gandolfi, M.T.; Prodi, L.; Venturi, M.; Philp., D.; Riclets, H.G-; Stoddart, J.F. (1993) *Angew. Chem., Int. Ed. Engl.*, **32**, 1301.

132 Shinkai, S. (1984) *Top. Curr. Chem.*, **121**, 67.

133 Shinkai, S.; Ishihara, M.; Ueda, K.; Manabe, O. (1984) *J. Chem. Soc., Chem. Commun.*, 727; see also Shinkai, S.; Yoshida, T.; Miyazaki, K.; Manabe, O. (1987) *Bull. Chem. Soc. Jpn.*, **60**, 1819.

134 Gromov S. P., Alfimov M. V., (1997) *Izv. Akad. Nauk, Ser. Khim.*, **4**, 641 [*Russ. Chem. Bull.*, (1997), **46**, 611 (Engl. Transl.)]; see also Alfimov, M.V.; Fedorov, Yu.V.; Fedorova, O.A.; Gromov, S.P.; Hester, R.E.; Lednev, I.K.; Moore, J.N.; Oleshko, V.P.; Vedernikov, A.I. (1996) *J. Chem. Soc., Perkin Trans.* **2**, 1441.

135 Balzani, V.; Scandola, F. in ref. 123, 687ff.

136 Hu, Y.-Z.; Loyen, D.v.; Schwarz, O.; Bossmann, S.; Dürr, H.; Huch, V.; Veith, M. (1998) *J. Am. Chem. Soc.*, **120**, 5822.

137 Ranjit, T.K.; Marx-Tibbon, S.; Ben-Dov, I.; Willner, I. (1997) *Angew. Chem., Int. Ed. Engl.* **36**, 147.

138 Plenio, H.; Aberle, C. (1998) *Angew. Chem., Int. Ed. Engl.*, **37**, 1397.

139 Fyles, T.M. (1996) in *Physical Supramolecular Chemistry*, Echegoen, L.; Kaifer, A.E. eds. (NATO ASI Ser. C), 39.

140 Lamb, J.D.; Christensen, J.J.; Izatt, S.R.; Bedke, K.; Astin, M.S.; Izatt, R.M. (1980) *J. Am. Chem. Soc.*, **102**, 3399.

141 Visser, H.C.; Reinhoudt, D.N.; de Jong, F. (1994) *Chem. Soc. Rev.* 75.

142 Tien, H.T. (1988) in *Thin Liquid Films, Surfactant Science Series, Vol. 29*, Ivanov, I.B., ed. Marcel Dekker, New York, 927.

143 Castaing, M.; Morel, F.; Lehn, J.-M. (1986) *J. Membr. Biol.*, **89**, 251.

144 Fyles, T.M.; James, T.D.; Kaye, K.C. (1993) *J. Am. Chem. Soc.*, **115**, 12315.

145 Menger, F.M.; Davis, D.S.; Persichetti, R.A.; Lee, J.J. (1990) *J. Am. Chem. Soc.*, **112**, 2451.

146 Gokel, G.W.; Murillo, O. (1996) *Acc. Chem. Res.*, **29**, 425.

147 Pregel, M.J.; Jullien, L.; Lehn, J.-M. (1992) *Angew. Chem., Int. Ed. Engl.*, **31**, 1637.

148 (a) Nakatsuju, Y.; Kobayashi, H.; Okahara, M. (1983) *J. Chem. Soc., Chem. Commun.*, 800; (b) Sakamoto, H.; Kimura, K.; Koseki, Y.;

Shono, T. (1987) *J. Chem. Soc., Perkin Trans.* **2**, 1181.

149 Kotera, M.; Lehn, J.-M.; Vigneron, J.-P. (1994) *J. Chem. Soc., Chem. Commun.*, 197 (for Problem 1)

150 Nolte, R.J.M.; van Nostrum, C.F.; Picken, S.J. (1994) *Angew. Chem. Int. Ed. Engl.* **33**, 2173 (for problem 3).

151 McCurdy, A.; Jimenez, L.; Stauffer, D.A.; Dougherty, D.A. (1992) *J. Am. Chem. Soc.*, **114**, 10314.

152 Hosseini, M.W.; Lehn, J.-M.; Jones, K.C.; Plute, K.E.; Mertes, K.B.; Mertes, M.P. (1989) *J. Am. Chem. Soc.*, **111**, 6330

153 Lehn, J.-M.; Sirlin, C. (1978) *J. Chem. Soc. Chem. Commun.*, 949

154 Tee, O.S.; Donga, R.A. (1996) *J. Chem. Soc., Perkin Trans.* **2**, 2763

155 Shinkai, S.; Nakaji, T.; Nishida, Y.; Ogawa, T.; Manabe, O. (1980) *J. Am. Chem. Soc.*, **102**, 5860.

156 see Grady, T.; Butler, T.; MacCraith, B.D.; Diamond, D.; McKervey, M.A. (1997) *Analyst*, **122**, 803.

157 Willner, I.; Rubin, S. (1996) *Angew. Chem., Int. Ed. Engl.*, **35**, 367; Willner, I. (1997) *Acc. Chem. Res.*, **30**, 347.

158 See Md. Hossain, A.; Schneider, H.-J. (1998) *J. Am. Chem. Soc.*, **120**, 11208.

159 see for detailed discussion: Fersht, A.R. (1985) *Enzyme Structure and Mechanism*, 2nd ed., Freeman, New York.

APPENDICES

Table I. Factor values for hydrogen bonds and other polar interactions: C increments after Raevsky *et al.* (Raevsky, O. A. *Russ. Chem. Rev. (Engl. Translation)*, **1990**, *59*, 219); α'^H/β'^H values after Abraham *et al.* (Abraham, M. H. *Chem. Soc. Rev.*, **1993**, *22*, 73). The C_a and β values refer to hydrogen bond acceptors or electron pair donors, C_d and α values *vice versa*. The values were calculated from the original tables for functions instead for whole molecules. Values denoted — are not available.

Nr.	Function	Substituents	C_a	C_d	α'^H	β'^H
1a	Ph–OH	—	—	-2.50	0.60	—
1b		2-Me	—	-2.27	0.55	—
1c		2,6-diMe	—	-1.58	0.39	—
1d		4-OMe	—	-2.41	0.58	—
1e		4-NO$_2$	—	-3.45	0.82	—
2a	Ph–NH$_2$	—	1.42	-1.04	0.26	0.38
2b		4-OMe	1.76	-1.00	0.25	0.46
2c		4-NO$_2$	—	-1.82	0.44	—
3	R–NH$_2$	R $= CH_3-(CH_2)_n-$ [a]	2.77 ± 0.05	—	0.16	0.70 ± 0.01
4a	RR'NH	R $=$ R' $=$ Ph	1.25	-1.26	0.31	0.34
4b		R $=$ R'$CH_3-(CH_2)_n-$ [b]	2.83 ± 0.03	—	—	0.71 ± 0.01
5a	R$_3$N	R $=$ Ph	1.11	—	—	0.31
5b		R $= CH_3-(CH_2)_n-$ [c]	2.51 ± 0.11	—	—	0.63 ± 0.02
6a	R–NH$_3^+$	R $=$ Ph	—	-3.55	—	—
6b		R $=$ Me	—	-2.50	0.64	—
7	RR'NH$_2^+$	R $=$ Et	—	-3.00	0.76	—
8a	RR'R''NH$^+$	R $=$ R' $=$ R'' $=$ Bu	—	-4.31	1.01	—
8b		R $=$ Bz, R' $=$ R'' $=$ Me	—	-4.82	1.19	—
9	R–CH$_2$–OH	R $= CH_3-(CH_2)_n-$ [a]	1.82	-1.42	0.35 ± 0.02	0.48
10	RR'–CH–OH	R $=$ R' $=$ Me	1.78	-1.06	0.27	0.47
11	RR'R''C – OH	R $=$ R' $=$ R'' $=$ Me	1.88	-1.28	0.32	0.49
12a	R–O–R'	R $=$ R' $=$ Ph	1.01	—	—	0.28
12b		R $=$ Et, R' $=$ Ph	0.96	—	—	0.27
12c		R $=$ R' $= CH_3-(CH_2)_n-$ [b]	1.62 ± 0.14	—	—	0.42 ± 0.03
12d		R $=$ Et, R' $=$ CMe$_3$	1.91	—	—	0.50
12e		$(-CH_2-CH_2-O-CH_2-O-)_{cycle}$	1.48	—	—	0.39
13a	R–CHO	R $=$ Ph	1.60			0.42
13b		R $=$ Et	1.50			0.40
14a	R–CO–R'	R $=$ R' $=$ Ph	1.73	—	—	0.45
14b		R $=$ R' $=$ Et	1.62	—	—	0.43
15	RCOO$^-$	R $=$ Ph	1.29	—	—	—
16a	R–COOH	R $=$ Ph		-2.50	0.59	0.40
16b		R $=$ Me		-2.58	0.61	0.44
17a	R–COOR'	R $=$ R' $=$ Ph	1.50	—	—	0.40
17b		R $=$ Ph, R' $=$ Et	1.64	—	—	0.43
17c		R $=$ R' $=$ Et	1.78	—	—	0.47
18a	R–CO–NR'R''	R $=$ Ph, R' $=$ Et, R'' $=$ H	2.53	—	—	
18b		R $=$ Ph, R' $=$ R'' $=$ Et	2.73	—	—	0.69

Table I.   (*continued*)

Nr.	Function	Substituents	C_a	C_d	α'^H	β'^H
18c		$R = R' = R'' = Et$	2.68	—	—	0.68
18d		$R = Me, R' = iPr, R'' = H$	2.67	—	—	0.68
19a	$R-SH$	$R = Ph$	—	-0.53	0.15	—
19b		$R = Me$	1.00	—	—	0.28
19c		$R = CH_3-(CH_2)_n-$ [b]	—	-0.16 ± 0.06	0.06 ± 0.01	—
20a	$R-S-R'$	$R = Ph, R' = Me$	0.50	—	—	0.16
20b		$R = R' = CH_3-(CH_2)_n-$ [d]	1.12 ± 0.12	—	—	0.31 ± 0.03
21a	$R-SO-R'$	$R = R' = Ph$	2.63	—	—	0.67
21b		$R = R' = CH_3-(CH_2)_n-$ [d]	3.19 ± 0.06	—	—	0.80 ± 0.02
22a	$R-SO_2-R'$	$R = R' = Ph$	1.93	—	—	0.50
22b		$R = R' = CH_3-(CH_2)_n-$ [d]	2.03 ± 0.02	—	—	0.53 ± 0.01
23a	$RO-SO_2-R'$	$R = R' = Ph$	1.21	—	—	0.33
23b		$R = Ph, R' = Et$	1.27	—	—	0.34
23c		$R = Me, R' = Ph$	1.52	—	—	0.40
23d		$R = R' = Me$	1.61	—	—	0.43
24a	$RO-SO_2-OR$	$R = Me$	1.30	—	—	0.35
24b		$R = Et$	1.43	—	—	0.38
25	$R-CO-SR'$	$R = R' = Me$	1.55	—	—	0.41
26	$R-CO-SH$	$R = Me$	—	-1.62	0.40	—
27a	R_3PO	$R = Ph$	3.65	—	—	0.91
27b		$R = CH_3-(CH_2)_n$ [d]	3.96 ± 0.10	—	—	1.00 ± 0.02
28a	$(RO)_3PO$	$R = Ph$	2.50	—	—	0.64
28b		$R = Et$	3.13	—	—	0.79
29	Benzene		0.08	—	—	0.03
30	Naphtalene		0.11	—	—	0.03
31a	$RCH=CH_2$	$R = H$	0.10	—	—	0.03
31b		$R = Ph$	0.15	—	—	0.04
31c		$R = CH_3-(CH_2)_n-$ [f]	0.15	—	—	0.05 ± 0.02
32a	$RCH=CR'-$ $CH=CH_2$	$R = Me, R' = H$	0.26	—	—	0.08
32b		$R = H, R' = Me$	0.21	—	—	0.07
33	$R-C\equiv C-R'$	$R = R' = Et$	0.50	—	—	0.14
34	$R-C\equiv C-H$	$R = CH_3-(CH_2)_n-$ [f]	0.41	—	—	0.11 ± 0.02
35a	$R-C\equiv C-H$	$R = Ph$	—	-0.12	0.05	—
35b		$R = CH_3-(CH_2)_n-$ [f]	—	-0.28	0.13 ± 0.02	—
36a	$R-NO$	$R = Ph$	1.65			—
36b		$R = Et$	1.65			—
37a	$R-NO_2$	$R = Et$	0.80	—	—	0.23
37b		$R = Ph$	1.23	—	—	0.33
38a	$R-CN$	$R = Ph$	1.55	—	—	0.41
38b		$R = CH_3-(CH_2)_n-$ [c]	1.55 ± 0.06	—	—	0.43 ± 0.02
39a	$R-N=C=S$	$R = Et$	0.92	—	—	0.26
39b		$R = Bu$	0.83	—	—	0.19
40a	$R-Hal$	$R = Ph, Hal = Cl, Br$	0.06 ± 0.01	—	—	0.02
40b		$R = CH=CH_2, Hal = Cl, Br$	0.05	—	—	0.02
40c		$R = CH_3-(CH_2)_n-$ [e] $Hal = F, Cl$	—			0.10
40d		$R = CH_3-(CH_2)_n-$ [e] $Hal = Br, I$	—			0.14 ± 0.01
41a	Hal^-	$(Hept)_4N^+ \ Cl^-$	3.50	—	—	0.88
41b		$(Hept)_4N^+ \ I^-$	3.24	—	—	0.81

Table II. Effective van der Waals Radii

H	1.2 Å					
N	1.5	O	1.40 Å	F	1.35 Å
P	1.9	S	1.85	Cl	1.80
As	2.0	Se	2.00	Br	1.95
Sb	2.2			I	2.15
CH_2, CH_3	2.0				
benzene	1.7				

Table III. Ionic radii [Å] (Wells. A. F. *Structural Inorganic Chemistry*, 5th Edition, Clarendon Press, Oxford, **1984**) [a]

Li^+	0.76	Mg^{2+}	0.72	Ni^{2+}	0.69	F^-	1.33
Na^+	1.02	Ca^{2+}	1.00	Cu^{2+}	0.73	Cl^-	1.81
K^+	1.38	Sr^{2+}	1.18	Zn^{2+}	0.74	Br^-	1.96
Rb^+	1.52	Ba^{2+}	1.35	Pb^{2+}	1.19	I^-	2.20
Cs^+	1.67	Fe^{2+} low spin	0.61	La^{3+}	1.03	OH^-	1.37
NH_4^+	1.50	Fe^{2+} high spin	0.78				
Ag^+	1.15	Co^{2+} low spin	0.65				
Tl^+	1.50	Co^{2+} high spin	0.75				

[a] Based on the value 1.40 Å for O^{2-} in 6-coordination; an alternative set of radii derived for alkali halides and extended on the basis of 1.19 Å for F^- has cation radii larger by 0.14 Å, and anion radii smaller by 0.14 Å.

Table IV. Overview of supramolecular structures on a website http://www.uni-sb.de/matfak/fb11/schneider (*status january 1999*).

The structures – or those from the Cambridge or Brookhaven databases etc. – can be used as training grounds for the identification and use of non-covalent interactions. Typical exercises would involve (a) evaluation of binding modes and possible alternatives; (b) lists of pairwise non-covalent interactions with relevant distances; (c) where possible estimation of corresponding free energy contributions and expected changes with solvent and salts; (d) where applicable, determinations of possible deviation of host or guest conformation from ideal (strainfree) torsional angles; (e) selection of methods for measuring affinities and conformations of the complexes, with lists of possible effects e.g. in NMR spectra etc; (f) possible modifications of host or guest structures (including e.g. alternative inhibitors for enzymes etc). The website contains in several cases more detailed exercises as well as some typical answers.

A) Ionophores

A) I) Crown Ethers
1) 18C6/K^+-complex
2) 18C6/Li^+-complex
3) 18C6/Na^+-complex
4) 15C5/Li^+-complex
5) 15C5/Na^+-complex
6) 12C4/Li^+-complex

7) 12C4/Na$^+$-complex
8) 18C6/NH$_4^+$-complex
9) 1,10-Diaza-18C6/K$^+$-complex: nitrogen lone pairs in axial position
10) 1,10-Diaza-18C6/K$^+$-complex: nitrogen lone pairs in equatorial position
11) Complexation of ANS and Zinc: an allosteric system

A II) Cryptands
1) [222] + K$^+$ complex
2) [222] + Na$^+$ complex
3) [221] + Li$^+$ complex
4) [221] + K$^+$ complex

A III) Natural Ionophores
1) Valinomycin + K$^+$ complex
2) Monensin B + Na$^+$ complex * H$_2$O

A IV) Anion Complexation
1) Complex of a urea/tren derivative with AMP
2) Complex of azide with a octa-aza-cryptand
3) Complex of chloride with a octa-aza-cryptand

B) Cyclophanes
1) Complex of the tetraazonia-cyclophane CP44 and benzene
2) Complex of the tetraazonia-cyclophane CP66 and AMP
3) Complex of the veratrole-based cryptand with tetramethylammonium

C) Cyclodextrins
1) Complex of heptakis-6-deoxy-heptakis-6-methylamino-β-cyclodextrin (**HMA**) and AMP
2) Complex of β-cyclodextrin and the fluorescence dye ANS
3) Mono-6-deoxy-mono-6-benzylamino-β-cyclodextrin: (**CDBenz**); with phenyl substituent inside the CD-cavity
4) Complexes of **CDBenz** with the **peptide** Ac-Gly-Phe-OH
5) Heptakis-6-deoxy-heptakis-6-methylamino-β-cyclodextrin (**HMA**) and complex of a **HMA** and the **peptide** Ac-Gly-Phe-OH

D) Calixarenes and Resorcarenes
1) Complex of *N*-permethylated (hexaaminomethyl) calix[6]arene and C60
2) Complex of a resorcarene and the chinuclidinium ion
3) X-ray structure of a complex consisting of six resorcarenes and eight water molecules

E) Peptides and Proteins
1) Pentaalanin as antiparallel and a parallel β-Sheet
2) Synthetic **Peptide Receptors**
      a) A enantioselective β-sheet model
      b) A length and sequence selective host
      c) A porphyrin-based system

Protein Complexes
3) Streptavidin/biotin complex
4) Alkaline phosphatase with Zn, Mg, phosphate
5) Carbonic anhydrase, complex with acetate
6) A γ-chymotrypsin complex
7) Camphor hydroxylase
8) LADH liver alcohol dehydrogenase with NADH and DMSO

F) Nucleic Acids and Derivatives
1) DNA (dodecamer, A-form)
2) DNA (dodecamer, B-form)/netropsine; groove complex
3) DNA (dodecamer, B-form)/chloroquine complex; intercalation
4) Single stranded DNA with a stacking ligand

G) Self Organisation/Materials
1) Zeolite A
2) Zeolite pentasil ZSM-5
3) A polycyclic iron chelate complex
4) A triplehelix from oligopyridyls and copper

INDEX